Microporous Framework Solids

RSC Materials Monographs

SERIES EDITOR

ADVISORY PANEL

The chemistry of materials will be the central theme of this Series which aims to assist graduates and others in the course of their work. The coverage will be wide ranging, encompassing both established and new, developing areas.

Although focusing on the chemistry of materials, the monographs will not be restricted to this aspect alone.

TITLES IN THE SERIES:

Polymer Electrolytes
F.M. Gray, *School of Chemistry, University of St Andrews, UK*

Flat Panel Displays: Advanced Organic Materials
S.M. Kelly, *Department of Chemistry, University of Hull, UK*

The Chemistry of Medical and Dental Materials
J.W. Nicholson, *Medway School of Science, University of Greenwich, UK*

Visit our website at http://www.rsc.org/Publishing/Books/MM

How to obtain future titles on publication
A standing order plan is available for this series. A standing order will bring delivery of each new volume immediately on publication.

For further information please contact:

Sales and Customer Care, Royal Society of Chemistry, Thomas Graham House, Science Park, Milton Road, Cambridge, CB4 0WF, UK
Telephone: +44 (0)1223 432360, Fax: +44 (0)1223 426017, Email: sales@rsc.org

RSC MATERIALS MONOGRAPHS

Microporous Framework Solids

Paul A. Wright
School of Chemistry, St Andrews University, Fife, UK

RSC Publishing

ISBN: 978-0-85404-812-0

A catalogue record for this book is available from the British Library

Published by The Royal Society of Chemistry,
Thomas Graham House, Science Park, Milton Road,
Cambridge CB4 0WF, UK

Registered Charity Number 207890

For further information see our web site at www.rsc.org

Preface

There are few areas of solid state chemistry that have received so much attention, and found such widespread application, as that of microporous solids. For over forty years there has been continued development of the field worldwide. The classic text on 'Zeolite Molecular Sieves' by Breck in 1974[1] described the structural types then known, their synthesis and characterisation and their properties of adsorption and ion exchange. Since that time research into what are more commonly referred to as microporous solids has burgeoned. New crystalline materials with compositions that include over two thirds of the elements of the Periodic Table have been prepared, with cages up to around 30 Å in diameter, and ordered mesoporous solids with pore sizes upwards of tens of Angstroms have added a new dimension. Over the same period, techniques have been developed that can image the structures to better than 2 Å point-to-point resolution, determine the local coordination geometry of atoms in the solids and monitor the position and chemical state of molecules adsorbed within the pores. The massive increase in available computational power and associated theory enables simulations to extend our understanding of these materials beyond that which we are able to observe by experimental methods. In parallel, the adsorptive, ion exchange and catalytic properties of these solids have found increasing use and research fields have opened up in membrane technology and in device-oriented applications.

Academically, the attraction of crystalline microporous solids lies in the regularity and the molecular dimensions of their pore structure. In this type of solid the pore system and active site are defined completely by the crystal structure, as are the protein arrangement and active site of an enzyme. Consequently, studies of these systems have provided the key to understanding solid acid catalysis, among other important properties. Experimental data obtained from vibrational and NMR spectroscopy and by studies of reaction kinetics and diffusion can be interpreted in terms of molecule–solid interactions at the pore walls and at the active site. These systems lend themselves perfectly to computer simulations of increasing complexity that are increasingly fundamental – and include *ab initio* quantum mechanical calculations. 'Zeolitic' solids have therefore provided the ideal systems, being both interesting and important, on which numerous physico-chemical techniques have been developed and refined.

This is not to say that all has been discovered about these materials. New solids are continually being synthesised, and the rapidly developing field of porous metal organic frameworks (MOFs) is a current example. The applicability of microporous solids is continually being extended within the traditional fields of adsorption, catalysis and ion exchange. One exciting application explores their use as catalysts in the manufacture of fine chemicals. Other emerging areas of research concern their potential use in medicine or in devices that require special electronic or optical properties.

I aim in this book to describe the fundamental principles and experimental practices of the synthetic chemistry and the characterisation of microporous solids, and to give clear accounts of the uses to which they are put. This leads on to a discussion of how these solids are being developed, for example as increasingly selective acid and oxidation catalysts and as advanced functional materials. The new family of mesoporous solids must be compared here, to put the relative attributes of the two families of solids into context.

This book is not meant to be an exhaustive text or compilation, examples of which are available elsewhere,[2,3] but rather to introduce the essential background to the science of molecular sieves, rather as Dyer's introduction to zeolites achieved previously,[4] and then to describe recent trends and likely areas for future development. In doing so I will provide a range of references for the interested reader, and these should give a cross-section of the research currently being pursued. My intention is to describe the research field as an open rather than a closed one, with important opportunities and challenges, without ignoring the tremendous advances and achievements that have made the current field both exciting and important. More than that, I want to convey my own enthusiasm for the way our understanding of the behaviour of these solids stems from a knowledge of their crystalline architecture.

1. D. W. Breck, *'Zeolite Molecular Sieves'*, Wiley, New York, 1974.
2. F. Schuth, K. S. W. Sing and J. Weitkamp, Eds., *'Handbook of Porous Solids'*, Wiley-VCH, New York, 2002.
3. S. M. Auerbach, K. A. Carrado and P. K. Dutta, *'Handbook of Zeolite Science and Technology'*, Marcel Decker, New York, 2003.
4. A. Dyer, *'An Introduction to Zeolite Molecular Sieves'*, Wiley, Chichester, 1988.

Contents

Chapter 10 Advanced Applications and Current Trends

Acknowledgements

I warmly appreciate the help of all my co-workers, friends and colleagues who have helped me in the preparation of this monograph by supplying material, giving permission to publish and proof reading. My thanks to David Cole-Hamilton for suggesting I write the text and encouraging me to do so and to Sir John Meurig Thomas for inspiration in all things zeolitic. Last but not least, I thank my family for all their patience and support.

CHAPTER 1
Introduction

1.1 Microporous Framework Solids: Definitions

First, a word on definitions. The materials I shall describe in this book will be those with ordered structures that are able to adsorb molecules reversibly and selectively based on differences in their size and shape. For a long time this definition would have applied almost exclusively to aluminosilicate zeolites, crystalline solids with pore sizes up to around 8 Å, and excluded those materials such as porous carbons or ceramics derived from sol-gel preparations that possess microporosity but no regularity of structure. Nowadays the term encompasses families of microporous solids with ever-increasing compositional variety, frameworks with metals in mixed coordination and hybrid metal-organic frameworks. The definition might also include some of the newly discovered class of mesostructured solids, discovered first by researchers at Mobil in the early 1990s. As molecular sieves, it is clear that the pore size of these solids, which extends from 10 Å upwards, enables the adsorption of much larger molecules than is possible for zeolites (Figure 1.1). I have excluded a consideration of layered solids such as pillared clays, which also show porosity, because their order is predominantly in two dimensions, rather than three.

A distinction should also be drawn between framework molecular sieves and the wealth of crystalline solids prepared in the presence of an organic species that becomes incorporated in voids within the structure but cannot be removed, thermally or otherwise, without structural collapse. These are better described as open framework solids. In practice the distinction is blurred and, although only materials that are thermally and hydrothermally robust are likely to be commercially exploited in traditional technologies, open framework solids containing organic species can possess properties that may make them suitable for more specialised use.

1.2 Historical Development of the Subject

There is great current interest in the use of combinatorial methods, or high throughput experimentation (in which a large number of experiments are performed to explore the different effects of many variables), to prepare new

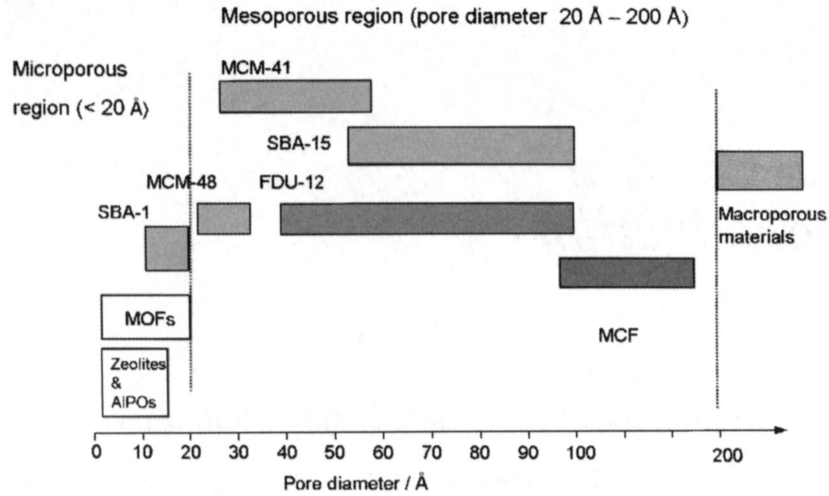

Figure 1.1 Schematic representation of the pore size ranges of microporous and mesoporous solids. Whereas crystalline materials have well-defined pore dimensions, those solids without atomic order on the long range (particularly the mesoporous solids) can be prepared with pore sizes that can vary widely, depending on synthetic and post-synthetic treatment conditions.

solids by optimised routes. I will give an example of this applied to zeolite synthesis in Chapter 5. In a sense, though, the range of geochemical situations that have existed naturally has acted in the same way, and solid state chemists have learnt much from observing and preparing laboratory analogues of natural minerals, including clays, zeolites and other porous silicates and phosphates. Natural zeolites have long been recognised as a class of solids with characteristic properties. The first recorded description of a zeolite mineral was of stilbite, by Cronsted in 1756.[1] Crystals of natural stilbite are shown in Figure 1.2: routes have since been developed for the laboratory synthesis of zeolites with the same framework structure of stilbite, and with a wide variety of compositions. Figure 1.2 also shows a scanning electron micrograph of a synthetic, high silica version of stilbite that has recently been prepared,[2] and a representation of its microporous structure. The property of zeolites to reversibly evolve adsorbed water when heated gives rise to their name (*Zeo* (*Greek*, boiling) *Lithos* (*Greek*, stone)). Zeolites are a class of 'tectosilicate' minerals that possess tetrahedrally linked, three-dimensional frameworks made up of corner-sharing aluminate and silicate tetrahedra that are sufficiently open to be able to reversibly adsorb molecules.

Under favourable geological conditions significant deposits of useful natural zeolites (such as clinoptililite and mordenite) have resulted and are mined and used in large quantities. More often, and certainly for many of the most important commercial zeolites, mineral forms of the zeolite types only occur in minute quantities. Very small amounts of natural analogues to the widely used zeolites Y, ZSM-5 and Beta have all been found, and are known as faujasite,

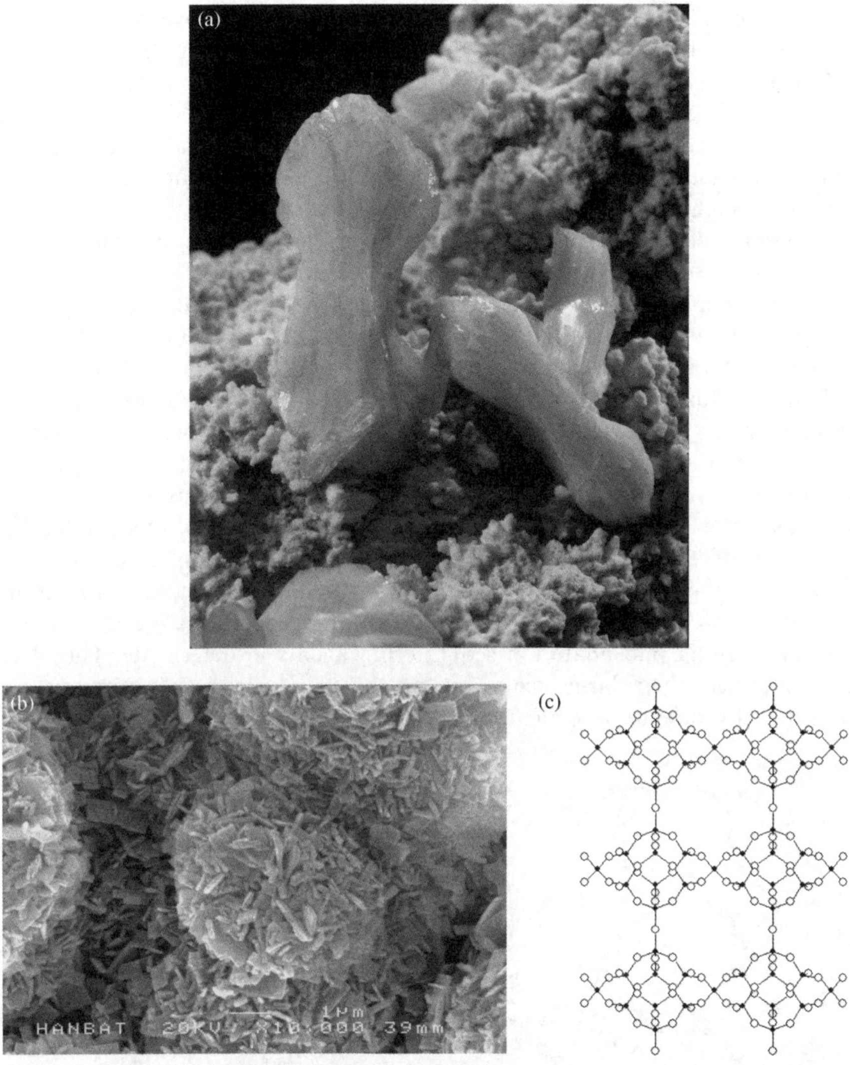

Figure 1.2 (a) Crystals of natural stilbite. Copyright Natural History Museum, London. (b) Scanning electron micrograph of microcrystals of synthetic high silica stilbite, prepared using an organic molecule as a template (see Figure 3.23). (Courtesy S.B. Hong) The framework structure of stilbite (c) has a system of parallel channels *ca.* 5 Å in free diameter, linked perpendicularly by smaller pores.

mutinaite[3] and czernickite[4], respectively. Such mineralogical occurrences (such as those found in Antarctica)[5] are more than interesting curiosities to the zeolite chemist: they were essential to establishing a structural basis for zeolite science (many zeolite structures have been solved from natural samples) and they point the way to unexpected structural possibilities. Two zeolites reported only as natural minerals are shown in Figure 1.3. The recently discovered structure of tschörtnerite,[6] for example, obtained from a few crystals from the Eifel region of Germany, includes the supercage shown, which has an internal free diameter of 17.3 Å. A synthetic tschörtnerite might be of considerable use in gas separation or in detergency. Boggsite,[7] on the other hand (also shown), possesses an intersecting channel system that would be of more interest in catalysis. Structures such as these are attractive targets for synthesis.

Zeolite mineralogy is certainly interesting and informative, but the development of zeolite chemistry was made possible by their large scale synthesis under laboratory conditions. It was recognised that zeolites could be crystallised hydrothermally from reactive precursor gels under alkaline conditions on timescales of a few days. Early synthetic work, by Richard Barrer, in academia, and industrialists such as Robert Milton, Donald Breck, George Kerr and Edith Flanigen, explored the use of alkali and alkali earth metals to direct the crystallisation of aluminosilicate gels from the 1940s onwards. These pioneering studies gave rise to many of the zeolites that are currently widely used, in particular the zeolites A, X, Y and mordenite. The characteristic zeolitic properties of these materials, such as adsorption and ion exchange, were established, their study receiving impetus with their adoption in gas drying and separation technologies and by the need for replacements for environmentally threatening phosphate ion sequestering agents in detergents. They have since retained their large scale use in these industries, with innovations prompted by new developments and considerations of interest to individual

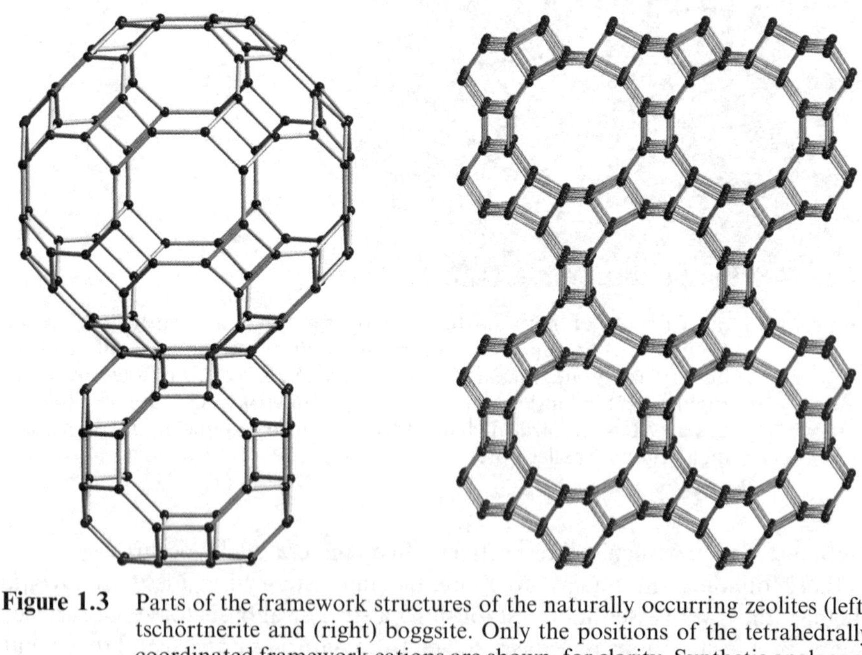

Figure 1.3 Parts of the framework structures of the naturally occurring zeolites (left) tschörtnerite and (right) boggsite. Only the positions of the tetrahedrally coordinated framework cations are shown, for clarity. Synthetic analogues of these minerals are attractive targets for zeolite chemists.

manufacturers (such as the need to prepare new products that are outside existing patent restrictions).

The compositional similarity of zeolites to silica-aluminas in widespread use as solid acid catalysts in catalytic cracking processes suggested that they could also be used in these processes if they were suitably modified. Chemical routes to the incorporation of acid sites were found, and the resulting catalysts were found to be strong solid acids (Chapter 8). In particular, zeolites based on zeolite Y are now widely used on a massive scale in catalytic cracking. Furthermore, the unique pore structure of zeolites was found to give unique and desired product selectivities. Further development in this area was made possible by the discovery by Barrer and Denny[8] that organic cations, particularly alkylammonium ions, could be used to direct the synthesis of new structures in a similar way to alkali and alkaline earth cations. The use of bulky organic cations was found to give zeolites with high silica-to-alumina ratios and consequently high thermal stability. In particular zeolite Beta[9] and the so-called pentasil zeolites ZSM-5[10] and -11 (pentasil refers to the predominance of five-membered rings in the frameworks) were discovered. These were found to possess unique activities and selectivity for a range of hydrocarbon conversions, in particular in the alkylation and isomerisation of aromatic molecules. The catalytic possibilities were greatly expanded and resulted in further penetration of zeolite catalysts into industrial processes, with increasing replacement of less selective and more corrosive acid catalysts such as supported phosphoric acid or liquid sulfuric acid. The trend continues to the present day, with new zeolite-catalysed hydrocarbon reactions continuing to meet the ever-changing demands of the petrochemical industry (Chapter 8). In parallel, programs of synthesis of novel solids using novel organic structure directing agents (or, more loosely, 'templates') continue today to widen the range of silicate-based solids (Chapters 2 and 5).

By the 1980s most of the aluminosilicate zeolites currently used industrially were known, and the emphasis shifted to the study of these materials using a range of powerful new techniques that came of age at this time. These included, in particular, solid state NMR, X-ray and neutron powder diffraction analysis, high resolution electron microscopy and computational methods. All were ideal for the study of structural details of solids that were rarely available, and never used in industrial applications, other than as microcrystalline powders. All these techniques are applicable to the bulk of the solid – this in turn makes up the (internal) surface, which is accessible to adsorbed molecules. Since the techniques are able to operate under any conditions of gas pressure, they may be used to extract structural details *in situ* under the operating conditions of ion exchange, adsorption and catalysis. In particular, zeolitic systems have proved ideal for the study, understanding and subsequent improvement of solid acid catalysts.

Already by 1980 it was known that molecular sieve frameworks of compositions other than aluminosilicate could be prepared. Pure silica polymorphs of ZSM-5 and ZSM-11, for example, (patented as silicalite-1 and silicalite-2) were of interest because their internal surfaces were hydrophobic rather than hydrophilic, and offered new possibilities in the separation of organics from aqueous solutions.[11] There was also compelling evidence that elements such as B, Fe, Ga and even P could be introduced, with consequent modification of the acidic properties. Indeed, the announcement by UOP in 1982 of a new series of microporous framework aluminophosphates stimulated a major expansion of the chemical compositions available in the form of molecular sieves.[12] Aluminophosphates possess a chemistry that is sufficiently different from that of zeolites to offer alternative catalytic possibilities, for example in selective oxidation. With a few exceptions, however, the stability of phosphate materials at high temperatures, particularly in the presence of water vapour, is lower than that of silicates, so that their applicability is more likely to be found in fields different from those exploited so successfully by zeolites. A major significance of the aluminophosphates, however, was to open up the possibility of making molecular sieves from an enormously enlarged range of building blocks. Current expression of this is to be found in the great diversity of novel microporous organic-inorganic hybrids described in Section 2.8.

The last decade has seen the increased application of zeolitic solids as catalysts outside the realm of petrochemicals, including the synthesis of fine chemicals and as potential materials for environmentally related catalysis. The development by Enichem of titanosilicate analogues of zeolites[13] as highly efficient catalysts for a range of epoxidations and oxidations is arguably the most significant development in this area, but a wide range of more specialised chemical conversions have also been achieved (Chapter 9). Considerable recent success has also been achieved in the fabrication and use of zeolite membranes in separation technology.

As described above, the field of microporous solids has developed in a series of steps. The discovery of mesoporous silicas,[14] with ordered pores of 20 Å and above, has stimulated a spectacular research effort in synthesis, characterisation and potential applications of these materials, such that the field has grown to a size (in terms of numbers of publications) similar to that of microporous solids. Although lacking the atomic order observed for zeolitic solids, many of the same techniques used to characterise microporous solids are also applicable to them. It is not yet apparent what the most important applications of these solids will be, but it is clear that they have caught the scientific imagination in the same way that microporous solids have, and I include a limited discussion of these materials in the appropriate chapters.

References

1. A. F. Cronsted, *Akad. Handl.*, 1756, **17**, 20.
2. S. B. Hong, E. G. Lear, P. A. Wright, W. Z. Zhou, P. A. Cox, C. H. Shin, J. H. Park and I. S. Nam, *J. Am. Chem. Soc.*, 2004, **126**, 5817.
3. G. Vezzalini, S. Quartieri, E. Galli, A. Alberti, G. Cruciani and A. Kvick, *Zeolites*, 1997, **19**, 323.

4. J. V. Smith, J. J. Pluth, R. C. Boggs and D. G. Howard, *J. Chem. Soc., Chem. Commun.*, 1991, 363.
5. A. Alberti, G. Cruciani, E. Galli, S. Merlino, R. Millini, S. Quartieri, G. Vezzalini and S. Zanardi, *Stud. Surf. Sci. Catal.*, 2001, **135**, 83.
6. H. Effenberger, G. Giester, W. Krause and H. J. Bernhardt, *Am. Mineral.*, 1998, **83**, 607.
7. J. J. Pluth and J. V. Smith, *Am. Mineral.*, 1990, **75**, 501.
8. R. M. Barrer and P. J. Denny, *J. Chem. Soc.*, 1961, 971.
9. R. L. Wadlinger, G. T. Kerr and E. J. Rosinski, 1967, US Patent 3,308,069.
10. R. J. Argauer and G. R. Landolt, 1972, US Patent 3,702,886.
11. E. M. Flanigen, J. M. Bennett, R. W. Grose, J. P. Cohen, R. L. Patton, R. M. Kirchner and J. V. Smith, *Nature*, 1978, **271**, 512.
12. S. T. Wilson, B. M. Lok, C. A. Messina, T. R. Cannan and E. M. Flanigen, *J. Am. Chem. Soc.*, 1982, **104**, 1146.
13. M. Taramasso, G. Perego and B. Notari, 1983, US Patent 4,410,501.
14. J. S. Beck, J. C. Vartuli, W. J. Roth, M. E. Leonowicz, C. T. Kresge, K. D. Schmitt, C. T. W. Chu, D. H. Olson, E. W. Sheppard, S. B. McCullen, J. B. Higgins and J. L. Schlenker, *J. Am. Chem. Soc.*, 1992, **114**, 10834.

CHAPTER 2
Families of Microporous Framework Solids

2.1 Introduction

As described in the first chapter, the continual discovery of novel structure types of microporous and open framework solids and their preparation within a wide compositional range has been one of the most striking features of research in this area. Figure 2.1 illustrates the chronological development of inorganic families of such solids, and emphasises their recent proliferation. The introduction of novel inorganic chemistry is particularly noticeable, and associated with this is the incorporation of framework-forming metal cations that possess five- or six-fold coordination rather than the tetrahedral coordination typical of zeolites and aluminophosphates.[1] Furthermore, novel families of inorganic-organic hybrid solids such as metal phosphonates with inorganic frameworks 'lined' with organic groups and metal-organic frameworks made up of coordination polymers have provided an exciting recent extension. The family of mesoporous solids (of which more than 10 different structure types have so far been identified) can usefully be included in the discussion. In this chapter I will describe the important structural features of these different families.

Considering only frameworks made up entirely of tetrahedral corner-sharing TO_4 species, full details of all the structure types are collected, refereed and published by the Structure Commission of the International Zeolite Association. The most recent publication indicates that around 170 framework types (each of which is given a unique three-letter code) have been unambiguously identified and both hardcopy publications (in particular the so-called '*Atlas of Zeolite Framework Types*'[2]) and the continuously updated structural summary on the web site (www.iza-structure.org) are indispensable resources for the researcher in this field.

Figure 2.2 illustrates the way in which new types of tetrahedrally connected frameworks have been discovered and their structures solved over the last forty years, and highlights at least two important trends. The first is the decline in the discovery of framework solids formed in the presence of inorganic cations alone

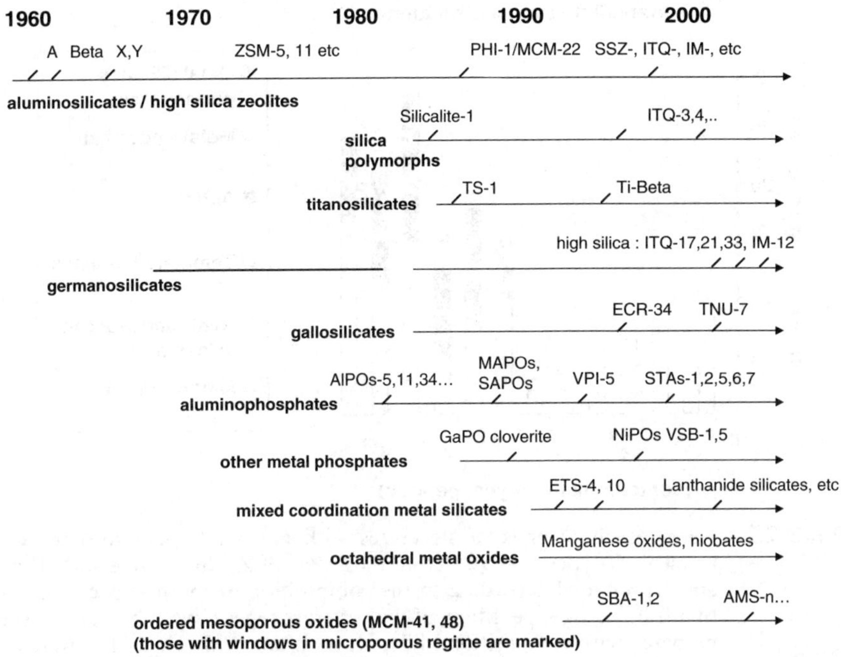

Figure 2.1 The chronological development of families of crystalline microporous and ordered mesoporous inorganic solids, marking on examples covered in this text.

as structure directing agents and without the use of organic cations. This decrease is a consequence of the limited choice of such cations that are soluble under alkaline conditions, and the tremendous efforts of exploratory synthesis already made. The major recent growth of structure types has therefore occurred in syntheses that have made use of organic 'templates' and novel framework compositions, often in tandem. This importance of templated syntheses is particularly important in both the high silica zeolites and in metal phosphates (aluminium, gallium, iron phosphates, *etc*.). In particular the designed synthesis of alkylammonium templates by the groups of Zones, at Chevron, and Corma, at the ITQ in Valencia, has provided a prolific route to the most recent tetrahedrally connected silicate structure types. The acceleration in the discovery of new structure types is therefore largely due to innovative synthetic work, including template design, an increase in the compositional variety and the use of mineralising agents such as fluoride ions (Chapter 5). In addition, improvements in the methods available for the structure solution of these solids (which are typically prepared as microcrystalline powders or as small, weakly diffracting crystals) have contributed to the increase in known tetrahedrally connected structure types. These methods include the remarkable development of methods of structure solution from powder data and also microcrystal diffraction at synchrotron X-ray sources (Chapter 3).

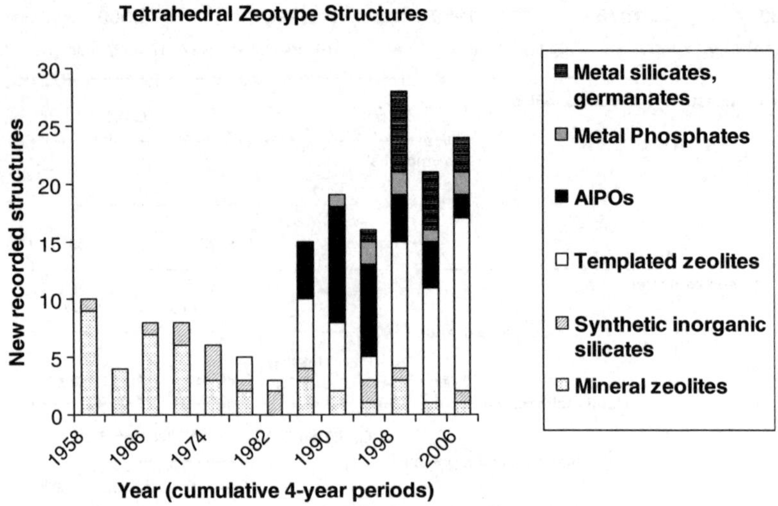

Figure 2.2 Reported discoveries of structures with novel tetrahedrally connected framework types, as reported in the 'Atlas of Zeolite Framework Types' and represented according to the composition of the first occurrence of that framework type. Many of the structure types have subsequently been prepared with compositions different from that with which they were first observed.

There is currently no comprehensive compilation of structure types that include framework cations with coordination environments other than tetrahedral. Among the more important of the inorganic-only solids are titanosilicates such as ETS-4 and -10 (containing octahedrally coordinated titanium), gallium and nickel phosphates, germanates and metal oxide molecular sieves that are described later in this chapter. More recently still, these have been augmented by many inorganic-organic hybrids (including the rapidly expanding field of porous metal-organic framework coordination compounds, or MOFs). These structures are also introduced later in the chapter. Finally, I include a brief description of the structure types of ordered 'mesoporous' solids. Although these are less well defined than truly crystalline solids, enough is now known (from X-ray diffraction, electron microscopy and solid state NMR) to establish their main structural features. These studies, in combination with adsorption measurements, have indicated that several are more accurately described as microporous, with windows less than 2 nm connecting larger cages.

Structural diagrams are of great importance in understanding crystalline structures, and indeed the combination of extended frameworks and pore space within these materials makes them particularly attractive for graphical representation. The most useful ways of displaying them are shown in Figure 2.3. These are: *line plots* showing only the bonds (with or without bridging oxygens); *ball-and-stick plots*, representing atoms and bonds; *ORTEP crystallographic plots*, where the atoms are represented by thermal ellipsoids that describe the volume occupied by atoms with a given percent of probability; plots of the *coordination polyhedra* (tetrahedra, octahedra, *etc.*); *space-filling* diagrams. In space-filling diagrams, the oxygen atoms are shown with their van der Waals radii of 1.35 Å. In this monograph, most structures are represented either as ball-and-stick models (to emphasise the pore space) or by use of polyhedra, when it is helpful to distinguish different cation coordination geometries within the framework.

2.2 Aluminosilicate Zeolites and Silica Polymorphs

2.2.1 Structural Chemistry

Zeolites may be considered to be made up of frameworks consisting of corner-sharing silicate and aluminate tetrahedra. Whereas pure silica frameworks are electrically neutral, the substitution of trivalent aluminium for tetravalent silicon imparts a negative charge to the framework that has to be balanced by positively charged extra-framework cations. The general formula of a zeolite is therefore:

$$M^{n+}_{x/n}Al_xSi_{1-x}O_2 \cdot yY$$

where M is the extra-framework charge-balancing cation and Y represents species such as H_2O included or adsorbed within the pores.

For an as-prepared zeolite M^{n+} may be inorganic or organic (depending on the composition of the gel during crystallisation) and can be replaced by other metal cations or even protons by suitable chemical conversions. Other species can also be introduced into the pore system by ion exchange or by post-synthesis treatments (Chapter 5). For zeolites where the charge-balancing cations are all inorganic (*e.g.* Na^+, K^+, Ca^{2+}) heating will initially remove weakly physisorbed water from the pores and then remove water of hydration from the cations. For divalent and trivalent cations there is good evidence that during this process the water of hydration can dissociate to give a hydroxyl species attached to the cation and a proton attached to the framework. The bare cations left after dehydration optimise their coordination with oxygen atoms of the zeolitic framework by migrating to favoured sites where possible. For zeolites A, X and Y and many other low silica zeolites a great deal is known about these cation locations because they control the important properties of gas adsorption and ion exchange. Recent studies have shown that they may move within the structure as a response to adsorption of molecules.

Many zeolites may be prepared in the acidic form by heating the ammonium-exchanged form at temperatures of 450 °C or above (see Chapter 6). For solids prepared with organic templates the protonic form can be obtained by removal of template molecules at high temperature in air ('calcination'). The protons in these solids are located on oxygen atoms that are bound to both an aluminium and a silicon atom, giving bridging hydroxyls. These protons are often very strong Brønsted acid sites of great importance in acid catalysis (see Chapter 8).

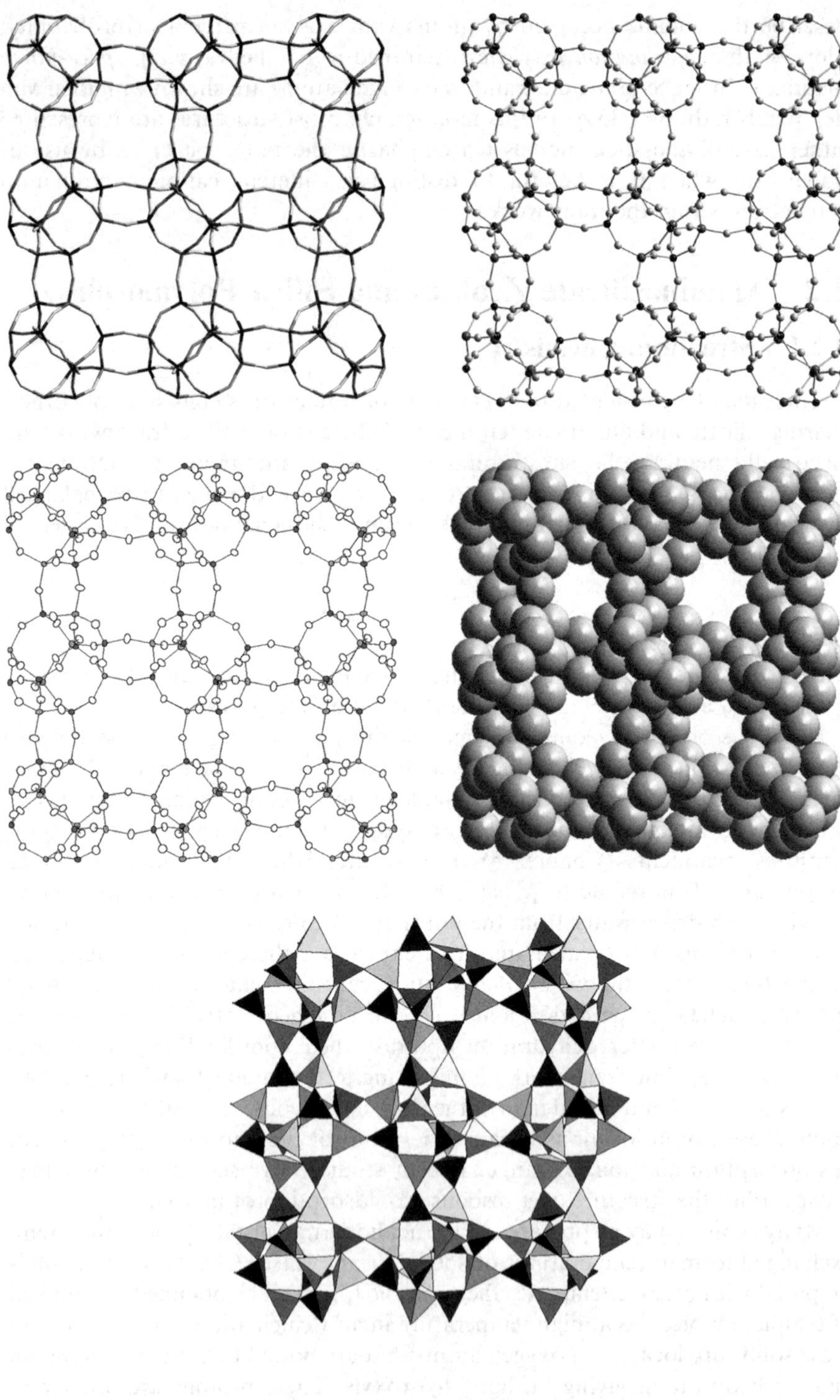

For some zeolites with high aluminium contents, such as zeolite A or X, however, the acid form is unstable.

The ratio of silicon to aluminium in the framework can vary between infinity (for the silica polymorphs) and 1. The lower limit of 1 arises because below this value it becomes impossible to avoid Al-O-Al linkages, which are electrostatically unfavourable because of the close approach of the negative charges associated with aluminium substitution for silicon in the lattice. This is Loewenstein's aluminium avoidance principle. For zeolites with Si/Al ratios close to 1, a high degree of local Al/Si order is achieved within the structure, and for a value of 1 (as observed, for example, for zeolite A) there is strict alternation of aluminium and silicon atoms between adjacent tetrahedral framework sites. Typically, zeolites prepared using organic cations as templates have low aluminium contents, because these cations have a higher volume-to-charge ratio than inorganic cations, and so fewer can be incorporated within the same pore space. As a result, fewer negative charges (and therefore fewer aluminium atoms) are required within the framework to balance the template charge. In general, zeolites with high Si/Al ratios possess high thermal stability and acid strength, and so post-synthesis methods have been developed to tailor Si/Al ratios to give these desired properties. Post-synthetic chemistry, along with a discussion of defects that it can introduce into the idealised zeolite structure described above, is discussed in Chapter 6.

2.2.2 Zeolite Framework Types

Zeolite structure types are commonly described in terms of identifiable structural units within the frameworks. These are often referred to as secondary building units (SBUs), although which, if any, of these are added intact to growing zeolite surfaces during crystallisation remains an open question. It is helpful to describe the SBUs and frameworks in terms of rings made up of alternating tetrahedral cations and oxygens. These are named in terms of the number of cations in the ring, so for instance, a six-membered ring (6MR) contains six cations and six oxygens, a twelve-membered ring (12MR) contains twelve cations, and so on. Small rings may themselves be considered SBUs. Other SBUs are most easily described as being built up of two rings linked together as a prism, so two four-membered rings linked together in this way are known as a double four-membered ring, D4R, two 6MRs as a D6R, *etc.* (Figure 2.4). Other SBUs that can be used to build up important structures are also given in the figure: these include the cancrinite cage and the sodalite cage.

Figure 2.3 The framework structure of the aluminophosphate STA-7 (framework type SAV) represented from top left to bottom as a line plot; a ball-and-stick plot; a plot in which the atom positions are represented by thermal ellipsoids that describe their likely location averaged over time and space; a space-filling representation, using van der Waals radii for the framework atoms and coordination polyhedra of the framework cations (in this case, tetrahedra).

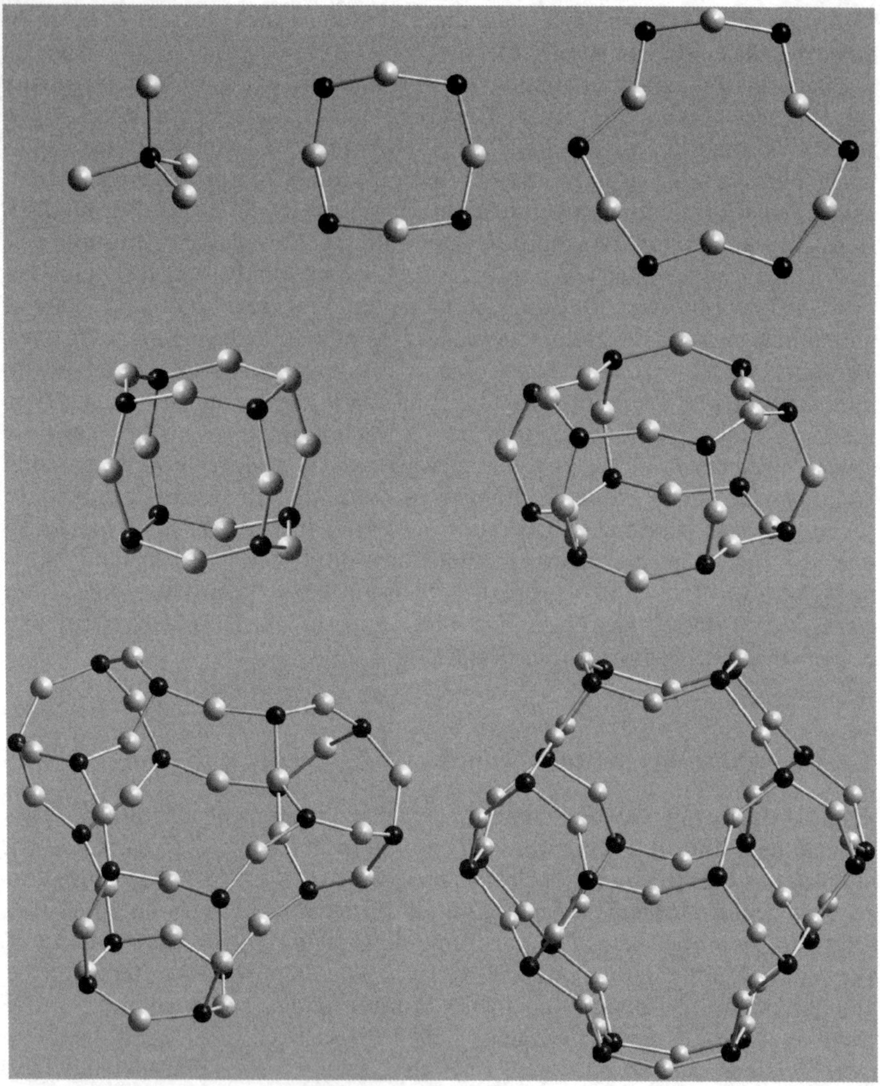

Figure 2.4 Building units commonly found in zeolite frameworks. From left to right, (top row) the tetrahedral primary building unit, followed by the secondary building units 4-membered ring (4MR); 6MR; (second row) double 4-membered ring (D4R); D6R; (bottom row) cancrinite cage (ε-cage); sodalite cage (β-cage).

Another convenient shorthand way to describe cages in the framework structure is to write them in terms of the rings that make up the faces of the cage. For example, a D6R is described $[4^6 6^2]$ and a sodalite cage $[4^6 6^8]$. There are many other such examples, and a wealth of such details on the nomenclature is to be found in the 'Atlas' on the IZC Structure Commission website.

Using a building unit approach, it is convenient to consider the structures of the important zeolites sodalite, zeolite A and the faujasitic zeolites X and Y to be built up from cuboctahedral cages of tetrahedra, known as sodalite cages (Figure 2.5). These are linked *via* shared four-membered ring faces in sodalite, *via* double four-membered rings in zeolite A and *via* double six-membered rings in Y. Whereas there is only one way of arranging the cages for sodalite and zeolite A, it is found that there are different ways of linking layers of cages through D6Rs. If the centre of the D6Rs possess inversion centres of symmetry,

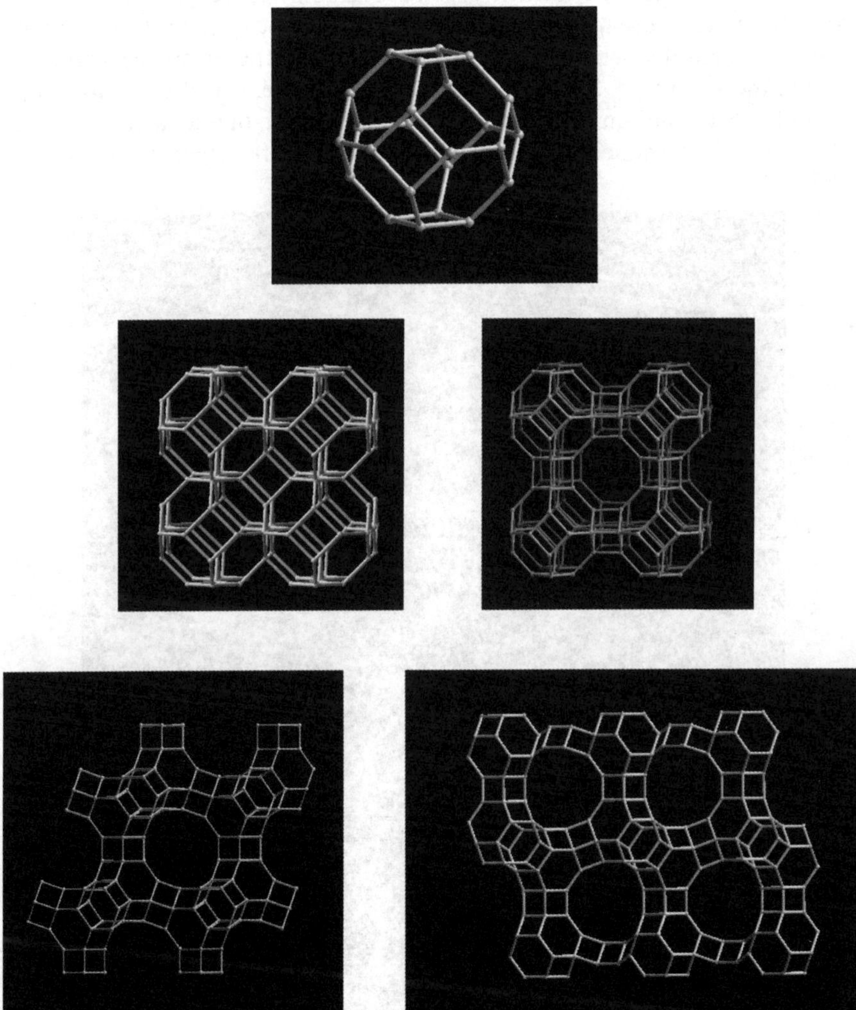

Figure 2.5 Framework structures built up from the sodalite cage, top. Middle left to bottom right: sodalite, zeolite A, zeolite Y and EMC-2. In each case only the connected locations of tetrahedral cations are illustrated. See text for a full explanation of the structures.

the cubic zeolite Y structure results, whereas if the D6Rs are at a mirror plane, hexagonal zeolite Y (EMT) is produced. Which of these two structure types forms is a consequence of different synthesis conditions during crystallisation, and polymorphism where different modes of stacking occur (polytypism) can by controlled by template choice (see Chapter 5).

The extra-framework cation sites present in zeolite Y are worthy of description here, since they have been the subject of extensive study, and are illustrated in Figure 2.6. The sites are given Roman numerals, starting from a site within the D6Rs, coordinated to six framework oxygen atoms (site I) and moving progressively through the sodalite cage towards the supercage: site I′ is threefold coordinated, close to site I, but within the sodalite cage; site II′ is within the sodalite cage but diametrically opposite to I′, close to the supercage; and site II is within the supercage. Site III is within the supercage, coordinated to oxygen atoms of a 4MR and on the inside of the cage. The site that a particular cation will occupy is determined by its size and valence, and the chemical history and

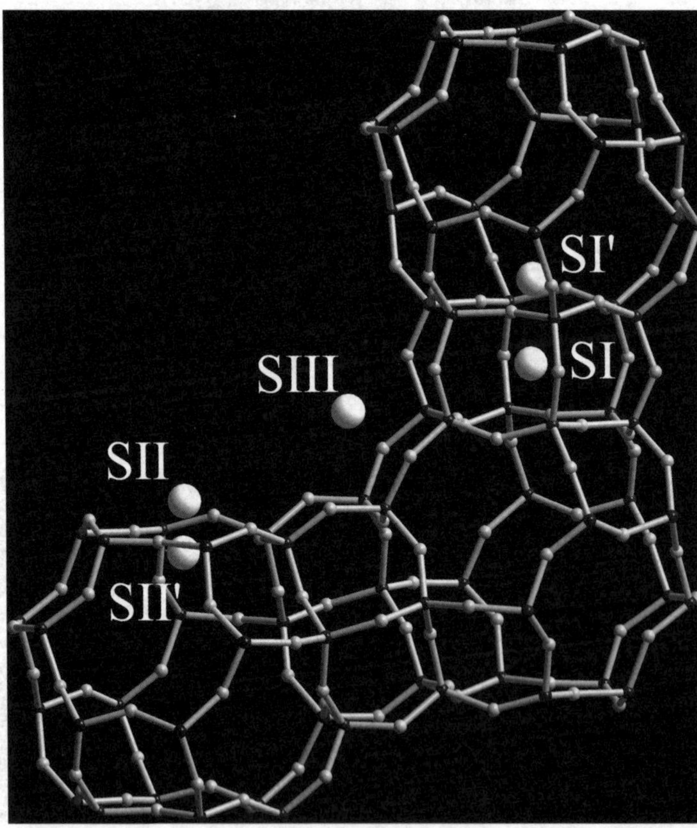

Figure 2.6 Position and nomenclature of the most important cation sites in zeolites X and Y (FAU type). SI is in the D6R, SI′ and SII′ in the sodalite cage, SII in the 6MR between the sodalite cage and the supercage and SIII is within the supercage.

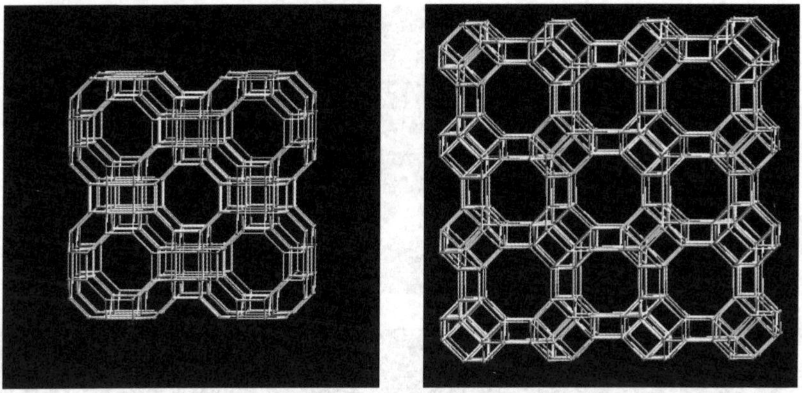

Figure 2.7 The framework types RHO (left) and KFI (right) – oxygen atoms not shown – both contain the α-cage that is also found in zeolite A. In addition, RHO also possesses D8Rs, whereas KFI also contains D6Rs and cages found in the MER structure type.

the presence or absence of adsorbed molecules within the pores. For example, hydrated nickel cations in the supercage migrate to site I upon dehydration,[3] but migrate back into the supercage as molecules are adsorbed into the pores.

The zeolite A structure, in addition to sodalite cages, possesses larger α-cages, or supercages, which possess six 8MR windows into adjacent supercages, and make up the accessible volume of the structure, since the sodalite cages are inaccessible to adsorbed molecules other than water. α-cages are also found in two other cubic small-pore (8MR) zeolites, zeolite Rho and ZK-5 (Figure 2.7). In Rho, the α-cages are linked *via* D8Rs, giving two interpenetrating but separate pore volumes, whereas in ZK-5 the basic building units are D6Rs, which are arranged to give α-cages linked *via* smaller MER cages.

Another important family of zeolite structures is made up of polytypic structures derived by different stacking sequences of layers of six-membered rings arranged in a hexagonal net (Figure 2.8). The six-membered rings can be stacked either one above another (at the A position), or at one of two other positions in a hexagonal unit cell (the B or C positions) and linked to the layers above and below by 4MRs. The stacking continues until a repeating sequence is obtained. Currently 16 members of this polytypic series have been discovered, either as zeolites or aluminophosphates (Table 2.1), including the important zeolites sodalite (repeat stacking sequence ABC), cancrinite (AB), chabazite (AABBCC) and erionite (AABAAC). Being polytypes, the framework densities (tetrahedral units per 1000 Å^3) are similar, and the unit cell repeat perpendicular to the layers is a multiple of *ca.* 2.5 Å.

In fact, there are many families of zeolite structures whose members contain similar structural units, such as chains, sheets or blocks, arranged in different, regularly repeating ways. The important zeolites ZSM-5 and ZSM-11 (ZSM = Zeolite Socony Mobil), where adjacent, similar layers are crystallographically related either across inversion centres (ZSM-5[4]) or by mirror planes

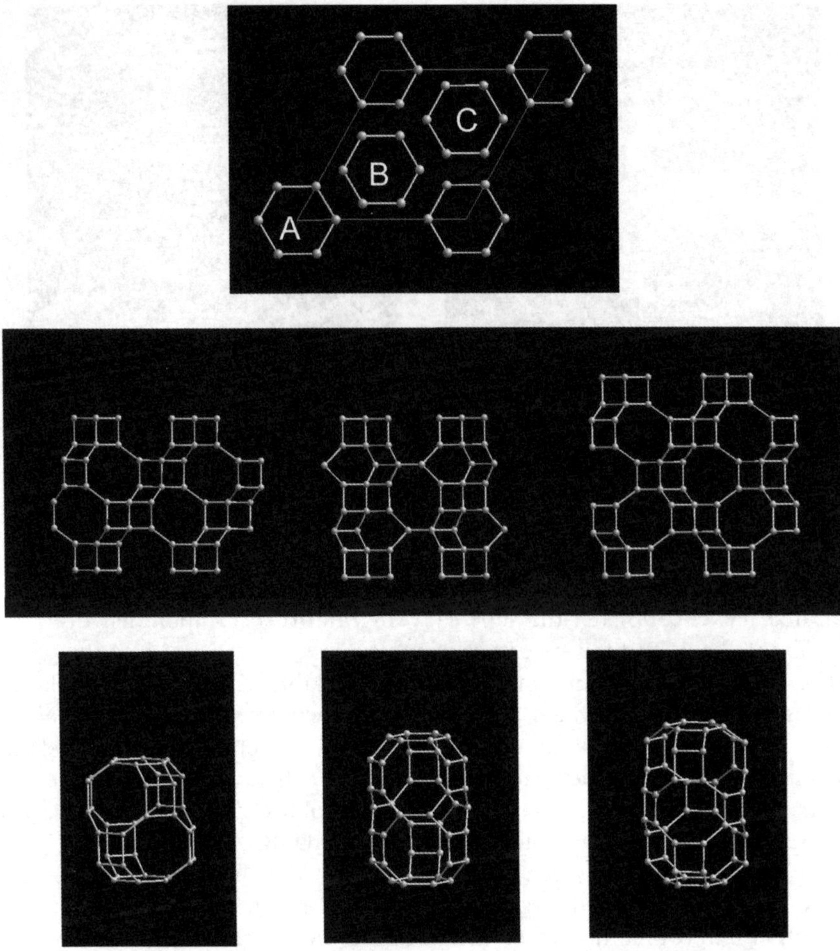

Figure 2.8 Many tetrahedrally connected frameworks are based on different stacking sequences of connected 6MRs. To construct these nets, 6MRs are placed at one of three different positions (A, B or C) within a hexagonal unit cell (top row). Only one of these positions can be occupied at any given height in the unit cell, allowing a large family of polytypic structure types, including, from left to right and viewed perpendicular to the direction of stacking, the zeolites chabasite (CHA), repeating sequence AABBCC; erionite (ERI), AABAAC and SSZ-16 (AFX), AABBCCBB (middle row). These three structures are characterised by cages of different sizes linked via 8MRs (bottom row).

(ZSM-11[5]) are the best known example of high silica zeolites related in this way (Figure 2.9). The basic building units of these zeolites are 5-5-1 units, which link together to form chains. These in turn link to give silicate sheets, which contain 10MR openings, and these can be connected across mirror planes or centres of symmetry to give the related structures. There are many other examples where two structures are related by different ways of linking of sheets. The zeolites

Table 2.1 Zeotypic polytypes derived by stacking 6MRs.

Name (code)	*Layers in repeat*	*Sequence*	*a/Å*	*c/Å*	*Framework density / 10^{-3} T/Å^3*
Cancrinite (CAN)	2	AB	12.8	5.1	16.7
Offretite (OFF)	3	AAB	13.3	7.6	15.5
Sodalite (SOD)	3	ABC	8.9	cubic	17.2
Gmelinite (GME)	4	AABB	13.8	10.0	14.6
Losod (LOS)	4	ABAC	12.9	10.5	15.8
Chabazite (CHA)[a]	6	AABBCC	13.2	15.1	14.6
EAB (EAB)	6	ABBACC	13.3	15.2	15.4
Erionite (ERI)[a]	6	AABAAC	13.3	15.1	15.6
Liottite (LIO)	6	ABABAC	12.8	16.1	15.7
Afghanite (AFG)	8	ABABACAC	12.8	21.4	15.9
SSZ-16[a] (AFX)	8	AABBCCBB	13.8	19.9	15.7
Levyne[a] (LEV)	9	AABCCABBC	13.3	23.0	15.2
Franzinite (FRA)	10	ABCABACABC	12.9	26.5	15.6
AlPO-52 (AFT)[b]	12	AABBCCAACCBB	13.7	29.7	15.2
STA-2 (SAT)[b]	12	ABAACACCBCBB	13.0	30.4	16.2
Giuseppettite (GIU)	16	ABABABACBABABABC	12.6	41.0	15.9

[a] Both aluminosilicate and aluminophosphate analogues are known.
[b] Only known as substituted aluminophosphate.
Structural references from the 'Atlas'.

Theta-1 and ZSM-23,[6] for example (Figure 2.10), contain the same sheets related either by translation or by mirror planes ('unit cell twinning'), as do RUB-13 and ITQ-3.[7]

As well as fully ordered end-members with different stacking sequences, there are many materials that possess disordered stacking sequences of common structural units. This is particularly true for intergrowths of the A- and B-polymorphs of the catalytically important zeolite Beta,[8] which are made of the same type of layers (each with four-fold symmetry) stacked layer to layer *via* one of four possible translations and with a high degree of disorder. This is illustrated more fully with the help of electron microscopy in Chapter 3. No fully ordered A- or B-polytype has yet been synthesised although a third, related, polymorph C has been prepared.[9] A database describing structural disorder in a number of zeolite families is being assembled and may be accessed on the IZA Structure Commission website.

On a related theme, there are also structures that are made up of more than one type of structural unit, which could be an SBU or a layer. There is growing evidence that such phases occur close to the boundaries between the synthesis fields of two different structures – the concept of boundary phases proposed by

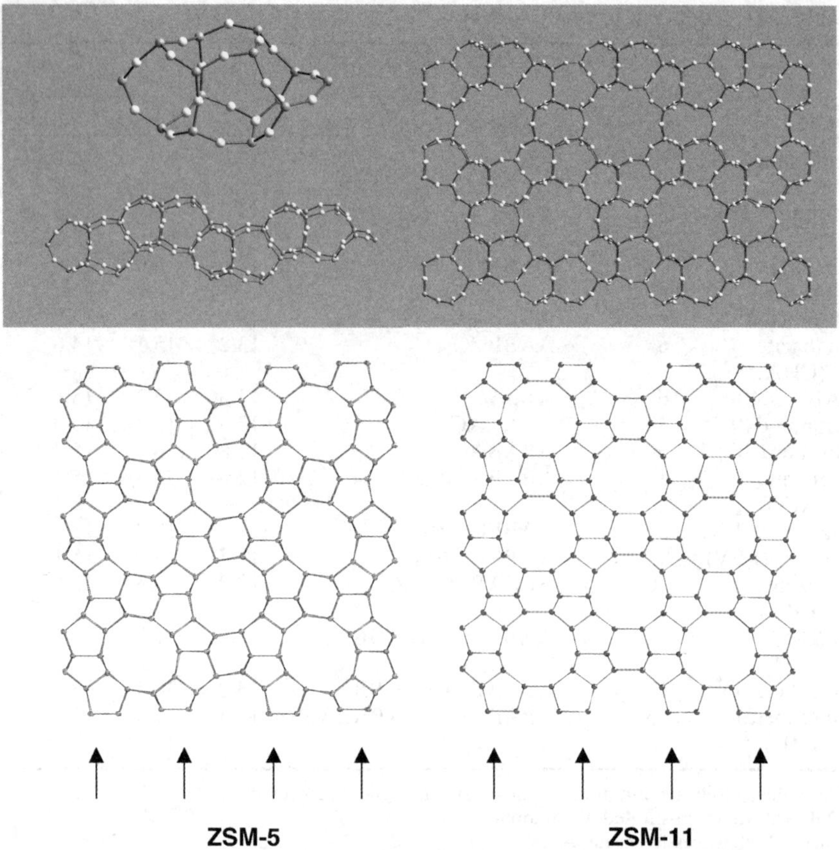

Figure 2.9 The frameworks of the 'pentasil' zeolites ZSM-5 and ZSM-11 are built up from SBUs containing predominantly 5MRs (top left), which assemble into chains and then sheets, the latter with 10MR openings (top right). These sheets (shown edge-on in the lower figures and indicated by arrows) can be connected to other identical sheets either across inversion centres to give the ZSM-5 structure or across mirror planes to give the ZSM-11 structure.

Vaughan.[10] One example of this is the zeolite ECR-1[11] and its gallosilicate analogue TNU-7,[12,13] which are made up of alternating mazzite (MAZ) and mordenite (MOR) sheets that have a close structural fit in two directions (Figure 2.11) and which crystallise at the boundary between the compositional fields responsible for the end-members. Synthetic Paulingite (ZSM-25) is an example where many different structural units are present in the structure, which crystallises at the boundary between phases containing smaller subsets of these SBUs.[10]

While the approach of considering structural units is helpful in the description and understanding of individual framework types, the categories of structure types become less clearly defined as zeolites with lower symmetry

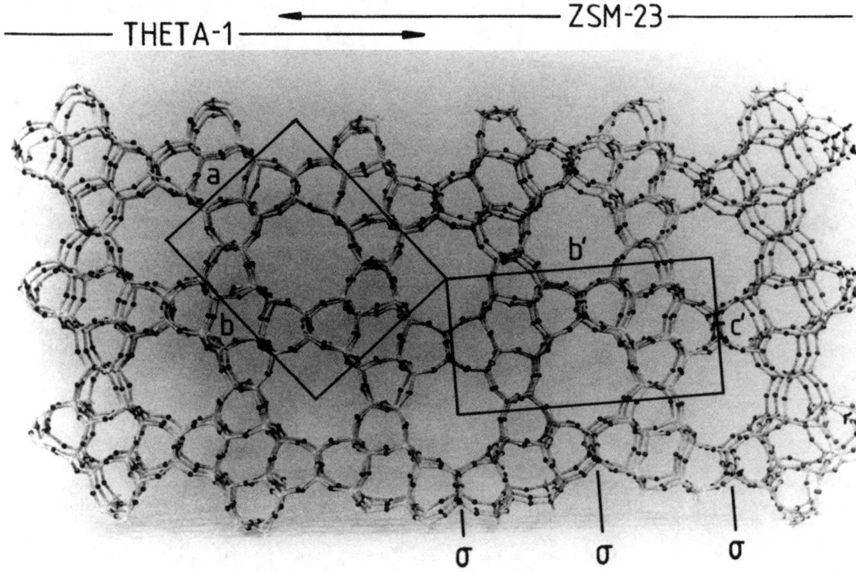

Figure 2.10 Comparison of the structures of the 10MR zeolites Theta-1 and ZSM-23 shows that ZSM-23 is related to Theta-1 by unit cell twinning. The photograph illustrates the unit cells of the two related structures.

and less regular arrangements are considered. A more practical classification is then in terms of the main features of the pores within the structure. The important criteria are the geometry of the pore windows that permit molecular access and transport, the size and shape of the cavities and/or tunnels that make up the pore space and the way in which they are connected. Structural details of the building units are of importance as they determine likely sites for cations and molecules and the acid strength of bridging hydroxyls.

The *pore size* determines which molecules can enter or leave the zeolite and is therefore responsible for its molecular sieving character. Pore sizes are determined primarily by the number of tetrahedral cations in the ring that bound the pore and also by the geometry that the ring adopts. The pore diameter is usually defined in terms of the free diameter, which corresponds to the space available for molecules to pass through the opening once a van der Waals radius of 1.35 Å has been added for each oxygen atom. 6MR pores are impermeable for all except the very smallest molecules (*e.g.* H_2O). For rings that adopt a configuration that is close to planar, the terms small, medium and large pore apply to openings delimited by planar 8MRs (*ca.* 4 Å), 10MRs (*ca.* 5.5 Å) and 12MRs (*ca.* 7.5 Å). This is only a guide, however, as structures with 7MRs, 9MRs, 11MRs and even 14MRs and 18MRs are known. UTD-1,[14] for example, possesses a 14MR elliptical pore with minor and major axis dimensions of 7.5 × 10 Å, and is therefore known as an extra-large-pore zeolite: CIT-5 is another 14MR zeolite.[15] The presence of cations near to the pore windows can modify the effective size for adsorption and this is particularly

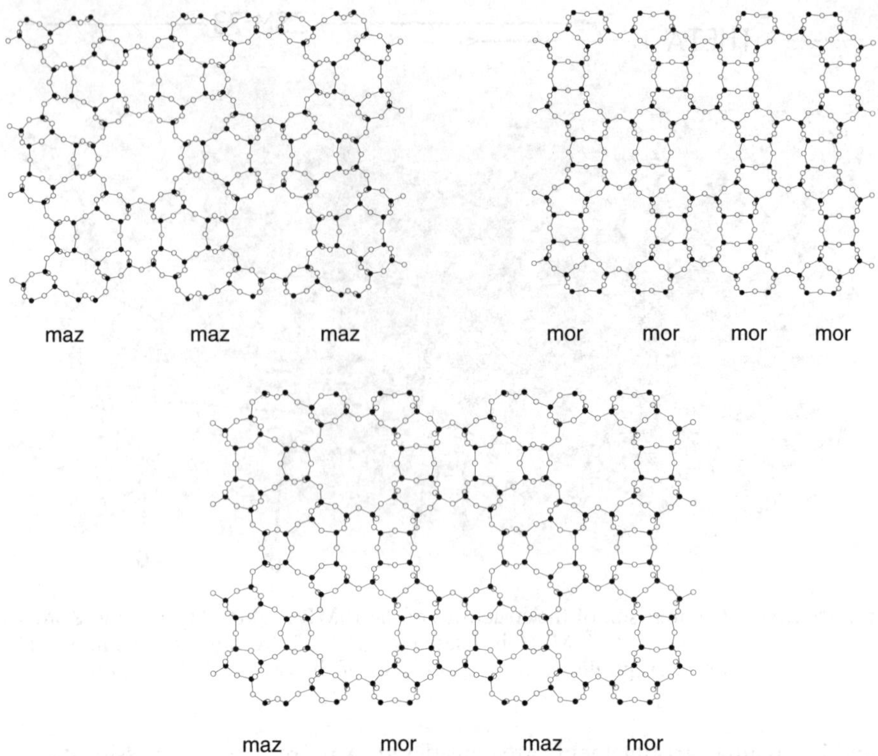

Figure 2.11 The structure of the gallosilicate TNU-7 (and the aluminosilicate ECR-1), shown below, projected down the large channels, can be thought of as being built up from strictly alternating sheets (*maz* and *mor*) that are found in the mazzite and mordenite structures, shown above left and right, respectively.

important in zeolites with high cation contents. For example, for zeolite A in the sodium form, cations near the windows restrict the openings, whereas for the calcium form fewer divalent cations are needed to balance the charge and there is consequently a larger effective pore size (See Chapter 7).

The *geometry of the pore space* may be described in terms of cages, as in zeolites A and X, where the internal free dimensions far exceed the pore window size, or in terms of channels, where the free dimensions are approximately constant and equal to the window size, as in SSZ-24,[16] for example. There may also be combinations of cages and channels within the same structure; two different, unconnected pore systems, as in the zeolite MCM-22;[17] or intersecting channel systems (such as observed in ZSM-5 and in TNU-9[18] (Figure 2.12)).

Connectivity of the pore space is important in applications. One-dimensional connectivity refers to access of a molecule being restricted to a single channel: the many examples include low silica zeolites such as zeolite L and high silica zeolites such as SSZ-24.[2] Two-dimensional connectivity requires a molecule to have access to any part of a planar dimensional arrangement of pore space

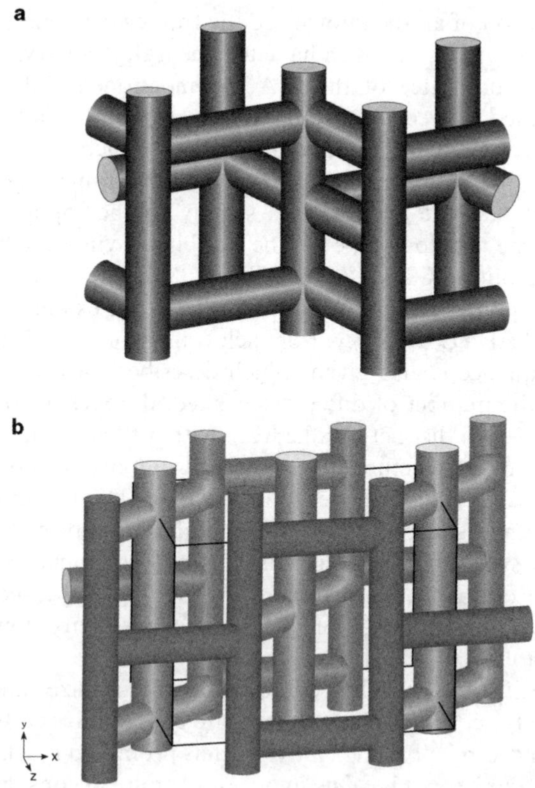

Figure 2.12 ZSM-5 has a channel system in which straight channels intersect with sinusoidal channels (a) whereas the new medium-pore zeolite TNU-9 has a complex pore structure, (b), consisting of two distinct channel systems linked perpendicularly, so that any part of the pore system is accessible from any other.

(as is observed for clays, for example, where molecules cannot pass between the layers). Examples of materials displaying two-dimensional connectivity include the high silica zeolite NU-87.[19] If a molecule can diffuse from a starting point in the pore system to any other point in the pore system without leaving the crystal the pore space may be described as three-dimensionally connected. Zeolites A and Y (and all other cubic zeolites) possess three-dimensional connectivity, as do the small-pore chabazite (or its high silica synthetic equivalent, SSZ-13), the medium-pore ZSM-5 and TNU-9 and the large-pore zeolite Beta. Note that by this definition three-dimensional connectivity does not require the connectivity at any one cage or channel intersection to be in all three dimensions – in ZSM-5, for example, access to adjacent channel systems is possible only by motion along two sets of channels, one straight and one sinusoidal. Zeolites whose pore space is three-dimensionally connected show a much lower tendency to become blocked by the deposition of reaction by-products in catalytic reactions.

A full compilation of all the known zeolitic framework types, based on four-connected nets, the nodes of which have tetrahedral geometry, is kept updated by the structure committee of the IZA. As mentioned earlier, each unique topology is assigned a three-letter code, based on the name of the type material (the framework topology of the mineral faujasite, which is identical to that of zeolites X and Y, for example, has been assigned the framework code FAU). It should be noted that the code applies strictly to the topology and not the material. The same topology type can be exhibited by materials with different compositions. Zeolites X and Y, as well as aluminophosphates, zincophosphates, gallophosphates and beryllium phosphates, for example, can all exist as structures with the FAU topology. For each framework topology a tetrahedral coordination sequence is also given, which describes for each distinct type of tetrahedral site the number of different tetrahedral nodes in consecutive coordination shells of each distinct tetrahedron, without 'counting back'. This acts as a fingerprint for a structure type and is an important aid in establishing whether a new material has a novel topology type. The maximum topological symmetry of the net is also given in the compilation, although real structures can have lower symmetry. The compilation of data includes full structural details for type examples of each framework topology type, including composition, space group symmetry, pore size and connectivity, unit cell size and atomic coordinates.

There are specific reasons why zeolites with window sizes greater than 10 Å would be of great use. That no aluminosilicate has so far been synthesised with an opening bounded by more than a 14 MR has prompted much speculation on the reasons why this should be. One important empirical consideration emerges from the plot of Meier[20] (Figure 2.13), in which minimum framework density is found to correlate with the smallest rings within the structure, so that the presence of 4MRs and even 3MRs can result in lower framework densities than is observed to be possible with materials that contain 5MRs or 6MRs as their smallest components. Examples of very open beryllo- and germanosilicates illustrate this principle (see Section 2.3.1).

Pure silica end-members may be considered as special cases of aluminosilicate zeolites. They may be prepared directly from hydrothermal synthesis and in some cases from aluminosilicates by post-synthetic treatment. For example, the pure silica analogue of ZSM-5 (Silicalite-1) is readily prepared by direct synthesis, whereas purely siliceous zeolite Y can only be obtained by post-synthetic treatment (Chapter 6). The microstructures present in these solids depend on the way in which they are prepared. For direct preparation routes the presence or absence of fluoride as a mineraliser in the preparation (see Chapter 5) determines whether the framework is prepared defect-free or with high concentrations of terminal silanol (SiOH) hydroxyls, where silicon is attached to three bridging oxygen atoms and a hydroxyl group. Post-crystallisation preparation of pure silica zeolites can be achieved by treatment of appropriate starting materials with silicon tetrachloride or by removal of aluminium from the aluminosilicate framework by heating the ammonium form in steam (Chapter 6).

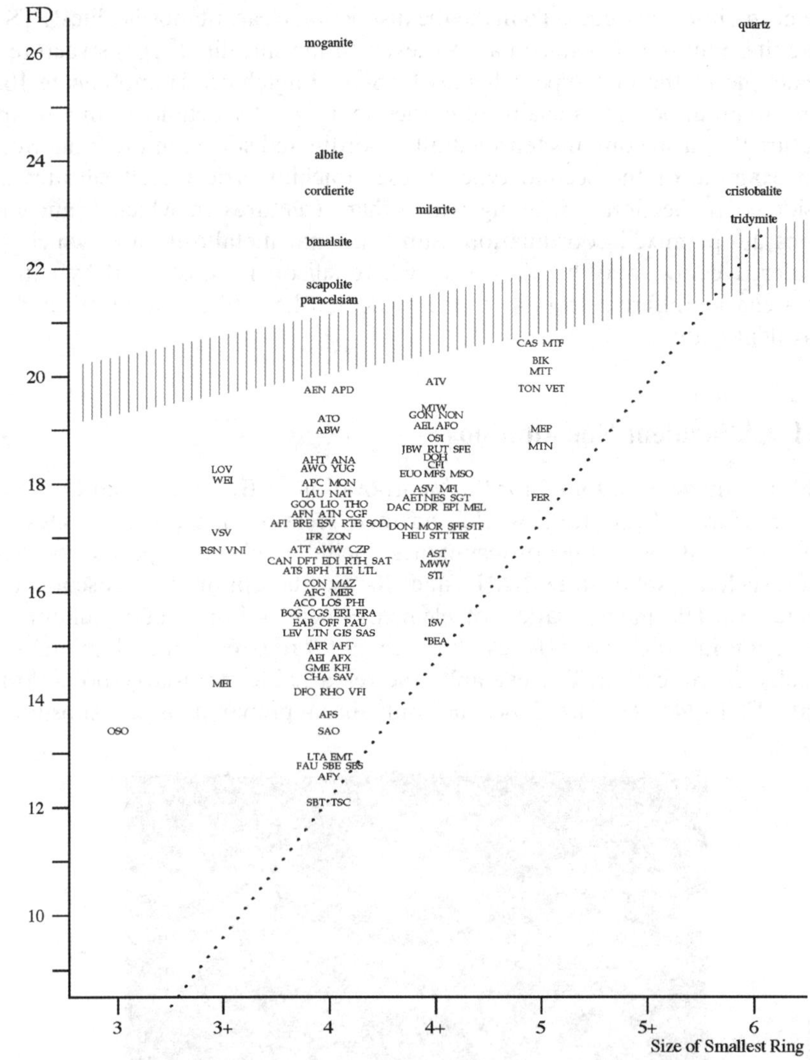

Figure 2.13 Meier's plot of the framework density (FD) of tetrahedral frameworks against the size of the smallest ring present in the structure indicates that the most open structures tend to contain small rings, particularly 4MRs and 3MRs. [Reproduced from reference 2 with permission. Copyright 2001, Elsevier.]

2.3 Substitutional Metallosilicates

For framework metallosilicates other than aluminosilicates we should make the distinction between those materials in which the heteroatom substitutes for silicon in a known zeolite-type framework and those in which it takes up a distinctive framework site in stoichiometric quantities within structures that have no zeolitic analogues. To make the distinction clear, titanosilicalite-1, TS-1, where titanium replaces silicon at low levels in the silicalite (MFI) structure, is an example of the first type, whereas ETS-10, Englehard TitanoSilicate-10,[21] where titanium occurs as chains of corner-sharing TiO_6 octahedra in a unique structure that also contains tetrahedrally coordinated silicon in the framework, is an example of the second type. These stoichiometric metallosilicates are considered in Section 2.5, along with other structures in which framework cations adopt mixed coordination. Substitutional metallosilicates, which are considered below, may be aliovalent, where silicon is substituted by cations with a charge different from 4+, or isovalent, where silicon is substituted by tetravalent cations.

2.3.1 Aliovalent Substitutions

Conditions have been found for the incorporation of B^{3+}, Ga^{3+} and Fe^{3+} into the tetrahedral silicate framework, and their location in tetrahedral sites has been confirmed by a range of techniques, including element specific spectroscopies such as solid state NMR and X-ray absorption spectroscopy (see Chapter 3). The incorporation of elements such as boron and gallium can also result in framework types that are not prepared with aluminium as the substituting cation. For example, the remarkable extra-large-pore 18MR silicate ECR-34[22] (Figure 2.14) has only been prepared as a gallosilicate.

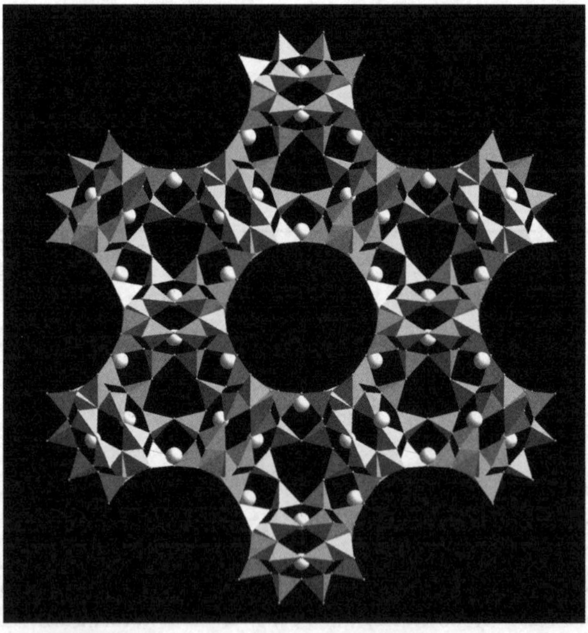

Figure 2.14 The gallosilicate ECR-34 was the first reported crystalline tetrahedrally connected framework silicate with 18MR channels.

The incorporation of trivalent cations of charge less than 4+ in place of silicon atoms imparts a net negative charge to the framework and gives the potential for properties of ion exchange and solid acidity, as does the inclusion of aluminium, but, in general, the solid acidity of such materials is weaker and the substituting cations are more likely than aluminium to leave the framework.

The substitution of divalent metals such as Be^{2+} and Zn^{2+} into tetrahedral positions has also been observed, often with the result of producing unique structures, such as the zincosilicate VPI-7[23] and the beryllosilicate OSB-1.[2] The OSB-1 framework shows remarkable openness, with a framework density of only 13.4 T/ 1000 Å^3, and this can be rationalised by the presence of a large proportion of 3MRs, which is predicted on the basis of Meier's plot (Figure 2.13) to permit less dense framework types. It is not thermally stable, however.

2.3.2 Isovalent Substitutions ($M^{4+} \Leftrightarrow Si^{4+}$)

The most important isovalent substitution in silicates is titanium for silicon. The resulting titanosilicate analogues of zeolites have particularly attractive catalytic properties for selective oxidation. The titanosilicalites-1 and -2 (TS-1 and TS-2) were the first solids of this kind to be prepared – TS-1 is the titanosilicate analogue of ZSM-5 and TS-2 is the titanosilicate analogue of ZSM-11. Titanium-Beta and other large-pore, titanium-containing high silica zeolites have also been synthesised. X-ray EXAFS spectroscopy has shown that in the as-prepared solids titanium is in five-fold coordination, and adopts tetrahedral coordination upon calcination and in the absence of adsorbed water molecules.[24] Of all the substitutions other than aluminium, titanium loading at low levels is the most important and the best defined. There is little loss of titanium from the framework sites upon activation or during catalysis and the system offers an ideal opportunity to study an active site for catalytic oxidation, as discussed further in Chapters 7 and 9.

Silicates can also be prepared with germanium in the framework cation sites. Recently, the inclusion of germanium has been found to direct the synthesis to novel types of high silica zeolite structures. In particular, the slightly larger size of germanium compared to silicon atoms (and the smaller $O\hat{T}O$ angles this permits) favours D4R building units and the synthesis of novel large-pore structures such as the C-polymorph of zeolite Beta[9] and the ITQ-21 structure[25] (Figure 2.15). In both of these cases there is a three-dimensionally connected large-pore system, with three channel systems running at right angles to one another. It is notable that a structure with the tetrahedral topology of polymorph C of zeolite Beta (BEC) was first synthesised phase pure as the germanium oxide analogue, FOS-5[26] from which the template could be thermally removed. In addition, the isostructural germanosilicates IM-12[27] and ITQ-15[28] have pore systems limited by 12MRs and 14MRs and the novel germanoaluminosilicate ITQ-33,[29] which contains both 3MR and D4R units, has a very open structure (12.3 T atoms per 1000 Å^3) with 18MR channels in one direction, linked *via* 10MR openings (Figure 2.16). This underlines the versatility of

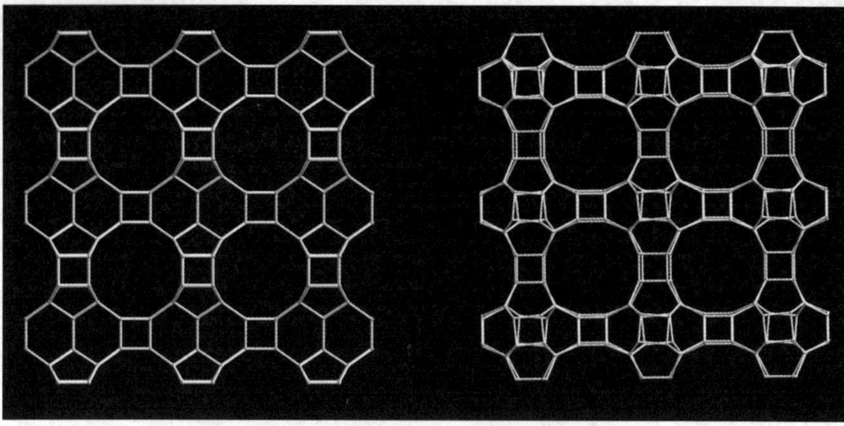

Figure 2.15 Framework structures of the germanosilicates Beta C (left) and ITQ-21 (right), each of which possesses three-dimensionally connected large pore (12MR) channel systems. These germanosilicates are characterised by the presence of D4Rs in their frameworks: D4Rs are favoured by the presence of germanium in framework cation positions.

germanium as a framework-forming element. Further work has yielded fascinating mixed coordination germanates (Section 2.5.3).

2.4 Metallophosphate Zeotypes and Related Materials

2.4.1 Aluminophosphates ($AlPO_4$s)

The discovery of the aluminophosphate ($AlPO_4$) molecular sieves by Flanigen, Wilson, Patton and others of Union Carbide[30] heralded a burgeoning of the chemistry of non-silicate microporous solids, which has formed part of the great increase in structure types shown in Figure 2.2. The $AlPO_4$s are a family of solids that typically possess a neutral framework, composition $AlPO_4$, made up of alternating corner-sharing aluminate and phosphate tetrahedra. Aluminophosphates are usually prepared in the presence of amines or quaternary alkylammonium ions as structure-directing agents. Some possess structural types closely similar to those of zeolites (analogues of zeolites sodalite, chabazite, A and Y are known, among many others); others possess structures unique to this composition. The most widely observed aluminophosphate structure type, readily prepared in the presence of a wide variety of organic additives, is $AlPO_4$-5, a large-pore solid with one-dimensional pores bounded by 12-membered rings (Figure 2.17).[31] This structure type was first observed as an aluminophosphate, but has since been prepared as a silicate (SSZ-24)[16]: the same applies to the $AlPO_4$-36 structure (Figure 2.17), more recently prepared as the silicate SSZ-55.[32] As a result of the strict alternation of aluminium and phosphorus cations in tetrahedral sites, only frameworks made entirely of even-numbered rings are observed, so there are no aluminophosphate

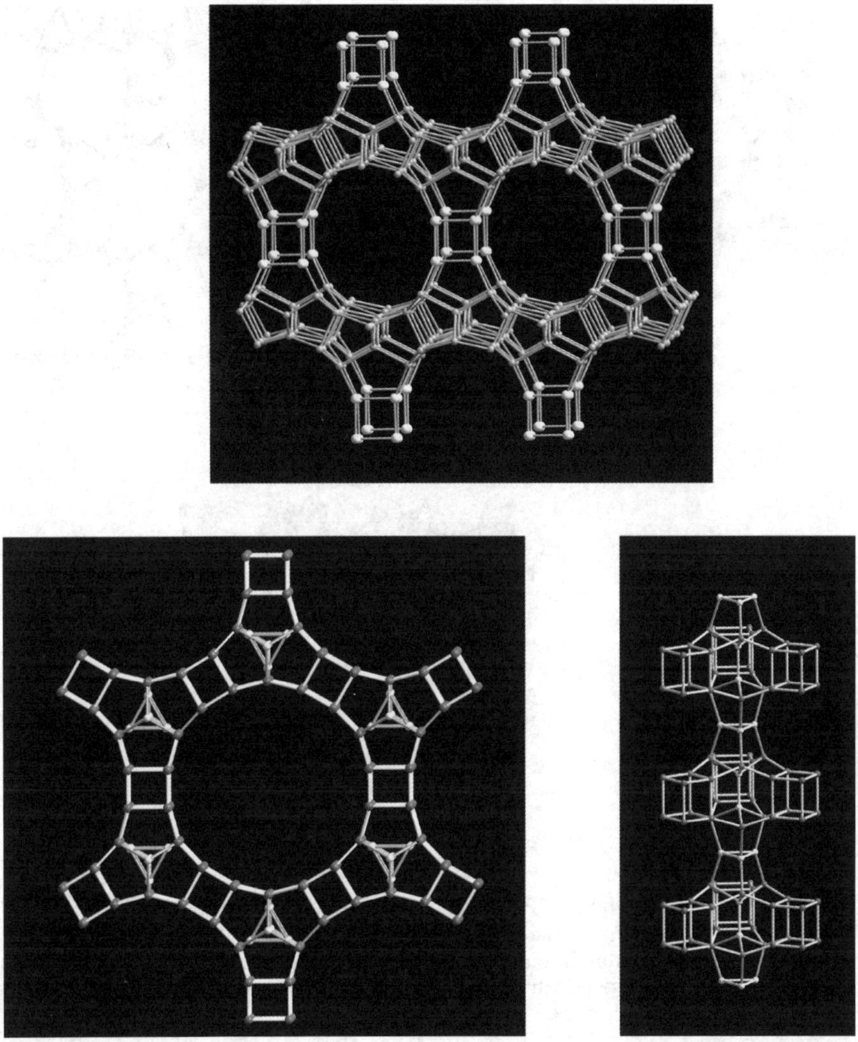

Figure 2.16 The germanosilicates IM-12 (above) and ITQ-33 (below, left and right) contain pores limited at their narrowest by 14MRs and 18MRs, respectively. ITQ-33 contains both 3MRs and D4Rs in its sub-units.

analogues of ZSM-5, for example. In addition, aluminophosphates with framework structures not yet observed as zeolites can be prepared – the most notable of which is the ultra-large-pore solid VPI-5 of Davis *et al.* (Figure 2.17), which possesses one-dimensional channels bounded by 18MRs.[33] In fact, in as-prepared VPI-5, aluminium in one of the sites is octahedrally coordinated with additional water molecules, but these can be removed to leave a tetrahedral framework. Although the channels of VPI-5 possess a larger free diameter than observed in any zeolite (12.7 Å free diameter), the utility of this solid is reduced by its limited hydrothermal stability, because it loses crystallinity at high

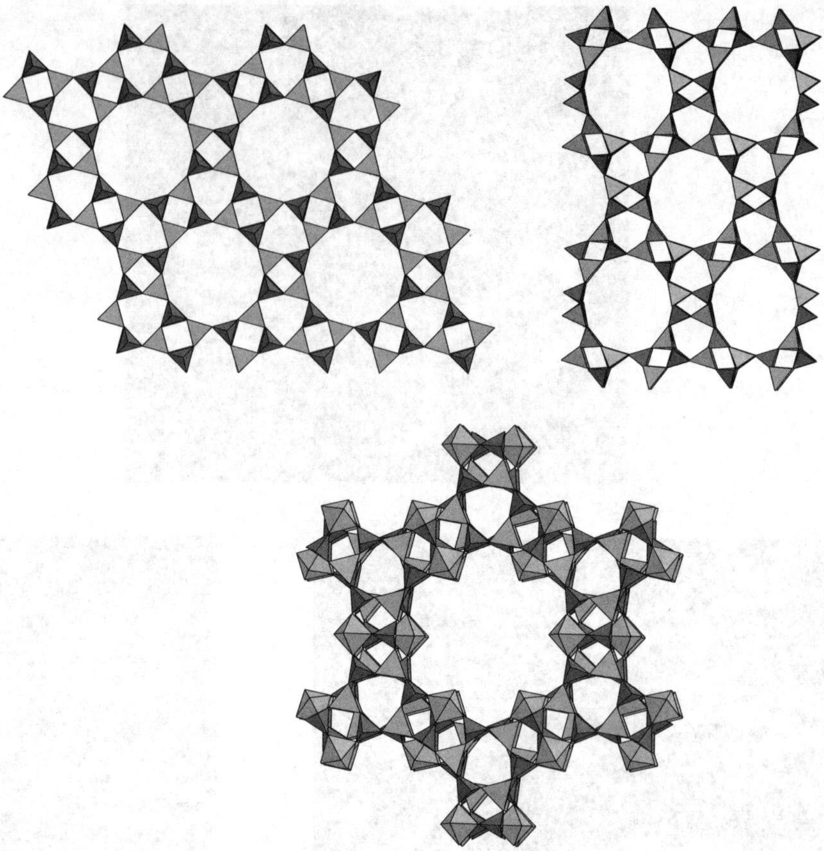

Figure 2.17 The large pore (12MR) aluminophosphate structures (top left) $AlPO_4$-5 and (top right) $AlPO_4$-36 and the extra large pore (18MR) VPI-5 (below) possess one-dimensional channel systems. In the as-prepared VPI-5 (shown) aluminium exists in both tetrahedral and octahedral coordination, the octahedral coordination being made up by two coordinated water molecules.

temperatures in the presence of water vapour. A full list of aluminophosphate structure types is given in the review of Patarin *et al.*[34] and reference to the more important solids is made in Table 5.2. Although of structural interest and adsorptive interest, the neutral aluminophosphates are not catalytically active, but the introduction of atoms other than aluminium or phosphorus into the tetrahedral sites has been shown to introduce both solid acidity and redox activity.

2.4.2 Substituted $AlPO_4$s

A wider and chemically different range of heteroatomic substitutions is possible in aluminophosphates than in zeolites. This is partly because the framework has a more ionic character, according to simulations (see Chapter 4) and is therefore able to take in a wider range of metal cations, and also because the conditions of synthesis (typically pH 7) enable more metal cations to be in solution during crystallisation. In addition, substitution can occur at either the aluminium or the phosphorus sites.[35] Well-characterised substitutions include the introduction of di- and trivalent cations in place of aluminium[36] and of silicon in place of phosphorus.[37]

Divalent metals that have been substituted for aluminium in the framework include Mg, Mn, Fe, Co and Zn. Typically the negative charge imparted to the framework is balanced in the as-prepared sample by protonated amines or alkylammonium cations incorporated during synthesis, and a typical composition can be described as

$$M_x^{2+}Al_{1-x}PO_4.R_{x/m}^{m+}$$

Removal of the templates from MgAPOs by calcination in oxygen can result in bridging hydroxyl groups or in the generation of Lewis acid sites which are thought to be Mg^{2+} cations that are not fully tetrahedrally coordinated within the framework.[38] In the case of transition metals such as Mn^{2+}, Fe^{2+} and Co^{2+}, calcination in oxygen results in the oxidation of some or all of the cations to the trivalent state.[39] These can then be reduced back to the divalent state and protons introduced (Chapters 3 and 9). For substituted metal cations that show no redox behaviour, such as Mg^{2+}, calcination results in solid acid catalysts, with both Brønsted and Lewis acidity, whereas for Mn-, Fe- and CoAPOs the solids have been shown to possess the ability to act as oxidation catalysts (Chapter 9).

Phosphorus has been most successfully substituted by silicon, although there are many reports that other elements, including titanium and vanadium, can be introduced. The direct replacement of isolated phosphorus atoms by silicon atoms results in a negatively charged framework and, upon calcination, an acid site. Since this can take the form of a bridging hydroxyl between aluminium and silicon atoms, it is similar in character to those found in aluminosilicate zeolites.

$$H_xAlP_{1-x}Si_xO_4$$

In some structures, silicon is incorporated by a second mechanism, where aluminosilicate 'islands' are developed within the frameworks by the substitution of both Al and P by Si, and the two mechanisms are clearly distinguished by NMR (Chapter 3). Silicon has never been observed to substitute into isolated aluminium sites where they would then be linked *via* bridging oxygens to phosphorus atoms. However, one interesting variation on the cation arrangement in aluminophosphates is exhibited by the silicoaluminophosphate ECR-40.[40] This has the same framework topology as the zeolite ZSM-18[41] (topology MEI) which contains three-membered rings, so that strict Al-O-P(Si) ordering cannot be maintained. In fact, the framework cations display a unique ordering scheme, in which Al-O-Al bonds occur.

Many of the structures made originally as aluminophosphates have been found to be able to incorporate heteroatoms, although the readiness with which they are able to do this varies strongly from one structure type to another, and between added elements. The $AlPO_4$-36 structure,[42,43] for example, readily takes in divalent cations for aluminium, but the silicoaluminophosphate SAPO-36 has not yet been prepared. In addition, the introduction of metal cations other than aluminium has given rise to structures that were not previously observed as pure aluminophosphates, although AlPO forms have in some cases later been prepared. This is due to the enhanced ability of such frameworks to accommodate charged templates. The substituted aluminophosphates DAF-1[44] and the STA-n family of solids (Figure 2.18),[45–49] prepared using dicationic, tricationic and azamacrocyclic templates (Chapter 5) are good examples of this. The structure of DAF-1 (see illustration on the front cover) is a rare example of a large-pore material with two different, interconnected pore systems.

Metal phosphates with distinctive, fully tetrahedrally coordinated frameworks have also been prepared for metals such as beryllium, cobalt, zinc and gallium, although gallophosphates often show mixed coordination geometry. Pure templated cobalt phosphates and cobalt aluminium phosphates with high

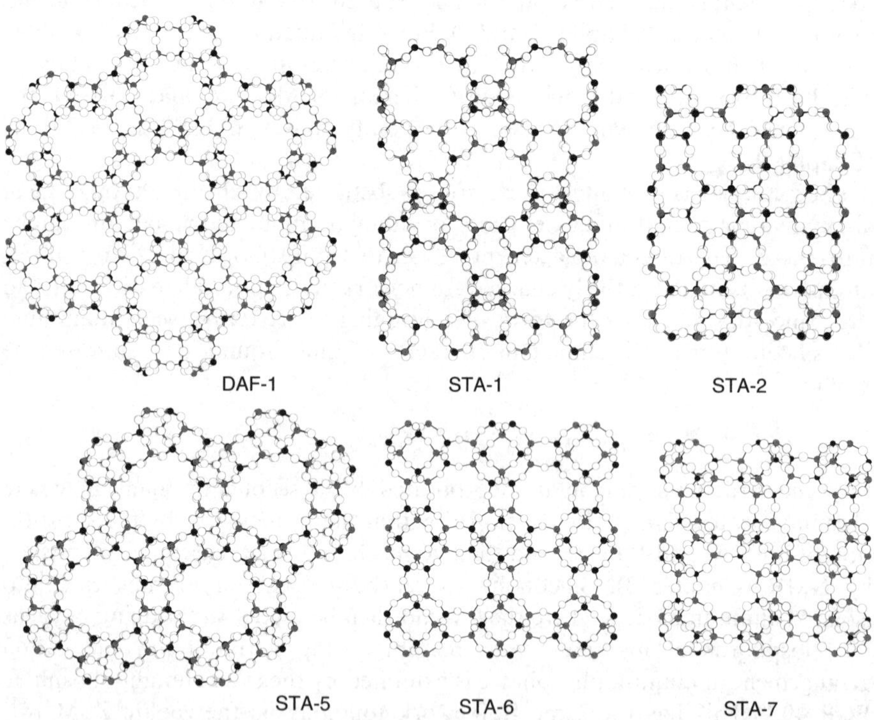

Figure 2.18 DAF-1 and the STA-n materials are examples of substituted aluminophosphates (MAPOs, SAPOs) that are not readily prepared as pure $AlPO_4$s.

Co/Al ratios that possess novel fully tetrahedral frameworks have been reported, including the UCSB materials of Bu *et al.*[50] Unfortunately the instability to template removal of this class of phosphate remains a severe limitation to their use.

2.5 Mixed Coordination Inorganic Frameworks

As well as the 170 or so tetrahedral framework structures currently described in the 'Zeolite Atlas', an increasing variety of porous framework structures are being determined in which the framework cations exhibit mixed coordination. Typically, this involves cations in both octahedral and tetrahedral coordination by oxygen, where the tetrahedral species may be silicates or phosphates and the octahedral cations are larger cations, such as first row transition metals, elements of Groups III and IV heavier than aluminium and silicon, and even lanthanides. Porous framework structures are also known that possess cations in entirely octahedral coordination, or in geometries that are neither tetrahedral nor octahedral. Aluminium, for example, can exist in four-, five- and six-fold coordination, and sometimes in a mixture of two different geometries. In the fluorinated aluminophosphate UT-6[51,52] aluminium exists both tetrahedrally and octahedrally coordinated, and in the octahedra the coordination includes two fluoride ligands left over from crystallisation (Figure 2.19). In as-prepared VPI-5 some of the aluminium cations achieve octahedral coordination with water molecules. In both these cases the extra ligands may be removed by heating to leave a porous tetrahedrally coordinated net but, in others, the mixed coordination is essential for the integrity of the framework. Mixed geometry inorganic frameworks are commonly encountered in solid state chemistry, but relatively few show porosity for molecules other than water. Representative examples are discussed below. A wide range of open framework phosphates, arsenates and sulfates with novel compositions can be prepared with amines or organic cations as templates, but only a small number can be rendered porous and crystalline.

2.5.1 Metallosilicates

Microporous framework silicates that contain metal cations in octahedral coordination remain relatively few in number, particularly those with pore sizes greater than 4 Å (small pore). The MO_6 octahedra (or tetragonal pyramids in some cases) can exist isolated within silicate frameworks, with all of the oxygens shared with silicate tetrahedra, connected in chains within a silicate network or as layers of edge-sharing octahedra, sandwiched between porous silicate layers. The attraction of such silicates is that they potentially provide an opportunity to combine shape selective adsorption with catalytic, optical, magnetic and electronic properties exhibited by transition and other metals that favour octahedral geometry. Examples of such materials, grouped according to the way the metal coordination octahedra are arranged, are given below.

Figure 2.19 The structure of the aluminophosphate fluoride UT-6 is based on the CHA framework but includes coordinated fluoride ions linking octahedrally coordinated aluminium cations (white octahedra).

2.5.1.1 Frameworks Containing Isolated Octahedra, Tetragonal Pyramids and Clusters of Octahedra

A wide structural and chemical variety of metal silicates possess porous frameworks with isolated octahedrally coordinated cations in a silicate network. Many of these are minerals, but in several cases their synthetic analogues have been prepared hydrothermally. Most are small pore solids which display ion exchange capacity and reversible water adsorption. Among the titanosilicates of this kind, the framework structure of $K_2TiSi_3O_9 \cdot H_2O$ (as in the mineral umbite – with synthetic analogues reported independently by several groups[53,54]) is made up of titanate octahedra and silicate tetrahedra, resulting in a small-pore one-dimensional channel system. Analogues of this structure containing zirconium[54] and tin[55] have also been prepared. The review of Rocha and Anderson[56] details other such materials. There are also vanadosilicate minerals, cavansite and pentagonite, in which VO_5 tetragonal pyramids (containing the VO^{2+} vanadyl grouping) link silicate sheets to give stable frameworks. Synthetic (small-pore) solids related to each of these have been prepared by the group of Jacobson.[57]

There are also metallosilicates in which the metal cations exist octahedrally coordinated within clusters. Two related small-pore structures with this arrangement are of considerable interest as cation exchangers. The first is a synthetic titanosilicate that is isostructural with the arsenate mineral pharmacosiderite ($KFe_4^{3+}(AsO_4)_3(OH)_4 \cdot 6H_2O$), the second a synthetic version of the mineral sitinakite ($Na_2Ti_2SiO_7 \cdot 2H_2O$). In each structure the octahedrally coordinated titanium atoms exist in clusters of four octahedra, $(Ti_4O_{16})^{16-}$, linked *via* their edges in a 'cubane-like' geometry. In the pharmacosiderite-like structure $M_4(TiO)_4(SiO_4)_3 \cdot 4H_2O$, where M is a univalent cation,[58] each titanate cluster is linked to six others *via* corner-sharing oxygen atoms belonging to silicate tetrahedra, making a simple cubic lattice (Figure 2.20). Charge-balancing cations occupy sites between the silicate tetrahedra and within the channel space. In the sitinakite structure the titanate clusters are linked to four others in the same plane *via* silicate tetrahedra in a similar way, but are connected above

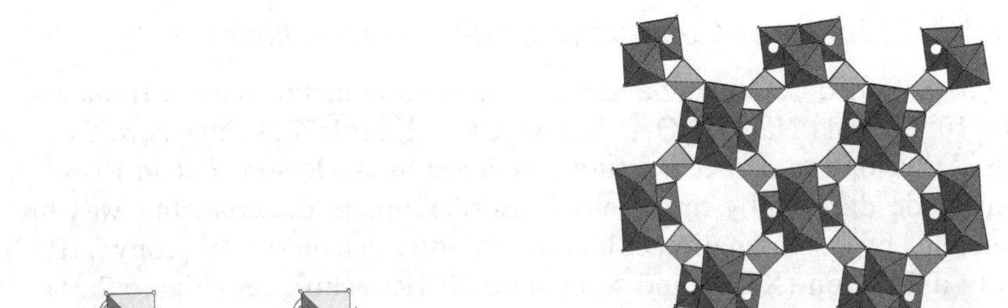

Figure 2.20 The pharmacosiderite structure possesses Ti_4O_{16} clusters of edge-sharing titania octahedra linked in three dimensions by silicate tetrahedra. Cations can reside in these channels (left). In the distorted protonic form (right, above) protons are attached to μ_3O oxygen atoms shared by three titanium cations (right, below).

and below this plane (along the crystallographic c-axis) directly to other titanate clusters.[59,60] In this crystalline silicotitanate, synthesised industrially as CST (IONSIV® IE-911 UOP Inc., Des Plains, IL) extra-framework sodium cations occupy window sites between silicate tetrahedra or within the 8MR channels running along the c-axis. All of the sodium ions can be exchanged. These structures display very wide chemical variation, by substitution in both octahedral (cluster) and tetrahedral sites within the framework and also by ion exchange of the charge-balancing cations.[61] CST can be ion-exchanged in acid to give the protonic form. Neutron diffraction indicates that the structural formula of the H-form can best be written $Ti_2(OH)_2SiO_5 \cdot 1.5H_2O$ because the protons reside on oxygen atoms of the Ti_4O_{16} clusters.[62] These hydroxyl groups are each bound to three titanium atoms of the cluster. The structural OH groups point into the channel space, where they form H-bonds with adsorbed water. The solids also show high selectivity for large alkali and alkaline earth cations whose radii closely match the size of the channel sites, as described in Section 6.3.

2.5.1.2 Materials Containing Chains of Octahedra

The best studied example of a mixed coordination metallosilicate framework is ETS-10, formula $Na_2TiSi_5O_{13}$, discovered by Kuznicki at Engelhard Corporation.[63] Elucidation of the structure, published by Anderson *et al.* in 1994,[21] had been made difficult by the level of disorder in the material, but was finally completed by a combination of high resolution electron spectroscopy (HREM), X-ray diffraction (XRD) and solid state NMR techniques (Figure 2.21). The structure is made up of chains of corner-sharing TiO_6 octahedra that share the four other oxygen atoms with SiO_4 tetrahedra that form part of a silicate framework. This results in a three-dimensionally connected large-pore solid with a channel system similar to that found in zeolite Beta. Like that of zeolites, the framework has a net negative charge and the as-synthesised form includes sodium ions to balance the framework charge.

In the cationic form, ETS-10 possesses potential as a gas adsorbent and as a basic catalyst, but cannot be converted to the acid form without structural collapse. Furthermore, because the titanium is already octahedrally coordinated, it is unable to expand its coordination sphere to accommodate reactants and products in the same way as titanium in TS-1 and Ti-Beta, and is therefore not catalytically active for selective oxidation. One of the most important structural features is that ETS-10 contains 'insulated' –Ti-O-Ti-O-chains arranged in layers. Chains of TiO_6 in stacked sheets run orthogonal to each other and are displaced by $\frac{1}{4}$ unit cell. Stacking sequence defects occur and give rise to extra large pores, which accounts for the wide-pore behaviour it displays. The original publication did not state the location of the cations in the framework, but the location of the cations has subsequently been determined to be principally in the sites adjacent to the –Ti-O-Ti-O-chains where they balance the charge on the titanium octahedra, which is -2 (6 shared O^{2-} per Ti).[64] Framework substitution into the Si and Ti sites of ETS-10 has been shown

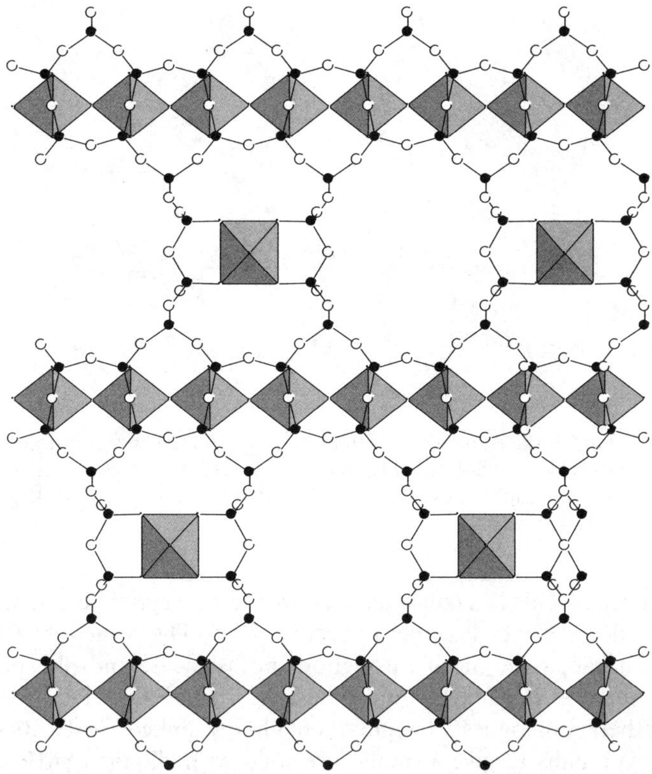

Figure 2.21 In the large-pore mixed coordination titanosilicate ETS-10, chains of corner-sharing TiO_6 octahedra are linked via silicate tetrahedra. The offset of the chains of octahedra between consecutive layers exhibits disorder on the long range, and end-member polytypes have not yet been prepared.

to be possible. Aluminium, gallium and boron can be substituted for silicon[65,66] (in sites that are not linked to titanium, see Section 3.4.1.2.6) and chromium, niobium[67] and vanadium[68] can be substituted for titanium.

Earlier than the discovery and structure solution of ETS-10, Chapman and Roe published the synthesis of a titanosilicate that gave a diffraction pattern which resembled that of the rare mineral zorite.[69] Independently, Kuznicki patented ETS-4, which is also closely related to zorite. The structure of the synthetic ETS-4 has been the subject of careful study, since it is complicated by disorder within an ordered framework.[70] The 'sub-structure', or basic unit from which the framework is built up, contains two kinds of titanium: chains of corner-sharing TiO_6 octahedra running in two directions, in a similar fashion to those in ETS-10, and isolated tetragonal pyramidal TiO_5 units (Figure 2.22). The basic unit contains both 12MR and 8MR channels, but frequent faulting in the structure results in the interruption of the 12MR channels, so that the 8MR windows control access of adsorbing molecules to the internal pore space. The structure is very flexible, particularly during dehydration, when cation

Figure 2.22 Part of the framework of the small-pore mixed coordination titanosilicate structure ETS-4, in which silicate 8MRs limit access to the internal pore space. (Light grey silicate tetrahedra are linked by dark grey titanate octahedra).

migration occurs. This has consequences for the adsorption and separation for small molecules, described further in Section 7.5.4. The corner-sharing titanate chain is a characteristic motif throughout both silicate and phosphate structures. The mineral nenadkevichite, (Na,Ca)(Nb,Ti)$Si_2O_7 \cdot H_2O$, for example, and its synthetic analogues[71] include these chains, linked *via* 4MRs of corner-sharing silicate units to give a small-pore solid with elliptical 8MR channels.

The group of Jacobson has reported a microporous copper silicate that contains silicate layers linked by $CuO_4(OH_2)_2$ groups, themselves arranged as corner-sharing chains.[72] The structure has a neutral framework, in which alkali metal hydroxide species reside after synthesis, and possesses pores bounded by large 12MRs. So far it has not been possible fully to remove species from within the pores, so that the potential porosity of the solid has not been achieved, but it can be reversibly dehydrated.

2.5.1.3 Materials Containing Sheets of Edge-sharing Octahedra

Another family of silicates groups together structures that contain edge-sharing octahedra arranged in sheets that are sandwiched by microporous layers composed of silicate tetrahedra. This class of materials includes yttrium, calcium, cerium, europium and terbium silicates. The synthesis and characterisation of the first microporous yttrium and calcium silicates was reported by Rocha and co-workers in 1998.[73] The yttrium and calcium silicates, $Na_4K_2Y_2Si_{16}O_{38}.10H_2O$ and $HKCa_2Si_8O_{19}.6H_2O$, possess the structures of the minerals monteregianite and rhodesite. The structure of these materials is indicative of the whole class of such solids to date; namely, the presence of sheets of octahedra – for sodium yttrium silicate these include YO_6 and $NaO_4(OH)_2$ octahedra – sandwiched between porous silicate layers of the apophyllite type

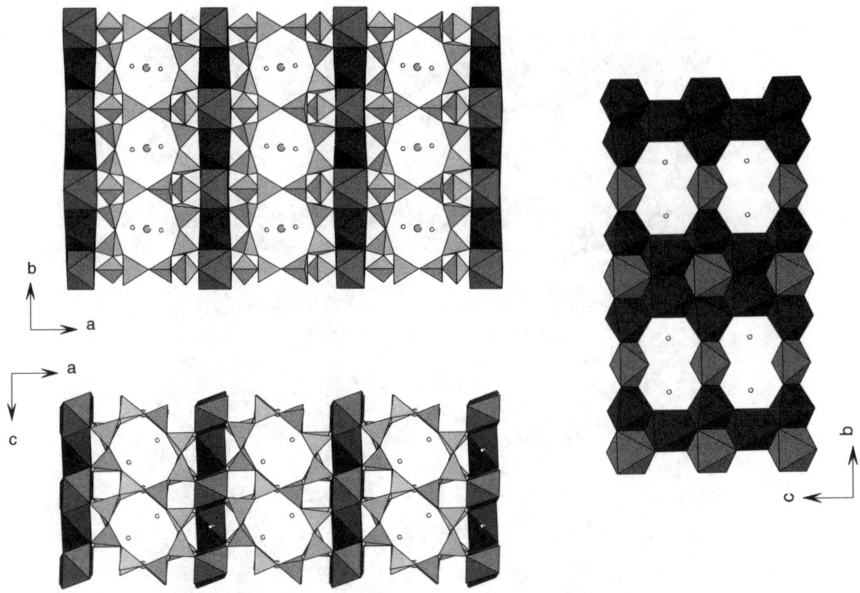

Figure 2.23 The microporous potassium sodium europium silicate AV-9 has a structure similar to the natural sodium calcium silicate monteregianite. Layers containing both edge-sharing NaO_6 (dark grey) and EuO_6 (medium grey) octahedra are separated by microporous sheets of silicate tetrahedra. The pore space is filled by extra framework potassium cations and zeolitic water.

(Figure 2.23). The monteregianite structure was again observed in a novel synthetic cerium (III) silicate reported by Rocha *et al.* in 2000.[74] This was the first lanthanide silicate synthesised and it displayed some of the optical properties known for Ce(III) compounds (*e.g.* lasing in the high energy blue spectrum). Two further lanthanide silicates were reported in 2001, europium and terbium silicates.[75] Both contained the characteristic layers of edge-sharing octahedra linked by slabs of porous silicate (see Figure 2.7) and both displayed photoluminescence.

2.5.2 Metallophosphates

One of the most notable mixed-coordinate metal phosphates is the gallophosphate cloverite, Figure 2.24, so-called because of the shape of the 20MR pore opening.[76] In fact, cloverite is a mixed-coordination interrupted structure, with most of the gallium cations surrounded by four oxygen atoms and a fluorine atom, and with some of the phosphate groups surrounded by three oxygens bridging to gallium and a terminal hydroxide group. Cloverite is truly a microporous solid, however, since the structure directing agent can be removed by heating, and it then adsorbs 1,3,5-trimethylbenzene, indicating its large-pore character.

Figure 2.24 The gallophosphate cloverite is an 'interrupted' framework structure, containing gallium in tetrahedral and octahedral coordination and PO_4 tetrahedra (dark grey) that share four or three oxygen atoms with gallium atoms. The protons of PO_4H groups point into the large cavities, giving the 'four-leafed clover' motif.

The nickel phosphates VSB-1[77] and VSB-5,[78] recently reported by Férey, Cheetham and co-workers, possess nickel in octahedral coordination as well as tetrahedral phosphate and hydrogenphosphate groups. The fully inorganic, large-pore VSB-5, for example, is prepared hydrothermally with composition $Ni_{20}[(OH)_{12}(H_2O)_6][(HPO_4)_8(PO_4)_4].12H_2O$, and contains channels that are bounded by 24 NiO_6 octahedra that share faces, edges and corners. The solid is stable at temperatures up to 400 °C and is the most porous mixed coordination metal phosphate structure.

The structures of many other metal phosphates have been reported in which the metal is in mixed or non-tetrahedral conformation, but without much information on the porosity or thermal stability. The vanadyl phosphates reported by Schindler and Baur,[79] for example, can be considered to be built up from planar $V_5O_9(PO_4)_4$ units that share phosphate tetrahedra. These have the same topology as 4MRs, so that analogue structures that are similar to zeolite structures that are made up completely of 4MRs such as sodalite and Rho can be prepared.

2.5.3 Germanates

Germanium is chemically similar to silicon, and as such is expected to show similar framework-building behaviour. The ability of germanium oxide to form zeolite-like tetrahedral lattices has already been described. In addition, the slightly larger size of germanium compared to silicon (Ge-O of 1.76 Å *cf.* 1.61 Å for Si-O) permits it to adopt four-, five- and six-fold coordination, rather than just tetrahedral coordination, and allows smaller O-M-O bond angles. This results in a higher tendency than silicates to form 3MRs, the incorporation of which tends to favour frameworks with low densities. As a consequence of this diverse crystal chemistry, a wide variety of open framework germanates and metallogermanates can be prepared hydrothermally in the presence of amines. Among the growing family,[80] the germanate fluoride ASU-16 is of particular interest because of its low framework density (only 8.6 cation sites per nm^3) and 24MRs but it has low thermal stability.[81] The material contains building blocks with germanium atoms in three coordination geometries, tetrahedral GeO_4, trigonal bipyramidal GeO_4F and octahedral GeO_5F. The stability can be enhanced by the partial substitution of germanium atoms by silicon atoms.

In practice, this ability to prepare large-pore germanates using mixed-coordination building blocks has turned out to be a 'curtain-raiser' for the crystalline mesoporous germanate, SU-M, of Zou *et al.*[82] of Stockholm University. They report the hydrothermal synthesis of a germanate with one of the largest unit cells of any crystalline porous solid (cubic Ia-3d, a = 51.3 Å, primitive unit cell volume 67, 640 $Å^3$) that is built from a single type of cluster, formula $Ge_{10}O_{24}(OH)_3$, which contains six GeO_4 tetrahedra and four GeO_6 octahedra (Figure 2.25). Each cluster is linked through shared oxygens to five others, leaving the three hydroxyl groups projecting into the mesopore volume. The linked building units fit on a cubic, periodic 'minimal surface' structure of the type seen elsewhere in amphiphilic surfactant liquid crystalline structures and in mesoporous silicas, such as MCM-48 (see Section 2.9). The structure contains two interpenetrating channel systems of opposite chirality, each linked *via* 30MR windows with dimensions of 10 × 22.4 Å. The amines present in the channels can at least partially be removed to give a porous structure. Remarkably, in a modified preparation, crystals have been found in which one of the enantiomeric channel systems is filled with additional germanate species, giving rise to a chiral crystalline mesoporous solid. This material therefore offers a range of opportunities for development, for example of crystallographically well-defined catalytic sites in a chiral environment.

2.6 Microporous Metal Oxides – Octahedral Molecular Sieves

Inorganic molecular sieves in which all of the framework cations are coordinated octahedrally comprise a small but significant family of microporous solids. The octahedral molecular sieves, or OMS materials, related to manganese oxide minerals of the hollandite family, are the most important of these. Examples have been prepared by Suib and co-workers through the hydrothermal treatment of layered manganese oxides. Careful choice of additional metal ion content of such preparations controls the inorganic phase that forms. The

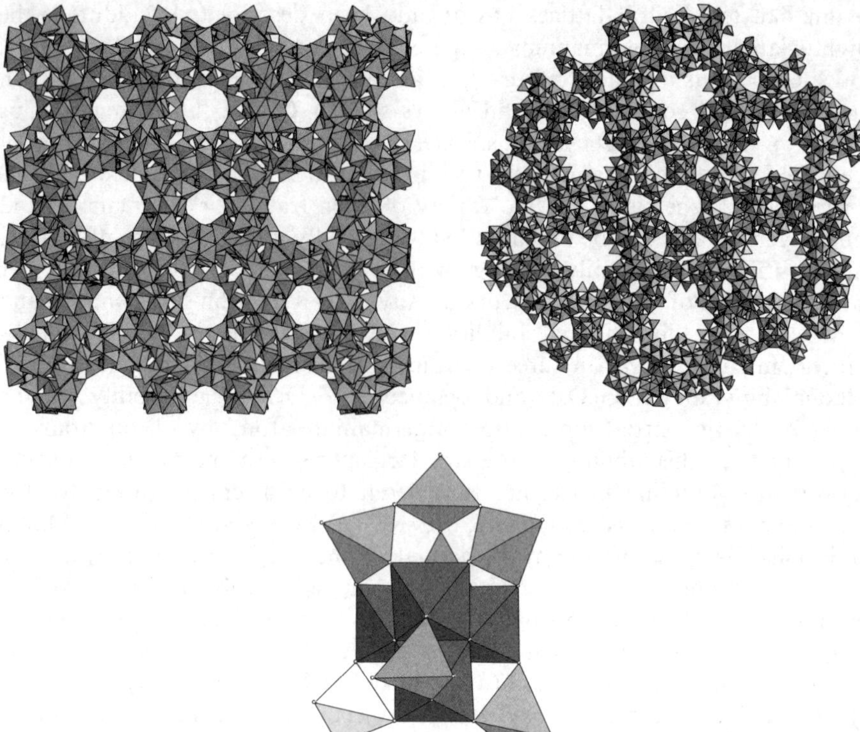

Figure 2.25 The germanate SU-M possesses cavities in the mesoporous regime (above). The framework is built from SBUs of formula $Ge_{10}O_{24}(OH)_3$ (below) containing corner-sharing tetrahedra and edge-sharing octahedra. The white tetrahedron belongs to an adjacent cluster.

structures are based on edge- and vertex-sharing MnO_6 octahedra, arranged to give tunnel structures (Figure 2.26) and are commonly described in terms of the size of tunnels that are made by the infinite slabs of edge-sharing octahedra.

For the mineral types hollandite and its synthetic equivalents, for example, the tunnels are delimited by four equivalent slabs, each two octahedra wide, giving a 2×2 tunnel structure. One such synthetic hollandite, OMS-2 (octahedral molecular sieve-2), has a chemical formula KMn_8O_{16}.[83] The manganese exists mainly as Mn^{4+} but also as Mn^{2+} and Mn^{3+} oxidation states, the distribution of which depends on which metal cations occupy cation sites in the tunnels or in the framework sites. Framework manganese cations have been substituted by other metal cations favouring octahedral sites, such as Cr^{3+}, whereas cations such as Li^+, NH_4^+ or Mg^{2+} may be found in the channels. The hollandite/OMS-2[83] structures are of interest as cathodes for rechargeable lithium batteries, but are not really microporous. By contrast, the 3×3 structure, observed in the mineral todorokite and in the synthetic OMS-1,[84,85] possesses pores large enough to adsorb cyclohexane. The material can be prepared by a 3-step procedure by which the sodium form of a layered

Figure 2.26 The manganese oxides hollandite (left) and todorokite (right) – synthetic analogues OMS-1 and OMS-2 – have frameworks of edge-sharing octahedra. Whereas the channel of hollandite is 2 × 2 octahedra in dimension, and is large enough to take in small cations, the 3 × 3 todorokite channel is large enough to take in small molecules.

manganese oxide is ion exchanged with an inorganic salt containing divalent cations and washed to give another layered manganese oxide, buserite. Hydrothermal treatment of this gives synthetic todorokite. The final chemical formula of the magnesium form is $Mg_{7.5}Mn_{12}O_{28}$ with the magnesium ions residing in the channel.

Further to the synthesis of the OMS materials, researchers at Sandia National Laboratories reported the synthesis of a family of microporous niobates, general formula $Na_2Nb_{2-x}M^{IV}{}_xO_{6-x}(OH)_x.H_2O$ (M^{IV} = Ti, Zr; x = 0.04–0.40), prepared from hydrothermal treatment of intimately mixed metal alkoxide precursors.[86] The structure is made up of layers of edge-sharing octahedra interleaved with double chains of edge-sharing niobate octahedra, in which additional, hydrated sodium cations reside. The niobate chains exhibit solid substitution. These 'SOMS' (Sandia octahedral molecular sieves) demonstrate appreciable ion exchange capacity and are selective for divalent cations.

2.7 Non-oxide Microporous Solids

The preceding discussion of inorganic microporous structures illustrates the great variety of metal cations that may be accommodated in porous frameworks. By contrast, relatively little progress has been made in the preparation of porous frameworks with anions other than oxide ions. The obvious candidates are nitrides and sulfides and there have been some recent advances in these directions.

Schnick and co-workers, for example, have prepared nitridophosphates with zeolite-like frameworks by high-temperature routes, including analogues of sodalite. Most of these structures are not microporous, however, as might be expected from the high temperatures of their synthesis, and they include anionic guests in the pores. One sodalite-related example which demonstrates reversible hydrogen adsorption has been given, however (formula $Zn_6P_{12}N_{24}$, which has vacancies in the sodalite cages).[87] Success has also been achieved in the synthesis of microporous metal chalcogenides. In a remarkable recent paper, Zheng *et al.*[88] discuss the synthesis of a family of M_1^{3+}/M_2^{4+} sulfides and selenides (M_1 = Ga or In, M_2 = Ge or Sn) that are based on corner-sharing metal sulfide or selenite tetrahedra, often arranged in supertetrahedral M_4S_{10} groups. They can formally be related to aluminosilicates by substitution of aluminium by M_1, silicon by M_2 and oxygen by the chalcogen. The mixture of tri- and tetravalent cations gives rise to structures of high porosity (particularly considering the high formula weight), large pores and thermal stability up to 300 °C.

2.8 Microporous Organic–inorganic Hybrids

One of the fastest-growing classes of microporous solids is that in which organic groups form part of the framework. These solids are attractive because the organic groups impart different adsorptive properties, and because it is possible to tailor these by modifying the organic group with different functional groups. In some of these structures the organic groups are connected to framework atoms (which may, for example, be Si or P) and so line the pores. Examples of organically–lined structures with inorganic skeletons include aluminium methylphosphonates, nickel carboxylates and organosilicates. Far more numerous, however, are coordination polymers, or metal-organic frameworks (MOFs), in which coordinated metal cations or metal oxide clusters are linked by bi-, tri- or even tetradentate ligands, such as amines, carboxylates and phosphonates. Among this class of solids are many highly porous solids that offer exciting new possibilities and applications.

2.8.1 Organically–lined Inorganic Frameworks

There are some examples of phosphonates (RPO_3, where R is an organic group) where the phosphonate is bound into an inorganic framework, and the R group lines the framework, projecting into the pores. The aluminium methylphosphonates discovered by Maeda are of this kind.[89,90] Two polymorphs, α and β, have been prepared, each possessing aluminophosphate frameworks lined with methyl groups. The two polymorphs are very similar, with hexagonal arrays of one-dimensional pores around 6 Å in free diameter (Figure 2.27). The β-polymorph is more readily formed, but converts to the thermodynamically more stable α-form upon heating in moist nitrogen (Chapter 5). The methyl groups rotate rapidly, giving the solids highly hydrophobic character. These aluminium methylphosphonates have proved ideal model solids to investigate adsorption behaviour on microporous 'organozeolites'

Figure 2.27 The aluminium methylphosphonate polymorphs α and β (above, left and right) and the scandium methylphosphonate (below) possess inorganic frameworks lined with rapidly rotating methyl groups.

(see later chapters) and the synthesis of more materials of this kind are to be expected (see the sodium scandium methylphosphonate framework of Figure 2.27 as another example[91]).

Structures in which the organic group decorates an inorganic framework are not restricted to phosphonates – the nickel succinate of Forster and Cheetham, for example,[92] possesses a three-dimensionally connected framework of NiO_6 octahedra that is decorated by the methylene groups of the succinate ligands, resulting in a thermally stable porous structure that is lined by the methylene groups. Some success has also been obtained in preparing silica forms of zeolites that contain organic siloxane ($RSiO_3$) groups, distributed throughout the framework. In contrast with the phosphonates, the structures are not ordered, because the siloxane group can only be incorporated at a defect site in the zeolite lattice. Nevertheless, there is good evidence that zeolite Beta can assimilate siloxane groups such as aminoethyl, phenethyl and mercaptopropylsiloxanes upon crystallisation, with retention of the overall framework structure, and that these can be further functionalised and used as shape selective catalysts.[93–95]

2.8.2 Porous Metal Organic Frameworks

Metal-organic frameworks (MOFs) are best considered as three-dimensionally connected networks comprising nodes, which are metal cations or di-, tri- or tetra-, or polynuclear metal cation clusters, linked by organic spacers, which are usually di-, tri- or tetradentate ligands. Where the coordination bonds between metals and ligands are strong, this gives rise to crystallographically well-defined open structures that are stable to the removal of guest molecules, giving permanent porosity. These porous MOFs are able to reversibly adsorb gases and vapours, sometimes in very large amounts that exceed by many times the specific adsorption capacities (*i.e.* uptake per gram) of even the most open zeolites. Porous MOFs therefore form a subset of the compounds known as coordination polymers, which include one-, two- and three-dimensionally connected arrays which may or may not be porous. In this text we will concentrate on 3D-connected arrays with porosity: it is also possible for 1D and 2D bonded arrays to adsorb molecules if they are stacked in arrangements (commonly by H-bonding) that generate intermolecular microporosity. The first examples of such porous arrays, including porphyrins,[96] were reported by the group of Robson. Subsequently the field has become very busy, and the groups of Yaghi, Férey, Kitagawa and Rosseinsky have made major contributions to novel structural chemistry in this area.

A metal-organic framework is represented schematically in Figure 2.28(a), in which metal-based nodes are connected to six others through organic linkers, generating a cubic cage. This simple structure represents a family of real materials (MOF-5[97] and related solids – see later) with the same topology and more generally illustrates several key features of MOFs. The first is that the topology of the network is fixed by the coordination geometry of the nodes and the rigidity of the linkers, rather than specific templating effects from species within the cage. Considering the nodes and linkers as secondary building units (SBUs, in the terminology of zeolites), the most common case is that the organic linkers are input directly into the synthesis as reactants, whereas the metal-based nodes assemble *in situ*. It is therefore possible in principle, and also in practice, to generate series of compounds with the same topology but different composition and dimensions by using linkers with the same geometric arrangement of bonding groups (most commonly carboxylates or amines) but different organic structure. Such series of compounds are described as being isoreticular (reticular = having the form of a (usually periodic) net).[98]

Such crystal chemistry imparts to MOFs the possibility of design of structures and properties to a greater degree than is possible for zeolites, where the principal building unit is the silicate tetrahedron rather than the SBUs. In zeolites, these SBUs (such as the sodalite cage) are more usually helpful in structure description than present as well-defined units within the synthesis mixture – this is not the case for MOFs. It is therefore possible to predict for MOFs the families of structures that can form for different combinations of metal-based nodes of different connecting geometries, with linkers of given shape. Furthermore, the ability to modify the organic linkers chemically, by

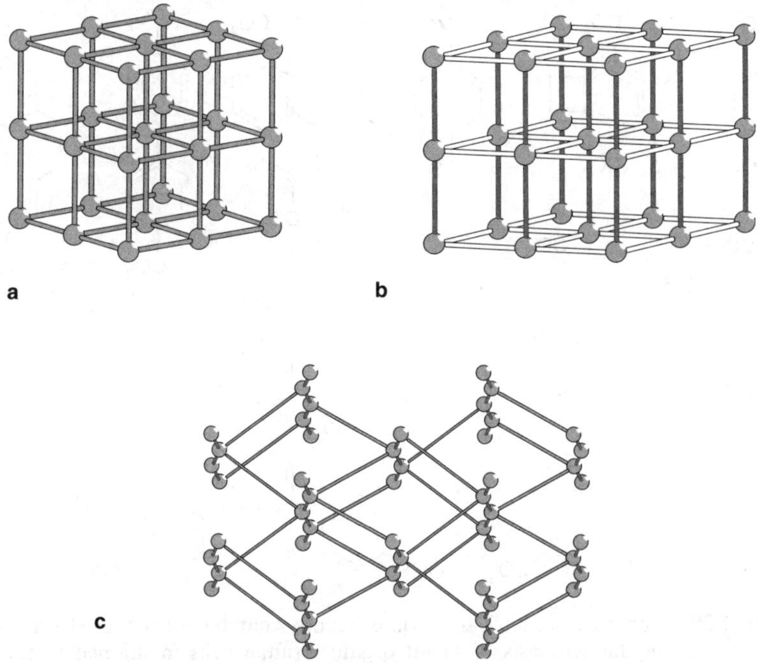

Figure 2.28 Examples of 'prototype' MOFs in which metallocentric nodes are connected via linear linkers. The simple cubic topology (a) is adopted by MOF-5, whereas in (b) there are two different sorts of linear linkers, as in the diamine pillared zinc terephthalate of Dybtsev.[104] Topology (c) is adopted by the terephthalate $Sc_2(O_2CC_6H_4CO_2)_3$.

organic synthesis, enables the predictable preparation of functionalised MOFs, the properties of which (for example in adsorption) can be calculated before they are made. Many of the crystallochemical principles that underlie 'reticulation' have derived from the consideration of theoretical nets, or tilings, of specified geometry. These are often inspired by crystal chemical features seen in simple inorganic structures or in three-dimensional coordination polymers with no porosity.[98–102] This is discussed further in Section 2.10.

The majority of the MOFs that show interesting porous frameworks are built from amine or carboxylate ligands, or combinations of the two. The ability of carboxylates to bind fully through two strong M–OC links rather than one M–N bond, tends to result in carboxylates making up the most thermally robust materials, but there are thermally stable microporous amine- and phosphonate- and even phosphine-based MOFs.

2.8.2.1 *Metal-carboxylate MOFs*

A very wide range of di-, tri- and tetra-carboxylic acids have been used in the synthesis of MOFs. Selected examples are shown in Figure 2.29, but a more

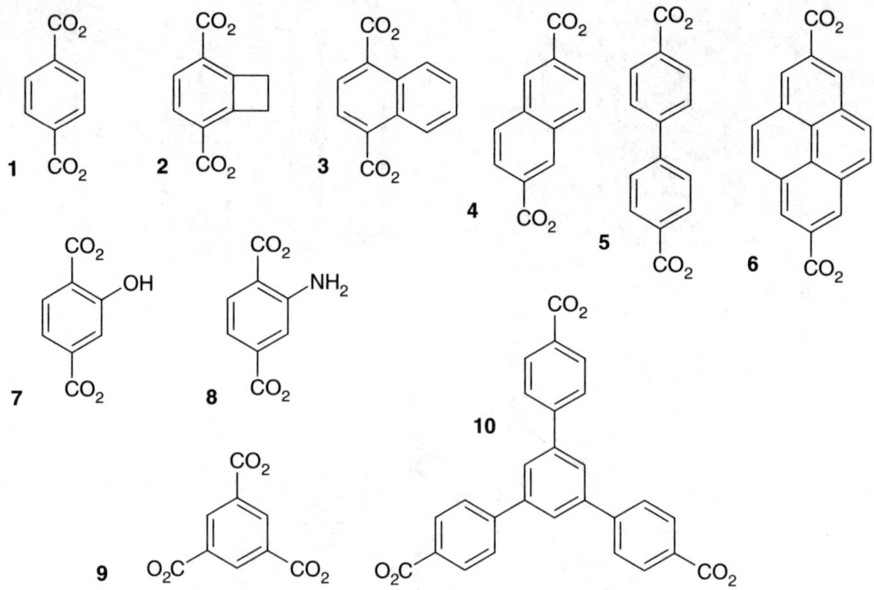

Figure 2.29 Some di- and tricarboxylate ligands that have been used successfully in the synthesis of metal organic frameworks mentioned in this text: **1** – MOF-5, MIL-53, MIL-68, MIL-101, $Sc_2(O_2CC_6H_4CO_2)_3$; **1**–**6** – structures with the same topology as MOF-5, but with extended linkers; **7**, **8** – functionalised linkers in MOF and MIL solids; **9** –HKUST-1, MIL-100; **10** – MOF-177.

complete summary is given in the review of Yaghi.[98] The corresponding acids (or esters) are commonly used directly as reagents in the syntheses and the metal-based units are typically generated *in situ* during crystallisation, as clusters characteristic of the synthetic conditions. Apart from isolated polyhedra, some examples of the types of clusters that act as nodes include the dinuclear 'paddle wheel' (M_2), trinuclear ($M_3(\mu_3O)$) clusters, tetranuclear ($M_4(\mu_4O)$) clusters and infinite chains (-M-O-), as illustrated in Figure 2.30. Examples of MOFs containing these types of units are discussed below to illustrate the features of some of the most fascinating MOFs so far discovered. No attempt is made to be comprehensive here – the research field is so dynamic that any such attempt would quickly be out of date.

2.8.2.1.1 Isolated polyhedra. Conceptually, the simplest type of MOF consists of a single type of metal coordination polyhedron as the node and a single type of dicarboxylic acid as a linker. In fact, there are relatively few porous carboxylate frameworks of this type. One example is the scandium terephthalate $Sc_2(O_2CC_6H_4CO_2)_3$ that displays a three-dimensional framework of ScO_6 octahedra and terephthalate (1,4-benzenedicarboxylate) groups (Figure 2.31).[103] This results in a small-pore solid with high thermal stability, relative to other MOFs.

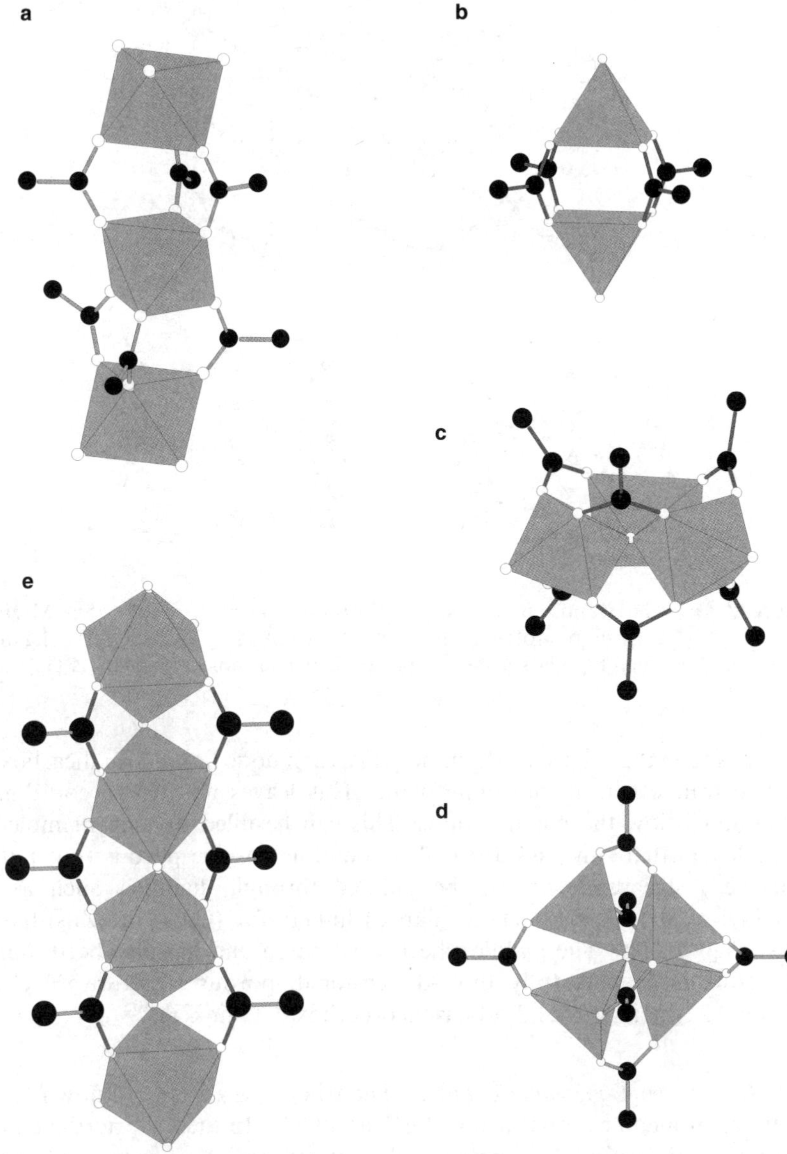

Figure 2.30 Some important metallocentric building units found in metal organic frameworks: (a) isolated octahedra; (b) dinuclear paddle wheel units; (c) trinuclear units; (d) tetranuclear units; (e) chains of corner-sharing octahedra.

2.8.2.1.2 Paddle wheel cluster. The paddle wheel cluster is known in many complexes and was found within several of the earliest MOFs to be prepared, typically as (Cu_2)- or (Zn_2)-based units. With 1,4-benzenedicarboxylate (BDC) or 2,6-naphthalene dicarboxylate (NDC) the paddle wheel complex typically

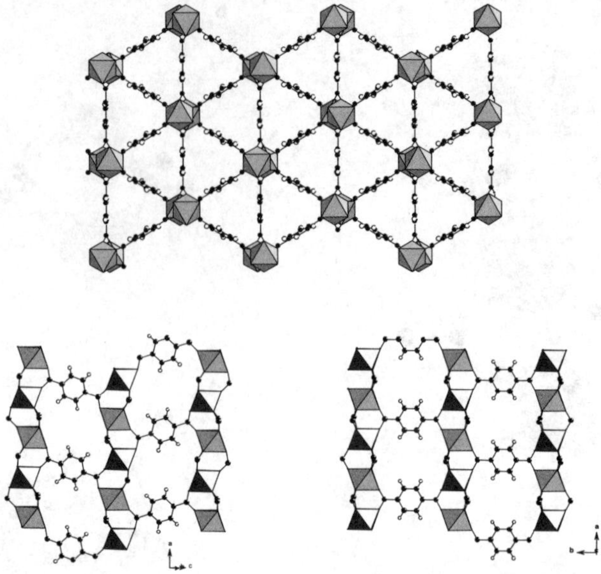

Figure 2.31 The scandium terephthalate structure is an example of a MOF that consists of isolated MO_6 octahedra linked by dicarboxylate ligands. It has a high adsorption capacity for small molecules such as H_2 and N_2.

combines to make square or rhombic grids, each node linking *via* dicarboxylate units to four others in the same plane. This leaves unsatisfied coordination above and below the metal cations. This can be filled by solvent molecules, giving layers that can pack through H-bonding to give porous (but not 3D-connected) structures, or can be pillared through diamines such as 4,4′-bipyridyl or diazabicyclooctane (dabco) linkers for (Cu_2)- or (Zn_2)-based[104] units (Figure 2.32). The paddle wheel (Cu_2)-based unit has also been found to act as the node in a fully three-dimensional, porous structure, HKUST-1 (Figure 2.32), joined by 1,3,5-benzenetricarboxylate linkers.[105]

2.8.2.1.3 $(M_3(\mu_3O))$ trimeric units. The trimer $(M_3(\mu_3O))$ is known to form for trivalent metal cations such as Fe^{3+} and Cr^{3+}. In these clusters each metal cation is octahedrally coordinated, with four equatorial oxygens that form part of the carboxylate units (that become linked into the framework), one terminal hydroxyl or fluoride or water ligand and a central oxygen shared between the three cations at the centre of the cluster. The group of Férey, at the Institut Lavoisier in Versailles, have made great use of this trimer in the synthesis of highly porous carboxylates. The iron fumarate, MIL-88A, for example, shows large pores and remarkable flexibility[106] – it is able to double its cell volume without losing structural connectivity as the dehydrated material adsorbs molecules (Figure 2.33). It is also possible to prepare the similar, isoreticular MIL-88 type structures with dicarboxylates other than fumarate, including

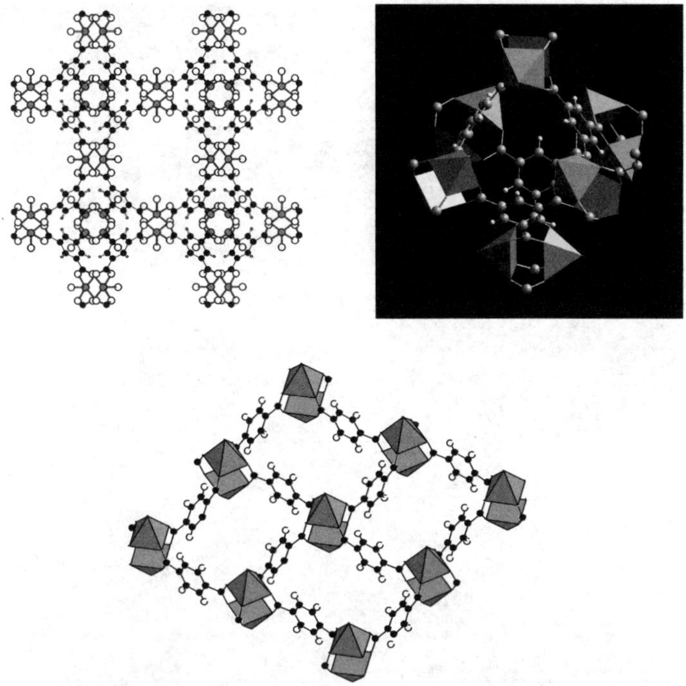

Figure 2.32 Examples of structures based on 'paddle wheel' dimer units. In the copper trimesate HKUST-1, (top, left and right), dimeric copper units link four trimesate ligands into a three-dimensional framework, leaving the apical coordination site accessible to other ligands (such as water molecules). In the pillared layered zinc terephthalate of Dybtsev *et al.*,[104] porous planar zinc terephthalate sheets (shown) are linked via the diamine diazabicyclo-octane ($N(C_2H_4)_3N$) coordinated at the apical sites.

terephthalate, 2,6-naphthalenedicarboxylate and 4,4′-biphenyldicarboxylate, showing the versatility of the approach.[107]

Most recently, the group has prepared two beautiful, highly open, porous hybrids, MIL-100 (Figure 2.34) and MIL-101,[108] under synthetic conditions that result in the *in situ* formation of the chromium versions of these trimeric clusters. MIL-100 forms with the benzenetricarboxylate linker ($Cr_3O(H_2O)_2(OH,F)BTC_2$) whereas MIL-101 is crystallised with the benzenedicarboxylate linker ($Cr_3O(H_2O)_2(OH,F)BDC_3$). Both structures can take up ferric ions in the metal nodes. These solids are prepared as powders, and their structure solution is a separate story (Chapter 4). They possess enormous cubic unit cells, both with space group Fd-3m, with unit cell dimensions of 72 and 89 Å, respectively. Both structures are built from what Férey describes as supertetrahedra, where each vertex of each supertetrahedron is occupied by a $Cr_3(\mu_3O)$ trimer.

For MIL-101, BDC linkers join trimers along the edges of the supertetrahedra, whereas for MIL-100, BTC linkers occupy the faces of the supertetrahedra, linking three trimers. The supertetrahedra are attached through shared trimers:

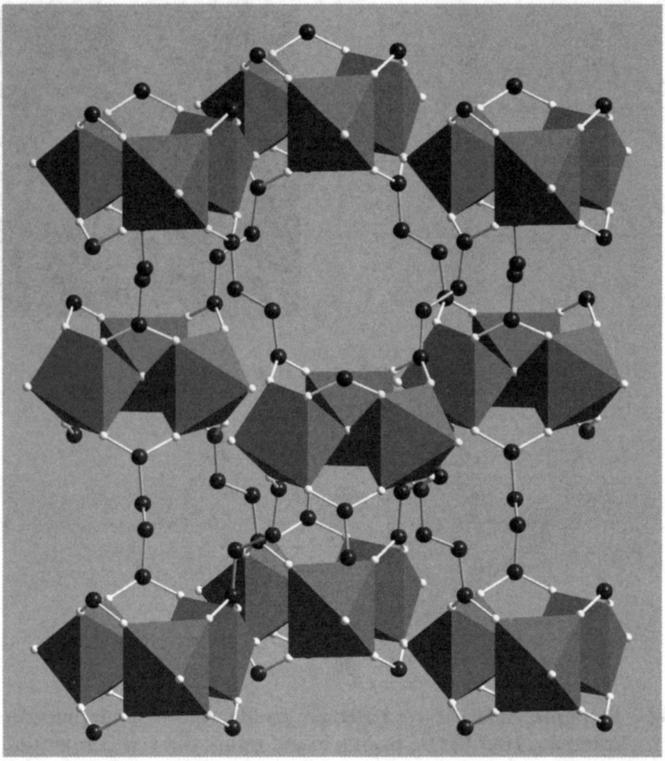

Figure 2.33 The iron (III) fumarate MIL-88 is built up from trimeric iron clusters linked by fumarate (*trans* $O_2CCH=CHCO_2$) linkers to give a highly flexible framework, shown here in the open form. Hydrogen atoms omitted for clarity.

the geometry of the trimer link must result in a local plane of reflection through the shared trimer (this was important in the structure solution). The way these tetrahedra link is then directly analogous to the way tetrahedra are joined in the tetrahedrally linked MTN zeolite topology, but with a much larger unit cell. Within each of the two structures there are three sets of cages: those within the supertetrahedra themselves, those within the structure that are made up of twelve 5-membered rings (pentagonododecahedra) and the largest cages, limited by 28 supertetrahedra with 12 pentagonal and 4 hexagonal windows. The apertures enabling access to these three cages and their free diameters are given for the two solids MIL-100 and MIL-101 in Table 2.2. It is clear that these solids demonstrate both microporosity and mesoporosity in the same structure and represent a breakthrough in solid state chemistry.

Yaghi *et al*[109] have taken the approach using trimeric building blocks a stage further with their synthesis of MOF-500, in which trimeric iron carboxylate units similar to those found in MIL-88, -100 and -101, but partially terminated by coordinated sulfate units that remain during the synthesis, are connected by coordination of the iron atoms in two further, hierarchical levels of complexity

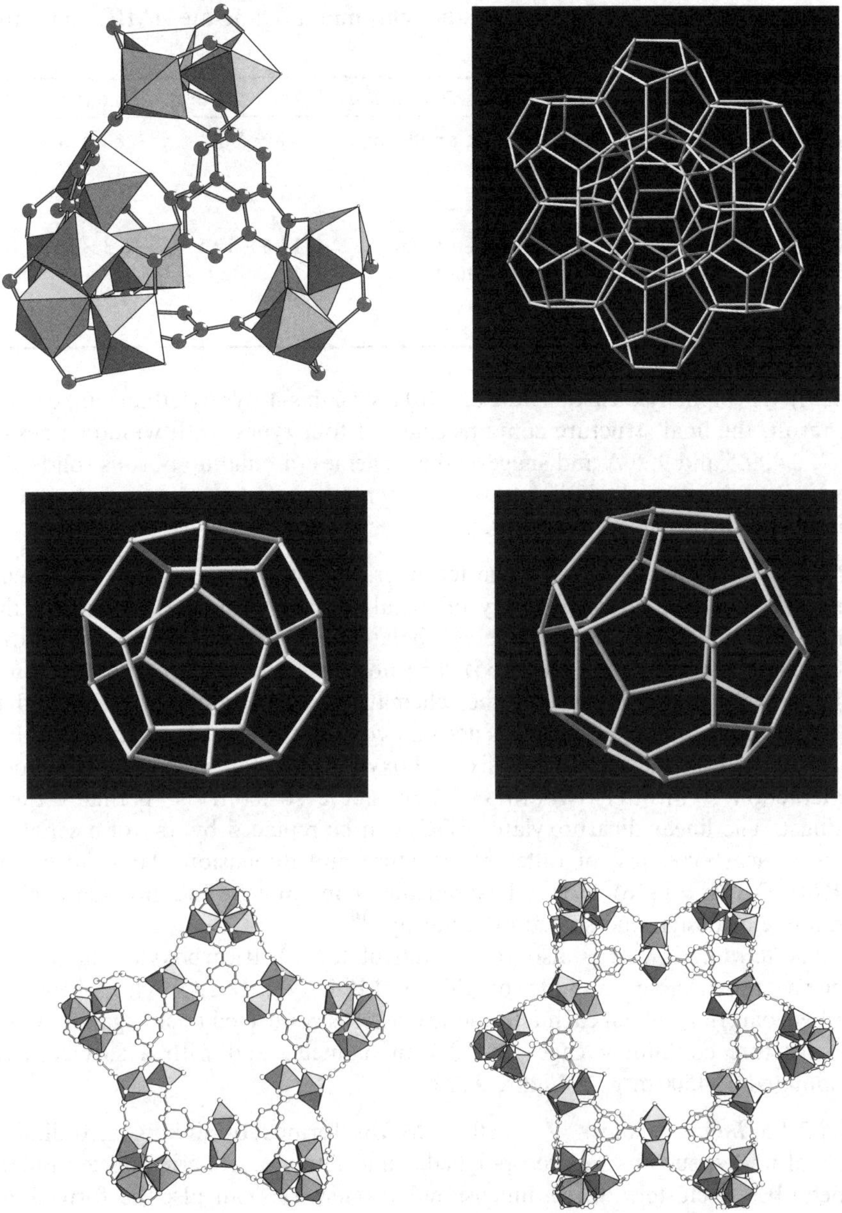

Figure 2.34 Trinuclear $Cr_3(\mu_3O)$ clusters connected by 1-,3-,5-benzenetricarboxylate (trimesate) ligands make up MIL-100. They are linked into supertetrahedra, with the benzene rings in the centres of the four faces (top left). These tetrahedra are then linked together in an analogous way to the linking of tetrahedra in the clathrate silicate, structure type MTN (top right). There are two cages in the structure, one built only of rings of five trimeric units, the other with openings bounded by both five and six trinuclear units (middle, left and right). These rings are shown in more detail below.

Table 2.2 Pore dimensions in the chromium trimesate (MIL-100) and terephalate (MIL-101).

Compound	*Opening/cage*	*Supertetrahedra*	*Dodecahedra*	*Hexadecahedra*
MIL-100	Window dimensions/Å	Not applicable	4.8 × 5.8	8.6 × 8.6
	Cage free diameter/Å	6.6	25	29
MIL-101	Window dimensions/Å	Open to small molecules	12.5 × 12.5	16.3 × 16.3
	Cage free diameter/Å	8.7	30	34

via (i) 4,4′-biphenyldicarboxylate and (ii) *cis*-1,2-bis-4-pyridylethane linkers. As a result, the final structure contains cages of four types, with window sizes of 3.4, 6.4, 9.5 and 9. 5 Å and suggests that schemes to build up porous solids can be rationally-devised, similar to organic syntheses. This synthetic approach is discussed further in Section 2.10.2.

2.8.2.1.4 ($M_4(\mu_4O)$) units. The tetrameric (Zn_4O)-based unit has a special place in the structural chemistry of metal-organic frameworks, forming the nodes for the prototypical zinc terephthalate (Zn_4O)(BDC)$_3$, MOF-5.[97] In this remarkable structure (Figure 2.35) each node is connected to six others, in a similar way to that shown for the schematic MOF. Examination shows that each carboxylate linkage bridges across two zinc atoms of the cluster, so that each cluster is the centre for six carboxylate linkers. As each linker-node interaction is through two M–OC bonds, the structure is thermally quite robust. The linear dicarboxylate (BDC) can be replaced by any of a series of linear dicarboxylates, of different structure and dimension. These form the IRMOF-n (n = 1–16) series of isoreticular compounds, and provide a clear example of design and structural tailoring.[110]

The tetrameric unit is also able to reticulate with tricarboxylate linkers. A recent, remarkable example of this is MOF-177, (Zn_4O) (1,3,5-benzene-tribenzoate)$_2$, in which each tetrameric cluster is connected to six ligands giving a structure containing cages 11–12 Å in diameter and with a surface area estimated at 4500 m^2g^{-1} (Figure 2.35).[111]

2.8.2.1.5 Infinite chains. As well as MOFs formed by linking zero-dimensional nodes, such as isolated polyhedra and dimeric, trimeric and tetrameric metal-based clusters, three-dimensional frameworks can also be formed by cross-linking chains of polyhedra by dicarboxylate linkers. The MIL-53 structure (Material Institut Lavoisier, Versailles) is a good example of such a structure (Figure 2.36).[112] Chains of chromium(III) centred corner-sharing octahedra, joined along the chain through hydroxyl oxygens, are linked perpendicular to the chains to four other chains by dicarboxylates, such as BDC linkers. The Al^{3+} and Fe^{3+} analogues of this structure have also been reported,[113,114] as has the vanadium analogue, MIL-47.[115] Projection of the structure parallel to the chains shows diamond-shaped channels similar in

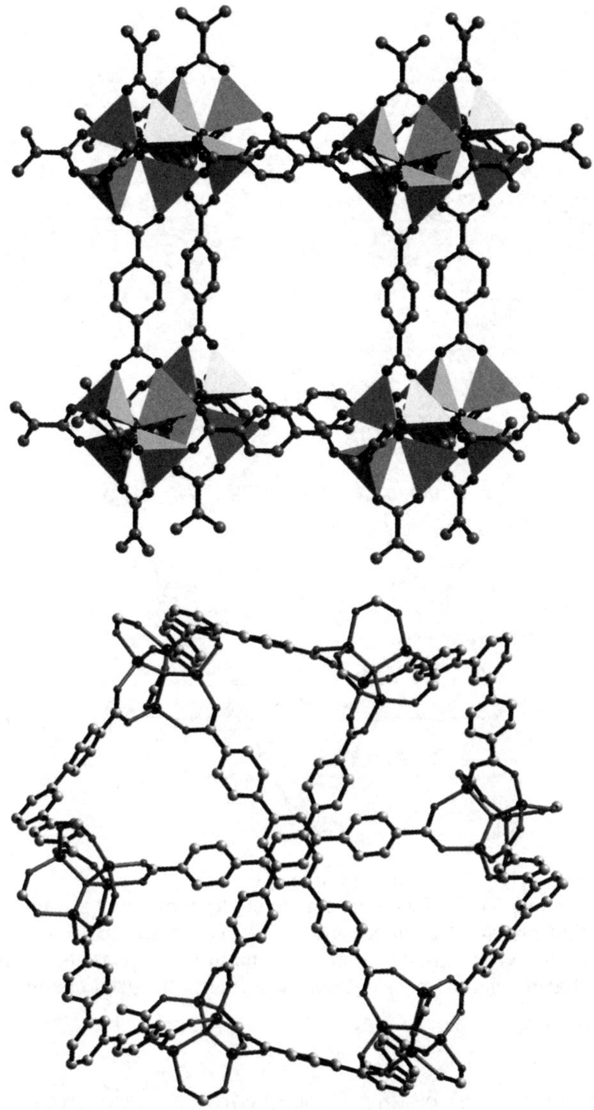

Figure 2.35 Two MOF structures based on the tetrameric Zn_4O unit. These units link octahedrally into the frameworks via six carboxylate groups. In MOF-5, above, the links are linear, via terephthalate groups, giving a cubic array, whereas in MOF-177, the links are through the much larger benzene-1,3,5-tribenzoic acid. Part of the MOF-177 structure is shown below.

cross-section to those seen for some of the layers produced by linked paddle wheel units. Like some of those structures, MIL-53 shows remarkable flexibility upon adsorption and desorption of molecules within the pores – the so-called breathing behaviour described further in Chapter 7. The vanadocarboxylate

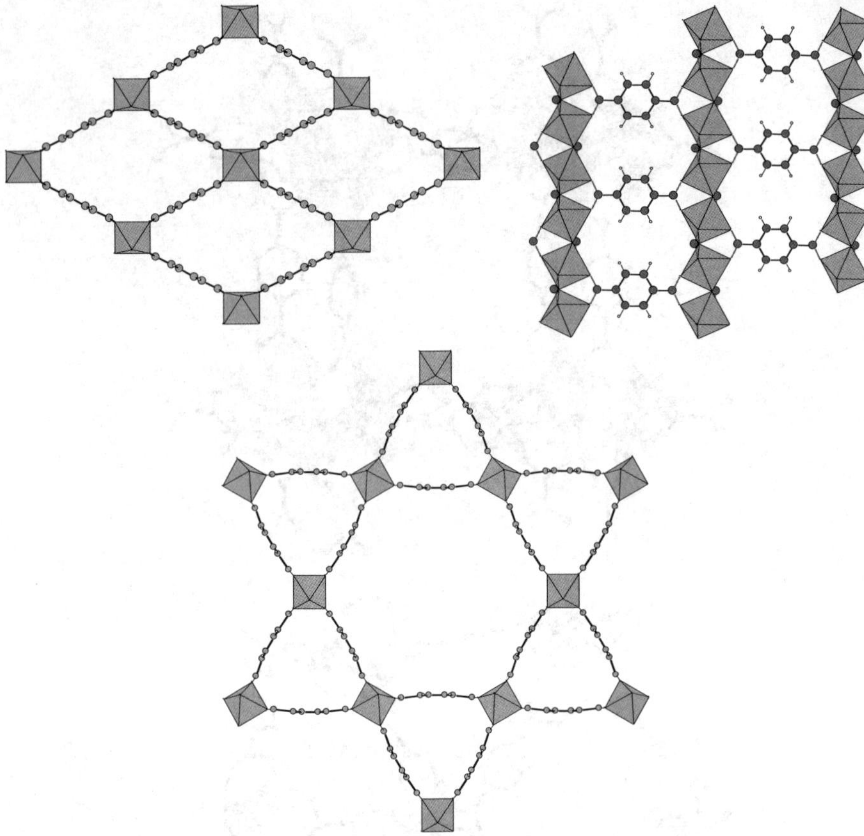

Figure 2.36 The MIL-53 structure (above, left and right) is based on corner-sharing chains of MO_6 octahedra linked via terephthalate groups giving channels that are rhombic in cross section. The linking oxygen atoms are part of hydroxyl groups. The MIL-68 structure (below) is also built from corner-sharing chains, but contains two differently sized channels.

MIL-68[116] also possesses chains of octahedra that are cross-linked by terephthalate groups, but in such a way that a rigid, large-pore structure is formed, giving channels with triangular and hexagonal cross sections when viewed down the *c* axis (Figure 2.36). The larger channels are some 16 Å in free diameter.

MOFs based on linked chains of metal-oxygen polyhedra are not restricted to those based on octahedra. For example, zinc-based MOFs based on chains of both tetrahedrally and octahedrally coordinated zinc, and lanthanide-based MOFs in which the larger lanthanide cations are coordinated by more than six oxygen atoms and linked by edge and corner-sharing, have also been reported. Rosi *et al.* describe some of these MOFs constructed from 'rod-shaped' secondary building units in reference.[117]

2.8.2.2 *Other MOFs – Metal Amines, Imidazolates, Phosphanes and Phosphonates*

The enormous variety of metal-ligand geometries available to coordination polymers results in countless possible open and porous frameworks. (There are few limitations to this approach, but when more than one apparently very open framework exhibits *interpenetration*, in which the free volume of one framework is filled by one or more similar frameworks, the result is often a non-porous solid (although even then some interpenetrating structures demonstrate small-pore microporosity between the different frameworks)).

Metal diamine frameworks with the 4,4′-bipyridyl ligand have a special place in this field,[118] because the $Ni_2(4,4'\text{-bipy})_3(NO_3)_4 \cdot 2EtOH$ structure was the first hybrid to show permanent porosity for nitrogen adsorption. Perhaps the most significant microporous framework structures based on metal-nitrogen bonds (other than diamine-pillared carboxylates) discovered so far are the metal imidazolates (see below). Microporous phosphanes have also been prepared, the 1,3,5-tris(diphenylphosphanyl)benzene complexing with silver to give a microporous solid with channels of 16 Å diameter.[119] Usually, though, M–N and M–P bonds are weaker than M–O bonds, so that porous frameworks based on metal-amine or metal-phosphane are in general likely to be less stable than carboxylates (which are also bidentate rather than monodentate). In this regard, metal phosphonates are also of interest, because the metal-O_3P bonds are strong (as seen in the aluminophosphonates) and the phosphonate group can be mono-, bi- or tridentate.

2.8.2.2.1 Metal imidazolates. Microporous metal imidazolates with framework topology types mathematically similar to those of zeolites were independently reported in 2006 by the groups of Chen[120] and of Yaghi.[121] Imidazole (IM) can lose a proton to give the negatively charged imidazolate anion (Scheme 2.1). Many members of this series of compounds are known, including methyl-, ethyl and phenyl derivatives (MeIM, EtIM and PhIM). Examination of the structures of dense metal imidazolates of cobalt and zinc showed that these tetrahedrally coordinated metals (M) coordinate to the imidazolate (IM) nitrogen atoms so that the <M-IM-M bridging angle is about 145°. This is very close to that of the T-Ô-T angle in zeolite structures, suggesting that zeolite analogues could be prepared in suitable synthetic conditions.

As a result of solvothermal syntheses using IM and MeIM, EtIM and PhIM derivatives, a family of zinc and cobalt imidazolates with porous topology types observed for zeolites has been discovered. Among the microporous solids

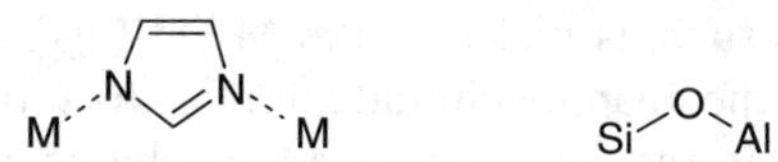

Scheme 2.1 The similarity between the geometry of metal-ligand bonds in metal imidazolate ZIFs and aluminosilicate zeolites.

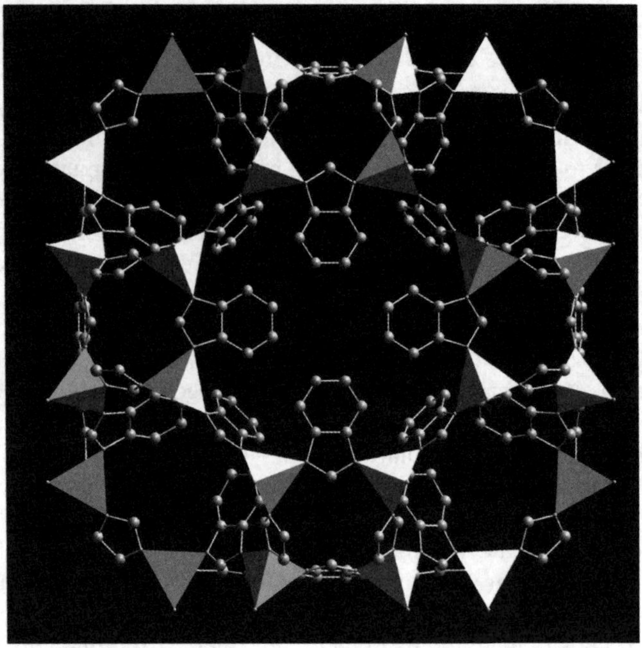

Figure 2.37 Part of the structure of the zinc phenylimidazolate ZIF-11, which has the same framework topology as zeolite Rho (Figure 2.7). The part of the structure shown corresponds to the decorated alpha-cage of the zeolite. The carbon atoms of the imidazolate group and the ZnN_4 tetrahedra are shown.

prepared so far are: $Zn(MeIM)_2$ with the SOD topology type; $Zn(IM)_2$, MER and $Co(PhIM)_2$, RHO (Figure 2.37). These 'ZIF' structures – 'Zeolitic Imidazolate Frameworks', as named by Yaghi – possess very high thermal stabilities, low densities and high surface areas. Structurally, the internal pores are lined by organic groups, which suggests these materials will have distinctive adsorption properties. The $Zn(MeIM)_2$, SOD structure ('ZIF-8'), for example, has a micropore volume of $0.64\,cm^3\,g^{-1}$. It is likely that these new materials will be the first of an important metal-nitrogen based MOF family.

2.8.2.2.2 Metal phosphonates. Many attempts have been made to prepare microporous bisphosphonates, but most crystalline solids of this type have small pores and few show adsorption of molecules other than water. More recently, though, use of the N,N′-piperazinebismethylenephosphonic acid (H_4L) has given microporous solids with 4 Å pores with titanium or aluminium in the MIL-91 structure[122] (*e.g.* $Al(OH)LH_2.xH_2O$) and large 10 Å pores with divalent metal cations such as nickel in the $M_2(H_2O)_2L.xH_2O$ structure.[123] In the former structure the bisphosphonate binds in a bidentate way by two phosphonate oxygen atoms between adjacent chains of apex-sharing MO_6 octahedra, whereas in the latter structure the acid binds *via* two phosphonate oxygen atoms and a nitrogen atom of the piperazine group to give helical

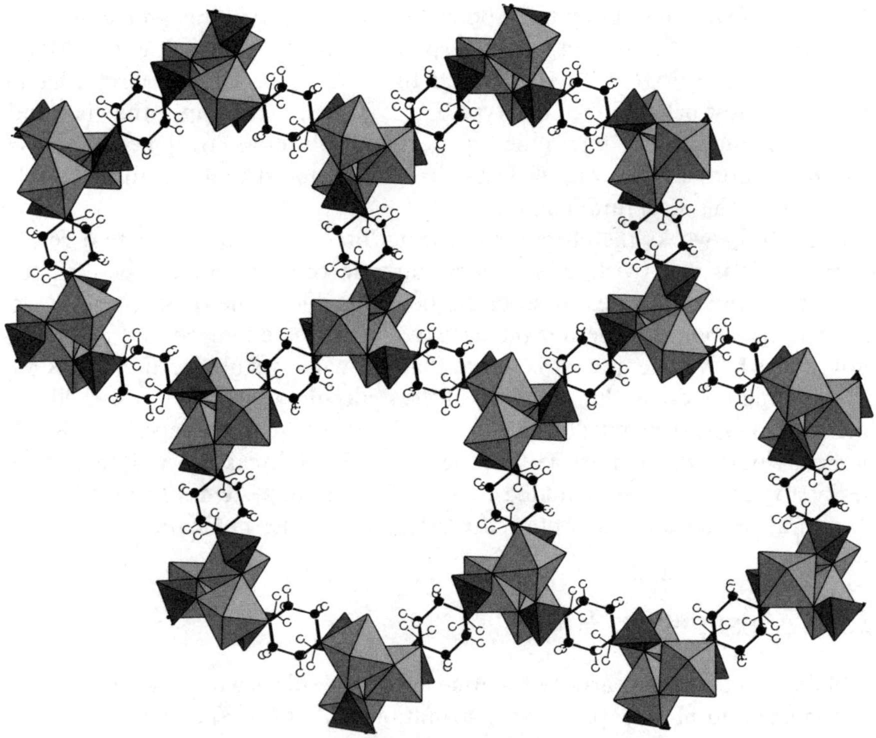

Figure 2.38 The nickel bisphosphonate $Ni_2(O_3PCH_2NC_4H_8NCH_2PO_3).2H_2O$ possesses channels 1 nm in free dimension. Edge-sharing helical chains of NiO_5N octahedra are linked by piperazine units: the bisphosphonic acid coordinates to the nickel by both oxygen and nitrogen atoms.

chains of edge-sharing MO_5N octahedra (Figure 2.38). These solids have permanent porosity upon dehydration and good thermal stability. In the latter structure, dehydration results in loss of a water molecule from the metal octahedral environment, giving a five-fold coordinated metal site.

2.8.2.3 Rigidity and Flexibility

The first generation of metal organic frameworks that were prepared tended to have relatively low stability upon guest removal, so that the synthesis of more robust porous solids was an important target. The discovery of MOF-5, the IRMOF series and most recently MOF-177,[111] MIL-100[124] and MIL-101[108] has gone a long way to meeting this challenge. Most of these are thermally stable and rigid frameworks with permanent porosity – they may be described as belonging to a second generation of MOFs. There is currently growing interest in preparing porous solids that possess flexible frameworks that respond to the uptake of molecules by structural adjustment. This behaviour, which is quite different from that displayed by more typical zeolite-like molecular sieves, has potential applications in the trapping and controlled release of gases, in sensing or as responsive materials. Among the MOFs described above, MIL-53, MIL-88 and the dabco-pillared zinc terephthalate structures show behaviour of this type. For MIL-53, for example, the diamond-shaped channels, which retain their shape upon solvent removal, become much flattened upon adsorption of H_2O, as additional strong bonding can be achieved by this structural change.

These frameworks therefore demonstrate 'breathing' behaviour that results from rotations and twists of structural units in order to achieve better coordination to bound species. A second type of flexible framework appears to be able to break bonds to permit the uptake of adsorbate molecules. In $Ni_2(4,4'\text{-bipy})_3(NO_3)_4$, for example, it is found that the windows observed in the X-ray crystal structure do not limit the size of the molecules that can be adsorbed into internal cavities: they must be able to open to permit guest passage.[125] This type of behaviour, which must require selective bond breaking and reforming, begins to resemble that exhibited by some biological systems. These effects of structural response to adsorption are discussed further in Chapter 7.

2.9 Mesoporous Solids

Mobil scientists in the early 1990s discovered that silicas with ordered arrays of mesopores and narrow pore-size distributions (MCM-41S) could be prepared using micelles composed of cationic surfactant molecules of the form $CH_3(CH_2)_nN^+(CH_3)_3$ (n typically 11–19) as 'liquid crystal-like' templates.[126] This opened up a new field of research that is currently comparable in activity to that of microporous solids. The MCM-41S materials are characterised by well-defined arrays of pores bounded by walls of highly condensed silica, itself without long-range order on the atomic scale. They are most clearly distinguished from high surface area silicas and controlled pore glasses by the regular arrangement of the pores and walls, which is beautifully demonstrated by high-resolution transmission electron microscopy (Figure 2.39).[127] Although they are not crystalline in the usual sense (atoms do not occur at the same coordinates in different unit cells) the structures possess lattice repeats and give well-defined diffraction maxima at low scattering angles. Under suitable conditions they form with well-defined 'crystal' morphologies.

There is great variability among mesoporous silicas, and even among different samples of nominally the same structure type. These differences arise because of different gel compositions, preparation conditions, surfactant templates and post-synthetic treatments, and manifest themselves as different unit cell repeats, pore and window sizes and wall thicknesses and silicon connectivities within the walls. In this respect they are very different from the crystalline zeolites. A mesoporous solid is usually identified primarily by its symmetry and its lattice dimensions. On this basis, at least seven different types have been synthesised, some with very different pore sizes. For example, whereas MCM-41 and MCM-48 have different symmetries (and structures), they have similar

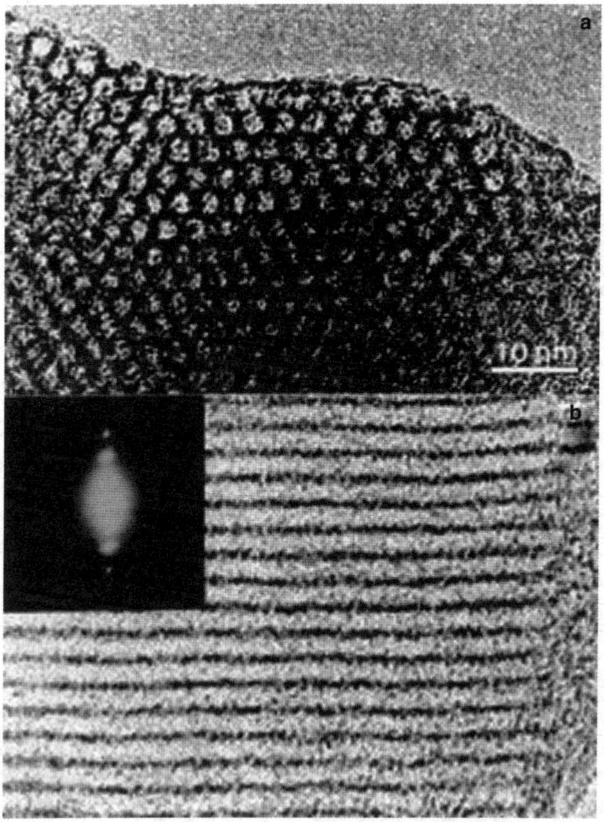

Figure 2.39 Electron micrographs of the mesoporous silica MCM-41 (a) parallel and (b) perpendicular to the channel axes reveal that the structure has periodicity in two dimensions only. [Reproduced from reference 127 with permission. Copyright 1997 Royal Society of Chemistry].

pore sizes and can be prepared using the same cationic surfactant template. SBA-15 and FDU-5 have similar structures (with the same space groups) to MCM-41 and -48, respectively, but are prepared using much larger non-ionic surfactants (such as the block copolymers P123 (EO_{20}-PO_{70}-EO_{20}) and F127 (EO_{106}-PO_{70}-EO_{106}), where EO is the more hydrophilic ethylene oxide and PO is the more hydrophobic propylene oxide) and consequently have larger dimensions and pore sizes. Other structures continue to be discovered, but unambiguous determination of their structure awaits careful analysis of HRTEM, low angle XRD analysis and porosimetry (Table 2.3).

The proposed mechanisms of synthesis are discussed in more detail in Chapter 5: at this stage the structures of the as-prepared inorganic-organic composites are simply considered as surfactant micelle arrays interpenetrated by amorphous and incompletely condensed silica walls, in which silicon has local coordination $Si(OSi)_{4-n}(OH)_n$, where n is mainly 0 (Q^4), 1 (Q^3) or 2 (Q^2).

Table 2.3 Mesoporous silicas and organosilicas, grouped according to symmetry and pore size, rather than synthesis conditions or the type of surfactant used to template them. The pore sizes vary with synthesis conditions.

Symmetry	*Materials; References*	*Structure/Pore space arrangement*	*Channel or cage diameter: window size of cages/Å*
p6mm	MCM-41,[127] FSM-16,[128] SBA-3,[129] AMS-3[130]	Hexagonal array of one-dimensional channels	30–50 Å channels
p6mm	SBA-15[131,132]	Hexagonal array of one-dimensional channels linked by irregular micropores	50 –100 Å channels
c2mm	SBA-8[133]	One-dimensional channel structure related to MCM-41, but distorted by extension along $[100]_{hex}$	*ca* 20 Å channels
Ia-3d	MCM-48,[134] AMS-4[130]	Pore structure comprises two interpenetrating but unconnected branching networks of cylinders/rods	Branching cylinders 20–40 Å in diameter
Ia-3d	FDU-5[135]	Pore structure comprises two interpenetrating but unconnected branching networks of cylinders/rods, linked by irregular micropores	Branching cylinders 40–80 Å in diameter
Pm-3n	SBA-1,[136,137] SBA-6,[136] AMS-2[130]	Pore structure contains two different types of cages with oblate spheroidal shape: intercage windows in the microporous regime	Cages of SBA-1 30–40 Å in diameter, separated by windows 10–20 Å in diameter
$P6_3$/mmc / Fm-3m	SBA-2,[138] SBA-12,[131] AMS-3[130]	Pore space based on close packed arrangement of spherical cages. These can be cubic close packed, hexagonally close packed or an intergrowth	Spherical cages 20–30 Å diameter: windows in microporous regime
Fm-3m	FDU-12[139]	Pore space based on cubic close packed array of large mesopores	Spherical cages 100 Å in diameter, windows 40 Å in diameter
Im-3m	SBA-16[131,136]	Body-centred array of connected cages	Spherical cages 90 Å in diameter, windows 20 Å
Fd-3m	AMS-8[140]	Face centred arrangement with two types of cages in the mesoporous region	Cages 60 and 80 Å in diameter. Windows between larger cages are 10 –20 Å, between small cages, windows are in the microporous regime
Pn-3m	AMS-10[141]	Bicontinuous cubic mesophase composed of an interwoven enantiomeric pair of 3D networks	Effective pore diameter 40 –50 Å

Removal of the surfactant by calcination results in further condensation of the silica in the walls and leaves highly porous solids.

The mesoporous solids that were originally discovered were MCM-41 and MCM-48. MCM-41, the most widely studied of the mesoporous solids, consists of a hexagonal array of one-dimensional channels ordered on the long range in two dimensions, but not ordered in the third (hence the two-dimensional space group symbol p6mm). The size of the pores (30–40 Å) and the thickness of walls depend on the surfactant used in the synthesis and the synthetic conditions and there is evidence to suggest that the pore walls may be flat or concave toward the pores, depending on the synthesis route. MCM-48 possesses two three-dimensionally connected channel systems that are interpenetrated but not connected, and which are separated by a silica wall a few angstroms thick.[134] This topology, space group Ia-3d, is well known in surfactant-water systems, where it results from a minimum energy arrangement of hydrophobic and hydrophilic regions, and the silicate occupies the position of the periodic bicontinuous gyroid structure (Figure 2.40). Both MCM-41 and MCM-48 possess larger pore analogues prepared under acidic conditions using non-ionic surfactants rather than cationic surfactants as structure directing agents. These larger-pore analogues, SBA-15 and FDU-5, possess channels around 50 –100 Å in diameter and thicker walls, and so they are still more different from the microporous solids that are of most interest in this text. It should be noted that these extra-large-pore mesoporous solids do possess considerable microporosity in the walls when they are calcined, due to the removal of non-ionic surfactant that is occluded there during synthesis. This microporosity is not well ordered, however.

Among the other mesoporous solids that have been reported, many have cage rather than channel structures (SBA-1, -2, -6, -12, -16, FDU-12) in which cavities are separated by restricted apertures that may be of sizes similar to those observed in typical microporous solids. SBA-1 is representative of this type of material, and has been the subject of careful structural study by the groups of Stucky, Terasaki and Anderson (Figure 2.41). The pore structure resembles that of mesoporous solids much more closely than it does the large-pore mesoporous solids, with cages up to 20 Å in diameter being separated by pore windows of *ca.* 1 nm. In fact, the structure of the hybrid chromium terephthalates, MIL-100 and 101 (Section 2.8), bear close similarity in pore size and overall symmetry. The differences are the degree of order in the walls, which for SBA-1 are amorphous, and variability in the pore sizes, which for mesoporous solids vary with the synthesis and post-synthetic treatment conditions. Other mesoporous silicas that possess mesocages include SBA-2 and related solids based on close packed arrays of spherical cages, and the body-centred cage structure SBA-16. SBA-2, for example, is made up of amorphous silica condensed around close packed arrays of spherical micelles, which may display cubic close packing, hexagonal close packing or intergrowth structures. The pore size of calcined SBA-2 can be very small, down to the small-pore equivalent of microporous solids. The definition between microporous and mesoporous solids in terms of pore sizes becomes blurred for such solids as these, but the structural and chemical differences remain.

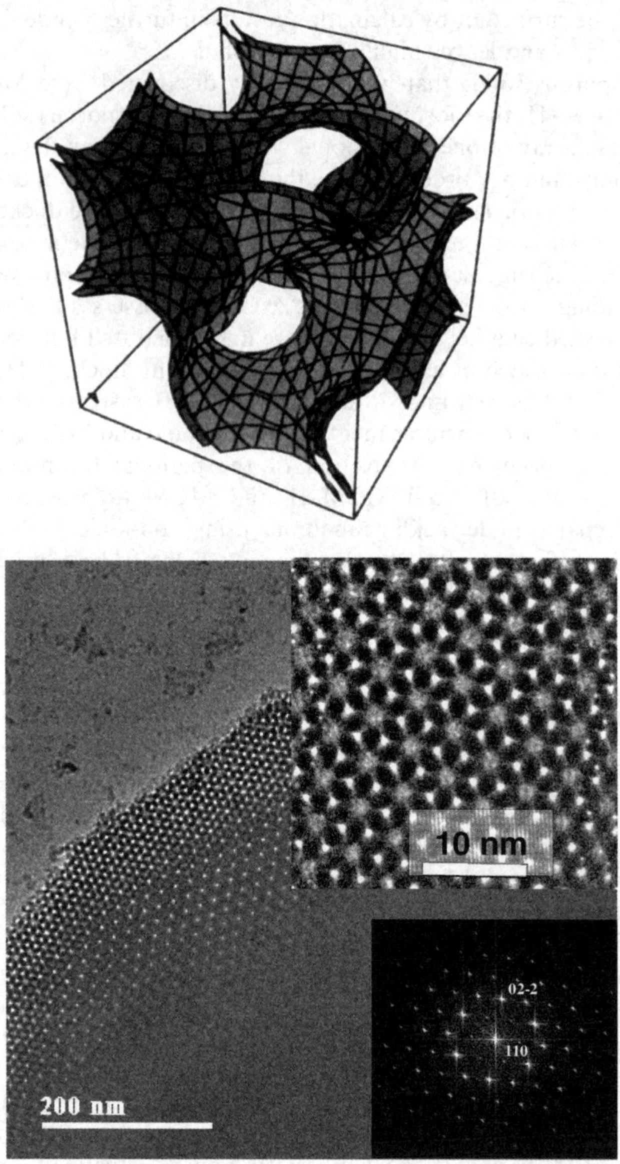

Figure 2.40 (Above) Representation of the mesoporous silica MCM-48, in which the mesopores are separated by walls made of amorphous silica between the two light and medium grey surfaces shown. [Reproduced from reference 134 with permission. Copyright 1997 Elsevier.] (Below) characteristic [111] electron microscope image of this structure, with the white spots indicating the positions of the channels in projection. The structure is based on a cubic bicontinuous gyroidal minimal surface structure, which has Ia-3d symmetry.

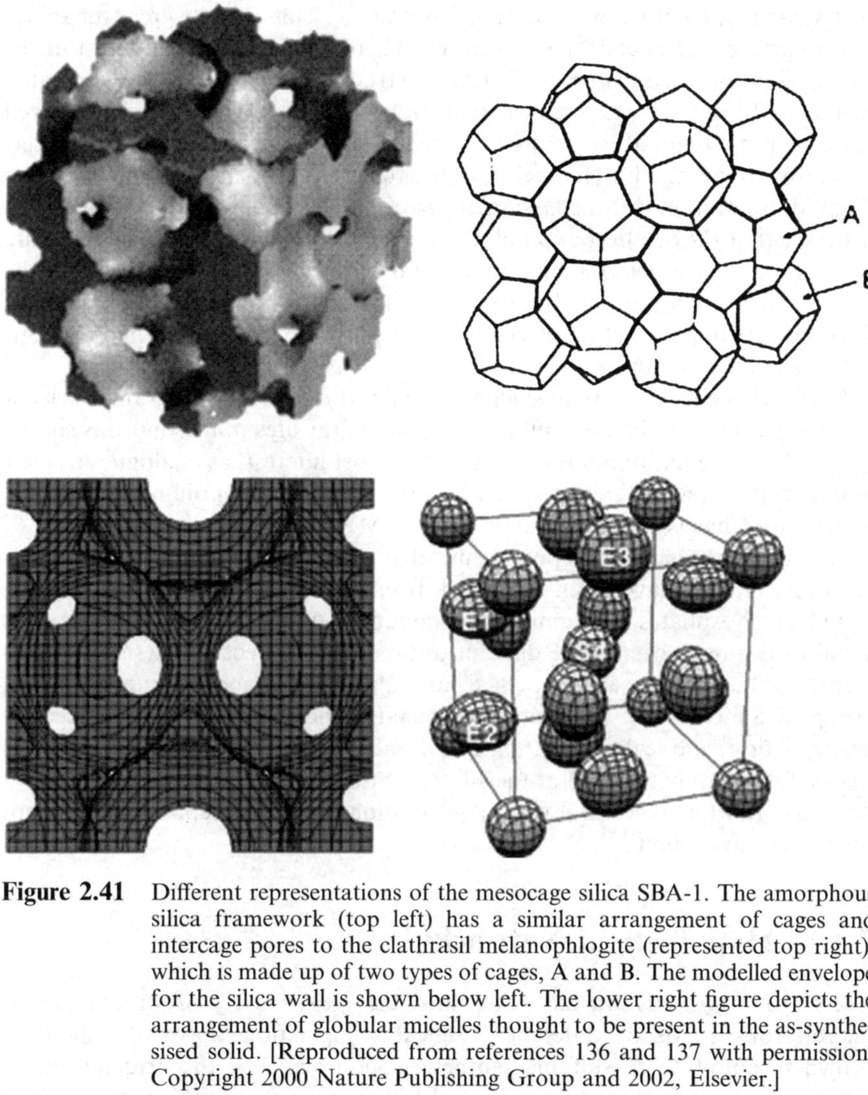

Figure 2.41 Different representations of the mesocage silica SBA-1. The amorphous silica framework (top left) has a similar arrangement of cages and intercage pores to the clathrasil melanophlogite (represented top right), which is made up of two types of cages, A and B. The modelled envelope for the silica wall is shown below left. The lower right figure depicts the arrangement of globular micelles thought to be present in the as-synthesised solid. [Reproduced from references 136 and 137 with permission. Copyright 2000 Nature Publishing Group and 2002, Elsevier.]

The AMS (anionic surfactant-templated mesoporous silicas) materials[130,140,142] are a family of well-ordered organosilicas prepared using anionic surfactant templates and including alkylammonium- or alkylamine-functionalised silica. The strong interactions between charged surfaces and surfactants in the synthesis results in a series of well-ordered organo-silica structures.

The chemistry of silicas templated by micelles is much closer to that of amorphous silicas prepared by sol-gel routes than to zeolites. In particular, the presence of surface hydroxyls permits the surfaces to be functionalised by organosiloxanes, of general formula $(RO)_3$-Si-X, where X may be alkyl, phenyl, alkylchloride, alkylthiol, aminoalkyl, *etc.* The groups can be incorporated post-synthesis, by grafting, or during synthesis.[143] This permits great tunability of the surfaces for chemical applications. There has been much interest in the preparation of organosilicas of regular structures by the micelle-mediated synthesis of bis-siloxanes, such as bis-benzene siloxanes, which Inagaki showed could be prepared with structures ordered in three dimensions.[144] The regular arrangement along the channel axis has been confirmed experimentally by X-ray diffraction and transmission electron microscopy, and interpreted by a structure that shows alternation of organic and inorganic parts of the structure parallel to the channel axis. Similar mesoporous organosilicas, but without the same degree of ordering within the walls, have been prepared with other bis-siloxanes. Examples include structures with pore windows in the microporous range, as in SBA-1 and -2.

Just as elements other than silicon can be introduced into the frameworks of zeolites, so they can be introduced into tetrahedral sites in mesoporous silicas. Aluminium, boron, titanium have attracted most interest, as analogues to their zeolitic counterparts: the framework incorporation of aluminium in MCM-41, for example, has been studied in detail by NMR and confirmed in this way.[145] Furthermore, ordered mesoporous metal phosphates with a very wide compositional range have been prepared from ethanolic sol-gel preparations, including phosphates of aluminium, titanium and zirconium.[146] Mesoporous metal oxides have been more difficult to prepare by direct synthesis, although there have been significant successes, particularly in mesoporous titanias by the group of Sanchez.[147,148] Mesoporous metal oxides can also be prepared by impregnation and calcination of metal salt precursors within mesoporous silicas, followed by removal of the silicate matrix to leave the metal oxide cast. A similar route is successful for the generation of inverse replicas of mesoporous silicas in carbon.[149]

2.10 Hypothetical Networks

The foregoing discussion has underlined the great variety of structures that microporous solids of different chemical compositions have already been shown to adopt given suitable synthetic conditions. For the original microporous solids, zeolites, there are around 170 topologies, and this increases by a few each year. Among the MOFs, by contrast, around 1200 three-dimensional periodic structures were deposited in the Cambridge Structure Database[150] by 2003, and the total number has doubled every 4 years or so since their discovery in 1978.[151] How can this accumulated experience of novel structures help us to understand the underlying principles of what is possible or probable in materials discovery and enable true design in the synthesis of materials with particularly valuable or desired properties? This is the 'Holy Grail' of materials synthesis, to enable us to compete with the organic chemist in programmed total synthesis.

To help answer this question, there has been a long history of geometric and mathematical analysis of observed and hypothetical periodic networks. As a result, a careful and precise vocabulary has been established, the details of which are beyond the scope of this book. The interested reader is referred to texts on crystal chemistry[152] and to recent articles on the subject.[153–155] One key term that is used is that of N-connected *nets* (the basic topologies that underlie most open structures), which are made up of *vertices* linked to N neighbours. Nets with one symmetrically distinct type of vertex are known as *uninodal*, those with two types *binodal* and so on. Nets are only topological abstractions – to arrive at a chemical structure these have to be embodied in terms of atoms or groups of atoms which have the appropriate coordination or connectivity. One additional term will be given here: a *tiling* is a description of a collection of tiles that fill space, so that the edges and vertices of the tile form a net. In two dimensions a tiling corresponds to an array of edge- and vertex-sharing flat tiles, whereas in three dimensions the tiles become polyhedra or cages and in *face-to-face* tilings each face is common to exactly two tiles. (It should be noted that the crystal symmetry of a structure is not necessarily that of the net, which displays the maximum possible symmetry. Distortions away from maximum symmetry can occur as frameworks relax to minimise energy, or if the building units have lower than maximum symmetry.)

2.10.1 Hypothetical Zeolites

Taking first the important example of zeolitic, tetrahedrally connected nets, many enumerations have been made. These include compilations based on geometric principles, observed structures and the assembly of observed structural building units (such as chains, cages or sheets) *via* different symmetry operations. Some of the best-known compilations are those derived by Smith.[156] The problem inherent to this approach is that the number of possible nets is infinite so that all compilations of this type will be incomplete. As a result, more systematic computer-based approaches have been taken to enumerate tetrahedrally coordinated crystalline networks.

Treacy *et al.* have built up a database of theoretical structures by the ‘symmetry-constrained intersite bonding search’ (SCIBS) method,[157,158] which involves systematic searching of all the possible arrangements of p unique vertices in specific space groups. For each of the 230 space groups, all of the combinatorial possibilities for uninodal ($p=1$) and binodal ($p=2$) nets have been determined, and for some space groups (such as P6/mmm) all nets up to $p=6$ have been computed. Once a net has been found, the associated structure can be predicted by energy minimisation by computer simulation (Chapter 4). The computing effort required increases massively as p increases: for $p=1$, around 40 000 possible arrangements were determined, whereas for $p=2$ over 440 million were found. Exponential increases are observed within each space group, the steepness of which depends on the symmetry. For this reason it seems unlikely that all possible zeolite structures can be predicted, given that the most complex zeolite structure yet determined, TNU-9, has $p=24$. Nevertheless, the approach has already had successes: the framework of ZSM-10,[159] for example, was unambiguously identified from its powder X-ray diffraction pattern by searching all 18 million nets in P6/mmm ($p = 6$).[160] Notably, among the hypothetical structures predicted in this same space group was the framework type of ITQ-33, before it was prepared and characterised.[158]

A parallel approach to the systematic enumeration of tetrahedrally coordinated networks has been developed by Delgado-Friedrichs *et al.*[161] They make use of recent advances in mathematical tiling theory to derive a mathematically rigorous partial solution to the problem rather than relying on empirical methods to enumerate structures. For simple tilings (which are based on closed polyhedra or cages) Delgado-Friedrichs *et al.* show that there are exactly 9 uninodal, 117 binodal and 1351 trinodal structures based on tetrahedra. Of the nine uninodal structures, for instance, six correspond to known structures, SOD, LTA, RHO, FAU, KFI and CHA. Examination of the chemical feasibility of these structures suggests as yet unprepared materials based on binodal and trinodal nets that are likely to have interesting properties if synthesised.[162]

In summary, then, the complexity of zeolitic structures is too great to allow a complete enumeration, so that this approach could not be used to identify all new zeolite structure types from a comparison of observed diffraction profiles and those predicted from energy minimisation. Nevertheless, for structures where the symmetry is quite high the method may find application in structure solution. Furthermore, novel hypothetical structures which are chemically feasible and which would possess important new properties if made would be attractive targets for synthesis *via* the methods of templating described elsewhere in this book.

2.10.2 Nets and MOFs

Upon initial consideration, comparison of metal organic frameworks, which can be considered as made up of metallocentric vertices and connecting ligands of very varied coordination geometry with zeolitic solids, which are entirely tetrahedrally coordinated, would suggest that the diversity of structures associated with MOFs would be bewilderingly large, especially because they may contain vertices with mixed coordination. The number of structures being deposited with the Cambridge Structure Database (CSD) appears to confirm this. However, closer structure analysis in terms of the underlying structural topologies or nets using the SYSTRE program[163] written for this purpose tells a rather different story.[151]

In this analysis, in which both metallocentric clusters and organic ligands are reduced to their essential coordination geometry – for example, in MOF-5 the Zn_4O cluster has octahedral coordination *via* the terephthalate ligands – Ockwig *et al.*[151] show that most of the 1127 three-dimensional MOF structures in the database up to 2003 can be described using a 'handful' of nets. O'Keeffe suggests that around a dozen such nets are of major importance.[153] Of the 1127 structures examined, 774 are reduced to topologies belonging to nets with vertices having one coordination number. For the 113 MOFs with triangular building units, 64% have the **srs** topology – each lower-case three-letter code represents a distinct net topology and **srs** refers to the net described by the Si atom positions in $SrSi_2$. For the 335 tetrahedral MOFs, 70% have the diamond **dia** topology and for 152 MOFs based on only octahedral building blocks 95% adopt the primitive cubic net **pcu**. The authors discuss this discrepancy between observed and theoretical structural diversity in terms of the components, which are rather simple and link non-directionally according to steric considerations to give high symmetry structures.

The discovery of this dominance of relatively simple structures that act as default structures can be exploited in synthesis, because there are enough nets of this type that new materials can be prepared by choosing metal/metal cluster-ligand combinations of pre-determined geometries that are expected to form open structures with attractive properties. MOF-5, for example, possesses the same net described by the boron atoms in CaB_6, and if not already made could have been designed from the principles outlined by Yaghi using topological considerations. O'Keeffe *et al.* give several examples of augmented nets that are interesting targets for designed synthesis.[153] Augmented nets are N-connected nets where the vertices are replaced by groups of N vertices. Suggested 'targets' of O'Keeffe include the augmented rutile net (Figure 2.42).

To access the very many hypothetical MOF nets that could display important novel features, it becomes apparent that additional features should be present during the syntheses that they might confer directionality and complexity in the crystallisation. In zeolite synthesis this is achieved by the inclusion of templating molecules. MOFs could also be prepared in the presence of organic templates, but because the templates must be removed to render the solids porous, and the MOFs are unlikely to withstand this process, alternative approaches are required. Yaghi suggests increasing the complexity of the components of synthesis, for example by introducing terminal ligands that compete with the coordinating linkers for coordination sites at the metal centre or by using organic linkers with low symmetry. The synthesis of MOF-500 is an example of the first of these approaches, which contains four levels of complexity in the structural units.[109] This is clearly an area in which synthetic methodology and the generation of hypothetical structures can together make great progress towards the true designed synthesis of particular structures.

It is also clear that the modular nature of MOFs permits the incorporation of complex organic units within the frameworks (the second approach). One exciting example of this is the incorporation of chiral building blocks to give inherently chiral solids. The chiral zinc-based solid POST-1 is a related example, where enantiomers of the ligand of Scheme 2.2 are incorporated *via* bonding of the deprotonated carboxylic acid group and the nitrogen on the pyridine group to trimeric $Zn_3(\mu_3O)$ building units (similar to those described in Section 2.8) into porous layers. These layers then stack to give large open channels.[164]

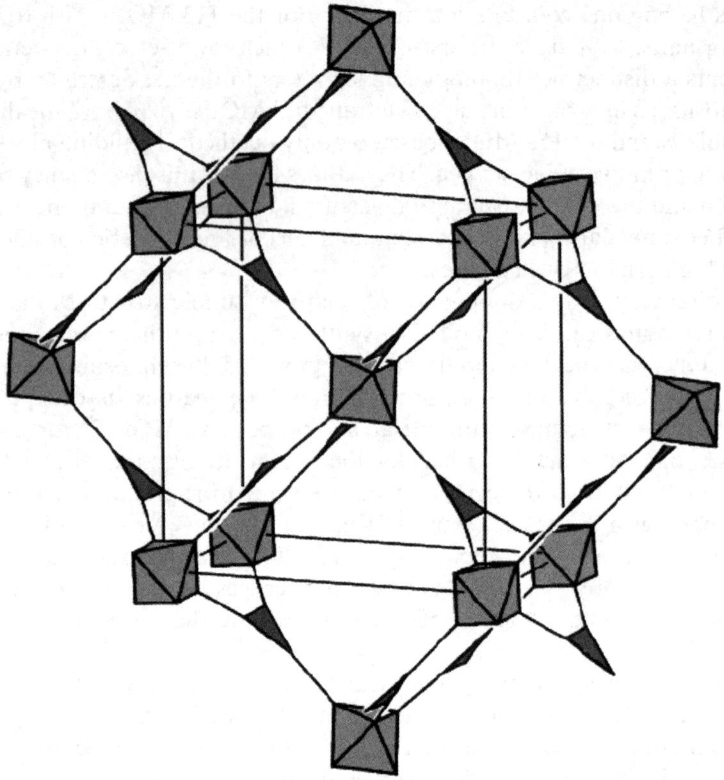

Figure 2.42 A decorated rutile net. This type of hypothetical structure could be formed by a combination of octahedral and trigonal building units. [Reproduced from reference 153 with permission. Copyright 2000 Elsevier.]

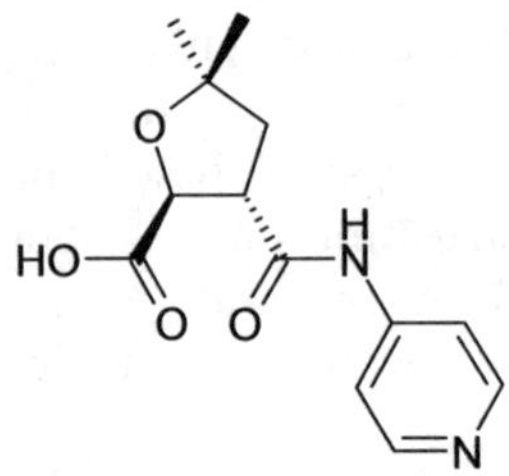

Scheme 2.2 The chiral ligand used in the synthesis of the porous solid POST-1.

This approach has been taken further, using as linkers diamine ligands that contain within them sites that themselves coordinate to catalytically active metals that are accessible to molecules from within the pore space.[165,166] This approach, which is discussed further in Chapter 9, has been used to include enantioselective catalytic sites into MOFs.

2.11 Summary

The diversity of ordered porous solids increases at an astonishing rate, particularly among the readily crystallised MOFs, and continues to offer novel materials properties. There is no obvious barrier to the synthesis of a myriad of new zeolite, zeotype or hybrid structures. Challenges remain, however. For zeolitic aluminosilicates, the 10 Å pore size restriction remains an important barrier, and an enantiomerically pure zeolite is still out of reach. For non-silicate crystalline microporous solids, thermal and hydrothermal stability, rather than framework geometry, limit their applicability, since fully crystalline germanates and carboxylates with pores in the mesoporous range now exist, and these solids have enormous specific surface areas. In these hybrid solids the ability to choose chirality in the building units indicates that it will be possible to prepare these in chiral form: the first examples have already been prepared.

Mesoporous silicas can possess intercage windows within the microporous region, but their stability and acid strength are low compared to those of zeolites. One current approach to this particular problem is the use of partially crystallised zeolite nuclei as starting materials in the synthesis of mesoporous solids, in an attempt to prepare a solid with structural features of both microporous and mesoporous solids. Such solids are said to possess hierarchical porosity, and are discussed further in Chapter 10.

References

1. A. K. Cheetham, G. Férey and T. Loiseau, *Angew. Chem. Int. Ed.*, 1999, **38**, 3268.
2. C. Baerlocher, W. M. Meier and D. H. Olson, *'Atlas of Zeolite Framework Types'*, 5th Ed., Elsevier, Amsterdam, 2001.
3. E. Dooryhee, C. R. A. Catlow, J. W. Couves, P. J. Maddox, J. M. Thomas, G. N. Greaves, A. T. Steel and R. P. Townsend, *J. Phys. Chem.*, 1991, **95**, 4514.
4. G. T. Kokotailo, S. L. Lawton, D. H. Olson and W. M. Meier, *Nature*, 1978, **272**, 437.
5. G. T. Kokotailo, P. Chu, S. L. Lawton and W. M. Meier, *Nature*, 1978, **275**, 119.
6. P. A. Wright, J. M. Thomas, G. R. Millward, S. Ramdas and S. A. I. Barri, *J. Chem. Soc., Chem. Commun.*, 1985, 1117.
7. M. A. Camblor, A. Corma, P. Lightfoot, L. A. Villaescusa and P. A. Wright, *Angew. Chem. Int. Ed.*, 1997, **36**, 2659.
8. J. M. Newsam, M. M. J. Treacy, W. T. Koetsier and C. B. Degruyter, *Proc. Roy. Soc. Lon. Ser. A.*, 1988, **420**, 375.
9. A. Corma, M. T. Navarro, F. Rey and S. Valencia, *Chem. Commun.*, 2001, 1486.
10. D. E. W. Vaughan and K. G. Strohmaier, *Micropor. Mesopor. Mater.*, 1999, **28**, 233.

11. M. E. Leonowicz and D. E. W. Vaughan, *Nature*, 1987, **329**, 819.
12. S. J. Warrender, P. A. Wright, W. Z. Zhou, P. Lightfoot, M. A. Camblor, C. H. Shin, D. J. Kim and S. B. Hong, *Chem. Mater.*, 2005, **17**, 1272.
13. B. Han, C. H. Shin, S. J. Warrender, P. Lightfoot, P. A. Wright, M. A. Camblor and S. B. Hong, *Chem. Mater.*, 2006, **18**, 3023.
14. C. C. Freyhardt, M. Tsapatsis, R. F. Lobo, K. J. Balkus and M. E. Davis, *Nature*, 1996, **381**, 295.
15. M. Yoshikawa, P. Wagner, M. Lovallo, K. Tsuji, T. Takewaki, C. Y. Chen, L. W. Beck, C. Jones, M. Tsapatsis, S. I. Zones and M. E. Davis, *J. Phys. Chem. B.*, 1998, **102**, 7139.
16. R. A. Van Norstrand, D. S. Santilli and S. I. Zones, *ACS Symp. Ser.*, 1988, **368**, 236.
17. M. E. Leonowicz, J. A. Lawton, S. L. Lawton and M. K. Rubin, *Science*, 1994, **264**, 1910.
18. F. Gramm, C. Baerlocher, L. B. McCusker, S. J. Warrender, P. A. Wright, B. Han, S. B. Hong, Z. Liu, T. Ohsuna and O. Terasaki, *Nature*, 2006, **444**, 79.
19. M. D. Shannon, J. L. Casci, P. A. Cox and S. J. Andrews, *Nature*, 1991, **353**, 417.
20. G. O. Brunner and W. M. Meier, *Nature*, 1989, **337**, 146.
21. M. W. Anderson, O. Terasaki, T. Ohsuna, A. Philippou, S. P. Mackay, A. Ferreira, J. Rocha and S. Lidin, *Nature*, 1994, **367**, 347.
22. K. G. Strohmaier and D. E. W. Vaughan, *J. Am. Chem. Soc.*, 2003, **125**, 16035.
23. M. J. Annen, M. E. Davis, J. B. Higgins and J. L. Schlenker, *J. Chem. Soc., Chem. Commun.*, 1991, 1175.
24. T. Blasco, M. A. Camblor, A. Corma, P. Esteve, J. M. Guil, A. Martinez, J. A. Perdigon-Melon and S. Valencia, *J. Phys. Chem. B.*, 1998, **102**, 75.
25. A. Corma, M. Diaz-Cabanas, J. Martinez-Triguero, F. Rey and J. Rius, *Nature*, 2002, **418**, 514.
26. T. Conradsson, M. S. Dadachov and X. D. Zou, *Micropor. Mesopor. Mater.*, 2000, **41**, 183.
27. J. L. Paillaud, B. Harbuzaru, J. Patarin and N. Bats, *Science*, 2004, **304**, 990.
28. A. Corma, M. J. Diaz-Cabanas, F. Rey, S. Nicolooulas and K. Boulahya, *Chem. Commun.*, 2004, 1356.
29. A. Corma, M. J. Diaz-Cabanas, J. L. Jorda, C. Martinez and M. Moliner, *Nature*, 2006, **443**, 842.
30. S. T. Wilson, B. M. Lok, C. A. Messina, T. R. Cannan and E. M. Flanigen, *J. Am. Chem. Soc.*, 1982, **104**, 1146.
31. J. M. Bennett, J. P. Cohen, E. M. Flanigen, J. J. Pluth and J. V. Smith, *ACS Symp. Ser.*, 1983, **218**, 109.
32. M. G. G. Wu, M. W. Deem, S. A. Elomari, R. C. Medrud, S. I. Zones, T. Maesen, C. Kibby, C. Y. Chen and I. Y. Chan, *J. Phys. Chem. B.*, 2002, **106**, 264.

33. M. E. Davis, C. Montes, P. E. Hathaway, J. P. Arhancet, D. L. Hasha and J. M. Garces, *J. Am. Chem. Soc.*, 1989, **111**, 3919.
34. J. Patarin, J. L. Paillaud and H. Kessler, in '*Handbook of Porous Solids*', ed. F. Schuth, K. S. W. Sing and J. Weitkamp, Wiley-VCH, New York, 2002.
35. E. M. Flanigen, B. M. Lok, R. L. Patton and S. T. Wilson, *Pure Appl. Chem.*, 1986, **58**, 1351.
36. S. T. Wilson and E. M. Flanigen, *ACS Symp. Ser.*, 1989, **398**, 329.
37. B. M. Lok, C. A. Messina, R. L. Patton, R. T. Gajek, T. R. Cannan and E. M. Flanigen, *J. Am. Chem. Soc.*, 1984, **106**, 6092.
38. P. A. Wright, C. Sayag, F. Rey, D. W. Lewis, J. D. Gale, S. Natarajan and J. M. Thomas, *J. Chem. Soc. Farad. Trans.*, 1995, **91**, 3537.
39. J. M. Thomas, G. N. Greaves, G. Sankar, P. A. Wright, J. S. Chen, A. J. Dent and L. Marchese, *Angew. Chem. Int. Ed.*, 1994, **33**, 1871.
40. M. Afeworki, D. Dorset, G. Kennedy and K. Strohmaier, *Stud. Surf. Sci. Catal.*, 2004, **154**, 1274.
41. K. D. Schmitt and G. J. Kennedy, *Zeolites*, 1994, **14**, 635.
42. P. A. Wright, S. Natarajan, J. M. Thomas, R. G. Bell, P. L. Gai-Boyes, R. H. Jones and J. S. Chen, *Angew. Chem. Int. Ed*, 1992, **31**, 1472.
43. J. V. Smith, J. J. Pluth and K. J. Andries, *Zeolites*, 1993, **13**, 166.
44. P. A. Wright, R. H. Jones, S. Natarajan, R. G. Bell, J. S. Chen, M. B. Hursthouse and J. M. Thomas, *J. Chem. Soc., Chem. Commun.*, 1993, 633.
45. G. W. Noble, P. A. Wright, P. Lightfoot, R. E. Morris, K. J. Hudson, A. Kvick and H. Graafsma, *Angew. Chem. Int. Ed*, 1997, **36**, 81.
46. G. W. Noble, P. A. Wright and A. Kvick, *J. Chem. Soc., Dalton Trans.*, 1997, 4485..
47. V. Patinec, P. A. Wright, R. A. Aitken, P. Lightfoot, S. D. J. Purdie, P. A. Cox, A. Kvick and G. Vaughan, *Chem. Mater.*, 1999, **11**, 2456.
48. V. Patinec, P. A. Wright, P. Lightfoot, R. A. Aitken and P. A. Cox, *J. Chem. Soc., Dalton Trans.*, 1999, 3909.
49. P. A. Wright, M. J. Maple, A. M. Z. Slawin, V. Patinec, R. A. Aitken, S. Welsh and P. A. Cox, *J. Chem. Soc., Dalton Trans.*, 2000, 1243.
50. X. H. Bu, P. Y. Feng and G. D. Stucky, *Science*, 1997, **278**, 2080.
51. S. Oliver, A. Kuperman, A. Lough and G. A. Ozin, *J. Mater. Chem.*, 1997, **7**, 807.
52. R. Garcia, I. J. Shannon, A. M. Z. Slawin, W. Z. Zhou, P. A. Cox and P. A. Wright, *Micropor. Mesopor. Mater.*, 2003, **58**, 91.
53. A. I. Bortun, L. N. Bortun, D. M. Poojary, O. Xiang and A. Clearfield, *Chem. Mater.*, 2000, **12**, 294.
54. Z. Lin, J. Rocha, P. Ferreira, A. Thursfield, J. R. Agger and M. W. Anderson, *J. Phys. Chem. B.*, 1999, **103**, 957.
55. Z. Lin, J. Rocha and A. Valente, *Chem. Commun.*, 1999, 2489.
56. J. Rocha and M. W. Anderson, *Eur. J. Inorg. Chem.*, 2000, 801.
57. X. Q. Wang, L. M. Liu and A. J. Jacobson, *Angew. Chem. Int. Ed*, 2001, **40**, 2174.
58. W. T. A. Harrison, T. E. Gier and G. D. Stucky, *Zeolites*, 1995, **15**, 408.

59. D. M. Poojary, R. A. Cahill and A. Clearfield, *Chem. Mater.*, 1994, **6**, 2364.
60. D. M. Poojary, A. I. Bortun, L. N. Bortun and A. Clearfield, *Inorg. Chem.*, 1996, **35**, 6131.
61. A. Tripathi, D. G. Medvedev, J. Delgado and A. Clearfield, *J. Solid State Chem.*, 2004, **177**, 2903.
62. P. Pertierra, M. A. Salvado, S. Garcia-Granda, A. I. Bortun and A. Clearfield, *Inorg. Chem.*, 1999, **38**, 2563.
63. S. M. Kuznicki, 1990, *US Patent* 4,938,989.
64. X. Q. Wang and A. J. Jacobson, *Chem. Commun.*, 1999, 973.
65. M. W. Anderson, J. Rocha, Z. Lin, A. Philippou, I. Orion and A. Ferreira, *Micropor. Mater.*, 1996, **6**, 195.
66. J. Rocha, P. Brandao, M. W. Anderson, T. Ohsuna and O. Terasaki, *Chem. Commun.*, 1998, 667.
67. J. Rocha, P. Brandao, A. Phillippou and M. W. Anderson, *Chem. Commun.*, 1998, 2687.
68. J. Rocha, P. Brandao, Z. Lin, M. W. Anderson, V. Alfredsson and O. Terasaki, *Angew. Chem. Int. Ed*, 1997, **36**, 100.
69. D. M. Chapman and A. L. Roe, *Zeolites*, 1990, **10**, 730.
70. S. Nair, H. K. Jeong, A. Chandrasekaran, C. M. Braunbarth, M. Tsapatsis and S. M. Kuznicki, *Chem. Mater.*, 2001, **13**, 4247.
71. J. Rocha, P. Brandao, Z. Lin, A. P. Esculcas, A. Ferreira and M. W. Anderson, *J. Phys. Chem.*, 1996, **100**, 14978.
72. X. Q. Wang, L. M. Liu and A. J. Jacobson, *Angew. Chem. Int. Ed.*, 2003, **42**, 2044.
73. J. Rocha, P. Ferreira, Z. Lin, P. Brandão, A. Ferreira and J. D. Pedrosa de Jesus, *J. Phys. Chem. B.*, 1998, **102**, 4739.
74. J. Rocha, P. Ferreira, L. D. Carlos and A. Ferreira, *Angew. Chem. Int. Ed.*, 2000, **39**, 3276.
75. D. Ananias, A. Ferreira, J. Rocha, P. Ferreira, J. P. Rainho, C. Morais and L. D. Carlos, *J. Am. Chem. Soc.*, 2001, **123**, 5735.
76. M. Estermann, L. B. McCusker, C. Baerlocher, A. Merrouche and H. Kessler, *Nature*, 1991, **352**, 320.
77. N. Guillou, Q. Gao, M. Nogues, R. E. Morris, M. Hervieu, G. Férey and A. K. Cheetham, *C.R. Acad. Sci. Paris*, 1999, **2**, 387.
78. N. Guillou, Q. Gao, P. M. Forster, J.-S. Chang, M. Nogues, S.-E. Park, G. Férey and A. K. Cheetham, *Angew. Chem. Int. Ed.*, 2001, **40**, 2831.
79. M. Schindler and W. H. Baur, *Angew. Chem. Int. Ed.*, 1997, **36**, 91.
80. Y. M. Zhou, H. G. Zhu, Z. X. Chen, M. Q. Chen, Y. Xu, H. Y. Zhang and D. Y. Zhao, *Angew. Chem. Int. Ed.*, 2001, **40**, 2166.
81. J. Plevert, T. M. Gentz, A. Laine, H. L. Li, V. G. Young, O. M. Yaghi and M. O'Keeffe, *J. Am. Chem. Soc.*, 2001, **123**, 12706.
82. X. D. Zou, T. Conradsson, M. Klingstedt, M. S. Dadachov and M. O'Keeffe, *Nature*, 2005, **437**, 716.
83. R. N. Deguzman, Y. Shen, E. J. Neth, S. L. Suib, C.-L. O'Young, S. Levine and J. M. Newsam, *Chem. Mater.*, 1994, **6**, 815.

84. Y. Shen, R. P. Zerger, S. L. Suib, L. McCurdy, D. Potter and C.-L. O'Young, *J. Chem. Soc., Chem. Commun.*, 1992, 1213.
85. Y. Shen, R. P. Zerger, R. N. Deguzman, S. L. Suib, L. McCurdy, D. Potter and C.-L. O'Young, 1993, **260**, 511.
86. M. Nyman, A. Tripathi, J. B. Parise, R. S. Maxwell, W. T. A. Harrison and T. M. Nenoff, *J. Am. Chem. Soc.*, 2001, **123**, 1529.
87. J. Weitkamp, S. Ernst, F. Cubero, F. Wester and W. Schnick, *Adv. Mater.*, 1997, **9**, 247.
88. N. Zheng, X. Bu, B. Wang and P. Feng, *Science*, 2002, **298**, 2366.
89. K. Maeda, J. Akimoto, Y. Kiyozumi and F. Mizukami, *Angew. Chem. Int. Ed.*, 1995, **34**, 1199.
90. K. Maeda, J. Akimoto, Y. Kiyozumi and F. Mizukami, *J. Chem. Soc., Chem. Commun.*, 1995, 1033.
91. S. R. Miller, E. Lear, J. Gonzalez, A. M. Z. Slawin, P. A. Wright, N. Guillou and G. Férey, *Dalton Trans.*, 2005, 3319.
92. P. M. Forster and A. K. Cheetham, *Angew. Chem. Int. Ed.*, 2002, **41**, 457.
93. K. Tsuji, C. W. Jones and M. E. Davis, *Micropor. Mesopor. Mater.*, 1999, **29**, 339.
94. C. W. Jones, K. Tsuji and M. E. Davis, *Micropor. Mesopor. Mater.*, 1999, **33**, 223.
95. C. W. Jones, M. Tsapatsis, T. Okubo and M. E. Davis, *Micropor. Mesopor. Mater.*, 2001, **42**, 21.
96. B. F. Abrahams, B. F. Hoskins, D. M. Michail and R. Robson, *Nature*, 1994, **369**, 727.
97. H. Li, M. Eddaoudi, M. O'Keeffe and O. M. Yaghi, *Nature*, 1999, **402**, 276.
98. O. M. Yaghi, M. O'Keeffe, N. W. Ockwig, H. K. Chae, M. Eddaoudi and J. Kim, *Nature*, 2003, **423**, 705.
99. J. L. C. Rowsell and O. M. Yaghi, *Micropor. Mesopor. Mater.*, 2004, **73**, 3.
100. M. J. Rosseinsky, *Micropor. Mesopor. Mater.*, 2004, **73**, 15.
101. D. Bradshaw, J. B. Claridge, E. J. Cussen, T. J. Prior and M. J. Rosseinsky, *Acc. Chem. Res.*, 2005, **38**, 273.
102. C. N. R. Rao, S. Natarajan and R. Vaidhyanathan, *Angew. Chem. Int. Ed.*, 2004, **43**, 1466.
103. S. R. Miller, P. A. Wright, C. Serre, T. Loiseau, J. Marrot and G. Férey, *Chem. Commun.*, 2005, 3850.
104. D. N. Dybtsev, H. Chun and K. Kim, *Angew. Chem. Int. Ed.*, 2004, **43**, 5033.
105. S. S. Y. Chui, J. P. H. Charmant, A. G. Orpen and I. D. Williams, *Science*, 1999, **283**, 1148.
106. C. Serre, F. Millange, S. Surblé and G. Férey, *Angew. Chem. Int. Ed.*, 2004, **43**, 2.
107. S. Surble, C. Serre, C. Mellot-Draznieks, F. Millange and G. Férey, *Chem. Commun.*, 2006, 284.
108. G. Férey, C. Mellot-Draznieks, C. Serre and F. Millange, *Acc. Chem. Res.*, 2005, **38**, 217.

109. A. C. Sudik, A. P. Côté, A. G. Wong-Foy, M. O'Keeffe and O. M. Yaghi, *Angew. Chem. Int. Ed*, 2006, **45**, 2528.
110. M. Eddaoudi, J. Kim, N. Rosi, D. Vodak, J. Wachter, M. O'Keefe and O. M. Yaghi, *Science*, 2002, **295**, 469.
111. H. K. Chae, D. Y. Siberio-Perez, J. Kim, Y. Go, M. Eddaoudi, A. J. Matzger, M. O'Keefe and O. M. Yaghi, *Nature*, 2004, **427**, 523.
112. F. Millange, C. Serre and G. Férey, *Chem. Commun.*, 2002, 822.
113. T. Loiseau, C. Serre, C. Huguenard, G. Fink, F. Taulelle, M. Henry, T. Bataille and G. Férey, *Chem.-Eur. J.*, 2004, **10**, 1373.
114. T. R. Whitfield, X. Q. Wang, L. M. Liu and A. J. Jacobson, *Sol. Stat. Sci.*, 2005, **7**, 1096.
115. K. Barthelet, J. Marrot, D. Riou and G. Férey, *Angew. Chem. Int. Ed*, 2001, **41**, 281.
116. K. Barthelet, J. Marrot, G. Férey and D. Riou, *Chem. Commun.*, 2004, 520.
117. N. L. Rosi, J. Kim, M. Eddaoudi, B. L. Chen, M. O'Keeffe and O. M. Yaghi, *J. Am. Chem. Soc.*, 2005, **127**, 1504.
118. M. Kondo, T. Yoshitomi, K. Seki, H. Matsuzaka and S. Kitagawa, *Angew. Chem. Int. Ed*, 1997, **36**, 1725.
119. X. L. Xu, M. Nieuwenhuyzen and S. L. James, *Angew. Chem. Int. Ed.*, 2002, **41**, 764.
120. X. C. Huang, Y. Y. Lin, J. P. Zhang and X. M. Chen, *Angew. Chem. Int. Ed*, 2006, **45**, 1557.
121. K. S. Park, Z. Ni, A. P. Côté, J. Y. Choi, R. D. Huang, F. J. Uribe-Romo, H. K. Chae, M. O'Keeffe and O. M. Yaghi, *Proc. Natl. Acad. Sci. U. S. A.*, 2006, **103**, 10186.
122. C. Serre, J. A. Groves, P. Lightfoot, A. M. Z. Slawin, P. A. Wright, N. Stock, T. Bein, M. Haouas, F. Taulelle and G. Férey, *Chem. Mater.*, 2006, **18**, 1451.
123. J. A. Groves, S. R. Miller, S. J. Warrender, C. Mellot-Draznieks, P. Lightfoot and P. A. Wright, *Chem. Commun.*, 2006, 3305.
124. G. Férey, C. Serre, C. Mellot-Draznieks, F. Millange, S. Surblé, J. Dutour and I. Margiolaki, *Angew. Chem. Int. Ed.*, 2004, **43**, 6296.
125. D. Bradshaw, J. B. Claridge, E. J. Cussen, T. J. Prior and M. J. Rosseinsky, *Acc. Chem. Res.*, 2005, **38**, 273.
126. C. T. Kresge, M. E. Leonowicz, W. J. Roth, J. C. Vartuli, J. S. Beck, *Nature*, 1992 **359**, 710: J. S. Beck, J. C. Vartuli, W. J. Roth, M. E. Leonowicz, C. t. Kresge, K. D. Schmitt, C. T. W. Chu, D. H. Olson, E. W. Sheppard, S. B. McCullen, J. B. Higgins and J. L. Schlenker, *J. Am. Chem. Soc.*, 1992, **114**, 10834.
127. C. Cheng, W. Z. Zhou, D. H. Park, J. Klinowski, M. Hargreaves and L. F. Gladden, *J. Chem. Soc., Farad. Trans.*, 1997, **93**, 359.
128. S. Inagaki, Y. Fukushima and K. Kuroda, *J. Chem. Soc., Chem. Commun.*, 1993, 680.
129. Q. S. Huo, D. I. Margolese, U. Ciesla, P. Y. Feng, T. E. Gier, P. Sieger, R. Leon, P. M. Petroff, F. Schuth and G. D. Stucky, *Nature*, 1994, **368**, 317.

130. A. E. Garcia-Bennett, O. Terasaki, S. Che and T. Tatsumi, *Chem. Mater.*, 2004, **16**, 813.
131. D. Y. Zhao, Q. S. Huo, J. L. Feng, B. F. Chmelka and G. D. Stucky, *J. Am. Chem. Soc.*, 1998, **120**, 6024.
132. D. Y. Zhao, J. L. Feng, Q. S. Huo, N. Melosh, G. H. Fredrickson, B. F. Chmelka and G. D. Stucky, *Science*, 1998, **279**, 548.
133. D. Y. Zhao, Q. S. Huo, J. L. Feng, J. M. Kim, Y. J. Han and G. D. Stucky, *Chem. Mater.*, 1999, **11**, 2668.
134. M. W. Anderson, *Zeolites*, 1997, **19**, 220.
135. X. Y. Liu, B. Z. Tian, C. Z. Yu, F. Gao, S. H. Xie, B. Tu, R. C. Che, L. M. Peng and D. Y. Zhao, *Angew. Chem. Int. Ed*, 2002, **41**, 3876.
136. Y. Sakamoto, M. Kaneda, O. Terasaki, D. Y. Zhao, J. M. Kim, G. Stucky, H. J. Shim and R. Ryoo, *Nature*, 2000, **408**, 449.
137. M. W. Anderson, C. C. Egger, G. J. T. Tiddy and J. L. Casci, *Stud. Surf. Sci. Catal.*, 2002, **142**, 1149.
138. Q. S. Huo, R. Leon, P. M. Petroff and G. D. Stucky, *Science*, 1995, **268**, 1324: Q. S. Huo, D. I. Margolese and G. D. Stucky, *Chem. Mater.*, 1996, **8**, 1147.
139. J. Fan, C. Z. Yu, T. Gao, J. Lei, B. Z. Tian, L. M. Wang, Q. Luo, B. Tu, W. Z. Zhou and D. Y. Zhao, *Angew. Chem. Int. Ed*, 2003, **42**, 3146.
140. A. E. Garcia-Bennett, K. Miyasaka, O. Terasaki and S. N. Che, *Chem. Mater.*, 2004, **16**, 3597.
141. C. B. Gao, Y. Sakamoto, K. Sakamoto, O. Terasaki and S. N. Che, *Angew. Chem. Int. Ed.*, 2006, **45**, 4295.
142. S. Che, A. E. Garcia-Bennett, T. Yokoi, K. Sakamoto, H. Kunieda, O. Terasaki and T. Tatsumi, *Nature Mater.*, 2003, **2**, 801.
143. H. H. P. Yiu, P. A. Wright and N. P. Botting, *J. Mol. Cat. B.*, 2001, **15**, 81.
144. S. Inagaki, S. Guan, T. Ohsuna and O. Terasaki, *Nature*, 2002, **416**, 304.
145. M. T. Janicke, C. C. Landry, S. C. Christiansen, D. Kumar, G. D. Stucky and B. F. Chmelka, *J. Am. Chem. Soc.*, 1998, **120**, 6940.
146. B. Z. Tian, X. Y. Liu, B. Tu, C. Z. Yu, J. Fan, L. M. Wang, S. H. Xie, G. D. Stucky and D. Y. Zhao, *Nature Mater.*, 2003, **2**, 159.
147. E. L. Crepaldi, G. Soler-Illia, A. Bouchara, D. Grosso, D. Durand and C. Sanchez, *Angew. Chem. Int. Ed.*, 2003, **42**, 347.
148. B. Smarsly, D. Grosso, T. Brezesinski, N. Pinna, C. Boissiere, M. Antonietti and C. Sanchez, *Chem. Mater.*, 2004, **16**, 2948.
149. R. Ryoo, S. H. Joo and S. Jun, *J. Phys. Chem. B.*, 1999, **103**, 7743.
150. F. H. Allen, *Acta Crystallogr. B*, 2002, **58**, 380.
151. N. W. Ockwig, O. Delgado-Friedrichs, M. O'Keeffe and O. M. Yaghi, *Acc. Chem. Res.*, 2005, **38**, 176.
152. M. O'Keeffe and B. G. Hyde, '*Crystal Structures I. Patterns and Symmetry*', Mineralogical Society of America, Washington, 1996.
153. M. O'Keeffe, M. Eddaoudi, H. L. Li, T. Reineke and O. M. Yaghi, *J. Solid State Chem.*, 2000, **152**, 3.
154. O. Delgado-Friedrichs and M. O'Keeffe, *J. Solid State Chem.*, 2005, **178**, 2480.

155. O. Delgado-Friedrichs, M. D. Foster, M. O'Keeffe, D. M. Proserpio, M. M. J. Treacy and O. M. Yaghi, *J. Solid State Chem.*, 2005, **178**, 2533.
156. J. V. Smith, *Chem. Rev.*, 1988, **88**, 149.
157. M. M. J. Treacy, K. H. Randall, S. Rao, J. A. Perry and D. J. Chadi, *Zeit. Krist.*, 1997, **212**, 768.
158. M. M. J. Treacy, I. Rivin, E. Balkovsky, K. H. Randall and M. D. Foster, *Micropor. Mesopor. Mater.*, 2004, **74**, 121.
159. J. B. Higgins and K. D. Schmitt, *Zeolites*, 1996, **16**, 236.
160. M. D. Foster, M. M. J. Treacy, J. B. Higgins, I. Rivin, E. Balkovsky and K. H. Randall, *J. Appl. Crystallogr.*, 2005, **38**, 1028.
161. O. Delgado-Friedrichs, A. W. M. Dress, D. H. Huson, J. Klinowski and A. L. Mackay, *Nature*, 1999, **400**, 644.
162. M. D. Foster, A. Simperler, R. G. Bell, O. Delgado-Friedrichs, F. A. A. Paz and J. Klinowski, *Nature Mater.*, 2004, **3**, 234.
163. O. Delgado-Friedrichs and M. O'Keeffe, *Acta Cryst. Sect. A*, 2003, **59**, 351.
164. J. S. Seo, D. Whang, H. Lee, S. I. Jun, J. Oh, Y. J. Jeon and K. Kim, *Nature*, 2000, **404**, 982.
165. S. H. Cho, B. Q. Ma, S. T. Nguyen, J. T. Hupp and T. E. Albrecht-Schmitt, *Chem. Commun.*, 2006, 2563.
166. C. D. Wu, A. Hu, L. Zhang and W. B. Lin, *J. Am. Chem. Soc.*, 2005, **127**, 8940.

CHAPTER 3

Structure Determination: Experimental Techniques

3.1 Introduction

The regular crystal structures of microporous solids have led, directly and indirectly, to their widespread study and application. The starting point in understanding their behaviour is the architecture of the repeat unit, or unit cell, which controls their properties. The well-defined pore size permits discrimination between molecules (and transition states in catalytic reactions) to better than 0.1 Å; activities for oxidation and solid acid catalysis are determined by the structural constraints on the local environments of framework cations and protons within the structure; and the regular distribution of extra-framework, charge-balancing cations is responsible for cation exchange and gas adsorption and separation behaviour.

A distinction can be drawn between structural features that may be determined by diffraction methods and those which may be measured spectroscopically. Whereas diffraction gives the time- and space-averaged structure over all unit cells, spectroscopy gives details of the local structure. For example, diffraction is indispensable in determining the framework structure and position of species within the pores, but it cannot in general give details of the distribution of silicon and aluminium atoms in the tetrahedral sites of zeolitic frameworks. NMR spectroscopy, however, readily reveals the degree to which they are ordered on the short range. Combination of experimentally determined details of the local environment of atoms and the overall framework structure, with computational modelling as appropriate, gives deep insight into the properties of the solid. A different situation exists for ordered mesoporous structures. In these materials atoms do not occupy well-defined sites in the unit cell and the walls between the pores are essentially amorphous, so that diffraction can only determine the distribution and thickness of pore walls in these structures and spectroscopy is required to give information on the type of species present and their local environments.

Diffraction-based techniques are widely applicable to crystalline molecular sieves. For samples available as crystals *ca.* 10 μm or more in size, X-ray single crystal diffraction methods are available to solve their structures *ab initio* (*i.e.* without any other information). Very often, however, new materials are only prepared as microcrystalline powders, often with structural defects, and single crystal methods are inapplicable. In these cases, a combination of X-ray and neutron powder diffraction and electron microscopy (EM) is required to solve unknown structures or establish structural details. Powder diffraction therefore occupies a special place among the structural techniques applied to microporous solids and is particularly relevant to zeolites, which are generally used in the form of microcrystalline powders.

Spectroscopic methods give structural details complementary to those available from diffraction-based techniques. The most generally applied methods continue to be NMR and vibrational spectroscopy (especially IR), but element specific techniques such as X-ray absorption spectroscopy are also of importance. As I will describe in Chapters 7 and 8, NMR and IR have also been used extensively for the study of the interactions of adsorbed molecules with molecular sieves and to investigate the nature of acidity in these solids.

Computer simulation of structures and interactions of molecular sieves with adsorbed molecules has become a third powerful approach. Simulation methods are now available that have been shown to model accurately structural details of framework geometry, cation location and the position and mobility of molecules within the pores. These approaches can now be applied to predict results for systems of interest (reducing expensive and time-consuming experimentation) and also where details cannot be determined by any other method.

In this chapter I will show how the structural chemistry of framework molecular sieves has been elucidated by diffraction-based techniques and spectroscopies. Computer simulation of structures will be discussed separately in Chapter 4 as an additional technique.

3.2 Diffraction-based Methods

The structures of many naturally occurring zeolites, and some important synthetic varieties, have been solved straightforwardly from laboratory single crystal diffractometry. This is possible when specimens of sufficient size and crystal quality are available. Crystals with volumes as small as *ca.* 10 000 μm^3 can now be studied using laboratory diffractometers operating with rotating anode X-ray sources, because detector technology has been greatly improved, for example by the development of very sensitive detectors based on charge coupled devices (CCDs). The advent of single crystal diffractometers sited at synchrotrons[1] reduces the crystal volume necessary for single crystal study still further, to less than *ca.* 1000 μm^3. This is because synchrotron X-ray fluxes are orders of magnitude higher than those available in laboratory diffractometers (even when a single X-ray wavelength is selected from the full spectrum by monochromation). These advances have stimulated single crystal structure determinations and structural studies of aluminophosphates and zeolites, as well as of open framework inorganic solids of an expanded compositional range. Many of these new framework types, such as metal phosphates and germanates, readily form crystals suitable for single crystal diffraction.

The majority of important synthetic novel zeolites, however, have not been prepared as crystals either large enough or sufficiently well ordered to be studied by available single crystal methods. Zeolites such as Beta, ZSM-12, NU-87, MCM-22 and ITQ-21, and related microporous solids such as the titanosilicate ETS-10, for example, were solved from X-ray powder diffraction and other methods (such as electron microscopy and computer simulations) rather than by single crystal diffraction.

Both single crystal and powder methods have been used for the structure solution of hybrid microporous solids, such as phosphonates and carboxylates, although crystals large enough for single crystal XRD are more generally available for these materials. For hybrid solids with exceptionally large unit cells, such as MIL-100 and -101 (described in Chapter 2) that do not form as single crystals, it has been necessary to develop auxiliary modelling approaches that can be combined with powder diffraction for structure solution (Chapter 4).

3.2.1 Single Crystal Diffraction

Single crystal diffraction of microporous solids using X-rays of a single wavelength gives time- and space-averaged positions of atoms within the unit cell, in the same way as it does for other crystalline solids.[2–5] For suitable crystals, enough reflection intensity data may be collected to solve structures through crystallographic 'direct methods' routines within programs such as SHELX[6] or SIR.[7] There are, however, several characteristic difficulties experienced with the structure solution of microporous solids that prevent it from being routine, even in the relatively rare cases where single crystals of suitable quality are available. These difficulties are usually associated with fractional occupancies and disorder in the position of framework and extra-framework cations and adsorbed molecules.

The first feature that is established by these studies is the *framework structure* such as is shown for many examples in Chapter 2. For aluminosilicate zeolites the structure can be described in terms of the way the tetrahedra are connected, which is unique for a given framework type (Chapter 2). The identity of the framework cations may be ambiguous in the case where two or more species are disordered over a given site (such as aluminium and silicon in zeolites). NMR spectroscopy is the preferred method in determining short-range ordering in such cases (Section 3.3) but a detailed examination of the bond angles and lengths of the framework structure is also informative. Typical bond lengths and angles of importance for zeolites and aluminophosphates are Si–O, 1.62 Å; Al–O, 1.73 Å; P–O, 1.52 Å; $T\hat{O}T$, 145°; $O\hat{T}O$, 109.7°, where T is a tetrahedral framework cation. These can vary in the presence of bridging hydroxyl groups.

In specific examples, the details of framework geometry can be used to establish the contents of individual framework cation sites (T sites) when the mixed cations that are present at the site possess different numbers of electrons or different ionic radii. Siting of the larger aluminium, gallium, germanium or titanium ions in silicon sites, for example, will be seen by lengthened average T–O distances. Similarly in aluminophosphates: for example, in crystals of the cobalt aluminophosphate form of STA-7, $Co_{0.22}Al_{0.78}PO_4$, average bond lengths from tetrahedral cations to oxygen atoms for the three distinct tetrahedral sites are determined by single crystal diffraction to be 1.735(4), 1.744(4) and 1.784(4) Å, indicating that cobalt, which has a larger ionic radius than aluminium (0.58 Å *cf.* 0.39 Å) is preferentially located in the third of these sites.

Single crystal measurements at different temperatures have also been made and analysed to explain in detail the remarkable negative thermal expansivity observed for many microporous solids.[8] Simplistically, enhanced transverse thermal vibration of oxygen atoms in T–O–T bonds can result in the reduction of T–T distances in the frameworks and the shrinkage of cell dimensions that depend on these T–T distances.[9]

The location of *extra-framework species* can be more difficult to determine than framework atom positions, and is best performed in conjunction with accurate chemical analysis. The usual approach is *via* difference Fourier syntheses, in which comparison of observed reflection intensities and those predicted on the basis of the 'nearly complete' structural model enables the residual electron density (in the case of X-ray diffraction) to be located. This can then be interpreted in terms of the species present and their expected distances from the framework. Extra-framework charge-balancing cations may occupy sites at less than unit occupancy, that is they may be either present or absent from that site in any given unit cell, and the situation is further complicated in mixed cation systems, where different cations may partition between different sets of extra-framework sites.

Templates occupy sites in the channels with close to full occupancy, and have in many cases been located unambiguously. In other cases, where the template possesses lower symmetry than the cage or pore space it occupies, it is likely to be statically disordered over several symmetry-related positions that are equivalent in energy and its position becomes difficult to determine. It may in some cases be possible to locate the position of mineralising anions, such as hydroxide or fluoride ions, that become included in the solids upon crystallisation. Fluoride-containing syntheses of pure silica polymorphs of zeolites, for example, give solids in which fluoride ions are left coordinated to silicon atoms to balance the positive charge on the alkyl ammonium templates and in aluminophosphates the fluoride ions remain coordinated to aluminium cations, increasing their coordination.[10,11] Figure 3.1 illustrates a single crystal structure determination of the template and F^- locations in the pure silica form of EU-1 templated by dibenzyldimethylammonium cations.[12] The template spans the channel and sidepocket in EU-1, and fluoride ions are coordinated to one of the silicon atoms in the characteristic $[4^1 5^4 6^2]$ silicate cage units that bridge 'layers' in the framework.

For the study of adsorbed molecules the situation is still more complicated. Cages may contain molecules at lower than unit occupancy or may contain more than one molecule, which can interact with each other as well as with the

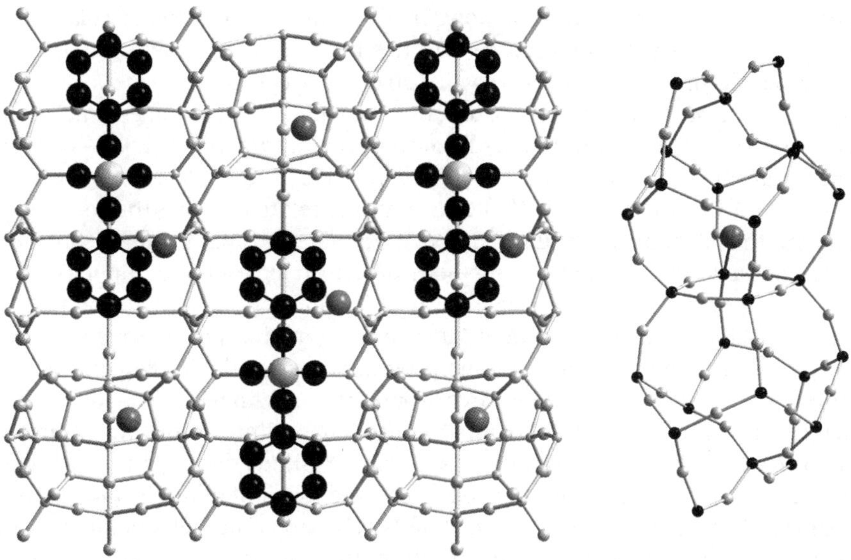

Figure 3.1 The pure silica EU-1 structure (type EUO), as determined by single crystal XRD (left). The silicate framework is templated by the singly fluorinated dibenzyldimethylammonium cation (shown, with the fluorine atom in medium grey, without H atoms). Fluoride ions are also found coordinated to silicon atoms of the framework, within silicate cage units (right – only one fluoride ion per two such cages). The configuration shown is one of a number of symmetrically equivalent ones and the relative positions of organic fluorine and coordinated fluoride is proposed to be the lowest energy configuration on the basis of computational atomistic simulation of the type described in Chapter 4. [Reproduced from reference 12 with permission. Copyright 2005 American Chemical Society.]

framework (and extra-framework) cations. As is the case for the templates, the molecules may have lower symmetry than the pores, and be disordered. Finally, there may be a number of adsorption sites and configurations that are similar in energy, and the molecules will be distributed and mobile between these sites except at very low temperatures. Despite these difficulties, some elegant single crystal diffraction studies of small adsorbed molecules have been performed on zeolites readily available as single crystals, such as zeolite A and silicalite. These are described in Chapter 7. They give a picture of the zeolite-sorbate complex, and can also be used to assess the reliability of energy minimisation calculations such as those described in Chapter 4.

3.2.2 X-ray and Neutron Powder Diffraction

X-ray powder diffraction is routinely applied to identify products from syntheses. For phases that are known, powder diffraction pattern databases are available and can rapidly be searched for matches of peak position and intensity. For zeolitic solids (with tetrahedrally connected frameworks), the most up-to-date compilation of powder diffraction data is the simulation of X-ray diffraction patterns prepared by the International Zeolite Association Structure Commission (http://www.iza-structure.org). More generally, if an experimental pattern can be indexed and the unit cell determined, the ICSD database (*e.g.* http://icsd.ill.fr/icsd) of reported inorganic crystal structure determinations should be searched to determine if the phase is known or related to a known material. If the structure is related to one that is already known, the structural details can be determined using profile refinement methods (such as the Rietveld method, described below) and for important cases the special properties of neutron powder diffraction can be exploited.

For novel phases, structure determination is sometimes possible from X-ray powder data alone. Crystallographic programs are available that use modifications of the structure-solving algorithms used in single crystal diffraction analysis, making use of additional structural constraints. Some very complex structures have been solved in this way, using methods designed to work with powder data (see later). Others have been solved by combining the information from the electron microscopy, solid state NMR and computer simulation with X-ray diffraction data. In particular, structures that possess high degrees of inherent disorder (Beta, ETS-10) have been solved in this way because they give poorly resolved X-ray diffraction profiles. The usual procedure involves establishing a model from electron microscopy using diffraction and high resolution electron microscopy (HRTEM) (see below) and confirming the model by matching its simulated X-ray powder pattern with the experimental one. Although powder diffraction is limited by giving many fewer data than single crystal diffraction it does possess some advantages, in particular for *in situ* studies under controlled environments. Most working zeolitic materials are employed in the form of powders, and so studies of microcrystalline powders give results that are more representative of the materials as used.

3.2.2.1 Structure Solution from Powder Diffraction Data

Structure determination from powders is much more difficult than from single crystal data because fewer resolved reflection intensities are available, and since direct methods routines for structure solution work by probability methods to solve the so-called 'phase problem', this is made much less reliable.[13] (The phases of diffracted beams, or reflections, are not directly accessible from the X-ray reflection intensities, but direct methods enable them to be determined from considering groups of reflections together.) Loss of data arises due to peak overlap because the diffraction data are collected over three dimensions in a single crystal diffraction experiment but are collapsed into a single dimension (the 2θ dimension) in powder diffraction. Use of high resolution synchrotron diffraction[14] reduces the overlap by giving better resolution of the diffraction peaks, and methods have also been developed to deconvolute overlapping peaks,[15] and even to permute relative intensities for peaks that overlap exactly. Such methods have found widespread use in solving the structures of molecular

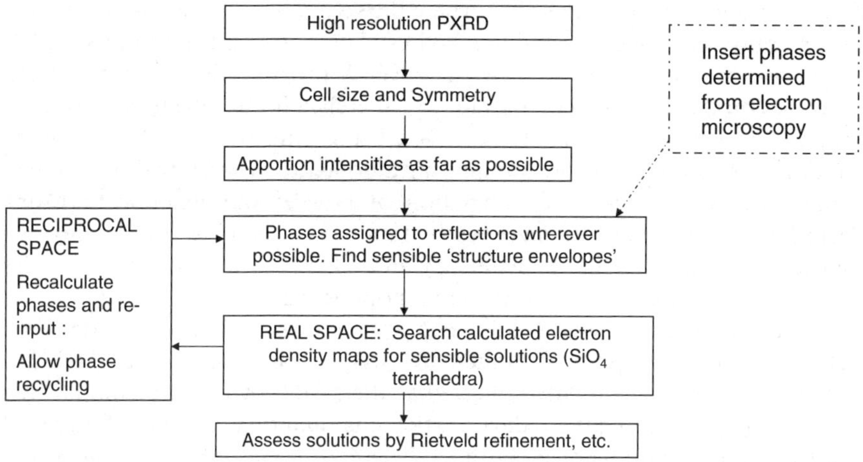

Scheme 3.1 The FOCUS program can use data from both real space and reciprocal (diffraction) space, and from electron microscopy as well as powder X-ray diffraction (PXRD).[17] It has successfully solved the structure of a zeolite with 24 independent silicon positions, TNU-9.

crystals or simple inorganic structures, where the approximate geometry of the molecules are already known, but have proved more restricted for framework solids, although successes have been achieved.[16]

One structure solution program for zeolites, FOCUS (Scheme 3.1),[17] has been written to incorporate the constraint of a tetrahedrally connected network in the trial models, and concentrates on scattering from the tetrahedral framework cations. Reflection intensities are extracted from the indexed pattern, and those of overlapping reflections are equipartitioned. Random phases consistent with the symmetry are assigned to the observed structure factors of these reflections and an electron density map calculated. From the resulting partial structure a new set of phases is generated and used in generating a new map, which is searched for a tetrahedrally connected framework. This procedure is repeated very many times until a candidate structure is obtained, and the process then starts again. The FOCUS program has solved structures with up to nine independent T-sites, and possesses the advantage that constraints and information from either direct space (chemical constraints) or reciprocal space (phase information) can be included to improve the chances of success. Other successful technical approaches include that of Rius,[18] in the data treatment, and Baerlocher, through experimental modification of the powder method to make use of the effects of preferred orientation.[17] The state-of-the-art of these approaches is represented by the structure solution of the zeolite TNU-9 by the group of McCusker, by combination of powder XRD and HRTEM data within the FOCUS program (Section 3.2.3). In every structure solution example, the final structure must be refined against the powder data.

Among the many successful structure solutions that have been achieved for tetrahedrally connected microporous solids from X-ray powder data alone are the aluminophosphate SAPO-40[19] (where the solution was constrained to give a tetrahedrally connected net) and the zeolites ITQ-4[20] (solved using the SIRPOW program) and SSZ-58[21] (solved by the ZEFSA program, in which initial trial structures obtained from the program were searched for tetrahedrally connected units). A recent application of the FOCUS program permitted the structure solution of a novel large-pore aluminosilicate zeolite, ITQ-27, with two dimensional pore connectivity.[22] Combination of powder diffraction with Monte Carlo-Simulated Annealing methods of randomly searching for tetrahedrally connected frameworks that fit symmetry and diffraction data has been successful for the structure solution of the aluminophosphate UiO-7.[23]

Hybrid organic-inorganic solids have also been solved from powder data, including the small-pore scandium methylphosphonate, $NaSc(CH_3PO_3)_2$, solved from synchrotron data collected at the ESRF by direct methods[24] (the final fit to synchrotron data, indicating the data quality, is shown in Figure 3.2) and the nickel N,N'-piperazinebismethylenephosphonate (Figure 2.38) solved from laboratory powder data by direct methods and computer modelling. For hybrid structures of this type, the building units present (metal clusters, ligands, *etc.*) are known, and this additional information can be input into 'real space' methods, in which the 'molecules' are moved around until promising fits of simulated and measured data are obtained. The FOX program of Favre-Nicolin is one such realisation of this methodology.[20,25,26] The remarkable structure solution of MIL-100, in which powder data were augmented by computer-assisted modelling, is described further in Chapter 4.

Advances in single crystal diffraction suggest that for structurally complex samples that do not show significant framework disorder, structure solution from microcrystals (or even nanocrystals – see the electron microscopy section) will be the most convenient route for solids that can be prepared as single crystals, even if they are less than 1000 μm^3 in volume. Many aluminosilicate zeolites, however, can only be prepared as micron-sized crystals or may include stacking faults. As a result, continued developments in structure solution from X-ray powder diffraction are of great value. (See, for example, the Special Issue of *Zeit. Krist.*, **219**, 2004.)

3.2.2.2 Structure Refinement from Powder Diffraction Data

Powder diffraction also has a special role to play in following structural details of microporous solids for which the basic structure is known. Laboratory and synchrotron X-ray powder diffraction and neutron powder diffraction can all be used to give information on changes in framework geometry, and on cation and adsorbate location, *via* the powder profile refinement method.

First developed by Rietveld in 1969 for neutron diffraction,[27] but now applied to both X-ray and neutron diffraction, the profile refinement method involves the gradual refinement of a starting model that is reasonably close to the correct structure.[28] The profile is calculated using two sets of parameters; structural and instrumental. The structural parameters include unit cell, symmetry, atomic coordinates and thermal displacements whereas the instrumental

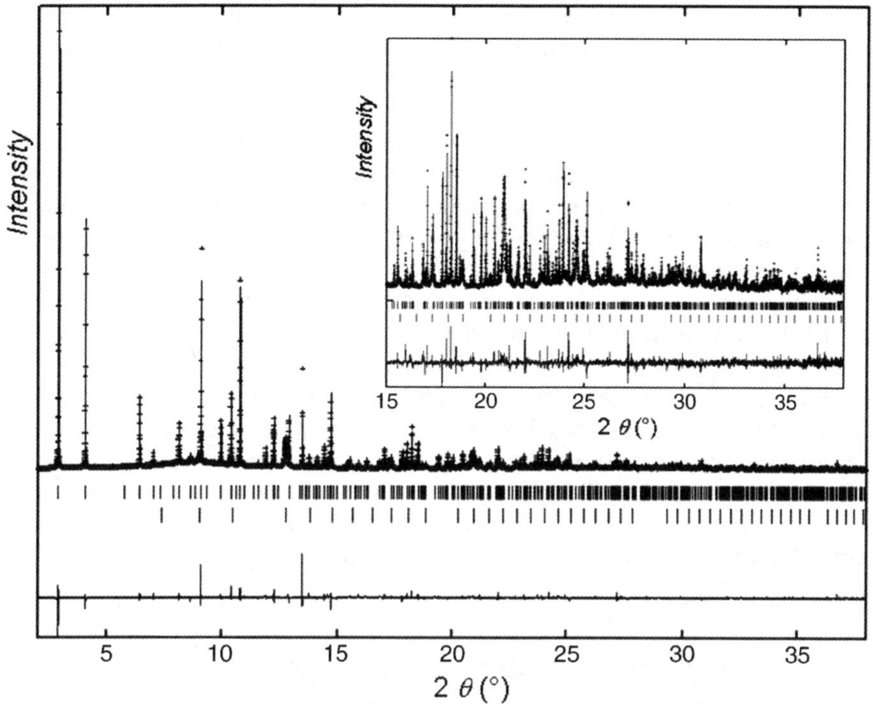

Figure 3.2 The final matched X-ray powder diffraction profile for $NaSc(CH_3PO_3)_2$. The high-resolution data were collected at beamline ID-31 at the ESRF, Grenoble, France, and the structure was solved by direct methods from these data. The structure of this phase is shown in Figure 2.27. The plot is typical for a Rietveld profile refinement: experimental data points given as a function of diffraction angle, 2θ, matched by a profile simulated from the final refined structural model, taking into account the instrumental parameters (solid line). Tick marks below the profile represent the positions of reflections predicted on the basis of unit cell geometry and symmetry – in this case there are two sets because the sample included a small amount of Sc_2O_3 impurity, the structure of which is known, and this necessitated a 2-phase refinement. Finally, the lowermost line is the difference between observed and simulated intensity and gives a visual guide to the correctness of the model. The goodness of fit parameter for this refinement, R_{wp}, is 8% (see text for more details). [Reproduced from reference 24 with permission. Copyright 2005 Royal Society of Chemistry.]

parameters include the type of radiation, scale factor, zero point, peak shape and background. The difference between the profile calculated from the model and that observed experimentally is quantified by the goodness of fit weighted profile R value, R_{wp},

$$R_{wp} = \left[\sum_i w_i (y_{io} - y_{ic})^2 / \sum_i w_i y_{io}^2 \right]^{1/2}$$

where y_{io} and y_{ic} are observed and calculated intensities and w_i is the weighting factor, which is inversely proportional to the value of y. R_{wp} is minimised by refinement of the instrumental parameters and chemically reasonable changes to the structural parameters of the model. This includes extra-framework species assigned on the basis of electron density peaks in the difference Fourier analyses. The structure refinement improves the fit until further modifications give no significant improvement. Unit cell dimensions, framework geometry and cation distributions are readily determined by this technique, details of which are available in the text edited by Young.[29] Figure 3.3 illustrates this for the dehydrated sodium form of the recently discovered gallosilicate TNU-7, studied by synchrotron X-ray powder diffraction at the ESRF in Grenoble, where refinement enables the sodium cations to be located.[30]

Very many Rietveld determinations of cation siting have been performed on zeolites. Mortier gave a snapshot of what was known in 1982,[31] many other studies have been carried out since then. The measurement of the migration of nickel cations within zeolite Na,Ni-Y during its dehydration and subsequent use in catalysis[32] is an excellent example of this approach applied *in situ*, particularly because dehydrated Ni-Y is an active catalyst in the trimerisation

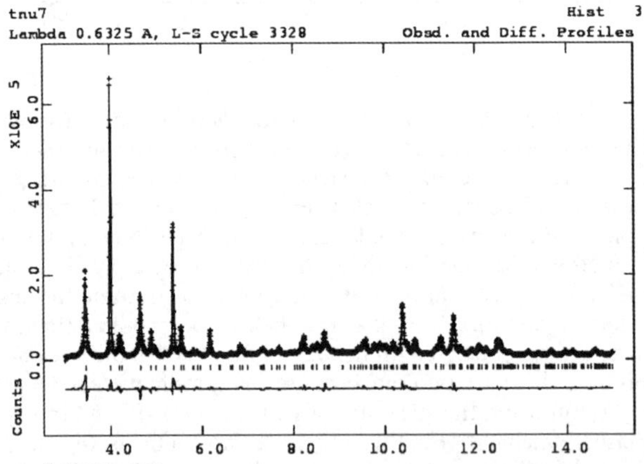

Figure 3.3 Rietveld refinement (top) of the structure of the dehydrated sodium form of the gallosilicate TNU-7 against synchrotron X-ray powder diffraction data, together with a view of the final structure, in which the sodium cation positions are indicated (bottom). An initial model for the framework was obtained from a description of the poorly ordered aluminosilicate ECR-1: cation positions were located from difference Fourier analyses and their positions and occupancies refined. The plot shows the observed data, the calculated profile and a difference plot. The final match is an acceptable fit ($R_{wp} = 6.5\%$). Electron diffraction (middle) along the [001] and [010] zone axes shows that this structure is a strictly alternating intergrowth of MOR and MAZ layers (see Figure 2.11) because no evidence of streaking in the diffraction maxima (characteristic of disorder) is observed. [Reproduced from reference 30 with permission. Copyright 2006 American Chemical Society.]

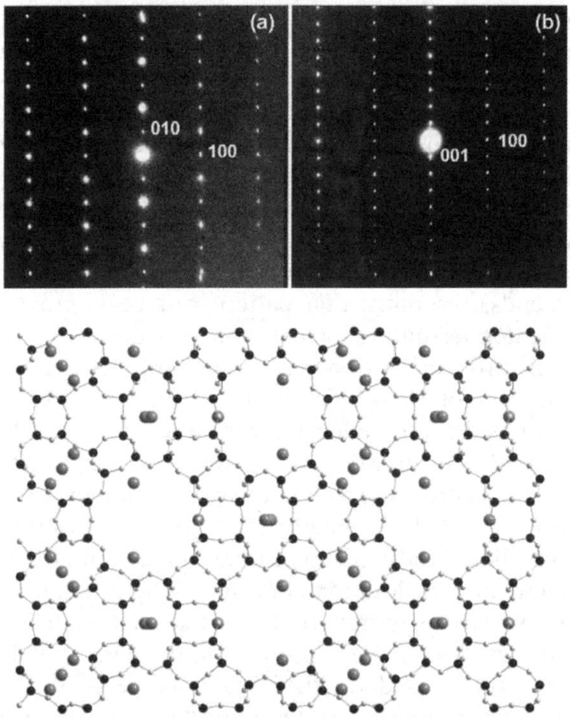

Figure 3.3 (*Continued*).

of acetylene to benzene. In addition, powder diffraction permits the study of incorporation of adsorbed molecules[33,34] and metallic and semi-conducting clusters ordered within the pores.[35–37]

Neutron diffraction has played a special role in the study of microcrystalline powders in general and of microporous solids in particular. Diffraction peaks using monochromatic neutrons from reactor sources give peaks with simple Gaussian lineshapes, and so neutron diffraction profiles were initially easier to match than X-ray profiles, the peaks of which are better fitted by convolution of Lorentzian and Gaussian functions (as the so-called Pseudo-Voigtian function). The use of reactor-based neutron diffraction therefore gave an important impetus to powder diffraction, although peakshape fitting of data from pulsed neutron sources, such as the ISIS spallation source at the Rutherford Laboratory near Oxford, is now routinely possible. (At pulsed neutron sources, scattered neutrons are detected as a function of their time-of-flight, which is a straightforward function of the associated d-spacing of the reflections.) The fall-off of scattering with increasing diffraction angle in diffraction using monochromatic radiation is far less marked for neutrons than X-rays (because scattering is from the nucleus rather than by the more dispersed electron cloud), and so high-quality diffraction data can be collected over a wide range of d-spacings. Before the widespread availability of intense synchrotron X-ray diffractometers, therefore, neutron diffraction gave the highest-quality powder diffraction data, and a series of important studies of low silica zeolites were performed using this method.

The high sensitivity of neutrons to hydrogen and deuterium nuclei makes the method of particular importance in locating proton (or deuteron) positions in dehydrated acidic zeolites. H-Y[38], H-SSZ-13[39] (a zeolite with the chabazite topology) and the lanthanum-exchanged form of zeolite Y, where hydrolysis results in $(LaOH)^{2+}$ species and protons[40] are good examples. It should be noted that since ^{1}H nuclei result in incoherent neutron scattering and high-profile backgrounds, the diffraction pattern can be improved by replacing protons with deuterons during preparation. Furthermore, combination of refinements of neutron diffraction data from both H- and D-forms, where the scattering lengths of H and D are –0.34 and 0.67, respectively, enables proton/deuteron positions to be located with more certainty. In addition, the greater sensitivity of neutron diffraction to oxygen atom positions compared to X-rays enables T–O distances to be measured accurately by neutron powder diffraction. Since the T–O bond lengths to oxygen atoms of bridging hydroxyl groups are longer than if there are no protons, inspection of T–O bond lengths can also be used to infer the location of protons. The structure of the acid form of zeolite Y is of particular importance: refinement of neutron diffraction data[38] indicated three positions for protons on the basis of refined occupancy of deuterons or protons and on the T–O bond lengths. The protons are associated with oxygen atoms O(1), O(2) and O(3) of the faujasite structure. The two most populated sites are indicated in Figure 3.4. One of these is associated with O(1), where the hydroxyl group points into the supercages of zeolite Y, whereas the second most abundant is associated with O(3) and is directed into the sodalite cages, where it is inaccessible to most adsorbed molecules.

In addition, the high neutron scattering of oxygen, carbon and nitrogen nuclei make neutron diffraction a suitable method to measure the minimum energy positions of adsorbed organic molecules within the pores at low temperatures, where there is little thermal motion. This was shown by early studies on the position of perdeuterated benzene in Na-Y[41] and of perdeuterated pyridine in K-Gallo-L,[42] the latter of which is illustrated later in this book (Chapter 7). Modern high resolution neutron powder diffractometers, coupled with enhanced detector facilities, continue to be of use for the study of microporous solids. Figure 3.5 shows the high resolution diffraction profile of the deuterated organozeolite aluminium methylphosphonate AlMePO-β, collected at the ISIS pulsed neutron source, and a calculated profile after structural refinement.[43] For this example, the positions of 49 atoms, including those of the methyl deuterons, were determined accurately without constraints from this refinement.

3.2.2.2.1 Advanced structure refinement. The Rietveld refinement structural method has been taken further in a number of directions. One of the most appealing is to combine it with the maximum entropy method (MEM), in which

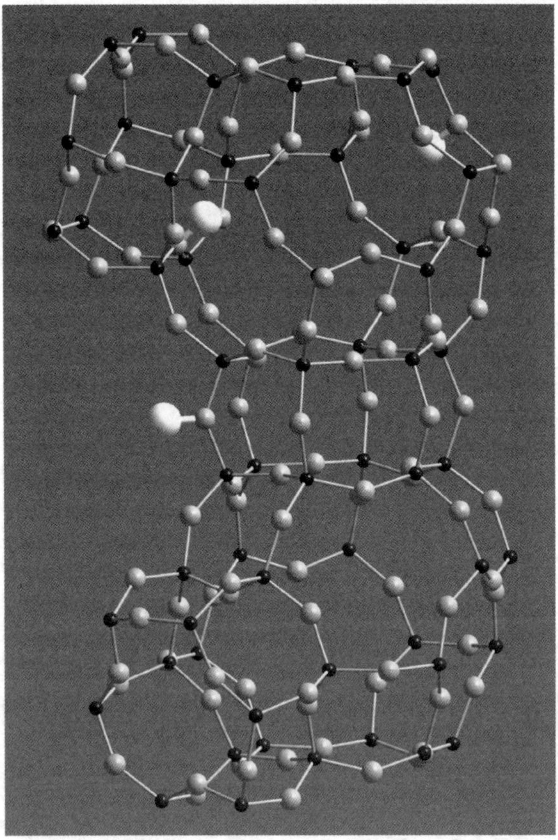

Figure 3.4 Proton positions (white spheres) in zeolite H-Y, as proposed on the basis of neutron powder diffraction of D-Y. Protons are located on three of the four different Si–O–Al bridging oxygens, with variable occupancies.

a model can be estimated from a limited amount of information by maximising information entropy under constraints consistent with observed physical quantities.[44] Put simply, given chemical constraints, the method is very effective in coming up with a sensible structural model that fits the experimental data. Using it in conjunction with Rietveld refinement, it is able to give a less noisy picture of the electron (or nuclear) density than is available *via* Fourier difference methods within the Rietveld method. As a result, the method has been used to great effect in determining electron density maps in microporous (and other) structures. These include the visualisation of hydrogen-bonding schemes of acetylene molecules held within a functionalised MOF material (see Section 10.3.3).[45] The 3D electron density approach is particularly powerful for the analysis of systems where disorder is not readily modelled by split atoms, such as the location of disordered potassium atoms in the K-loaded K-A zeolite or of residual water molecules in Na-A zeolite.[44]

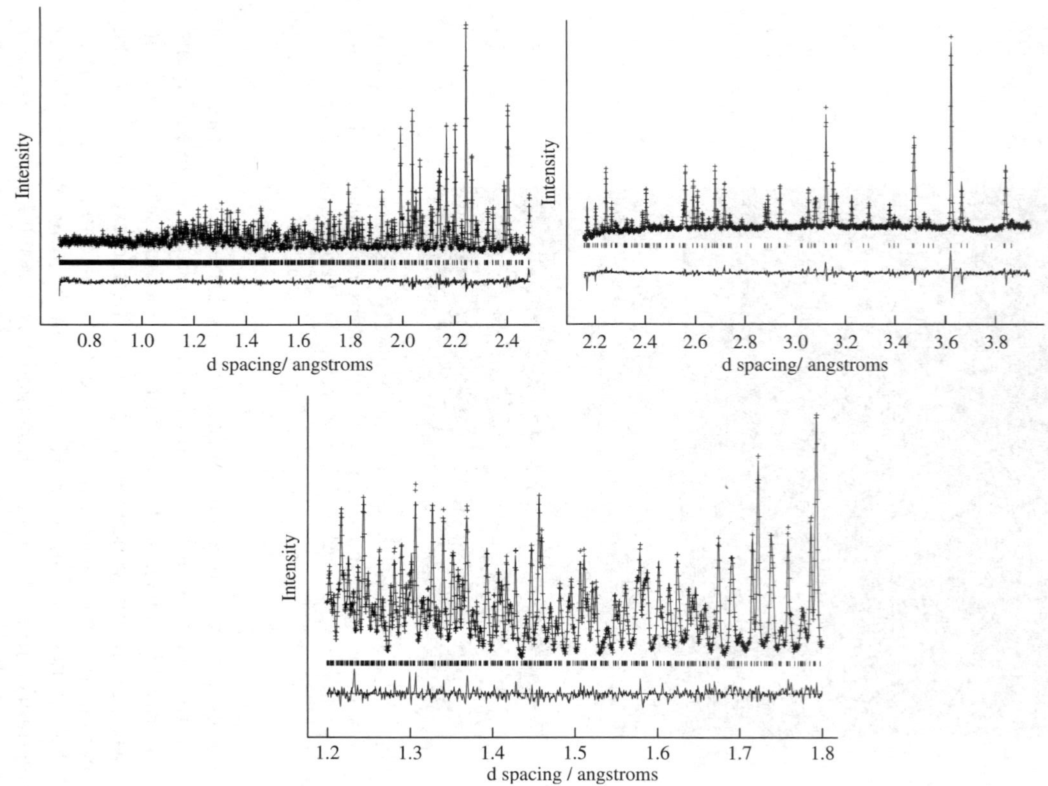

Figure 3.5 High resolution time-of-flight neutron powder diffraction collected on deuterated AlMePO-β ($Al_2(CD_3PO_3)_3$) at station HRPD at the pulsed neutron source ISIS, Oxfordshire, UK. The diffraction is measured at detectors at scattering angles of 168° (above left and, expanded, below) and 90° (above right). The profile has been fitted using Rietveld profile analysis in which the positions of 49 atoms were refined. [Reproduced from reference 43 with permission. Copyright 1999 Elsevier.]

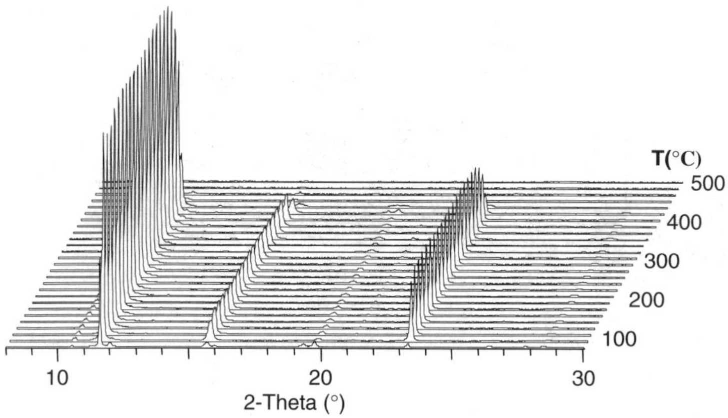

Figure 3.6 Variable temperature X-ray powder diffractometry of $Sc_2(O_2CC_6H_4CO_2)_3$ (see Figure 2.31) ramped from room temperature to 500 °C. Crystallinity is lost above 400 °C. [Reproduced from reference 103, chapter 2, with permission. Copyright 2005 Royal Society of Chemistry.]

3.2.2.3 *In situ* Diffraction Studies

X-ray diffraction can readily be applied *in situ* to follow structural changes associated with thermal transformations, cation migrations and crystallisation from solution. One simple example of the first application is to determine the temperature at which a material loses crystallinity, as illustrated for the scandium terephthalate $Sc_2(O_2CC_6H_4CO_2)_3$ in Figure 3.6. Another application is to measure the kinetics of cation migration throughout zeolites, as previously described for Na,Ni-Y.

The measurement of crystallisation kinetics *in situ* is more difficult, being complicated by X-ray absorption and scattering by the container and the solution. For this reason, the extremely high intensity of synchrotron sources is required for these experiments, where strong signal attenuation can be tolerated. In studies where the crystallising phases in the reaction have already been identified, and time resolution is more important than peak resolution, energy dispersive XRD (EDXRD) at synchrotron sources can be used.[46–48] In this technique, X-rays of the entire spectrum produced by the synchrotron are used in the diffraction, and the X-rays diffracted at a fixed angle are analysed as a function of their energy, which data can be converted to a diffraction pattern of intensity versus d-spacing. Because all available X-rays are used, the diffraction patterns can be collected very quickly and give acceptable diffraction on the timescale of minutes (Figure 3.7).[48] By following the diffraction intensity of characteristic peaks as a function of time, the kinetics of nucleation and growth can be measured *in situ* for those materials that crystallise over relatively short time periods, giving information on the mechanisms of these processes. Crystal growth kinetics can be determined in this way, typically being analysed in terms of the Avrami equation:

$$\alpha(t) = 1 - \exp(-k(t - t_o)^n)$$

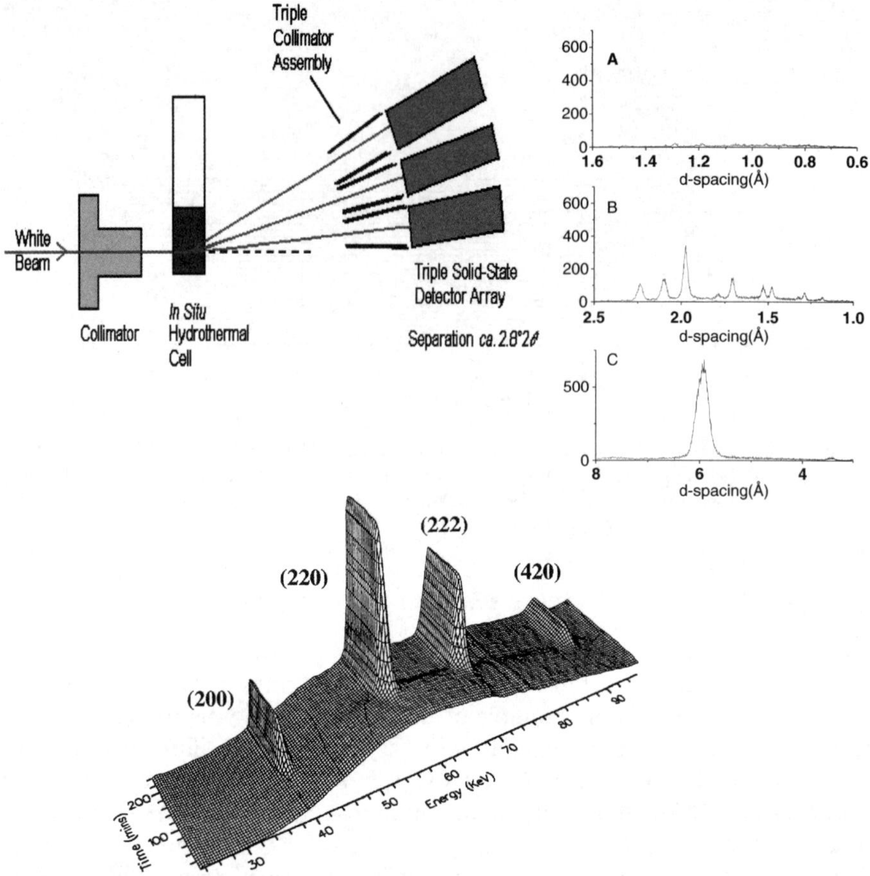

Figure 3.7 *In situ* time-resolved energy dispersive synchrotron X-ray diffraction studies of the crystallisation of zeolite A at 90 °C. In the experimental setup (above) an intense white beam (*i.e.* with a spectrum of wavelengths) is incident upon the crystallising gel and diffracted X-rays collected as a function of energy at different (fixed) detector angles (right, A, B, C). These diffraction patterns may be expressed as a function of d-spacing (above right). Crystallisation is followed as a function of heating time, as shown in the stacked energy dispersive diffraction patterns (below). [Figure courtesy of G. Sankar.][48]

where α is the degree of crystallisation, k is the rate constant, t_o is the nucleation time and n is the order. If k is measured using a series of isothermal crystallisations at different temperatures, an apparent activation energy of crystallisation can be obtained. Although the physical meaning of these energies is obscure, the parameters can be used to compare similar systems. Rapidly crystallising zeolites or phosphates have been investigated in this way. Metal aluminophosphates (these crystallise over periods of hours) have been studied[49] as have other metal phosphates, including gallophosphates. In the crystallisation of gallium phosphates, for example, O'Hare and Walton have observed crystalline intermediates not seen in quenched experiments.[50]

As well as analysing the wide angle scattering (WAXS) it is possible to measure simultaneously small angle X-ray scattering (SAXS), which consists of broad peaks that arise from non-crystallographic short-range ordering features, such as nuclei, with large associated characteristic lengths. From this it is becoming possible to relate the processes of agglomeration and nucleation to subsequent crystal growth (Chapter 5).[51,52]

3.2.3 Electron Diffraction and Transmission Electron Microscopy

Transmission electron microscopy and electron diffraction, taken together, enable structural information to be obtained on crystalline samples under high vacuum. Thin samples are required because electrons interact very strongly with matter and are multiply and incoherently scattered by thick samples. Electron microscopy is a challenging field for framework solids, compared, for example, to ceramic metal oxides or alloys, because electron beam radiation causes damage to the frameworks and loss of crystallinity on the timescale of seconds. However, careful sample preparation, reduction of beam intensities, advances in electron sources and the sensitivity of detectors are permitting significant progress for the imaging of microporous and mesoporous solids.[53]

Electron diffraction occurs when the electron beam is incident at an angle that satisfies the Bragg condition for diffraction. A consequence of the samples being thin and the electron wavelength very small is that many electron diffraction maxima can be measured simultaneously. By careful tilting of the sample and measuring series of electron diffraction patterns it is possible to establish the unit cell and the space group symmetry (see Figure 3.8 for the sequence of electron diffraction patterns obtained and used by Ohsuna *et al.*, for the determination of the unit cell and symmetry of the zeolite Beta C).[54] This is possible because the experiment is essentially a single crystal diffraction experiment on a crystal of much less than a micron in dimension.

The analogy to single crystal X-ray diffraction can be further extended by measuring many diffraction intensities in this way, and using these intensities in modified direct methods programs to give structure solutions. This is much more difficult for electron diffraction than for X-ray diffraction, because of the effects of multiple scattering, but if sufficiently thin samples are examined then it is possible. In fact, usual application of this electron crystallography technique makes use of high resolution lattice images that are formed by recombining diffracted beams. Phase information is retained in the diffracted beams, so that their recombination by focusing using electromagnetic fields as lenses gives images with down to better than 2 Å point resolution in a process known as phase contrast imaging. These images give structural information directly (see below) but by a process of selected area Fourier transformation a diffractogram can be generated mathematically for a region of interest, giving diffraction maxima and their intensities and phases. The retention of phase

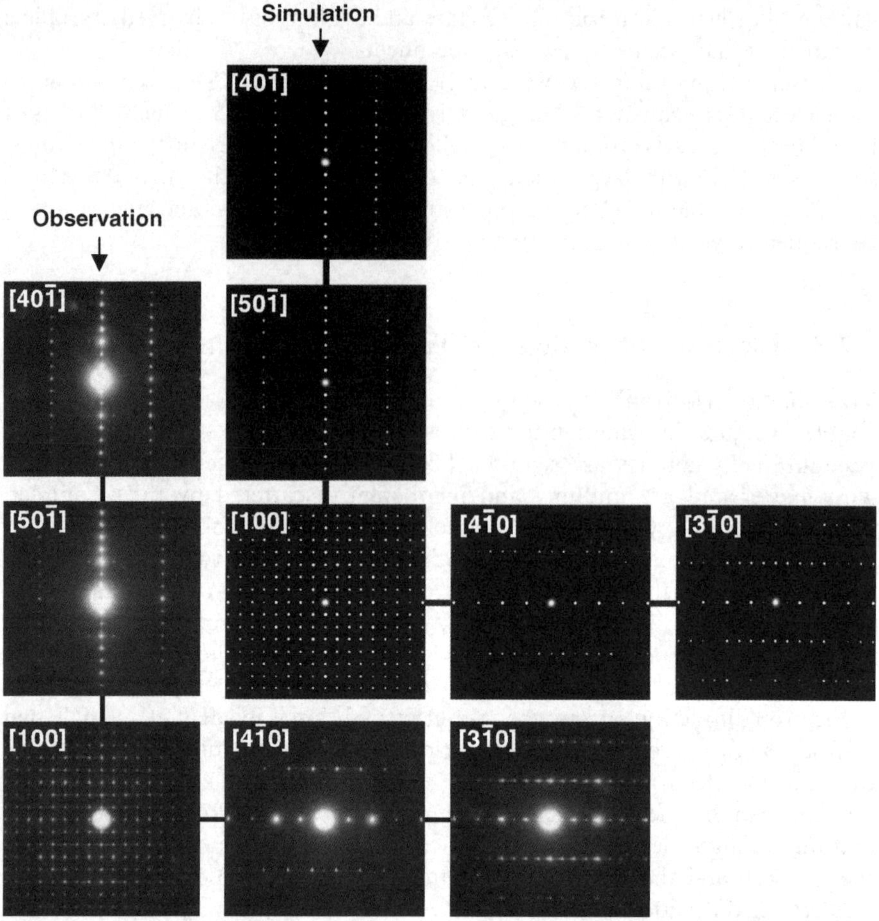

Figure 3.8 In their solution of the structure of pure silica zeolite Beta C (see Figure 2.15 for the framework), Ohsuna *et al.*[54] first determined the unit cell size and symmetry from a series of electron diffraction patterns, taken down different directions of the unit cell. These are shown above: they compare closely with patterns simulated using the determined cell ($P4_2/mmc$, a = 13.1 Å, c = 13.8 Å). [Figure courtesy O. Terasaki.]

information is an important advantage over X-ray diffraction. The process can be repeated down several crystal directions, so that a set of amplitudes and phases for the reflections is obtained and can be used in structure solution. Electron crystallography has been used to solve the structures of the novel high-silica zeolites SSZ-48[55] and the C-polymorph of zeolite Beta[54] and is an area of developing research. A series of papers by Dorset[56–58] demonstrates the possibilities and limitations of this technique for a series of known structures.

The combination of phase data derived from electron microscopy and intensity information from high resolution X-ray diffraction offers the most powerful approach to structure solution of zeolites from powder data.

The scheme for this approach is given in Scheme 3.1, where the FOCUS program is used in conjunction with phase data obtained from electron microscopy. In the solution of TNU-9, phase data were obtained from the electron micrographs shown in Figure 3.9 and combined with XRD data from station ID-31 at the ESRF synchrotron facility to give models of the electron density. These were then searched using the FOCUS program for tetrahedrally connected frameworks, as previously described, ultimately giving the framework structure of TNU-9, Figure 3.10, which with 24 distinct tetrahedral sites is the most complex zeolite structure solved so far.[59] The final structure was refined against high-resolution X-ray powder diffraction data using the Rietveld method to confirm the structure (Figure 3.10).

More commonly in the study of zeolites, high-resolution transmission electron microscopy (HRTEM) has been used to establish structural models of novel materials by identification of important structural features. The images typically represent a projection of electron density of the sample, and are of particular interest for zeolitic samples when the image is collected viewed down aligned channels, or pores (see Figure 3.9(a) for the HRTEM image of zeolite TNU-9 taken looking down the two sets of straight 10MR channels). The electron density contrast is clear in these cases, and it becomes possible to pick out arrangements of pores and building units within the structure and so to propose models. These are then refined against X-ray powder diffraction data. The framework geometries of ZSM-23,[60] MCM-22[61] and UZM-5,[62] for example, have been established with the help of electron diffraction and high-resolution imaging, followed by structure refinement against X-ray diffraction data.

Electron microscopy is invaluable in structure solution of samples that possess inherent disorder, which cannot therefore be solved by other diffraction methods. Zeolite Beta is an excellent example.[63,64] The stacking of layers in Beta is never fully ordered over the long range, so that the X-ray diffraction profiles are broadened and give little detailed information. Viewed along the *a*-axis in an electron microscope the alternation of stacking sequences of layers is clearly visible (Figure 3.11). Similar images enabled structural models incorporating stacking disorder to be established and X-ray diffraction patterns calculated from these disordered models using the program DIFFAX[65] were in close agreement with those observed experimentally. Notably, extensive investigation of the properties of this solid only occurred once the structure had been determined. The zeolite has since been found to be of great use, particularly in catalysis and fine chemicals syntheses, because it is the most readily synthesised of the high-silica zeolites with a three-dimensionally connected large-pore structure. The permeability of the structure is not strongly affected by the disorder. Recently, we have been able to obtain very high resolution images of zeolite Beta, including pore defects that result from different stacking orientations originating in the same layer (Figure 3.11(c).[66] These serve to confirm the structural model as well as suggesting a mode of crystal growth (Figure 3.12) and an explanation for the high concentration of silanol groups observed in zeolite Beta. Very similar stacking disorder features have been observed in the three-dimensionally-connected large-pore titanosilicate ETS-10, the structure of which was also solved by a combination of HRTEM and XRD simulation using DIFFAX.[67] ETS-10 also contains defects that result from disorder in the stacking of layers that give rise to 'extra-large' pores irregularly distributed throughout the structure.

For known frameworks, then, the main role of electron microscopy is to image defects in the structure that could affect diffusion or porosity. The images of Millward and Thomas showing intergrowths of ZSM-5 and ZSM-11 were early examples of this, in which the mode of stacking of the structural layers (Figure 2.9 – *via* either mirroring or inversion) could be determined directly.[68,69] The defect structure of the one-dimensional medium-pore zeolite ZSM-48[70] has also been studied by combined electron microscopy and X-ray diffraction.

Electron microscopy has therefore been of widespread use in structural studies of zeolites. Although lattice images have been obtained for aluminophosphate-[71] and carboxylate[72]-based microporous solids, their lower stability in the electron beam makes them more difficult to study, and usually results in images of lower resolution.

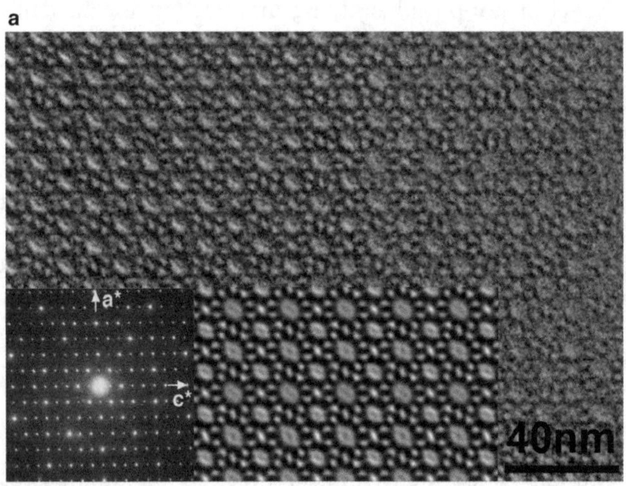

Figure 3.9 High-resolution transmission electron micrographs and corresponding diffraction patterns (insets) of the high silica zeolite TNU-9 (C2/m, a = 28.22 Å, b = 20.01 Å, c = 19.49 Å, β = 92.3°), down three different directions in the unit cell; (a) [010], (b) [001] and (c) [$\bar{1}$10]. In each case the experimental image is compared with an image simulated using the structural model of TNU-9 and appropriate instrumental and imaging conditions: 300 keV electron microscope. In each case, close agreement of observed and simulated images confirms the model is correct. The image down [010] shows the projected distribution of 10MRs, 6MRs and 5MRs of TNU-9 down that axis. Close examination even shows the difference between the two different 10MR channels parallel to the b-axis (see Figure 3.10). [Reproduced from reference 59, with permission. Copyright 2006, Nature Publishing Group.]

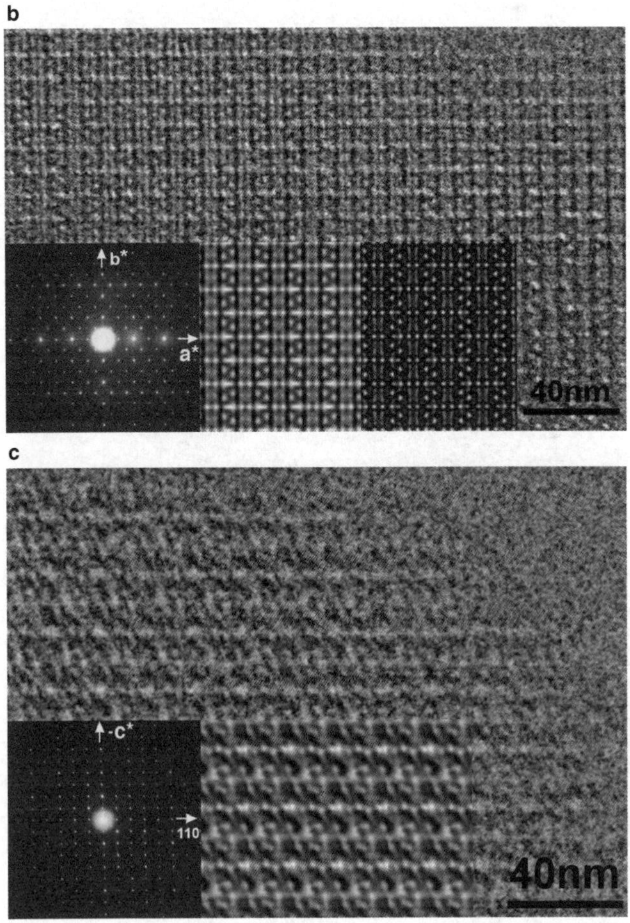

Figure 3.9 (*Continued*).

3.2.4 Structural Studies of Mesoporous Solids

Mesoporous solids prepared by liquid crystal templating routes possess properties that are in many respects similar to those of microporous solids. Although these solids do not possess crystallinity as it is usually defined (*i.e.* atoms at certain positions in each unit cell) they do have a regular structure of pores and pore walls. Each mesoporous structure gives a characteristic X-ray diffraction pattern, which comprises in each case a small number of diffraction maxima at low scattering angles with an absence of high-angle reflections. Although these patterns can be indexed according to crystal system and, more ambiguously, space group symmetry, they possess insufficient information to permit structure solution. It therefore falls to electron microscopy, and in particular imaging and electron crystallography, to provide structural information on them.[53]

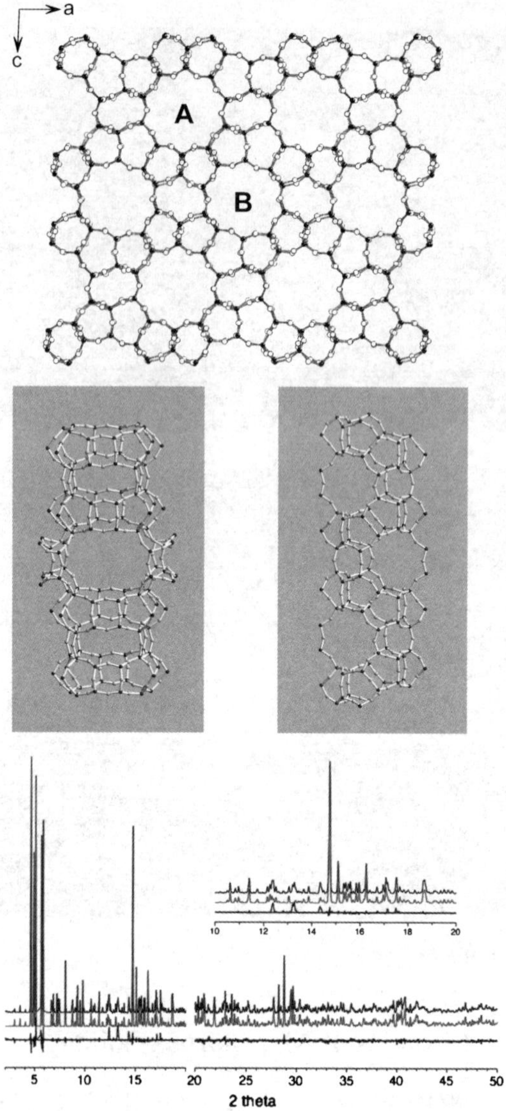

Figure 3.10 The framework of TNU-9 (above, projected down the *b* axis and directly comparable with the TEM image in Figure 3.9; middle, showing the wider A channel (left) and the narrower B channel (right), both viewed perpendicular to the channel axis). The complex structure possesses two straight channels (A and B) bounded by 10MRs and three-dimensional connectivity via 10MR channels (Figure 2.12). The structure was determined by combining synchrotron X-ray powder diffraction with electron microscopy. The Rietveld refinement (below) is shown with experimental, simulated and difference profiles. [Reproduced from reference 59, with permission. Copyright 2006 Nature Publishing Group.]

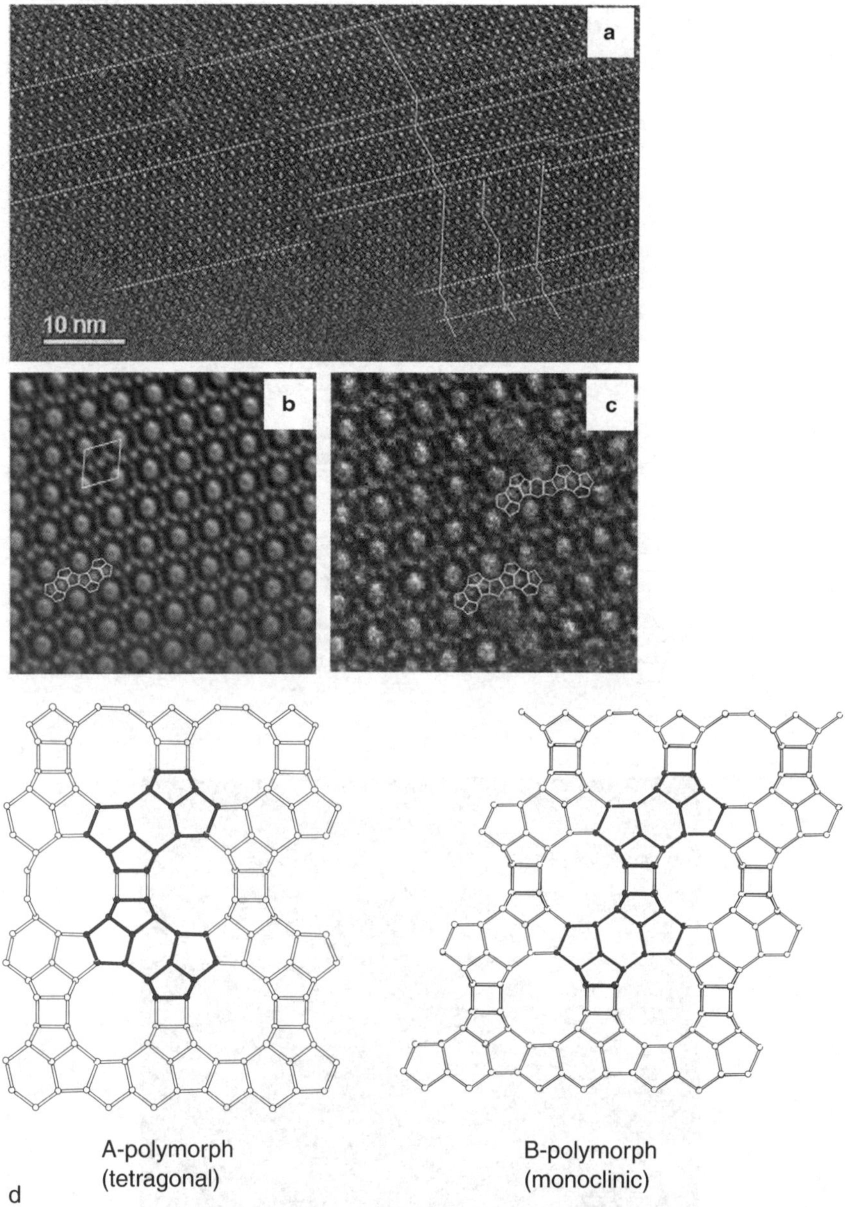

Figure 3.11 (a) Transmission electron micrograph of a particle of zeolite Beta showing stacking disorder. The particle has both well-ordered regions (shown with noise reduction by Fourier filtering, (b)) that reveal the 12MRs, 6MRs and 5MRs in projection and also regions (c) where different stacking directions starting from the same layer result in 'double pore' defects that heal after three layers. The end member A- and B-polymorph structures of zeolite beta are given in projection below. [Reproduced from reference 66 with permission. Copyright 2005 American Chemical Society.]

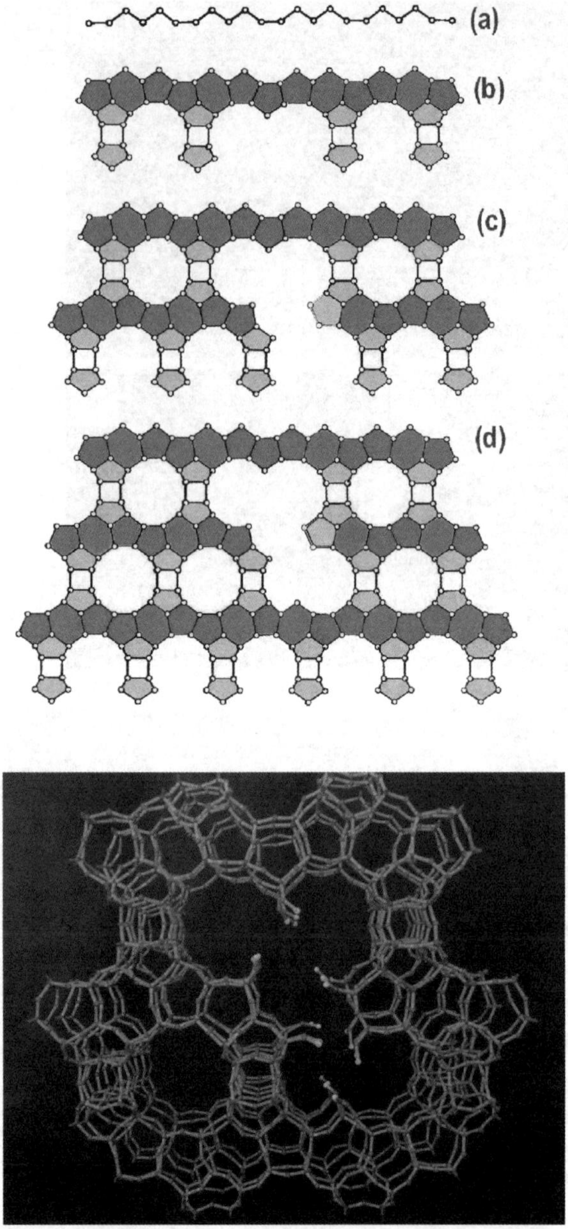

Figure 3.12 A model for the formation of defects in zeolite Beta. Growth onto a surface can take place in (at least) two different configurations (a to b). If the two stacking directions nucleate on the same plane, they cannot join in the next layer, and result in a double pore. Continued growth results in the defect healing in the third layer (c, d). A physical model of the defect (including silanol groups) is shown (below). This is in close agreement with the 'double pore' defects observed by electron microscopy in Figure 3.11.

For mesoporous solids of the MCM-41 type the structure is best considered as a hexagonal array of cylindrical channels, with order only in two directions. Images taken parallel to the channel axes therefore show a hexagonal arrangement of pores. The images can be used to give directly details on the spacings between pores, wall thickness and deviation of the pores from cylindrical, and compared with adsorption and X-ray data.[73] For three-dimensionally ordered mesoporous silicas such as those with space group symmetries Ia-3d (MCM-48),[74] P 6_3/mmc (SBA-2, SBA-12[75]) and Pm-3n (SBA-1, AMS-2, SBA-6[76]) careful electron microscopic studies have revealed intricate geometries. In each case a combination of electron diffraction and high-resolution imaging followed by extraction of phase data and image simulation by recombination of diffraction data collected down different axes has given a model for the structure.[76] Figure 3.13 shows an HRTEM image of AMS-2 (closely similar to SBA-1) taken down [100] and Figure 2.41 shows the model for SBA-1 produced by combining information from several such images. In SBA-2, which is based on the close-packed arrangement of spherical micelles, intergrowths of hexagonal close-packed and cubic close-packed domains are readily identified by electron microscopy (Figure 3.14).[77]

Similar high-resolution structural work has been performed to determine the structures of the functionalised hybrid mesoporous silicas prepared using anionic surfactants and incorporating cationic siloxanes (the AMS materials)[78] and to confirm the three-dimensional ordering of the benzene silicate of Inagaki, which TEM shows to have an MCM-41-like structure but with ordering along the channel axis due to the regular alignment of the aromatic groups.[79]

3.2.5 3D TEM–Electron Tomography

The preceding sections on electron microscopy have referred to studies that investigate the periodic structure of microporous and mesoporous solids by electron crystallography and imaging. Scanning transmission electron microscopy can also be used to give images of non-periodic features. In the simplest cases this may show details of sample thickness and shape and the inhomogeneous distribution of particles of different electron scattering power. However, when series of images are taken at different tilt angles in the microscope and then appropriately recombined, 3D TEM images, or electron tomographs can be produced of non-periodic features in microporous solids, such as the distribution of irregular mesopores or of metal particles, that can strongly influence their performance in adsorption and catalysis.[53] Research in this area has been pioneered by two groups, those of de Jong in Utrecht,[80,81] and of Midgley, Weyland and Thomas in Cambridge.[82–84]

Electron tomography requires that an extensive series of micrographs (typically > 100) be obtained over an angular range of around $\pm 70°$ *via* automated data collection with auto-compensation for shifts in the image position and of variations in focus. The individual images have to be aligned prior to

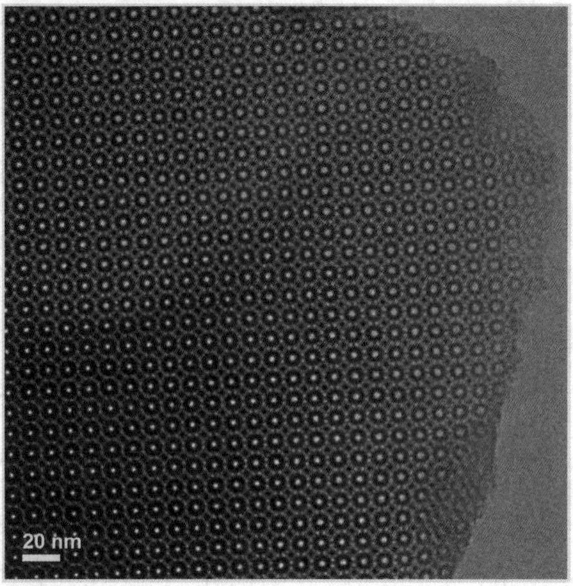

Figure 3.13 Electron micrograph of the mesocage organosilica AMS-2 taken along the [100] axis, courtesy of A. Garcia-Bennett. AMS-2 is similar in structure to SBA-1 (Figure 2.41) but is prepared with a high percentage of functionalised siloxane (O_3Si-R-X, X = $N(CH_3)_3^{3+}$) and using the anionic surfactant N-lauroylglutamic acid. The TEM image was taken on a JEOL 3010 microscope, resolution 1.7 Å, working at an accelerating voltage of 300 kV and at a magnification of 60 k. A Gatan CCD camera was used to record the image.

recombination to give a 3D image. The resolution of the image, d, is approximated by $d = \pi(T/N)$, where N is the total number of projections and T is the sample thickness. The emphasis of the electron tomography of the two groups has been different. Whereas de Jong has concentrated on the use of the more conventional 'bright field' imaging, using a parallel incident electron beam and obtaining the image from the electrons scattered forward at low angles, the Cambridge group have made use of high-angle annular dark field imaging 'HAADF', using a convergent incident beam and detecting the electrons scattered at high angles. In the bright field method the image results from phase contrast, which is strongly affected by thickness, lattice orientation and also the optics of the electron microscope. In HAADF the contrast from the Rutherford scattered electrons at high angle is strongly dependent on the atomic number, Z, of the element present (it is proportional to Z^2). As a result, 'Z-contrast tomography', as it is known, is highly sensitive to the distribution of heavy metal particles on supports of low atomic number.

Bright field electron tomography has been used by de Jong to determine the three-dimensional distribution of mesopores within dealuminated zeolite Y (see also Section 6.2.3). These studies show the geometry of secondary mesopores formed by different steaming methods (Figure 3.15),[80] and can therefore be

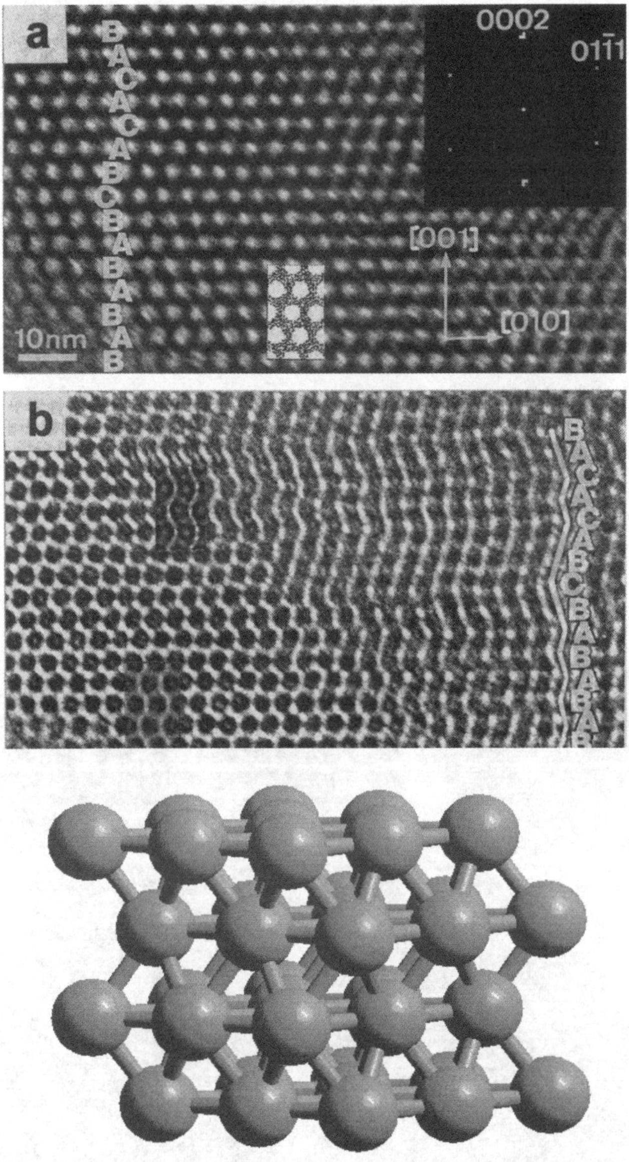

Figure 3.14 Electron micrographs, taken at different focus conditions, (a) and (b), of the mesocage silica SBA-2, which is synthesised around the close packing of spherical micelles (below). Disorder is commonly observed in the stacking sequence of pores in this material, so that hexagonal (ABAB), cubic (ABCABC) and also random sequences may be observed. [Reproduced from reference 77 with permission. Copyright 1998 American Chemical Society.]

used to optimise the stabilisation process. A proprietary steaming method has been shown to produce interconnected cylindrical mesopores that will enhance the transport of molecules to and from active sites in the microporous regions, and therefore improve the catalytic performance. In parallel, Z-contrast tomography has been used to determine the distribution of bimetallic Pd-Ru particles within a mesoporous support, which cannot be achieved by a single transmission image, because it is usually not clear whether the particles are on the surface or within the pores. A comparison of the two methods is given by Thomas *et al.*[85]

3.3 Scanning Electron Microscopy and Scanning Probe Microscopy

In scanning electron microscopy, images are obtained from the low-energy secondary electrons (< 50 eV) that are emitted from the sample in response to a narrow electron beam rastered across the surface. The intensity of the secondary electrons is sensitive to the orientation of the surface, so that the resulting image indicates its topography. The technique is widely used to determine

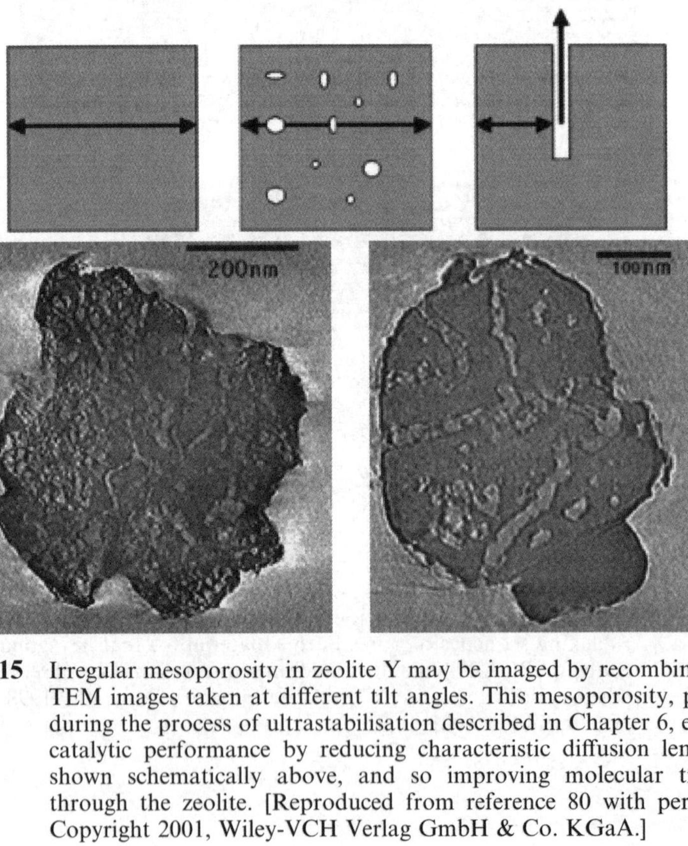

Figure 3.15 Irregular mesoporosity in zeolite Y may be imaged by recombination of TEM images taken at different tilt angles. This mesoporosity, prepared during the process of ultrastabilisation described in Chapter 6, enhances catalytic performance by reducing characteristic diffusion lengths, as shown schematically above, and so improving molecular transport through the zeolite. [Reproduced from reference 80 with permission. Copyright 2001, Wiley-VCH Verlag GmbH & Co. KGaA.]

crystal morphology and size and also, using this information, the phase purity of samples. Crystals of the aluminophosphate DAF-1, for example, while exhibiting broadly hexagonal morphology, also show details of overgrowths and dissolution (Figure 3.16) and SEM of the surfaces of mesoporous silicas can give remarkable detail on the arrangement of channels in the structure (Figure 3.17).[86] In conjunction with EDX analysis (Energy Dispersive analysis of characteristic X-rays emitted following ionisation events under electron irradiation) it is possible to determine the composition of selected crystals. Also, with suitable specimen treatment (*i.e.* the preparation of sections through particles) it is possible to observe compositional zoning, if present. For ZSM-5, for example, different zoning patterns of the Si/Al ratio have been observed. In direct synthesis using the tetrapropylammonium cation as a template, for example, the cores tend to be enriched in silica,[87] the explanation for which is given in Section 5.4.3.

Recent high-resolution SEM studies[86] on modern instruments have shown that, if care is taken to avoid charging effects by working at low currents and accelerating voltages, SEM (and STEM) images with a resolution of around 1 nm can be obtained without coating the surface by a layer of conducting metal (which is typically used to make the surface conducting and avoid charging). Under these conditions the micrographs show surface details as fine as the tubular structure of the channels in SBA-15 and the nature of random interconnections between the channels. At this resolution, SEMs are competitive with Atomic Force Microscopes (AFMs) for imaging the surfaces of microporous and mesoporous solids.

Figure 3.16 Typical scanning electron micrograph of crystals of the magnesioaluminophosphate DAF-1. The crystals reflect the hexagonal symmetry of the structure and show growth features at the surface of the main crystal.

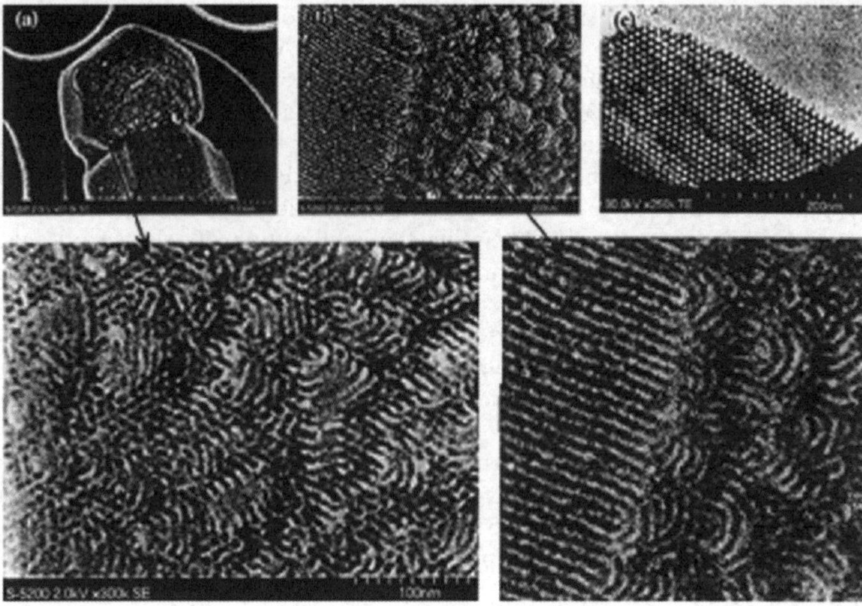

Figure 3.17 Scanning electron micrographs at high resolution of the surface of MCM-41 reveal the surface expression of the mesopores at the surface. [Reproduced from reference 86 with permission. Copyright 2003, Wiley-VCH Verlag GmbH & Co. KGaA.]

Atomic Force Microscopes operate by rastering a narrow tip on a cantilever over a surface, allowing the tip to move vertically to keep the required distance from the surface to feel a fixed force. In this way it is possible to determine the topology of well-oriented surfaces. A resolution of *ca.* 1 nm is obtained by this method, which is in general preferred over scanning tunnelling microscopy for the study of zeolites, due to their insulating character. The study of the surfaces of microporous solids by AFM and the subsequent quantitative analysis of the images has revealed important details of their crystallisation mechanisms.[88–90] An AFM image of the surface of zeolite Y, for example, showing the development of successive layers by growth at kink and edge sites, is given in Figure 3.18:[88a] the insights such images have given for the crystallisation of microporous solids are discussed in Chapter 5.

3.4 Spectroscopic Methods in Structural Studies

3.4.1 Solid State NMR Spectroscopy

3.4.1.1 The Method

The widespread applicability and availability of solid state NMR spectroscopy have established it as an indispensable structural tool. Furthermore, it is complementary to diffraction methods, since it gives information on local

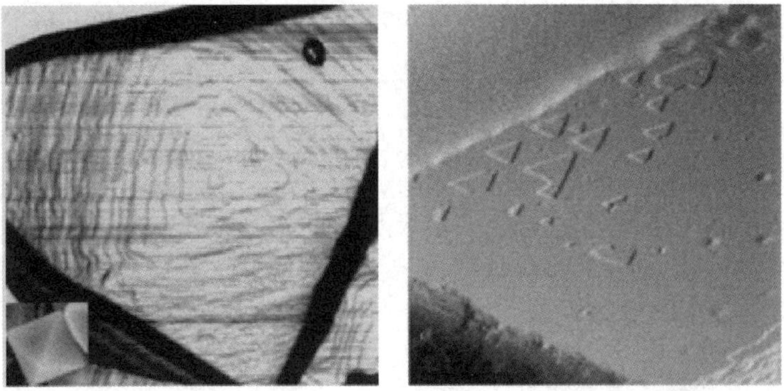

Figure 3.18 Atomic force microscopy of the <111> surfaces of zeolite Y and a zinc phosphate prepared under specific 'reverse micelle' synthetic conditions. Both have the FAU structure type. Examination of images such as these are enabling details of the crystal growth process to be established. [Reproduced from references 88 (a) and (b) with permission. Copyright 1996 Wiley-VCH Verlag GmbH & Co. KGaA and 2002 American Chemical Society.]

environments and mobility that are not available from X-ray diffraction, which yields only the space- and time-averaged positions of atoms. The method permits identification and resolution of atoms in different environments, their coordination geometry and relevant internuclear distances and molecular motion. The versatility of the technique, in terms of the types of measurement that are possible, coupled with the very large number of elements (and isotopes) that can be studied make it difficult to summarise in a single book, let alone a section. I will in this chapter concentrate on the application of NMR to structural studies of molecular sieves, and go on to discuss the information that can be obtained on the environment and mobility of adsorbed molecules in Chapter 7 and on molecular processes in catalysis in Chapter 8. Clear and detailed texts on the application of solid state NMR are available, some with detailed sections on porous solids.[91] Table 3.1 summarises the properties of NMR active nuclei that have been of most interest in studies of microporous solids.[92]

Solid state NMR of static samples gives relatively little structural information, because the broad lines that result from anisotropic dipolar interactions and chemical shift anisotropy prevent the separation of individual resonances. Dipolar interactions (which may be homo- or heteronuclear) are through-space interactions with nearby spins, whereas chemical shift anisotropy arises from the modification of the applied magnetic field experienced at the nucleus by the surrounding electrons. The line-broadening effects of both dipolar interactions and chemical shift anisotropy can be removed by sample spinning to give higher-resolution spectroscopy, but under certain circumstances, static (also known as wideline) spectroscopy is of use. One example is in the study of the mobility of adsorbed molecules by ^{2}H NMR where the adsorbate, rather than the solid, is moving (described in more detail in Chapter 7).

Table 3.1 NMR properties of commonly-studied nuclei in microporous solids.[92]

Nucleus	*Natural abundance %*	*Spin*	*NMR freq. at 23.487 KG*	*Receptivity rel. to ^{1}H*	*Receptivity rel. to ^{13}C*	*Quadrupole moment (fm^2)*
^{1}H	99.9885	1/2	100	1	5.87×10^{3}	–
^{2}H	0.0115	1	15.35	1.11×10^{-6}	6.52×10^{-3}	0.286
^{6}Li	7.59	1	14.72	6.45×10^{-4}	3.79	−0.081
^{7}Li	92.41	3/2	38.86	0.271	1.59×10^{3}	−4.01
^{11}B	80.1	3/2	32.07	0.132	777	4.059
^{13}C	1.07	1/2	25.14	1.7×10^{-4}	1	–
^{14}N	99.63	1	7.23	1×10^{-3}	5.90	2.044
^{15}N	0.37	1/2	10.14	3.8×10^{-6}	0.0225	–
^{17}O	0.037	5/2	13.56	1.1×10^{-5}	0.065	−2.558
^{19}F	100	1/2	94.09	0.83	4.9×10^{3}	–
^{23}Na	100	3/2	26.45	9.27×10^{-2}	545	10.4
^{27}Al	100	5/2	26.06	0.207	1.22×10^{3}	14.66
^{29}Si	4.7	1/2	19.87	3.68×10^{-4}	2.16	–
^{31}P	100	1/2	40.48	6.65×10^{-2}	391	–
^{45}Sc	100	7/2	24.29	0.302	1.78×10^{3}	−22.0
^{69}Ga	60.1	3/2	24.00	4.2×10^{-2}	246	17.1
^{71}Ga	39.9	3/2	30.50	5.7×10^{-2}	335	10.7
^{129}Xe	26.4	1/2	27.81	5.7×10^{-3}	33.6	–

In general, though, the spectra from Magic Angle Spinning (MAS) experiments, in which dipolar interactions and chemical shift anisotropy are averaged to zero, are much more informative. In MAS NMR, the sample is typically spun at 10–30 kHz about an axis that makes an angle of 54° 44′, the Magic Angle, to the magnetic field. This averages dipolar interactions and chemical shift anisotropy to zero for nuclei with spin $I = \frac{1}{2}$, giving narrow resonances. If dipolar coupling to protons is important, the signals can be further narrowed by the removal of X-^{1}H interactions (where X is the dipolar nucleus under study) by simultaneous irradiation of the sample at the proton resonance frequency to 'decouple' the protons. It is not possible, however, to average the very strong ^{1}H-^{1}H homonuclear dipolar interactions at accessible MAS spinning speeds, so that other methods are required to narrow resonances where these interactions dominate.

MAS NMR spectroscopy of the spin $\frac{1}{2}$ nuclei ^{1}H, ^{13}C, ^{29}Si and ^{31}P has been particularly useful in studies of microporous solids but many other nuclei of interest in the area of microporous solids are quadrupolar, and indeed 74% of NMR-active nuclei have a spin greater than $\frac{1}{2}$. Among these, ^{2}H is the only nucleus of major chemical interest with an integer spin ($I = 1$), and is of particular relevance to porous solids for studies of mobility. By contrast, many quadrupolar nuclei with non-integral spin are important, including ^{11}B, ^{17}O, ^{23}Na, ^{27}Al and ^{71}Ga. The central ($m_I = +\frac{1}{2} \rightarrow -\frac{1}{2}$) transition of quadrupolar nuclei is to first order unaffected by quadrupolar coupling, and since dipolar coupling and chemical shift anisotropy are removed by MAS, reasonable linewidths can be obtained. However, second order quadrupolar interactions remain. Further line narrowing can be achieved by working at high fields, because the second order quadrupolar coupling is inversely proportional to field strength. The resulting lineshapes depend strongly on the site symmetry of the quadrupolar nucleus, so that spherically symmetrical environments give narrow, symmetrical lines, whereas others, such as distorted tetrahedral or octahedral environments, or atoms in five-fold coordination, give broadened resonances with characteristic quadrupolar line shapes. In extreme cases the resonances may be so broad as to be unobservable.

While important structural information has already been obtained by MAS studies of quadrupolar nuclei such as ^{27}Al and ^{11}B, a great deal of effort has been expended in devising methods of improving resolution of spectra of quadrupolar nuclei. These involve techniques for removing the second-order quadrupolar broadening. Three ingenious experimental approaches have been developed to achieve enhanced resolution in the spectra of quadrupolar nuclei. These are Double Rotation (DOR),[93] Dynamic Angle Spinning (DAS)[94,95] and Multiple Quantum MAS NMR.[96] In double rotation NMR the sample is spun simultaneously at two angles, the magic angle and a second angle (30.6° or 70.1°). As might be imagined, this method, in which one rotor is spun within another, is experimentally challenging, but has been used with conspicuous success. In dynamic angle spinning, the experiment involves spinning the sample at two different angles alternatively over different time periods, in order to refocus the signal. In this way a two dimensional spectrum is obtained, with isotropic signals (quadrupolar broadening removed) in one dimension and quadrupolar broadened anisotropic patterns in the other. The 2 D spectrum therefore permits improved resolution in one dimension while retaining symmetry information in the other.

In the MQ MAS pulse sequence a time period of multiple quantum excitation (for ^{27}Al, $I = 5/2$, this can involve 5 quantum or 3 quantum transitions) is followed by a period where the coherence is converted to a single quantum ($m_I = +\frac{1}{2} \rightarrow -\frac{1}{2}$) transition and the signal measured in the normal way. Multiple Quantum MAS NMR gives two dimensional spectra that can be represented either by plotting the MAS NMR against the MQ MAS spectrum 'F_2 *vs.* F_1' or, by mathematical treatment (shearing), with isotropic spectra in one dimension. The isotropic projection is much more highly resolved than the 1D MAS NMR spectrum. One advantage of MQ MAS over the other methods is that it is not so challenging in terms of the physical procedure of performing the experiment, because this only involves magic angle spinning. This method offers great promise for the study of quadrupolar nuclei. MQ MAS has been used to great effect to resolve complex signals and to determine true isotropic chemical shifts of sites, but the quantification of signals remains a significant challenge.

Cross Polarisation is a widely used solid state NMR technique that enhances signal from one type of nucleus by the transfer of magnetisation from another, typically ^{1}H. It is applicable to both spin $\frac{1}{2}$ and quadrupolar nuclei, though the high natural abundance of NMR active nuclei such as ^{31}P and ^{27}Al tend to mean it finds most use for dilute spin $\frac{1}{2}$ nuclei such as ^{13}C and, especially in zeolites, for ^{29}Si. The transfer of magnetisation from the abundant, fast relaxing proton population to the nucleus under examination is achieved by matching the energies of the two spin systems at what is known as the Hartmann–Hahn energy matching condition (see for example, ref. 91). As a result of the technique, acquisition times are reduced and signal-to-noise ratios improved. Magnetisation transfer via the dipolar interaction is only efficient over short distances, so that the ratios of signals in CP NMR spectra are not quantitative. While this can be a drawback, it can also be utilised to act as an aid to signal identification, when comparing between environments with or without associated protons. For example, the signal from Q^3 silicon atoms attached to a hydroxyl group, $(SiO)_3SiOH$, found in some defective pure silica zeolite polymorphs or in high abundance in pure silica mesoporous solids, is preferentially enhanced over Q^4 silicon atoms, $(SiO)_4Si$, by CP MAS.

Two dimensional correlation spectroscopy has revolutionised solution state NMR in the last two decades, enabling the structure of complex biomolecules in solution to be determined. Correlation spectroscopy (COSY) gives information on the connectivity of atoms by measuring the strength of internucleus interactions. This is possible because the interactions either are strongly dependent on internuclear separations and therefore fall off rapidly with increasing distance (eg. NOESY) or make use of through-bond interactions (COSY). High resolution 2 D spectra are usually represented so that they indicate which nuclei are close to one another, and additional experimental methods are available to give accurate internuclear distances. Although the inherently lower resolution available from solids has held back the development of such methods in solid state NMR, elegant studies have been published that describe the location of adsorbed species in pure silica zeolite polymorphs by methods of this type. Pure silica polymorphs are amenable to these experiments because they give very well resolved ^{29}Si MAS NMR spectra. In addition, careful measurement of signal intensities and internuclear distances can, in collaboration with diffraction data, be used to determine structures ('NMR crystallography'). This approach is finding widespread use for studies of molecular crystals and shows promise for microporous solids. Examples of two dimensional correlation spectroscopies that have been applied to zeolites are homonuclear J-coupling correlation spectroscopy (COSY) and related methods probing spins of the same type, such as ^{29}Si–^{29}Si INADEQUATE experiments, and methods which can be applied to systems with heteronuclear coupling, *via* dipolar couplings (the TEDOR (transferred Echo Double Resonance) experiment), or *via* through-bond J couplings (the INEPT experiment). These are described in later sections.

A very wide range of solid state NMR experiments is therefore available for the study of the structural chemistry of microporous framework solids. Some of these are listed in Table 3.2, together with a summary of the information they give. In this chapter I will discuss the application of solid state NMR spectroscopy to the elucidation of the structural chemistry of zeolites, other inorganic molecular sieves and organic-inorganic hybrids.

Table 3.2 Solid state NMR experiments and the information they give on microporous solids.

NMR technique	*Description of experiment*	*Information*
MAS NMR	Sample spinning at the magic angle to the external magnetic field, typically up to 30 kHz, removes dipolar coupling, chemical shift anisotropy and first-order quadrupolar interactions	High-resolution spectra of solids, giving detailed information on number and type of different chemical environments
DOR, DAS, MQ MAS	Methods suitable for high-resolution spectra of half integer quadrupolar nuclei, because magic angle spinning alone is unable to remove second order quadrupolar interactions	High resolution of spectra of half integer quadrupolar nuclei such as ^{27}Al, ^{17}O, ^{23}Na, *etc.*, giving information on number and geometry of sites
Heteronuclear decoupling	Irradiation of sample at frequency of the nucleus under investigation and a second nucleus that couples strongly to it (such as ^{1}H, ^{19}F) to remove dipolar broadening	Improved resolution of spectra of nuclei coupled directly to protons, fluorine (or other suitable cases, such as ^{27}Al-^{31}P)
Cross polarisation	Transfer of magnetisation from abundant, high γ, fast-relaxing nuclei to adjacent non-abundant, more slowly relaxing nuclei (^{1}H-^{29}Si, ^{1}H-^{13}C, *etc.*), enhancing signal	Non-quantitative spectra with greatly enhanced signal-to-noise. Comparison of CP and non-CP can help assign resonances where there are differences in the proximity of the more abundant nucleus (protons, for example)
Cross polarisation with variable contact time	CP enhancement in isolated spin pairs exhibits an oscillatory signal as a function of contact time due to magnetisation transfer back and forth between the two heteronuclear spins	Analysis gives internuclear distances in isolated spin pairs such as ^{1}H-^{31}P, ^{1}H-^{29}Si, ^{19}F-^{29}Si, *etc.*
REDOR (Rotational Echo Double Resonance) and TEDOR	Double resonance heteronuclear (nuclei I, S) spectroscopies. Dephasing of signal from spin I nuclei through pulses applied to spin S. Spins with strong dipolar interactions decrease in intensity. The	These techniques enable the measurement of interatomic distances from heteronuclear spin pairs, particularly when these are isolated. Examples include ^{1}H-^{13}C, ^{27}Al-^{31}P, *etc.*

(*Continued*)

Table 3.2 (*Continued*).

NMR technique	*Description of experiment*	*Information*
	pulse sequences are designed to prevent dipolar averaging and to enable measurement of the dipolar coupling. The TEDOR technique is REDOR with a transfer of magnetisation, allowing 2 D correlation, also *via* dipolar coupling	
COSY, NOESY & INADEQUATE 2 D spectroscopies	Homonuclear correlation spectroscopies. COSY and INADEQUATE 2 D spectroscopies probe the through-bond J coupling. NOESY spectroscopy is homonuclear 2 D correlation spectroscopy through dipolar coupling	2 D spectra of both labelled and natural abundance materials give information on homonuclear connectivities (^{29}Si-^{29}Si; ^{13}C-^{13}C)
INEPT	INEPT measures 2 D heteronuclear correlation via J coupling.	Through-bond connectivities

3.4.1.2 Zeolites and Mixed Coordination Silicates

Solid state NMR studies, particularly of silicon and aluminium, occupy a special place in the history of zeolite science, partly because the method was developed and became widely available in the 1970s and 1980s – at the same time as there was an upsurge of industrial and academic interest in these solids.

^{29}Si is a spin $\frac{1}{2}$ nucleus with attractive NMR properties. Although its low natural abundance (5%) and slow relaxation necessitate long acquisition times, the low abundance also means that homonuclear ^{29}Si-^{29}Si interactions are not important and if the recycle delay is suitably chosen, the spectra are quantitative. Furthermore, ^{29}Si exhibits an acceptably large chemical shift range over different environments: typical aluminosilicates give ^{29}Si shifts between –80 and –110 ppm with respect to tetramethylsilane. The resolution of ^{29}Si MAS NMR of zeolites improves with field strength up to around 4.7 Tesla, after which no further narrowing is observed. Relaxation times in zeolites, which limit the rate at which signal can be measured and therefore determine spectral acquisition times, are typically shorter (5–30 s) than those of amorphous or dense silicates and there is strong evidence to suggest that proximity to adsorbed dioxygen is a key relaxation mechanism.[97]

3.4.1.2.1 Pure Silica Polymorphs. In the case of pure silica polymorphs of zeolites, such as silicalite-1, the ^{29}Si MAS NMR spectrum consists of a number of resonances at different chemical shift values: these arise from crystallographically distinct, tetrahedrally connected silicon atoms, known as Q^4 silicons ($Si(OSi)_4$).[98] Silicalite-1 is known to possess monoclinic symmetry, with 24 crystallographically distinct sites. The ^{29}Si spectrum of silicalite can be resolved into 20 lines, the total intensity of which is 24 times that of the least intense peak (Figure 3.19). The linewidth of these resonances is extremely narrow (5 Hz).[99] The ^{29}Si chemical shift therefore depends on the local structural environment (as well as the number of aluminium atoms in the first cation coordination shell, which has a large effect). Through studies of this kind on a wide range of pure silica zeolite polymorphs of known structure, it is possible to derive a linear relationship between the ^{29}Si (Q^4) chemical shift and the mean Si-Si non-bonded distance or the mean bond angle (θ)[100] of the four Si Ô Si angles around each silicon atom, e.g.,

$$\delta = -25.44 - 0.5793 \times \theta$$

Another relationship, proposed by Engelhardt and Radeglia,[101] relates the shift to the individual Si-Ô-Si angles (α) *via* the expression

$$\delta = -247.05 \times \left(\overline{\cos\alpha/(\cos\alpha - 1)}\right) + 2.19$$

Once such relationships were established and better understood,[102] it became possible to assign resonances of zeolites of known structure to particular crystallographic silicon sites. The relationship has since been extended to tetrahedral silicates containing aluminium in framework positions, so that a single relationship relating the ^{29}Si chemical shift to the sum of the four T–T bond distances, and taking into account an additional term for the paramagnetic contribution of each of n aluminium nearest neighbours to the shift (7.95 n ppm) can be used for all signals:[103]

$$\delta = 143.03 - 20.34 \sum_{1}^{4} d_{T-T}/\text{Å} + 7.95\,n$$

^{29}Si MAS NMR is therefore sensitive to structural modifications and symmetry changes that involve changes in bond angles, and this is particularly noticeable for pure silica polymorphs because the ^{29}Si resonances are so well resolved. For example, silicalite-1 undergoes a phase transition from monoclinic to orthorhombic upon heating which results in a decrease in the number of crystallographically distinct sites from 24 to 12, and this is readily followed by NMR. Addition of organic adsorbates also has a major effect on the chemical shifts of the different sites as a result of structural adjustments.

In many syntheses of pure silica zeolite polymorphs, alkylammonium cations are used as structure directing agents. The positive charge on the template may be balanced in syntheses performed under alkaline pHs by the presence of defects, which after template removal can leave vacant silicon sites surrounded by hydroxyl groups attached to neighbouring silicon atoms. For samples with a high level of Q^3 silicon ($Si(OSi)_3OH$), which results from template removal from silicas prepared from alkaline media, the ^{29}Si spectrum is broadened and

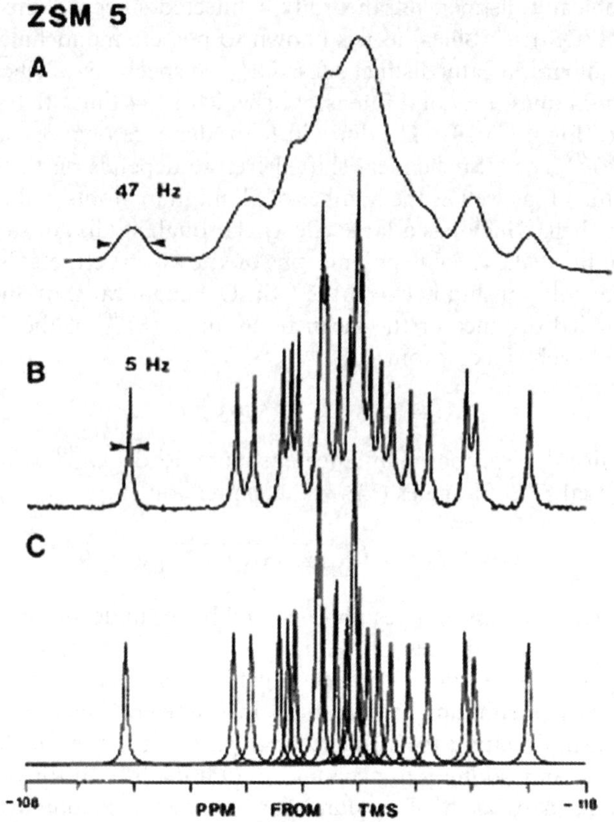

Figure 3.19 High-resolution ^{29}Si MAS NMR of silicalite (B) compared with that from a ZSM-5 (containing aluminium but with the same framework structure). (C) shows the deconvoluted spectrum for comparison, and indicates the 24 different resonances. [Reproduced from reference 99 with permission. Copyright 1988 American Chemical Society.]

the resolution between crystallographically different silicon species is usually lost. Q^3 peaks resonate *ca.* 5 ppm downfield of Q^4: this assignment is easily verified by ^1H-^{29}Si cross-polarisation measurements that indicate strong enhancement of peaks in this shift region, because of the proximity to those silicon atoms of protons of the attached hydroxyl group. In solids crystallising from fluoride-containing syntheses (see Chapter 5), fluoride ions can remain bound to silicon atoms, leaving five-coordinate SiO_4F species with overall negative charge. ^{29}Si MAS NMR can be used to detect these different species. Q^4 silicon with bound fluoride resonates upfield of normal Q^4 silicon, giving signals that are clearly visible in the ^{29}Si spectra. These fluoride ions are mobile at higher temperatures in some structures, so that SiO_4F signals are observed more clearly at low temperatures, when the fluoride ions become localised. Complementary ^{19}F ($I = \frac{1}{2}$) MAS NMR reveals a wide range of chemical shifts of fluoride ions bound to silicon atoms in small silicate cages, from –38 ppm in [4^6] cages (D4Rs), for example in octadecasil, to –64 ppm in D6Rs in chabasite and –76 ppm in [$4^1 5^4 6^2$] (nonasil-type) cages.[104]

3.4.1.2.2 Aluminosilicate Zeolites. The presence of aluminium within the silicate framework has a strong effect on the ^{29}Si spectra. The first effect, even at low framework Al concentrations (Si/Al > 100), is broadening of the resonances. In most cases, this results in overlap of the Q^4 silicons surrounded by four silicons ($Si(OSi)_4$). The second result is that different Q^4 silicon environments are produced, with different numbers of silicon and aluminium atoms in their second coordination shell. Increasing the concentration of aluminium atoms results in systematic downfield shifts of around 10 ppm for each extra aluminium in the nearest cation coordination shell. The effect is most clearly seen for zeolites that have only one crystallographically distinct site, such as those with the FAU structure type (zeolites X and Y) (Figure 3.20[105]). Zeolites A, Rho and ZK-5 are other examples. For these zeolites, each different peak must be from silicon atoms with different second-nearest-neighbour environments. For pure silica polymorphs there is only a single Q^4 peak. Increasing the aluminium content results in spectra with changing intensity ratios for Q^4 peaks with different second-shell compositions, until at Si/Al = 1 there is again a single peak, this time for a fully ordered structure in which silicon and aluminium atoms alternate strictly in the tetrahedral sites, obeying the Loewenstein aluminium avoidance principle. In fact, it is possible in the case of such single T-site zeolites to calculate the Si/Al ratio of the framework using the formula:[106]

$$\left(\frac{Si}{Al}\right)_{framework} = \frac{I_4 + I_3 + I_2 + I_1 + I_0}{I_4 + 0.75I_3 + 0.50I_2 + 0.25I_1}$$

where I_n denotes the intensity of the NMR resonance from the $Si(OAl)_n(OSi)_{4-n}$ environment (denoted Si(nAl)). The numerator is proportional to the total signal from silicon, while the denominator is proportional to the aluminium. This is measured indirectly, as each of the links to aluminium from a silicon atom corresponds to 0.25 Al (because of the tetrahedral coordination). This is of particular importance because the framework Si/Al ratio of zeolites may be very different from the overall composition, for example where framework dealumination occurs in the technologically important process of ultrastabilisation (see Chapter 6). The ^{29}Si NMR spectrum depends on framework composition in a similar way when gallium, rather than aluminium, is included in the framework, and so Si/Ga framework ratios can be determined.

For aluminosilicate zeolites that have more than one T-site in the unit cell (by far the majority), signals from all the Q_4 silicons with the same number of second coordination shell aluminium atoms may overlap, and then the formula given previously for the determination of framework composition still holds. Further details of the distribution of aluminium (or gallium) within a zeolite framework can also be obtained from the ^{29}Si spectrum in this case. For aluminium-rich zeolites the aluminium may order on the short range in

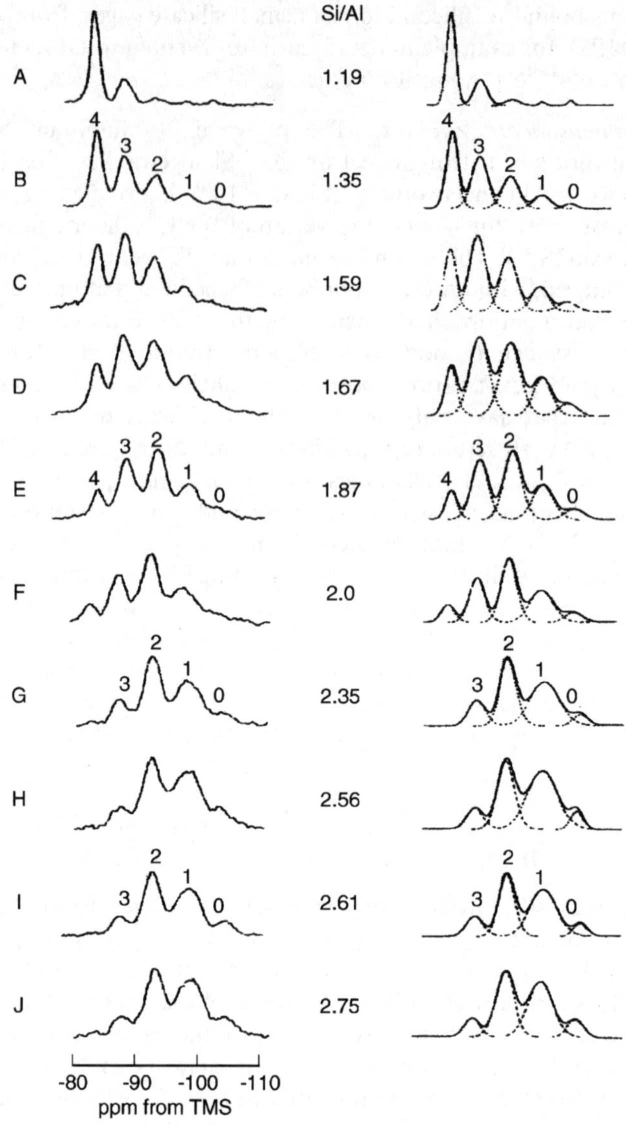

Figure 3.20 The ^{29}Si MASNMR spectra (A–J) of faujasitic zeolites X and Y with framework Si/Al ratios from 1.19 to 2.75. Deconvolutions of the spectra in terms of resonances due to silicon surrounded by 0 to 4 second nearest aluminium neighbours are given to the right of each spectrum. [Reproduced from reference 105 with permission. Copyright 1982 Royal Society of Chemistry.]

response to Loewenstein's principle or processes in the synthesis, so that the distribution over different sites will not be statistical, and the ^{29}Si NMR places limits on the nature of this ordering. The work of Melchior for zeolite Y samples is described in Section 5.4.3. In some cases, however, the mean T-Ô-T angles, and consequently the ^{29}Si chemical shifts from the crystallographically distinct silicon sites are so different that the observed resonances cannot unambiguously be attributed to silicon atoms with particular numbers of aluminium atoms in their second coordination shell. Zeolite mazzite (ZSM-4) is such a case:[107] more careful deconvolution is then required to obtain the framework Si/Al ratio.

3.4.1.2.3 ^{19}F MAS NMR Studies. Fluoride can play an important part in hydrothermal syntheses, where it can act as a mineralising agent in the syntheses of both silicates and aluminophosphates, and in many cases can remain attached to the framework cations such as silicon or (in the case of AlPOs, aluminium) in the as-prepared material. The ^{19}F nucleus has attractive NMR properties (abundant, spin $\frac{1}{2}$, large chemical shift range) and is readily studied to give information on its presence in different environments, the type of sites it adopts, and its mobility at different temperatures. Examples are given in Section 5.4.2 of the study of fluorine in as-prepared pure silica polymorphs of zeolites.

3.4.1.2.4 2 D MAS NMR. High resolution 1D, single resonance NMR has given information on the number of different silicon sites in a structure and how these may be assigned to a structural model. It can therefore be used to support or refute assigned crystal symmetries and as an accurate monitor of phase changes. It is not able to give atom-atom connectivities, however, and it is this that is needed to derive structural information and ultimately to solve structures. In a series of careful studies, the group of Fyfe has elegantly developed two-dimensional and double-resonance techniques by which framework connectivities in microporous silicas, zeolites and aluminophosphates can be established *via* experiments in which magnetisation is transferred from one set of nuclei to another.[108–111]

For investigation of ^{29}Si–^{29}Si connectivities in pure silica zeolite polymorphs the coupling is homonuclear and the COSY and INADEQUATE through-bond pulse sequences are used. The study of the pure silica polymorph of ZSM-12 is a good example. The structure has seven non-equivalent cation positions, which give clearly distinguishable resonances. COSY and INADEQUATE experiments are able to determine the connectivities unambiguously, and thereby enable the resonances to be assigned on the basis of the known structure (Figure 3.21).[112] A similar approach was successfully taken to establish the connectivity between silicon atoms giving resolved NMR resonances for the more complex ZSM-5 structure.[113] (For $^{27}Al \rightarrow {}^{29}Si$ and $^{27}Al \rightarrow {}^{31}P$ connectivities, heteronuclear couplings between the quadrupolar and the spin $\frac{1}{2}$ nuclei are determined through double-resonance methods that include those based on cross polarisation, REDOR (rotational echo double resonance) and TEDOR (transferred echo double resonance).)

NMR methods making use of connectivities also offer possibilities in structure solution: two dimensional ^{29}Si–^{29}Si methods are the most advanced for zeolites. The approach of Fyfe has been taken further by the group of Levitt. In place of the INADEQUATE sequence a new method ('Symmetry-based ^{29}Si

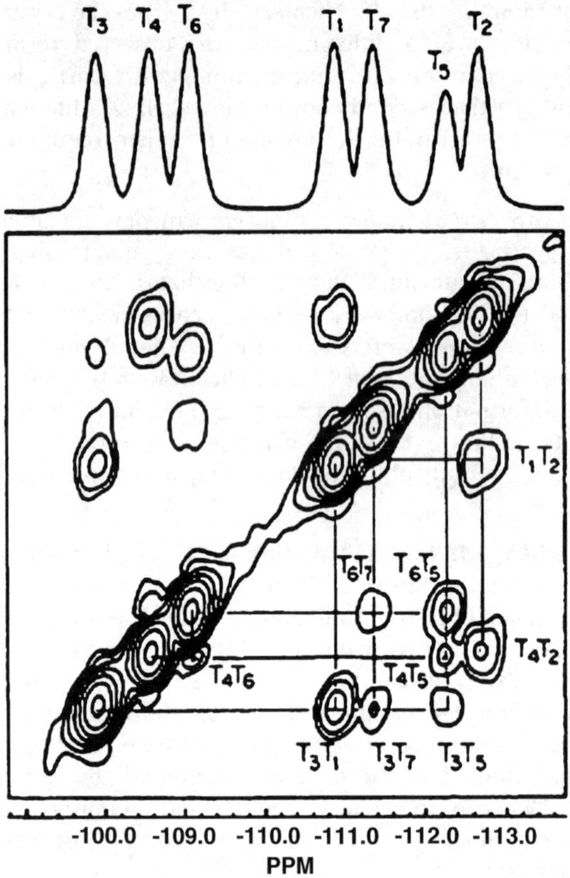

Figure 3.21 Two-dimensional correlation spectroscopy of a highly siliceous ZSM-12 zeolite enables all seven different sites to be resolved, and for the Si–O–Si site connectivity to be determined by looking for Si–Si cross peaks. [Reproduced from reference 112 with permission. Copyright 1990 American Chemical Society.]

Dipolar Recoupling') has been developed that gives full connectivities of resolved sites and also 'build-up curves', which reveal the distinctive radial distribution of silicon sites around each crystallographically distinct type of silicon.[114] Taken together with space group and symmetry information available from diffraction data, this has been shown to be sufficient to obtain the structures of pure silica zeolite polymorphs[115] and could conceivably offer an interesting alternative approach to direct methods associated with X-ray diffraction for structure solution. It is only applicable to solids giving highly resolved spectra, however, so that aluminosilicate zeolites would require conversion to the pure silica form to be amenable to the method.

The determination by NMR of internuclear distances in microporous and mesoporous solids (such as Si-H and Si-F[116] and also Al-H, P-H, *etc.*) can also be achieved by double resonance methods involving magnetisation transfer between the different nuclei, and can be useful for structures where structural details are not available from diffraction, for example in structures displaying disorder or fractional occupancies.

3.4.1.2.5 MAS NMR of ^{27}Al and Other Quadrupolar Nuclei. ^{27}Al NMR studies of zeolites have also been carried out extensively, because it is the incorporation of aluminium in, and in some cases its loss from, tetrahedral framework sites that determine the ion exchange and adsorption properties, catalytic activity and hydrothermal stability. The quadrupolar character of the ^{27}Al nucleus ($I = 5/2$) results in broader peaks than those obtained for ^{29}Si, but because aluminium atoms can exist in very different environments (four-, five- and six-fold coordination to oxygen) these resonances appear separated in the spectra. Typical approximate chemical shifts in zeolites are: tetrahedral aluminium at 50–60 ppm, five-fold coordinated species at 30 ppm and octahedral aluminium at *ca.* 0 ppm. Indeed, the study of the chemistry of aluminium in zeolites by observing its environment as a function of treatment has been a very important subject of ^{27}Al NMR (Chapter 6). Within this context, the most problematic aspect is the broadening of the lineshape as the symmetry of the environment is reduced. Undistorted tetrahedral and octahedral environments possess high, cubic symmetry, and therefore relatively narrow linewidths, but when the aluminium occupies distorted tetrahedral sites (for example, next to a bridging hydroxyl group or in five-fold coordination or when octahedral, extra-framework aluminium is dehydrated) the signal is strongly broadened by quadrupolar effects sometimes to the point where it becomes unobservable, and so the literature refers in places to 'hidden' aluminium. This has been detected indirectly by double resonance NMR in which the ^{1}H signal is reduced by the dephasing effect of 'hidden' ^{27}Al nuclei[117] (transfer of population *via* double resonance – TRAPDOR) and also directly by ^{27}Al MAS NMR and MQ MAS NMR of steamed, ultrastabilised Y (see Chapter 6) at very high magnetic fields,[118] where the quadrupolar broadening is strongly reduced (Figure 3.22). In the latter studies the five-fold coordinated aluminium is clearly visible and MQ MAS NMR reveals two different tetrahedral environments as well as octahedral sites.

It is sometimes possible to differentiate between aluminium atoms occupying crystallographically distinct T sites in zeolites, but to a much smaller degree than is possible for silicon. For zeolite ZSM-4, for example, the two crystallographic sites are readily resolved,[119] but for ZSM-5 the (orthorhombic) aluminosilicate version of silicalite-1, the best separations, even using 2 D MQ MAS NMR, only resolve two groups of peaks, rather than 12 separate peaks.[120] Continued development is needed to increase resolution, to give quantitative information on whether all T-sites are occupied in an ordered way, or whether they are distributed randomly, details that have important implications for catalytic performance. There is evidence, for example, that the type of template used in a synthesis could affect this distribution (see Chapter 5).

Specialised techniques such as multiple quantum NMR and the analysis of spinning sidebands are of promise in this area.

Several other species that can be included into the zeolitic framework are amenable to solid state NMR. ^{71}Ga (I = 3/2) is, like ^{27}Al, quadrupolar, so that the first information available is whether it remains tetrahedral in the framework or leaves the framework to adopt another geometry.[121] The smaller boron (B^{3+})

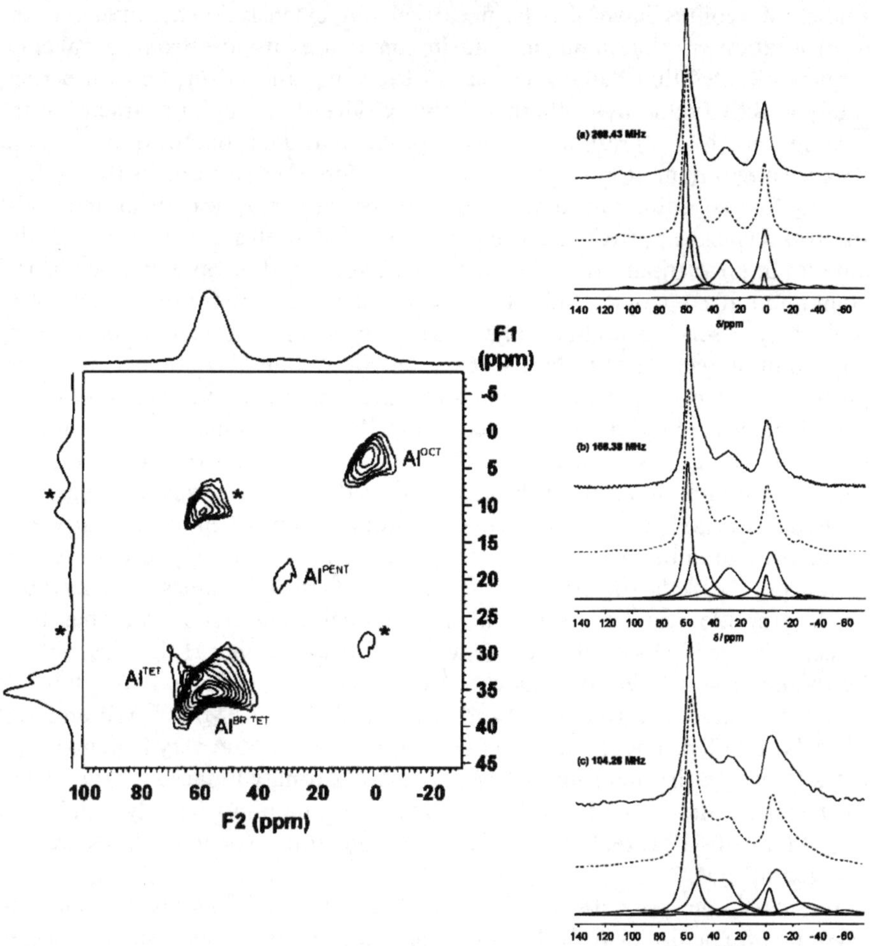

Figure 3.22 High-resolution multiple quantum (MQ) ^{27}Al MAS NMR of steamed zeolite Y, collected at 600 MHz, with satellite peaks asterisked (left) and 1D ^{27}Al MAS NMR of the same sample collected at three different fields (right). Taken together, these spectra indicate the presence of aluminium in octahedral (Al^{OCT}), five-fold (Al^{PENT}) and two kinds of tetrahedral (Al^{TET}, $Al^{BR\ TET}$) coordination. The two different tetrahedrally coordinated sites are best resolved at lower field in the 1D MAS NMR: their different lineshapes indicate different site symmetries. [Reproduced from reference 118 with permission. Copyright 2001 American Chemical Society.]

cation is shown by NMR (^{11}B, I = 3/2) to adopt either tetrahedral or trigonal geometry in silicates. A series of elegant experiments by Fild *et al.*[122–124] show, for example, that in a protonated boron-containing zeolite the bridging hydroxyl oxygen moves away from the boron atom, leaving it trigonally coordinated, with a characteristic change in the NMR lineshape. The crystal chemistry of boron zeolites is discussed in more detail in Chapter 6.

The final component of the zeolite framework that has been studied by NMR is the lattice oxygen. ^{17}O is quadrupolar, I = 5/2, with low natural abundance (0.04%). It frequently adopts highly asymmetric environments, so its study requires specific quadrupolar techniques for high resolution. For example, dynamic angle spinning experiments on a pure silica zeolite Y have resolved all four crystallographically distinct oxygen atoms.[125] MQ MAS NMR also enables resolution to be improved, and to resolve Si-O-Al and Si-O-Si environments. Such studies suggest the ^{17}O chemical shift correlates with bridging T-Ô-T angle in a similar way to the ^{29}Si chemical shift correlation with bridging angle described earlier.

^{1}H MAS NMR is of particular use in the study of the types of proton present in zeolites. Considering the protons of hydroxyl groups of dehydrated zeolites, where the protons are relatively dilute, the signals occur over a relatively narrow chemical shift range (1–5 ppm). Silanol hydroxyls resonate at around 1.3–2.3 ppm and groups on extra-framework aluminium are observed at 2.6–3.6 ppm. The bridging hydroxyl groups give signals over the narrow 3.8–4.5 ppm range, so that in many cases it is not possible to distinguish between protons in different crystallographic environments. It is possible, however, to distinguish between protons at their two crystallographic sites in zeolite Y, which resonate at 3.8–4.4 ppm (O_1) and *ca.* 5 ppm (O_3).

Solid state NMR has also been performed on charge-balancing cations in extra-framework positions, such as lithium, sodium and ammonium, primarily to investigate whether they are located in different sites. For as-prepared, templated solids, ^{13}C MAS NMR is used to determine whether alkylammonium cations in the original syntheses are retained intact in the crystalline solid. Typically, the solid state NMR of the alkylammonium chloride or bromide is measured for comparison (see Figure 3.23 as an example[126]). For these purposes ^{1}H-^{13}C cross polarisation is commonly used, because the signal intensity is strongly increased in this way. ^{1}H MAS NMR of these systems is also possible, but usually gives broad lines because the very strong ^{1}H-^{1}H homonuclear dipolar coupling present in such systems can only be removed by very fast spinning or advanced pulse techniques.

3.4.1.2.6 Mixed Coordination Silicates. As described in the second chapter, there is a small group of microporous metallosilicates in which the metals possess non-tetrahedral coordination. ETS-10 is the best known of these, in which tetrahedral silicon has two types of chemical environments, $Si(OSi)_3(OTi)$ and $Si(OSi)_4$. The titanosilicate end-member is able to take up a range of trivalent cations into the framework upon synthesis, including boron, aluminium and gallium.[127] Solid state NMR reveals the nature of this substitution:

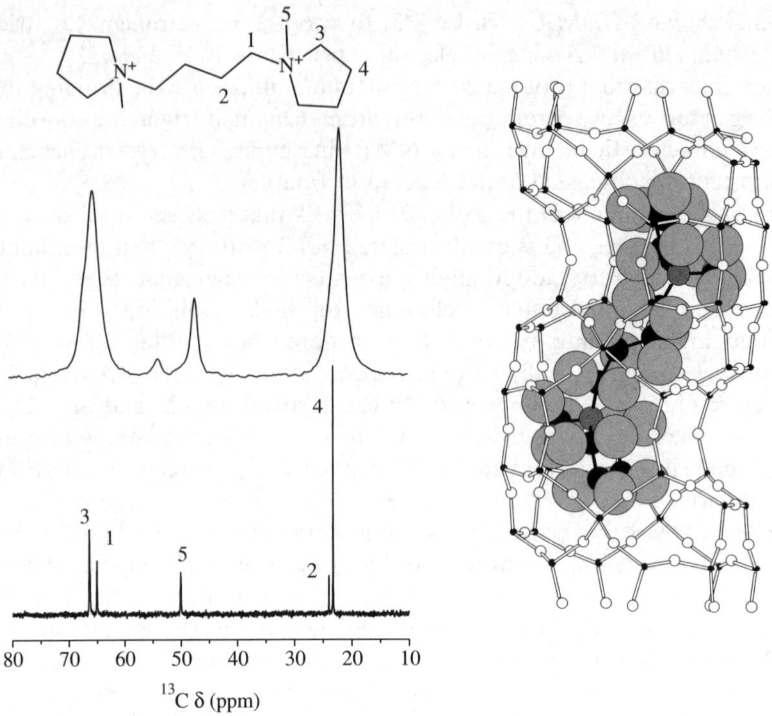

Figure 3.23 ^{13}C MAS NMR (left, middle) of the bis N-methylpyrrolidiniumbutane cation templating the synthetic stilbite TNU-10 (modelled position shown right). The MAS NMR spectrum is compared with the solution spectrum of the cation (shown left, below). [Reproduced from reference 126 with permission. Copyright 2004 American Chemical Society.]

that is that the trivalent cations only substitute into silicon sites without titanium next nearest neighbours, in a variation of Loewenstein's principle (in this case M(III)-Ti avoidance).

3.4.1.3 Aluminophosphates

Aluminophosphates are highly amenable to MAS NMR: as well as being ideally suited for study by ^{27}Al NMR (I = 5/2, 100% abundant), MQ MAS ^{27}Al NMR[128] and ^{31}P MAS NMR ($I = \frac{1}{2}$, 100% abundant) the substitutional chemistry of silicon and non-paramagnetic metal cations into tetrahedral sites in the framework can be studied by the technique.

For fully connected, unsubstituted framework aluminophosphates, $AlPO_4$, all phosphorus is tetrahedral and surrounded by four aluminium second nearest neighbours, $P(OAl)_4$, giving ^{31}P peaks in the chemical shift region – 20 to –40 ppm. Any hydrogenphosphate groups that might be present, for example in interrupted frameworks or layered aluminophosphates, are readily identified on the basis of their more positive chemical shift and their enhanced response to ^1H-^{31}P cross polarisation MAS NMR. For framework aluminophosphates, ^{31}P MAS NMR can in favourable cases be used in a quantitative way to identify different crystallographic sites if their signals can be resolved. The large-pore aluminophosphate VPI-5, for example, was originally solved as a tetrahedral framework with two distinct tetrahedral phosphorus sites, in the ratio of 2:1, but the ^{31}P NMR indicated that there were really three sites, in the ratio 1:1:1, so that the structure needed to be revised. Complementary ^{27}Al NMR indicated the presence of octahedral as well as tetrahedral aluminium, and the final crystallographic structure had to agree with these features. Furthermore, it was possible by comparing ^{31}P-^{31}P dipolar interactions measured by NMR (using different rates of spinning) with the crystal structure to assign the three ^{31}P resonances to particular sites.[129]

^{27}Al MAS NMR has revealed the important tendency of aluminium in $AlPO_4$s to increase coordination without leaving the framework. For as-prepared solids, the coordination can be increased to five or six by coordinated water or hydroxyl or fluoride ions, which may be terminal or linking. These anions can be removed by calcination, together with organic templates, giving fully tetrahedral networks. Upon calcinations, $AlPO_4$s become fully tetrahedrally coordinated, but some of their framework aluminium cations coordinate water molecules upon rehydration, which is readily observed by ^{27}Al NMR. Hydrated calcined VPI-5, for example, has three crystallographically distinct aluminium sites, one of which is octahedrally coordinated, with two water molecules in its coordination sphere.[130]

High resolution ^{27}Al MAS NMR spectra, including Multiple Quantum MAS experiments, can readily resolve Al with different coordination in aluminophosphates, and in many cases, resolves crystallographically different sites with the same coordination. The ^{27}Al MAS NMR of $AlPO_4$-14 in the as-prepared templated form is shown in Figure 3.24.[131] As-prepared $AlPO_4$-14 is known from single crystal diffraction to possess aluminium cations in tetrahedral, five-fold and octahedral coordination. ^{27}Al MAS NMR gives rather broad signals with quadrupolar lineshapes, with the five-fold resonance scarcely visible, but the isotropic projection of the MQ MAS NMR easily resolves the four sites. Once calcined and dehydrated the structure becomes fully tetrahedral, as shown by the MAS NMR and (with higher resolution in the isotropic projection) by MQ MAS NMR (Figure 3.24), so that all four distinct sites are readily distinguished.

Heteronuclear NMR spectroscopy – ^{27}Al and ^{31}P MAS NMR correlation spectroscopy – in which dipolar (CP, TEDOR experiments) or J-coupling (the INEPT experiment) is used to determine the correlation between different Al and P sites,[132–134] can assist in the NMR crystallography of aluminophosphates. For $AlPO_4$-14, for example, Wiench and Pruski[135] describe a series of NMR experiments that use polarisation transfer from the quadrupolar ^{27}Al to spin-$\frac{1}{2}$ ^{31}P *via* J-coupling to confirm the connectivity between aluminium and phosphorus sites in the structure.

The open framework aluminophosphate $AlPO_4$-14A containing isopropylamine as a template (originally an impurity in the $AlPO_4$-14 synthesis) has also

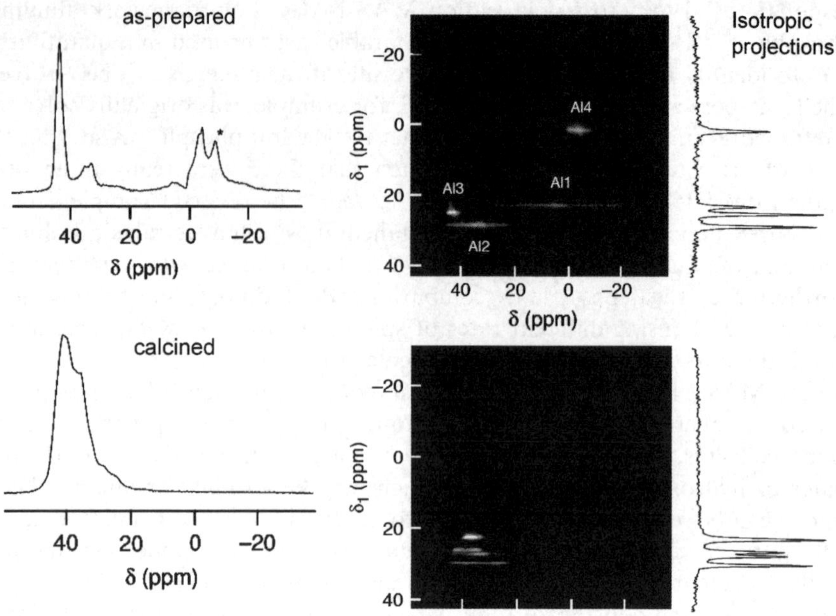

Figure 3.24 MAS NMR (left) and sheared MQ MAS NMR (right) of $AlPO_4$-14 in the as-prepared form (above) and when calcined and dehydrated (below). [Reproduced from reference 131 with permission. Copyright 2006 American Chemical Society.]

been studied in depth by 2 D NMR techniques by the group of Fyfe.[136] The structure is known from single crystal diffraction, with four different P and four different Al sites, and has proven to be an excellent model system to develop and assess the NMR methods. Unlike most of the $AlPO_4$s described in this text, $AlPO_4$-14A does not have a tetrahedrally connected network, either in the as-prepared or the calcined form, but instead contains tetrahedral and octahedral aluminium cations, Al-OH-Al linkages and aluminium cations linked *via* oxygens to five phosphorus atoms. In this work, ^{27}Al MQ MAS NMR enables one octahedral and three tetrahedral signals to be resolved; ^{1}H(^{27}Al) TRAPDOR and $^{1}H \rightarrow {}^{27}Al$ TEDOR enable the assignment of the OH proton signal and its location between Al(1) and Al(4); ^{1}H MAS NMR indicates the amine is protonated and $^{1}H \rightarrow {}^{27}Al$ TEDOR and $^{1}H \rightarrow {}^{31}P$ CP correlation indicates that the template nitrogen is located near the Al(1)-OH-Al(4) cluster and P(4); $^{27}Al \rightarrow {}^{31}P$ 2 D INEPT experiment unambiguously identifies a resolved ^{31}P signal in the spectrum. This study underlines what details of connectivity and internuclear distances can now be obtained from complex but well-ordered aluminophosphate structures.

The substitutional chemistry of aluminophosphates by non-paramagnetic cations, particularly Mg and Zn, can be examined by ^{31}P NMR. Typically, divalent cations (M = Mg, Zn ...) substitute only for aluminium, and the framework aluminium remains tetrahedral (as shown for STA-2). For MAPOs, therefore, substitution results in $P(OM)_n(OAl)_{4-n}$ environments with different chemical shifts (separated by about 7 ppm for MgAPO-20). Framework M/Al ratios can be calculated from ^{31}P NMR by using a related approach to that taken to determine framework Si/Al ratios in zeolites, and assuming that peaks are separated only on the basis of the number of second nearest neighbours of type M:

$$\left(\frac{M}{Al}\right)_{NMR} = \frac{\sum_{n=0}^{4} I_{P(nAl)}}{\sum_{n=0}^{4} 0.25nI_{P(nAl)}} - 1$$

where $I_{P(nAl)}$ is the intensity of the NMR signal of the P(nOAl, (4-n) OM) unit.

Also,

$$\left(\frac{M}{P}\right)_{NMR} = \frac{\sum_{n=0}^{4} \frac{(4-n)}{4} I_{P(nAl)}}{\sum_{n=0}^{4} I_{P(nAl)}}$$

Studies of this type have shown that incorporation of magnesium into the CHA aluminophosphate structure can be random, within the 'Al' site.[137] In the aluminophosphate form of sodalite, (AlPO-20) magnesium substitutes in an ordered way for aluminium, and Barrie and Klinowski interpret the observed ^{31}P NMR intensities in terms of an ordered distribution where there are never two magnesium cations in the same 4MR.[138]

Figure 3.25 shows the ^{27}Al MQ NMR of the as-prepared aluminophosphate fluoride and magnesioaluminophosphate forms of STA-2. The framework requires a negative charge to balance the charge of the diquinuclidinium cation that templates the structure: in the aluminophosphate fluoride form this is supplied by fluoride ions coordinated to the aluminium, whereas in the MgAPO form the negative charge is supplied by the substitution of divalent Mg^{2+} for Al^{3+} in the framework. The framework structure has two crystallographically distinct tetrahedral sites – these are well resolved for both materials by the MQ NMR. The additional peak in the NMR spectrum of the aluminophosphate at 10 ppm is consistent with aluminium in higher coordination, with additional fluoride coordinated.

The substitutional chemistry of silicon into $AlPO_4$s is more complex. Whether silicon can replace phosphorus directly or forms aluminosilicate islands is readily observed by ^{29}Si NMR, *via* measurement of the distribution and chemical shift of ^{29}Si resonances. This can clearly be seen by a comparison of samples of SAPO-34 and SAPO-18 described by Chen *et al.*[139] (Figure 3.26). Although the structures are very closely related, silicon is found to substitute exclusively for P in the SAPO-34 samples (single peak, $\delta^{29}Si = -92$ ppm) whereas in SAPO-18 it both replaces phosphorus ($\delta^{29}Si = -92$) and also forms aluminosilicate islands ($\delta^{29}Si = -111$, –105, –100, –96 ppm corresponding to

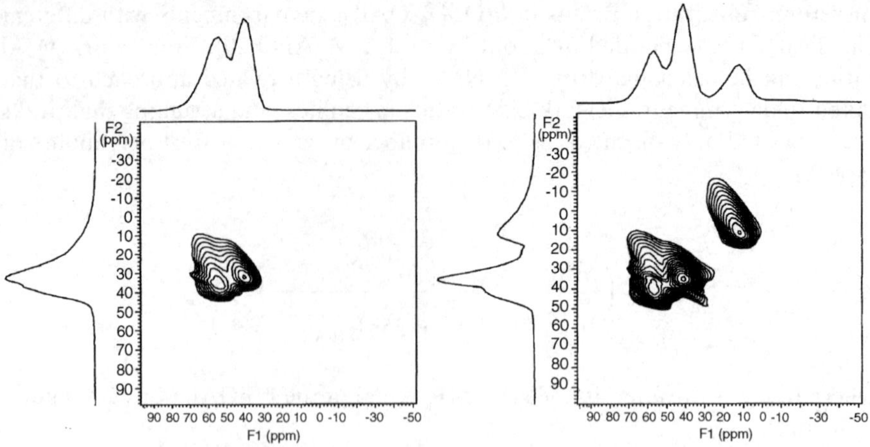

Figure 3.25 ^{27}Al MQ MAS NMR of as-prepared MgAPO (left) and $AlPO_4$ (right) versions of STA-2 (framework type SAT). In the MgAPO, only tetrahedral aluminium (in two crystallographically distinct sites) is observed (30–40 ppm) whereas in the $AlPO_4$, aluminium exhibits in both tetrahedral and higher (δ 12 ppm) coordination. The higher coordination is attributed to coordination by F^- or OH^- ions that are required for charge balance of the positively charged template.

Si in $Si(OSi)_n(OAl)_{4-n}$ environments, with n = 4 to 1). In another well-characterised example, SAPO-35, a small-pore aluminophosphate with two different tetrahedral sites, incorporation of silicon at low levels ($Si/(Al+Si+P) < 0.1$) is found to be preferred in one of the sites. At higher silicon levels aluminosilicate islands are formed.[140]

Solid state NMR is similarly informative for other microporous phosphates, such as those of gallium. For example in the gallophosphate cloverite, resonances at shifts of δ –9.7 and –11.2 can be attributed to $P(OGa)_4$ species, whereas a signal at –0.8 ppm is assigned to $P(OGa)_3OH$ species,[141] confirming the interrupted nature of the framework (Section 2.5.2). For those phosphates containing paramagnetic ions, such as iron, cobalt or nickel, however, the signals from phosphorus atoms close to paramagnetic cations are very greatly broadened (due to rapid relaxation) and become invisible to NMR. Notably, $AlPO_4$s containing paramagnetic complexes of transition metals in their pores can give well-resolved spectra.[11]

3.4.1.4 Organic-inorganic Hybrids

Microporous metal phosphonates have been studied by NMR in order to obtain crystallographic information and to determine the response of coordination geometry to dehydration and adsorption. The method is well suited, for example, to aluminium phosphonates, in which aluminium is able to exhibit a range of coordination geometries within the framework. In aluminium methylphosphonates -α and -β, for example, aluminium adopts tetrahedral and

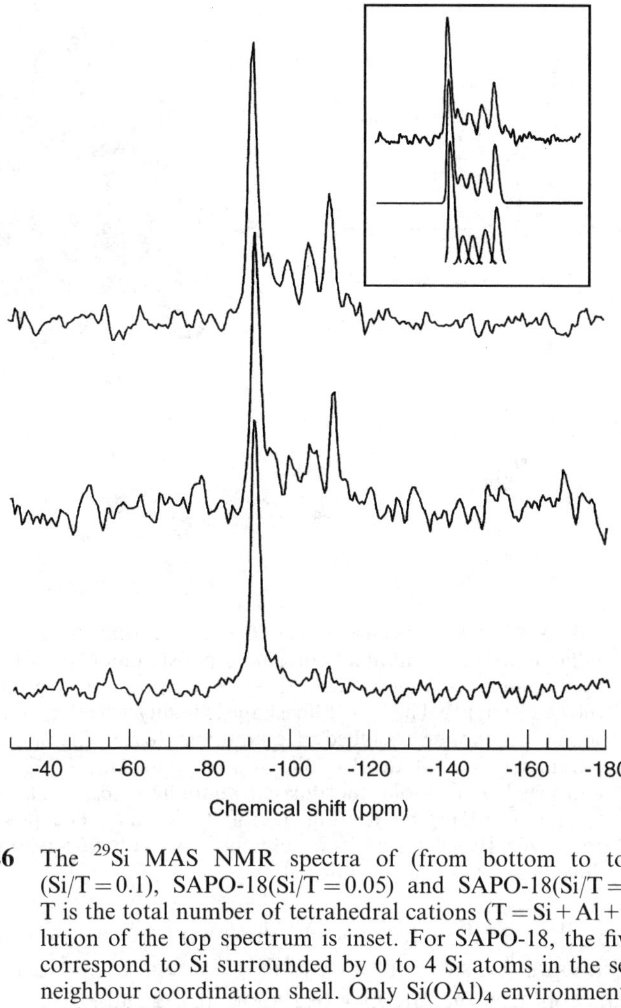

Figure 3.26 The ^{29}Si MAS NMR spectra of (from bottom to top) SAPO-34 (Si/T = 0.1), SAPO-18(Si/T = 0.05) and SAPO-18(Si/T = 0.09), where T is the total number of tetrahedral cations (T = Si + Al + P). Deconvolution of the top spectrum is inset. For SAPO-18, the five resonances correspond to Si surrounded by 0 to 4 Si atoms in the second nearest neighbour coordination shell. Only $Si(OAl)_4$ environments are present in the SAPO-34 sample. [Reproduced from reference 139 with permission. Copyright 1994 American Chemical Society.]

octahedral coordination. In the β–polymorph, there are three crystallographically distinct aluminium environments, and this provided an excellent demonstration of the enhanced resolution available from the multiple quantum technique, in which 5Q ^{27}Al NMR was able to resolve signals from each of these sites.[142] In the pillared microporous aluminium ethylenebisphosphonate $Al(OH)(O_3PCH_2CH_2PO_3).H_2O$, the aluminium exhibits octahedral geometry, with one of its coordination sites occupied by the water molecule. This may be removed at low temperature, giving aluminium in five-fold square pyramidal geometry with retention of the overall framework. The dehydration (and rehydration) process is readily followed by *in situ* ^{27}Al MAS NMR (Figure 3.27).[143] The ^{31}P NMR of phosphonates gives similar information to that

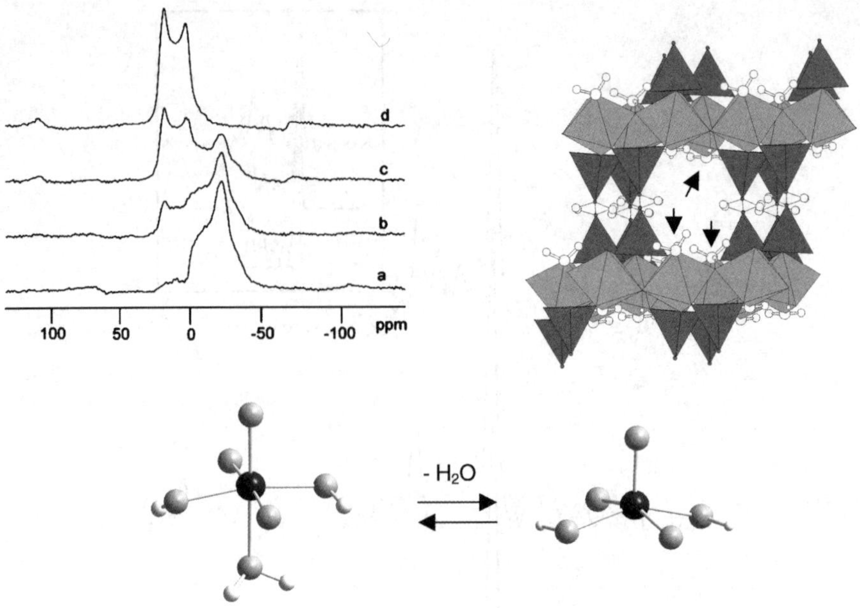

Figure 3.27 ^{27}Al MAS NMR spectra measured *in situ* during the dehydration of a pillared layered aluminium ethylenebisphosphonate, (a – d). The aluminium changes symmetry as the structure loses a coordinated water molecule (right). The NMR lineshape is highly sensitive to the symmetry of the aluminium: octahedral aluminium ($\delta_{iso} = -7.1$ ppm; asymmetry parameter, $\eta = 1.0$; quadrupolar coupling constant, QCC = 5.4 MHz) is converted to five-fold coordinated aluminium ($\delta_{iso} = 20.7$ ppm; $\eta = 0$; QCC = 5.8 MHz) by heating at 130 °C for 1 h. The process is fully reversible. [Reproduced from reference 143 with permission. Copyright 2004 American Chemical Society.]

available from ^{31}P NMR in framework phosphates (*i.e.* resolution between different crystallographic sites; presence of POH groups, *etc.*). ^{13}C NMR on the same materials confirms the integrity of the organic groups.

Although very many carboxylate and amine-based hybrid frameworks have been prepared, solid state NMR has not been used as a characterisation tool in the same way as it has for inorganic zeolitic frameworks. This is probably because they are commonly prepared as crystals suitable for single crystal diffraction, and often with paramagnetic metals. It is likely, however, that NMR will reveal important structural details as these materials are studied more carefully, particularly in monitoring interactions and framework rearrangements upon adsorption of molecules.[144] Studies of framework and sorbate motion by deuterium NMR are also highly relevant here (Chapter 7).

3.4.1.5 *Mesoporous Solids*

Mesoporous silicas, including those with pore windows in the 4–15 Å range, have been characterised extensively by NMR. Indeed, the lack of atomic long range order means that NMR is the most important local structural method for these solids. ^{29}Si NMR gives information on the relative abundances of Q^1–Q^4 silicon environments, because the amorphous silica walls are not fully tetrahedrally connected. This enables the processes of hydrolysis and condensation, described in Chapter 5, to be followed. For periodic organosilicas, in which organically functionalised siloxanes are included into the framework either by co-condensation or by post-synthesis grafting (see Sections 5.7.1 and 6.7.1) the presence of so-called T^n silicon environments (R-Si-$(OSi)_n(OH)_{3-n}$) are detected downfield from the Q^n signals. The number, n, of $(SiO)_n$-Si-RX links made by the tether to the internal surface is given directly by the ^{29}Si chemical shift. Typically, the incorporation of siloxanes during the synthesis gives a higher proportion of T^3 environments than when the siloxanes are incorporated post-synthesis by grafting onto a calcined sample, because it is more likely that the groups will achieve a higher degree of incorporation when the silicate is able to adjust during the synthesis process. For the remarkable benzene-silica of Inagaki, all silicons are of this type.

^{13}C MAS NMR also finds widespread use in the analysis of these solids. The silicas are templated by surfactants, so that NMR can be used to detect the presence of these, for example after attempted extraction by solvent washing. For organosilicas, where organically modified siloxanes are included in the framework, the presence of organic groups, and any subsequent reactions they undergo, can be followed by ^{13}C MAS NMR. In a similar way, it is possible to confirm the presence of such organic groups attached by 'grafting' organosiloxanes to the silanol-covered surface.

Solid-state NMR has been used extensively to identify the environment of cations such as aluminium that are introduced into the silicate walls.[145] When the mesoporous silicate is synthesised under alkaline conditions, for example, aluminium is readily incorporated into tetrahedral sites in the silicate network. Subsequent calcination and hydrothermal treatment can result in some of this aluminium changing coordination from tetrahedral to five- and six-fold, with chemical shift vales of ~ 30 and ~ 0 ppm. Similar studies have been performed on other substituting elements.

3.4.2 X-ray Absorption Spectroscopy (XANES and EXAFS)

X-ray absorption spectroscopy can be applied generally to give information on the local environment of atoms. It is an element-specific technique that gives information on the oxidation state and the number and type of neighbours that surround the absorbing atom. The need for tunable, intense X-radiation means that X-ray absorption spectroscopy is usually restricted to synchrotron X-ray sources.[1]

As the energy of X-rays incident on a sample containing the element of interest is increased, at a certain value (the absorption edge) the energy is sufficient to excite a photoelectron out of a core level to an unbound state and there is a sharp increase in absorption. This is illustrated in Figure 3.28 for the

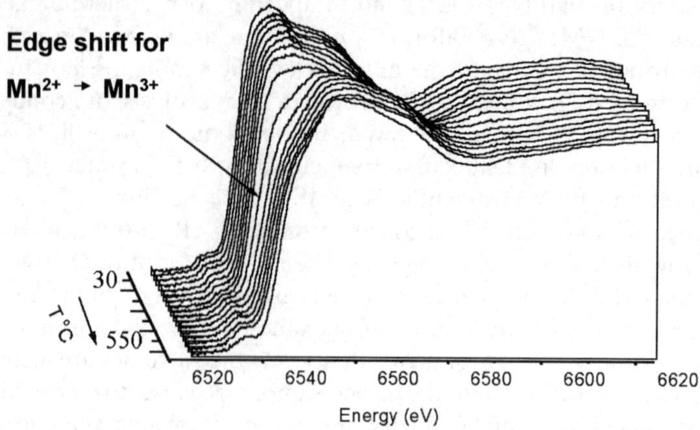

Figure 3.28 Mn K-edge X-ray absorption measurements as MnAPO-18 is calcined in oxygen. The change in edge position edge shift at 450 °C corresponds to the oxidation of framework manganese cations from 2+ to 3+. [Figure courtesy of G. Sankar.]

Mn K-edge of a MnAPO that is being calcined. The structure of the absorption spectrum close to the edge (XANES – X-ray Absorption Near Edge Structure) often gives important information on the local geometry and the oxidation state, so that the Mn K-edge moves to higher energy as the manganese undergoes oxidation. Pre-edge peaks in XANES (at energies less than the edge) result from excitations within energy levels, and can give information on local geometry. The pre-edge peak for tetrahedrally coordinated titanium atoms, for example in Figure 3.29, is a characteristic feature of activated titanosilicate catalysts. Increasing the incident X-ray energy above the edge increases the energy of the emitted photoelectron. At each value of the incident X-ray energy, the emitted photoelectron, acting as a wave, is backscattered by atoms surrounding the X-ray absorbing atom, so that the backscattered electron wave interferes with electron density at the absorbing atom and modulates the absorbance.

An X-ray absorption spectrum, which is obtained by increasing the incident X-ray energy, takes the form of an oscillation that may extend 0.5–1 keV above the edge, superposed on the raised absorption background. This is the Extended X-ray Absorption Fine Structure or EXAFS. The EXAFS spectrum is usually plotted and fitted as a k^3-weighted EXAFS function, where k is the wave vector ($2\pi/\lambda$) of the emitted photoelectron, which magnifies the oscillations at higher energies, giving higher sensitivity to the analysis. Fourier transforming the spectrum (with appropriate phase angles for the backscattering) then gives a representation of the distribution of backscattering atoms around the central absorbing atom. This is illustrated in Figure 3.30 for the Ni EXAFS of nickel complexed in a templing macrocycle in STA-6 and released into the same structure as an extra-framework cation by calcination.

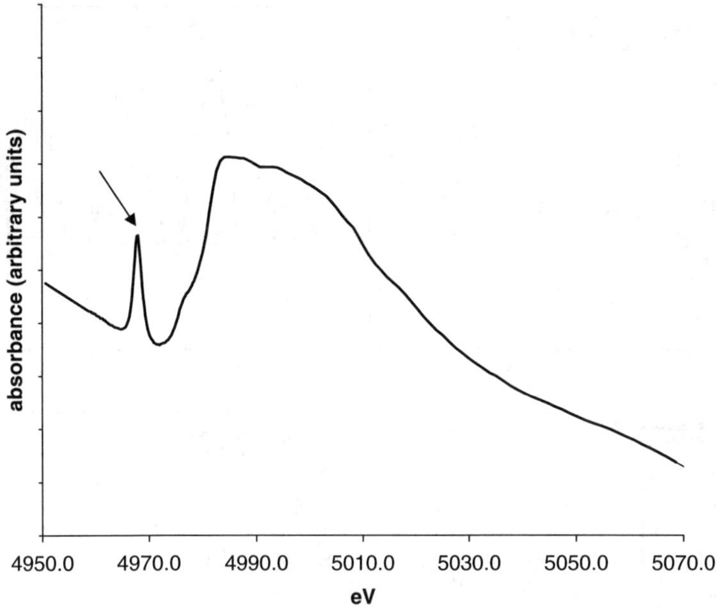

Figure 3.29 Ti K-edge spectra of calcined TS-1. The high pre-edge peak is characteristic of titanium in tetrahedral coordination.

The EXAFS spectrum may be analysed by refining proposed structural models to give important details of local structure that are not readily accessible from X-ray diffraction or other techniques. In particular, it is possible to model the absorption spectrum in terms of the number and types of atoms surrounding the absorbing atom, and the distances between the central and backscattering atom. Bond distances are accurately measured in this process (to 0.02 Å). During this process, the calculated Fourier transforms of measured data are compared with those predicted from the model, as an additional guide to the fit. Nanoparticles of metals or metal oxides are readily studied, because they give characteristic EXAFS spectra which are highly sensitive to particle size. It is also possible to measure elements present at small concentration by performing the experiments in fluorescence, rather than transmission. The technique does not rely on special electronic or magnetic properties, unlike NMR and ESR spectroscopies, and can be performed under variable environments, so that it can be used for *in situ* catalytic and sorption studies. For reliable results, it does require that all of the atoms of the element under study have similar environments – heterogeneity of sites is very difficult to treat unambiguously in the analysis.

Descriptions of the theory and application of X-ray absorption, and XANES and EXAFS spectroscopy, are given in excellent texts,[146,147] and detailed examples of their application to microporous solids in the review of Bordiga *et al.*[148]

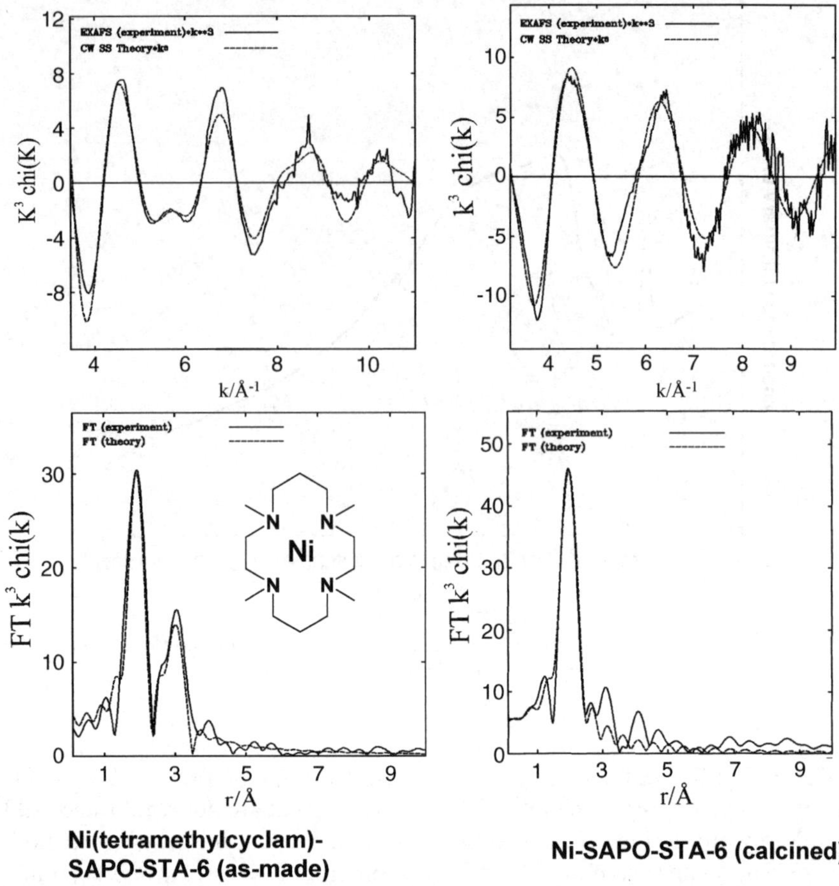

Figure 3.30 Ni K-edge k^3-weighted, background subtracted , EXAFS spectra (above) and their Fourier transforms (below) of (left) Ni(cyclam)-SAPO STA-6, as made, and (right) calcined to give Ni-SAPO STA-6, right. (The Fourier transform distances are not the atom–atom distances between shells, because of phase shifts in the scattering process.)

3.4.2.1 EXAFS of Elements in the Framework

The study of the environment of transition metal cations that substitute for aluminium in the framework of aluminophosphates is a good example of the application of EXAFS spectroscopy. Extensive studies of such materials are able to follow redox cycling between the divalent and trivalent forms of manganese and cobalt in the structure.[149] Mn^{2+} and Co^{2+} are oxidised to Mn^{3+} and Co^{3+} in tetrahedral coordination in these materials upon calcinations of the as-prepared solid, so the absorption edge is shifted to higher energies and the metal-oxygen bond lengths decrease. This edge shift is illustrated in Figure 3.28 for the calcination of MnAPO-18. Careful analysis of the

EXAFS spectra shows that the average bond length decreases from 2.0 Å to 1.86 Å. For cobalt in CoAPO-18 the average Co-O bond length decreases from 1.95 Å to 1.83 Å upon oxidation, and during subsequent reduction of the cobalt in hydrogen gas, the bond length increases again, to 1.90 Å, and bridging hydroxyl groups (Co-OH-P) are observed by infrared spectroscopy.

The method has also been applied highly successfully to the study of the environment of framework titanium(IV) in the catalytically active titanium silicates (Chapters 2 and 9). For the as-prepared and the de-templated structures of titanosilicalite-1, extensive study by X-ray absorption spectroscopy, NMR and IR has confirmed that titanium occupies a framework site and does not leave it upon calcination. In the calcined state, the XANES spectra are characterised by an intense pre-edge peak at 4967 eV (Figure 3.29) resulting from 1s → 3pd electronic transitions of titanium in tetrahedral coordination.[150] (The mixing of p and d orbitals arising in tetrahedral geometry results in the 'forbidden' 1s → 3d transitions becoming allowed, whereas the 1s → 3d transition is forbidden in octahedral geometry.) The tetrahedrally coordinated titanium readily coordinates ligands such as water or ammonia to become octahedral, whereupon the pre-edge peak decreases. In fact, recent studies suggest the situation is a little more complex. Notably, as-prepared silicalite samples without titanium already possess, in addition to a majority of $Si(OSi)_4$ silicon sites, internal defect sites where missing silicon atoms are replaced by nests of hydroxyls attached to neighbouring silicons. These silicon vacancy sites also appear to be clustered and associated with particular crystallographic sites. On the basis of EXAFS studies of samples including titanium, which suggest that the titanium in as-prepared, dehydrated TS-1 samples can be included up to the normal silicon vacancy level and has a coordination number around 4.5, Prestipino *et al.* suggest that in TS-1 titanium is incorporated in these vacancy sites upon synthesis, and that some of this exists in five- or six-fold coordinated environments. In the de-templated solids the titanium is the active site for selective oxidations, and is active because the titanium can expand its coordination from tetrahedral (dehydrated framework) to accommodate the hydrogen peroxide reactant in its local coordination environment to give five- or six-fold coordination, and this can be observed by careful analysis of the EXAFS spectra (Chapter 7).

Most recently, aluminium EXAFS has been used to investigate the local geometry of aluminium atoms in zeolites. Aluminium is difficult to study by EXAFS because the X-ray energies at the edge are quite low and easily absorbed by any form of matter, including gases. Nevertheless, two groups have been able to show that the geometry of aluminium associated with zeolitic acid sites is different from that in non-protonated samples.[151,152] It appears that one of the four Al-O bonds is significantly longer than the others (at around 1.9 Å), in remarkable agreement with the geometry inferred from NMR studies which suggest the bond to the oxygen atom of the SiOH group is lengthened (see Section 8.4.1). Similar studies also indicate that dehydroxylation of the H-form of zeolites above 400 °C leads to trigonal aluminium, presumably still within the framework.

EXAFS of the trivalent metals Fe and Ga substituted into zeolitic silicates show that while these are included into tetrahedral positions, they are removed from the framework upon calcinations. This is seen as a rapid fall-off in the intensity of the first nearest neighbour coordination shell as the metals adopt less regular environments in the pores.

3.4.2.2 *EXAFS of Extra-framework Species*

The coordination of extra-framework cations can readily be determined. Figure 3.30 shows the Ni K-edge k^3-weighted EXAFS spectra and their Fourier transforms for Ni^{2+} within tetramethylcyclam amine acting as a template in SAPO STA-6, and as an extra-framework cation liberated once the organic has been removed by calcination. The first spectrum is matched by nickel surrounded by a first shell of 4 nitrogen atoms at 2.08 Å, 8 carbon atoms at 2.82 Å, four C at 2.96 Å and 2 at 3.35 Å, consistent with the structure of the complex. The second is matched by 6 oxygen atoms at 2.06 Å, where the nickel is coordinated octahedrally by framework oxygen atoms and water molecules.

$$[NiN_4(cyclam)]^{2+} - SAPO \rightarrow Ni^{2+} - SAPO$$

EXAFS is also well suited for the study of finely divided metal (or metal oxide or metal sulfide) clusters supported within the pore structure (see Chapter 6). These particles are readily observed by X-ray spectroscopy, even if they are disordered throughout the solid. Analysis can even determine the average particle size of such clusters, which is of vital importance in catalytic preparation. Typically, for example, platinum supported on zeolites (and other solid acids) is a highly effective catalyst in the reforming of hydrocarbons.

3.4.3 Vibrational Spectroscopy

Infrared spectroscopy is a widely available technique and has been applied extensively in the study of microporous solids. Using Fourier Transform analysis, sensitive detectors and operating either in transmission or in diffuse reflectance (DRIFT) mode, powders can give spectra with high resolution and sensitivity. The method is most valuable when analysing the interaction of molecules with adsorption sites (acid or base) – this is described in Chapters 7 and 8. It does give some structural insights, however, for example on the environment of protons and on the presence of framework and non-framework cations.

A typical infrared spectrum is collected between 400 and 4000 cm^{-1}, which enables most of the fundamental bands from the framework and hydroxyls to be measured. Some spectrometers permit measurement in the near IR (up to 7000 cm^{-1}), so that overtones and combination bands (much weaker than the fundamental resonances) can also be measured. For porous inorganic solids empty of adsorbed species, fundamental hydroxyl stretches are observed in the range 3200–3800 cm^{-1} and framework vibrations are observed in the range 400–1200 cm^{-1}. Organic species present (such as templates or the organic parts of hybrid organic-inorganic solids) give characteristic vibrational modes typical of, for example, C–H, C–C, C–N and C–O bonds.

For microporous solids, the bands from framework vibrations depend on both the structure type and the composition. For zeolites, substitution of framework aluminium results in broadening of these bands and a shift to lower wavenumber, which is attributable to the Al–O bonds being weaker than Si–O bonds. As a result, and once the variation of frequency with composition of a certain framework stretching vibration (symmetric T–O–T or asymmetric T–O–T, for example) has been determined for a given structure, measurement of the stretching frequency can be used as an inexpensive way to estimate the framework composition of a zeolite (*e.g.* for zeolite Y[153]). Examination of framework vibrations has also been used to determine whether added cations substitute for silicon within the tetrahedrally coordinated silicate framework. In some cases, where the local geometry, mass and/or bond strength varies strongly from the silicate, characteristic new bands are observed.[154] The most commonly quoted example is that for isomorphously substituted titanosilicates, for which a band at 960 cm^{-1} has been shown to be quantitatively dependent on framework titanium content by complementary X-ray absorption and Raman spectroscopic studies.[155] Simulation studies reported in the same work indicate that the band is due to the out-of-phase antisymmetric vibration of the four connected Ti–O–Si oscillators – the four Si–O bonds pointing towards the Ti atom. Care should be taken, however, to rule out interference from other structural features that have been shown to give rise to absorbances at closely similar frequencies in some samples. In particular, hydroxyl nest defects at silicon vacancies, $(\text{-Si-OH})_4$, being rather similar in structure, also show a related resonance, although it is observed at a slightly higher wavenumber (975 cm^{-1}). Other framework substitutions (such as Fe) result in similar features.

Vibrational spectroscopy is a very powerful method for the study of hydroxyl species present. For zeolitic solids, these divide into three broad categories:[156] silanol groups, $(SiO)_3SiOH$, hydroxyl groups bound to metal cations such as aluminium, AlOH, and bridging hydroxyls, Si-OH-Al.

Isolated silanol groups are present on the external surfaces or at internal structural defects, and have stretching and in-plane bending vibrations in the ranges 3740–3745 cm^{-1} and 795–835 cm^{-1}, respectively. Defects can occur in as-prepared materials where there are isolated silicon vacancies (often observed for high silica zeolites prepared under alkaline conditions) or structural defects that result from stacking faults (such as those observed at defects in zeolite Beta or ETS-10). Internal silanol groups are also generated by dealumination, for example during 'ultrastabilisation' by steaming (Chapter 6). Hydrogen bonding can occur in clusters of silanol groups, resulting in broad resonances with observed wavenumbers of 3450–3550 cm^{-1}. Hydroxyl groups on aluminium cations at the surface or bound to extra-framework species give stretching bands between 3770 and 3800 cm^{-1}.

Bridging hydroxyls, which are responsible for Brønsted acidity, have stretching and in-plane bending frequencies in the ranges 3250–3660 cm^{-1} and

990–1055 cm^{-1}, respectively. The exact position depends on framework composition and local structural environment. Quantification is possible from calculated extinction coefficients, but is not routinely performed, partly because there is considerable variation in the values of these coefficients in the literature. (^1H NMR is more accurate for measurement of proton concentrations, but there is less resolution between similar species.) There has been much discussion of bridging hydroxyl resonances in zeolites and zeotypes. In zeolite H-Beta, for example, a single band for Brønsted protons is observed at 3614 cm^{-1}, whereas for zeolite H-Y, two bands are observed, at 3554 (sometimes referred to as the low frequency, LF, band) and 3627 cm^{-1} (high frequency, HF) associated with the bridging hydroxyls at oxygens O(3) (in the sodalite cage) and O(1) (in the supercage) that have been observed by neutron diffraction (Figure 3.4).[38] As expected from the structure, the LF band is not affected by the adsorption of molecules such as O_2 and N_2, which cannot gain access to the sodalite cages, but the HF band is perturbed.

Parallel investigations of aluminophosphates show that bridging hydroxyls with similar IR frequencies can be introduced by substitution of silicon for phosphorus.[139,157] In SAPO-34, for example, two bands at 3626 and 3599 cm^{-1} are unambiguously assigned to Brønsted acidic hydroxyls. Comparison of CoAPO-18 and SAPO-34[158] also shows that, in CoAPO-18 at least, IR gives clear evidence of bridging Co-OH-P species, although it should be noted that for many metal-substituted aluminophosphates vibrational modes related to bridging hydroxyls are not observed. To explain these cases, structural schemes that include lattice defects, such as missing lattice oxygens, have been proposed,[159] although the full picture remains unclear.

Raman spectroscopy is another form of vibrational spectroscopy that is subject to different selection rules from IR spectroscopy and therefore complementary to it. Raman spectroscopy has, for example, been used to fingerprint the framework region of zeolites (interpreting spectra in terms of characteristic building units, for example) and to investigate the incorporation of transition metals in the framework, such as titanium.[37,103,155] Raman spectra of titanosilicates give characteristic resonances at 1125 and 960 cm^{-1}, for example.

Inelastic neutron scattering (INS) is a specialised technique of vibrational spectroscopy that is not limited by the selection rules that apply to either IR or Raman spectroscopy. INS is particularly sensitive to vibrations of bonds containing deuterium and, as a result, O-D resonances are readily observed, and bending as well as stretching vibrations of hydroxyl groups can be distinguished.

3.4.4 Other Spectroscopies: UV-visible, Electron Spin Resonance

In addition to NMR and vibrational spectroscopies such as IR and Raman, which are by far the most generally applied, more specific UV-visible, ESR and Mossbauer spectroscopies are of value for particular investigations. Of these,

Mossbauer is the most restricted in applicability, because only a few nuclei are amenable (among them Fe and Sn).

UV-visible spectroscopy, typically measured in diffuse reflectance mode, gives broad resonances from electronic transitions. Examples of its application include the examination of transition metal complexes incorporated within microporous solids, either as structure directing agents or *via* post-synthetic procedures. The spectra indicate the oxidation state and coordination of the transition metal, particularly by reference to known spectra of the complexes. For example, the UV-visible spectra of aluminophosphates such as STA-6 and STA-7 (topology types SAS and SAV) prepared using a nickel cyclam complex as a structure-directing agent show that the nickel remains within the azamacrocycle, in square planar geometry and without axial ligands.[160] This confirms magnetic measurements that indicate that the composite solids are diamagnetic, as expected for a square planar d^8 complex. Investigation of the coordination and oxidation state of transition metals included within the frameworks of microporous solids is also possible by UV-visible spectroscopy although, when access to a synchrotron is possible, XANES and EXAFS spectroscopy is the preferred method for quantitative studies of these systems.

ESR spectroscopy of microporous solids is dominated by the investigation of transition metal cations with unpaired electrons. ESR is then sensitive to the symmetry and geometry of the local cation environment and its changes upon dehydration and adsorption. Cu^{2+} (d^9), Co^{2+} (d^7) and Mn^{2+} (d^5) are examples of suitable cations, and the investigation of their behaviour as charge-balancing cations or coordinated by amines in complexes are good examples of the use of ESR in the study of microporous solids. Schoonheydt and co-workers, for example, have over a number of years examined the geometry of Cu^{2+} in extra-framework sites in zeolites, using *ab initio* calculations on model clusters to interpret ESR spectra in terms of the coordination geometry of the cation sites.[161] ESR has also been used to follow quantitatively the geometry of Co^{2+} in aluminophosphate frameworks upon various treatments, indicating that the cobalt in these structures remains tetrahedral upon calcinations of the as-prepared material but only *ca.* 30% undergoes oxidation to the (ESR invisible) Co^{3+}.[162] Such studies complement EXAFS and UV-visible spectroscopies in building up a full picture of chemical changes upon calcination of AlPOs.[161,163] Complications inherent to ESR include the loss of signal as oxidation or reduction of the cation occurs, and difficulty in attributing signals where independent models of the environments are not available.

3.5 Summary

The study of the structure of microporous solids is both fascinating and important: important because their many uses depend on properties related directly to their structure and fascinating because of the challenge in determining structure on the long and short range in architectures of great chemical variety and geometric beauty. The sodalite cage-based zeolites A and Y are the most important microporous solids from the point of view of applications, and have a rich structural chemistry that has been and continues to be studied in detail by the techniques outlined in this chapter. By comparison, metal organic frameworks have mainly been studied by conventional diffraction, so that enormous scope exists for elucidation of their structure–property relationships by other diffraction and spectroscopy-based methods.

Most of the applications of these porous solids depend on interactions of adsorbed species with the internal surface. This internal surface is in fact the periodic bulk, so that methods that probe the whole material are directly relevant to understanding these interactions, a situation that is quite unlike that for surface-molecule interactions on non-porous metal oxides or metals. Diffraction-based methods give the basic framework structure, including details of framework geometry, pore size and connectivity that determine possible locations for extra-framework species and diffusion rates for adsorbed molecules. In many catalytic applications, however, the active site is best thought of as a type of point defect: a proton on a bridging oxygen, for example, or a titanium atom substituting for silicon in a framework site. In these cases, local spectroscopic probes (such as NMR, EXAFS, *etc.*) are required. Crucially, though, the results of such spectroscopic methods are most usefully interpreted in the context of the overall periodic structure. The likely geometry of titanium in TS-1, for example, must be studied by considering the titanium atoms distributed in some way over the 24 crystallographically distinct sites in the monoclinic MFI structure.

Defects of other types at scales quite different from that of the unit cell can also be important. The surface structure represents both the crystal face during growth and the first structure presented to adsorbing molecules, so that surface study, for example by AFM and HRTEM, can give important clues to these processes. Furthermore, the structure of mesopores generated by post-synthetic treatment, or the size and dispersion of metal particles in bifunctional catalysts are both examples of non-periodic features important in catalysis, so that recently developed 3D TEM methods to study them have particular significance.

The resolution at which microporous materials can be studied is continually improved by developments in analytical methods. X-ray synchrotron facilities enable ever smaller single crystals to be studied, giving improved structural detail over that available from powders, while improvements in resolution of powder diffraction also enable more complex structures to be solved. Synchrotron X-ray sources also enable better time resolution of *in situ* diffraction and EXAFS experiments of relevance in catalysis. Electron crystallography is also coming of age for structure solution, for example of mesoporous solids and of zeolites such as the complex TNU-9. NMR methods of studying quadrupolar nuclei and of determining connectivities all continue to make exciting progress.

That these solids are amenable to so many characterisation methods usually means that a combination of methods will ultimately reveal important structural details. For example, combined IR, UV, Raman and EXAFS together established the nature of titanium substitution into TS-1, whereas the nature of aluminium atoms at acid sites in zeolites has been established by X-ray and neutron diffraction, NMR and most recently EXAFS. Against this background, these structures are ideal for computational studies described in the next chapter. Starting from experimentally verified examples, simulation can extend understanding to those aspects not currently accessible to experimental methods.

References

1. G. Margaritondo, *'Introduction to Synchrotron Radiation'*, Oxford University Press, New York, 1988.
2. W. Clegg, '*Crystal Structure Determination*', Oxford University Press, 1998.
3. M. F. C. Ladd and R. A. Palmer, *'Structure Determination by X-ray Crystallography'*, 4th Ed., Kluwer Academic/Plenum, 2003.
4. G. H. Stout and L. H. Jensen, *'X-ray Structure Determination'*, 2nd Ed., Wiley, New York, 1989.
5. C. Giacovazzo, *'Fundamentals of Crystallography'*, Oxford University Press, New York, 1991.
6. G. M. Sheldrick, *'SHELXL97'*, Gottingen, Germany, 1997.
7. M. C. Burla, R. Caliandro, M. Camalli, B. Carrozzini, G. L. Cascarano, L. De Caro, C. Giacovazzo, G. Polidori and R. Spagna, *J. Appl. Cryst.*, 2005, **38**, 381.
8. P. Lightfoot, D. A. Woodcock, M. J. Maple, L. A. Villaescusa and P. A. Wright, *J. Mater. Chem.*, 2001, **11**, 212.
9. L. A. Villaescusa, P. Lightfoot, S. J. Teat and R. E. Morris, *J. Am. Chem. Soc.*, 2001, **123**, 5453.
10. S. Oliver, A. Kuperman, A. Lough and G. A. Ozin, *J. Mater. Chem.*, 1997, **7**, 807.
11. R. Garcia, I. J. Shannon, A. M. Z. Slawin, W. Z. Zhou, P. A. Cox and P. A. Wright, *Micropor. Mesopor. Mater.*, 2003, **58**, 91.
12. M. Arranz, J. Pérez-Pariente, P. A. Wright, A. M. Z. Slawin, T. Blasco, L. Gomez-Hortiguela and F. Cora, *Chem. Mater.*, 2005, **17**, 4374.
13. C. Baerlocher and L. McCusker, *Zeit. Krist.*, 2004, **219**, 782.
14. A. N. Fitch, *J. Res. Nat. Inst. Stand. Technol.*, 2004, **109**, 133.
15. A. LeBail, H. Duroy and J. L. Fourquet, *Mat. Res. Bull.*, 1988, **23**, 447.
16. A. W. Burton, *Zeit. Krist.*, 2004, **219**, 866.
17. L. B. McCusker, C. Baerlocher, R. Grosse-Kunstleve, S. Brenner and T. Wessels, *Chimia*, 2001, **55**, 497.
18. J. Rius, *Powder Diffr.*, 1999, **14**, 267.
19. M. A. Estermann, L. B. McCusker and C. Baerlocher, *J. Appl. Crystallogr.*, 1992, **25**, 539.
20. P. A. Barrett, M. A. Camblor, A. Corma, R. H. Jones and L. A. Villaescusa, *Chem. Mater.*, 1997, **9**, 1713.
21. A. Burton, S. Elomari, R. C. Medrud, I. Y. Chan, C. Y. Chen, L. M. Bull and E. S. Vittoratos, *J. Am. Chem. Soc.*, 2003, **125**, 1633.
22. D. L. Dorset, G. J. Kennedy, K. G. Strohmaier, M. J. Diaz-Cabanas, F. Rey and A. Corma, *J. Am. Chem. Soc.*, 2006, **128**, 8862.

23. D. E. Akporiaye, H. Fjellvag, E. N. Halvorsen, J. Hustveit, A. Karlsson and K. P. Lillerud, *J. Phys. Chem.*, 1996, **100**, 16641.
24. S. R. Miller, E. Lear, J. Gonzalez, A. M. Z. Slawin, P. A. Wright, N. Guillou and G. Férey, *Dalton Trans.*, 2005, 3319.
25. R. Cerny and V. Favre-Nicolin, *Powder Diffr.*, 2005, **20**, 359.
26. M. Edgar, V. J. Carter, D. P. Tunstall, P. Grewal, V. Favre-Nicolin, P. A. Cox, P. Lightfoot and P. A. Wright, *Chem. Commun.*, 2002, 808.
27. H. M. Rietveld, *J. Appl. Cryst.*, 1969, **2**, 65.
28. C. Giacovazzo, in *'Fundamentals of Crystallography'*, ed. C. Giacovazzo, Oxford University Press, New York, 1992, 61.
29. R. A. Young, *'The Rietveld Method'*, Oxford University Press, New York, 1996.
30. B. Han, C. H. Shin, S. J. Warrender, P. Lightfoot, P. A. Wright, M. A. Camblor and S. B. Hong, *Chem. Mater.*, 2006, **18**, 3023.
31. W. J. Mortier, '*Compilation of Extra-Framework Sites in Zeolites*', Butterworth, 1982.
32. E. Dooryhee, C. R. A. Catlow, J. W. Couves, P. J. Maddox, J. M. Thomas, G. N. Greaves, A. T. Steel and R. P. Townsend, *J. Phys. Chem.*, 1991, **95**, 4514.
33. J. B. Parise, J. A. Hriljac, D. E. Cox, D. R. Corbin and V. Ramamurthy, *J. Chem. Soc., Chem. Commun.*, 1993, 226.
34. C. Eylem, J. A. Hriljac, V. Ramamurthy, D. A. Corbin and J. B. Parise, *Chem. Mater.*, 1996, **8**, 844.
35. P. A. Anderson, A. R. Armstrong and P. P. Edwards, *Angew. Chemie Int. Ed.*, 1994, **33**, 641.
36. K. Moller, M. M. Eddy, G. D. Stucky, N. Herron and T. Bein, *J. Am. Chem. Soc.*, 1989, **111**, 2564.
37. J. E. Readman, I. Gameson, J. A. Hriljac, P. P. Edwards and P. A. Anderson, *Chem. Commun*, 2000, 595.
38. M. Czjzek, H. Jobic, A. N. Fitch and T. Vogt, *J. Phys. Chem.*, 1992, **96**, 1535.
39. L. J. Smith, A. Davidson and A. K. Cheetham, *Cat. Lett.*, 1997, **49**, 143.
40. A. K. Cheetham, M. Eddy and J. M. Thomas, *J. Chem. Soc., Chem. Commun.*, 1984, 1337.
41. A. N. Fitch, H. Jobic and A. Renouprez, *J. Phys. Chem.*, 1986, **90**, 1311.
42. P. A. Wright, J. M. Thomas, A. K. Cheetham and A. K. Nowak, *Nature*, 1986, **318**, 611.
43. V. J. Carter, J. P. Kujanpaa, F. G. Riddell, P. A. Wright, J. F. C. Turner, C. R. A. Catlow and K. S. Knight, *Chem. Phys. Lett.*, 1999, **313**, 505.
44. F. Izumi, *Solid State Ionics*, 2004, **172**, 1.
45. R. Matsuda, R. Kitaura, S. Kitagawa, Y. Kubota, R. V. Belosludov, T. C. Kobayashi, H. Sakamoto, T. Chiba, M. Takata, Y. Kawazoe and Y. Mita, *Nature*, 2005, **436**, 238.
46. G. Muncaster, A. T. Davies, G. Sankar, C. R. A. Catlow, J. M. Thomas, S. L. Colston, P. Barnes, R. I. Walton and D. O'Hare, *Phys. Chem. Chem. Phys.*, 2000, **2**, 3523.

47. R. I. Walton, F. Millange, D. O'Hare, A. T. Davies, G. Sankar and C. R. A. Catlow, *J. Phys. Chem. B,* 2001, 105, 83; R. I. Walton and D.O'Hare, *J. Phys. Chem. B* 2001, **105**, 91.
48. A. T. Davies, G. Sankar, C. R. A. Catlow and S. M. Clark, *J. Phys. Chem. B*, 1997, **101**, 10115.
49. A. N. Christensen, T. R. Jensen, P. Norby and J. C. Hanson, *Chem. Mater.*, 1998, **10**, 1688.
50. C. Livage, F. Millange, R. I. Walton, T. Loiseau, N. Simon, D. O'Hare and G. Férey, *Chem. Commun.*, 2001, 994.
51. P. de Moor, T. P. M. Beelen, R. A. van Santen, K. Tsuji and M. E. Davis, *Chem. Mater.*, 1999, **11**, 36.
52. P. de Moor, T. P. M. Beelen, B. U. Komanschek, L. W. Beck, P. Wagner, M. E. Davis and R. A. van Santen, *Chem.-Eur. J.*, 1999, **5**, 2083.
53. M. W. Anderson, T. Ohsuna, Y. Sakamoto, Z. Liu, A. Carlsson and O. Terasaki, *Chem. Commun.*, 2004, 907.
54. T. Ohsuna, Z. Liu, O. Terasaki, K. Hiraga and M. A. Camblor, *J. Phys. Chem. B*, 2002, **106**, 5673.
55. P. Wagner, O. Terasaki, S. Ritsch, J. G. Nery, S. I. Zones, M. E. Davis and K. Hiraga, *J. Phys. Chem. B*, 1999, **103**, 8245.
56. D. L. Dorset, *Zeit. Krist.*, 2003, **218**, 458.
57. D. L. Dorset, *Zeit. Krist.*, 2003, **218**, 525.
58. D. L. Dorset, *Zeit. Krist.*, 2003, **218**, 612.
59. F. Gramm, C. Baerlocher, L. B. McCusker, S. J. Warrender, P. A. Wright, B. Han, S. B. Hong, Z. Liu, T. Ohsuna and O. Terasaki, *Nature*, 2006, **444**, 79.
60. P. A. Wright, J. M. Thomas, G. R. Millward, S. Ramdas and S. A. I. Barri, *J. Chem. Soc., Chem. Commun.*, 1985, 1117.
61. M. E. Leonowicz, J. A. Lawton, S. L. Lawton and M. K. Rubin, *Science*, 1994, **264**, 1910.
62. C. S. Blackwell, R. W. Broach, M. G. Gatter, J. S. Holmgren, D. Y. Jan, G. J. Lewis, B. J. Mezza, T. M. Mezza, M. A. Miller, J. G. Moscoso, R. L. Patton, L. M. Rohde, M. W. Schoonover, W. Sinkler, B. A. Wilson and S. T. Wilson, *Angew. Chem. Int. Ed.*, 2003, **42**, 1737.
63. M. M. J. Treacy and J. M. Newsam, *Nature*, 1988, **332**, 249.
64. J. M. Newsam, M. M. J. Treacy, W. T. Koetsier and C. B. Degruyter, *Proc. Roy. Soc. London A*, 1988, **420**, 375.
65. M. M. J. Treacy, J. M. Newsam and M. W. Deem, *Proc. Roy. Soc. London A*, 1991, **433**, 499.
66. P. A. Wright, W. Z. Zhou, J. Perez-Pariente and M. Arranz, *J. Am. Chem. Soc.*, 2005, **127**, 494.
67. M. W. Anderson, O. Terasaki, T. Ohsuna, A. Philippou, S. P. Mackay, A. Ferreira, J. Rocha and S. Lidin, *Nature*, 1994, **367**, 347.
68. J. M. Thomas and G. R. Millward, *J. Chem. Soc., Chem. Commun.*, 1982, 1380.
69. G. R. Millward, S. Ramdas, J. M. Thomas and M. T. Barlow, *J Chem. Soc. Farad. Trans. I*, 1983, **79**, 1075.

70. R. F. Lobo and H. van Koningsveld, *J. Am. Chem. Soc.*, 2002, **124**, 13222.
71. P. L. Gai-Boyes, J. M. Thomas, P. A. Wright, R. H. Jones, S. Natarajan, J. S. Chen and R. R. Xu, *J. Phys. Chem.*, 1992, **96**, 8206.
72. O. I. Lebedev, F. Millange, C. Serre, G. Van Tendeloo and G. Férey, *Chem. Mater.*, 2005, **17**, 6525.
73. M. Kruk, M. Jaroniec, Y. Sakamoto, O. Terasaki, R. Ryoo and C. H. Ko, *J. Phys. Chem. B*, 2000, **104**, 292.
74. A. Carlsson, M. Kaneda, Y. Sakamoto, O. Terasaki, R. Ryoo and S. H. Joo, *J. Electron Micr.*, 1999, **48**, 795.
75. Y. Sakamoto, I. Diaz, O. Terasaki, D. Y. Zhao, J. Perez-Pariente, J. M. Kim and G. D. Stucky, *J. Phys. Chem. B*, 2002, **106**, 3118.
76. Y. Sakamoto, M. Kaneda, O. Terasaki, D. Y. Zhao, J. M. Kim, G. Stucky, H. J. Shim and R. Ryoo, *Nature*, 2000, **408**, 449.
77. W. Z. Zhou, H. M. A. Hunter, P. A. Wright, Q. F. Ge and J. M. Thomas, *J. Phys. Chem. B*, 1998, **102**, 6933.
78. A. E. Garcia-Bennett, O. Terasaki, S. Che and T. Tatsumi, *Chem. Mater.*, 2004, **16**, 813.
79. S. Inagaki, S. Guan, T. Ohsuna and O. Terasaki, *Nature*, 2002, **416**, 304.
80. A. H. Janssen, A. J. Koster and K. P. de Jong, *Angew. Chemie Int. Ed.*, 2001, **40**, 1102.
81. A. H. Janssen, A. J. Koster and K. P. de Jong, *J. Phys. Chem. B*, 2002, **106**, 11905.
82. P. A. Midgley, M. Weyland, J. M. Thomas and B. F. G. Johnson, *Chem. Commun.*, 2001, 907.
83. M. Weyland, P. A. Midgley and J. M. Thomas, *J. Phys. Chem. B*, 2001, **105**, 7882.
84. P. A. Midgley, M. Weyland, J. M. Thomas, P. L. Gai-Boyes and E. D. Boyes, *Angew. Chem. Int. Ed.*, 2002, **41**, 3804.
85. J. M. Thomas, P. A. Midgley, T. J. V. Yates, J. S. Barnard, R. Raja, I. Arslan and M. Weyland, *Angew. Chem. Int. Ed.*, 2004, **43**, 6745.
86. S. N. Che, K. Lund, T. Tatsumi, S. Iijima, S. H. Joo, R. Ryoo and O. Terasaki, *Angew. Chem. Int. Ed.*, 2003, **42**, 2182.
87. K. J. Chao and J. Y. Chern, *Zeolites*, 1988, **8**, 82.
88. (a) M. W. Anderson, J. R. Agger, J. T. Thornton and N. Forsyth, *Angew. Chem. Int. Ed.*, 1996, **35**, 1210; (b) R. Singh, J. Doolittle, M. A. George and P. K. Dutta, *Langmuir*, 2002, **18**, 8193.
89. J. R. Agger, N. Hanif and M. W. Anderson, *Angew. Chem. Int. Ed.*, 2001, **40**, 4065.
90. J. R. Agger, N. Hanif, C. S. Cundy, A. P. Wade, S. Dennison, P. A. Rawlinson and M. W. Anderson, *J. Am. Chem. Soc.*, 2003, **125**, 830.
91. M. J. Duer, *in 'Solid State NMR Spectroscopy: principles and applications'*, Blackwell Science, Oxford, 2002.
92. R. K. Harris, E. D. Becker, S. M. Cabral de Menezes, R. Goodfellow and P. Granger, *Solid State Nucl. Mag. Res.*, 2002, **22**, 458.
93. A. Samoson, E. Lippmaa and A. Pines, *Mol. Phys.*, 1988, **65**, 1013.
94. A. Llor and J. Virlet, *Chem. Phys. Lett.*, 1988, **152**, 248.

95. K. T. Mueller, B. Q. Sun, G. C. Chingas, J. W. Zwanziger, T. Terao and A. Pines, *J. Magn. Reson.*, 1990, **86**, 470.
96. L. Frydman and J. S. Harwood, *J. Am. Chem. Soc.*, 1995, **117**, 5367: A. Medek, J. S. Harwood and L. Frydman, J. Am. Chem. Soc. 1995, **117** 12779.
97. J. Klinowski, T. A. Carpenter and J. M. Thomas, *J. Chem. Soc., Chem. Commun.*, 1986, 956.
98. C. A. Fyfe, G. C. Gobbi, J. Klinowski, J. M. Thomas and S. Ramdas, *Nature*, 1982, **296**, 530.
99. C. A. Fyfe, J. H. O'Brien and H. Strobl, *Nature*, 1987, **326**, 281; C.A. Fyfe *et al.*, *J. Am. Chem. Soc.*, 1988, **110**, 3373.
100. J. M. Thomas, J. Klinowski, S. Ramdas, B. K. Hunter and D. T. B. Tennakoon, *Chem. Phys. Lett.*, 1983, **102**, 158.
101. G. Engelhardt and R. Radeglia, *Chem. Phys. Lett.*, 1984, **108**, 271.
102. R. Radeglia and G. Engelhardt, *Chem. Phys. Lett.*, 1985, **114**, 28.
103. S. Ramdas and J. Klinowski, *Nature*, 1984, **308**, 521.
104. L. A. Villaescusa and M. A. Camblor, *Rec. Res. Devel. Chem.*, 2003, **1**, 93.
105. J. Klinowski, S. Ramdas, J. M. Thomas, C. A. Fyfe and J. S. Hartman, *J. Chem. Soc. Farad. Trans. I*, 1982, **78**, 1025.
106. G. Engelhardt, U. Lohse, E. Lippmaa, M. Tarmak and M. Magi, *Zeit. Anorg. Allg. Chemie*, 1981, **482**, 49.
107. J. Klinowski and M. W. Anderson, *J. Chem. Soc. Farad. Trans. I*, 1986, **82**, 569.
108. C. A. Fyfe, Y. Feng, H. Grondey, G. T. Kokotailo and H. Gies, *Chem. Rev.*, 1991, **91**, 1525.
109. C. A. Fyfe, H. Grondey, K. T. Mueller, K. C. Wongmoon and T. Markus, *J. Am. Chem. Soc.*, 1992, **114**, 5876.
110. C. A. Fyfe, K. T. Mueller, H. Grondey and K. C. Wongmoon, *J. Phys. Chem.*, 1993, **97**, 13484.
111. C. A. Fyfe, K. C. Wongmoon, Y. Huang, H. Grondey and K. T. Mueller, *J. Phys. Chem.*, 1995, **99**, 8707.
112. C. A. Fyfe, Y. Feng, H. Gies, H. Grondey and G. T. Kokotailo, *J. Am. Chem. Soc.*, 1990, **112**, 3264.
113. C. A. Fyfe, H. Grondey, Y. Feng and G. T. Kokotailo, *J. Am. Chem. Soc.*, 1990, **112**, 8812.
114. D. H. Brouwer, P. E. Kristiansen, C. A. Fyfe and M. H. Levitt, *J. Am. Chem. Soc.*, 2005, **127**, 542.
115. D. H. Brouwer, R. J. Darton, R. E. Morris and M. H. Levitt, *J. Am. Chem. Soc.*, 2005, **127**, 10365.
116. C. A. Fyfe, D. H. Brouwer, A. R. Lewis, L. A. Villaescusa and R. E. Morris, *J. Am. Chem. Soc.*, 2002, **124**, 7770.
117. C. P. Grey and A. J. Vega, *J. Am. Chem. Soc.*, 1995, **117**, 8232.
118. C. A. Fyfe, J. L. Bretherton and L. Y. Lam, *J. Am. Chem. Soc.*, 2001, **123**, 5285.
119. J. Klinowski, M. W. Anderson and J. M. Thomas, *J. Chem. Soc., Chem. Commun.*, 1983, 525.

120. P. Sarv, C. Fernandez, J. P. Amoureux and K. Keskinen, *J. Phys. Chem.*, 1996, **100**, 19223.
121. C. R. Bayense, A. P. M. Kentgens, J. W. Dehaan, L. J. M. Vandeven and J. H. C. Vanhooff, *J. Phys. Chem.*, 1992, **96**, 775.
122. C. Fild, H. Eckert and H. Koller, *Angew. Chem. Int. Ed.*, 1998, **37**, 2505.
123. C. Fild, D. F. Shantz, R. F. Lobo and H. Koller, *Phys. Chem. Chem. Phys.*, 2000, **2**, 3091.
124. H. Koller, C. Fild and R. F. Lobo, *Micropor. Mesopor. Mater.*, 2005, **79**, 215.
125. L. M. Bull, A. K. Cheetham, T. Anupold, A. Reinhold, A. Samoson, J. Sauer, B. Bussemer, Y. Lee, S. Gann, J. Shore, A. Pines and R. Dupree, *J. Am. Chem. Soc.*, 1998, **120**, 3510.
126. S. B. Hong, E. G. Lear, P. A. Wright, W. Z. Zhou, P. A. Cox, C. H. Shin, J. H. Park and I. S. Nam, *J. Am. Chem. Soc.*, 2004, **126**, 5817.
127. M. W. Anderson, J. Rocha, Z. Lin, A. Philippou, I. Orion and A. Ferreira, *Micropor. Mater.*, 1996, **6**, 195.
128. C. Fernandez and J. P. Amoureux, *Chem. Phys. Lett.*, 1995, **242**, 449.
129. W. Kolodziejski, H. Y. He and J. Klinowski, *Chem. Phys. Lett.*, 1992, **191**, 117.
130. L. B. McCusker, C. Baerlocher, E. Jahn and M. Bulow, *Zeolites*, 1991, **11**, 308.
131. S. Antonijevic, S. E. Ashbrook, S. Biedasek, R. I. Walton, S. Wimperis and H. X. Yang, *J. Am. Chem. Soc.*, 2006, **128**, 8054.
132. L. Delevoye, C. Fernandez, C. M. Morais, J. P. Amoureux, V. Montouillout and J. Rocha, *Solid State Nucl. Mag. Reson.*, 2002, **22**, 501.
133. C. Fernandez, C. Morais, J. Rocha and M. Pruski, *Solid State Nucl. Mag. Reson.*, 2002, **21**, 61.
134. C. A. Fyfe, H. M. Z. Altenschildesche, K. C. Wong-Moon, H. Grondey and J. M. Chezeau, *Solid State Nucl. Mag. Reson.*, 1997, **9**, 97.
135. J. W. Wiench and M. Pruski, *Solid State Nucl. Mag. Res.*, 2004, **26**, 51.
136. D. H. Brouwer, J. M. Chezeau and C. A. Fyfe, *Micropor. Mesopor. Mater.*, 2006, **88**, 163.
137. F. Deng, Y. Yue, T. C. Xiao, Y. U. Du, C. H. Ye, L. D. An and H. L. Wang, *J. Phys. Chem.*, 1995, **99**, 6029.
138. P. J. Barrie and J. Klinowski, *J. Phys. Chem.*, 1989, **93**, 5972.
139. J. S. Chen, P. A. Wright, J. M. Thomas, S. Natarajan, L. Marchese, S. M. Bradley, G. Sankar, C. R. A. Catlow, P. L. Gaiboyes, R. P. Townsend and C. M. Lok, *J. Phys. Chem.*, 1994, **98**, 10216.
140. H. J. Jung, C. H. Shin and S. B. Hong, *J. Phys. Chem. B*, 2005, **109**, 20847.
141. S. M. Bradley, R. F. Howe and J. V. Hanna, *Solid State Nucl. Mag. Reson.*, 1993, **2**, 37.
142. J. Rocha, Z. Lin, C. Fernandez and J. P. Amoureux, *Chem. Commun.*, 1996, 2513.
143. R. N. Devi, P. Wormald, P. A. Cox and P. A. Wright, *Chem. Mater.*, 2004, **16**, 2229.
144. T. Loiseau, C. Serre, C. Huguenard, G. Fink, F. Taulelle, M. Henry, T. Bataille and G. Férey, *Chem. Eur. J.*, 2004, **10**, 1373.

145. M. T. Janicke, C. C. Landry, S. C. Christiansen, D. Kumar, G. D. Stucky and B. F. Chmelka, *J. Am. Chem. Soc.*, 1998, **120**, 6940.
146. D. C. Koningsberger and R. Prins, *in 'X-ray Absorption'*, Wiley, New York, 1988.
147. J. W. Niemantsverdriet, *'Spectroscopy in Catalysis'*, VCH, Weinheim, 1993.
148. S. Bordiga, C. Prestipino, F. Bonino, A. Damin, E. Groppo and C. Lamberti, *Phys. Chem. Chem. Phys.*, in the press.
149. J. M. Thomas, G. N. Greaves, G. Sankar, P. A. Wright, J. S. Chen, A. J. Dent and L. Marchese, *Angew. Chem. Int. Ed.*, 1994, **33**, 1871.
150. C. Prestipino, F. Bonino, S. Usseglio, A. Damin, A. Tasso, M. G. Clerici, S. Bordiga, F. D'Acapito, A. Zecchina and C. Lamberti, *Chem. Phys. Chem.*, 2004, **5**, 1799.
151. R. W. Joyner, A. D. Smith, M. Stockenhuber and M. W. E. van den Berg, *Phys. Chem. Chem. Phys.*, 2004, **6**, 5435.
152. J. A. van Bokhoven, A. M. J. van der Eerden and D. C. Koningsberger, *J. Am. Chem. Soc.*, 2003, **125**, 7435.
153. H. Fichtner-Schmittler, U. Lohse, H. Miessner and H.-E. Maneck, *Z. Phys. Chem.*, 1990, **271**, 69.
154. D. Scarano, A. Zecchina, S. Bordiga, F. Geobaldo, G. Spoto, G. Petrini, G. Leofanti, M. Padovan and G. Tozzola, *J. Chem. Soc. Farad. Trans.*, 1993, **89**, 4123.
155. G. Ricchiardi, A. Damin, S. Bordiga, C. Lamberti, G. Spano, F. Rivetti and A. Zecchina, *J. Am. Chem. Soc.*, 2001, **123**, 11409.
156. B. Zibrowius and E. Loffler, in *'Handbook of Porous Solids'*, ed. F. Schuth, K. S. W. Sing and J. Weitkamp, Wily-VCH, New York, 2002.
157. S. Bordiga, L. Regli, C. Lamberti, A. Zecchina, M. Jorgen and K. P. Lillerud, *J. Phys. Chem. B*, 2005, **109**, 7724.
158. L. Marchese, J. S. Chen, J. M. Thomas, S. Coluccia and A. Zecchina, *J. Phys. Chem.*, 1994, **98**, 13350.
159. M. P. J. Peeters, J. H. C. van Hoof, R. A. Sheldon, V. L. Zholobenko, L. M. Kustov and V. B. Kazansky, in *'Proceedings of the 9th IZC'*, Butterworth-Heinemann, Washington DC, 1993, 651.
160. R. Garcia, E. F. Philp, A. M. Z. Slawin, P. A. Wright and P. A. Cox, *J. Mater. Chem.*, 2001, **11**, 1421.
161. A. Delabie, K. Pierloot, M. H. Groothaert, R. A. Schoonheydt and L. G. Vanquickenborne, *Eur. J. Inorg. Chem.*, 2002, **515**.
162. B. M. Weckhuysen, A. A. Verberckmoes, M. G. Uytterhoeven, F. E. Mabbs, D. Collison, E. de Boer and R. A. Schoonheydt, *J. Phys. Chem. B*, 2000, **104**, 37.
163. S. Detavernier and R. A. Schoonheydt, *Zeolites*, 1991, **11**, 155.

CHAPTER 4

Computer Modelling

4.1 Introduction and Definitions

The experimental methods of diffraction and spectroscopy are uniquely applicable to the study of crystalline microporous solids and their chemistry. Nevertheless, there are important aspects of zeolite science that are not readily accessible to these techniques: the species involved in nucleation and crystal growth, the structure of sites (often present at low concentration) that are active for adsorption and catalysis or the reaction intermediates present in catalysis. In these cases computational atomistic simulation offers great possibilities for improved understanding. Furthermore, many experimental measurements, such as calorimetric studies of heats of adsorption, and NMR or neutron scattering studies of dynamics, may be very expensive and time-consuming. Computer simulation methods, which promise to predict the performance of materials as adsorbents and catalysts rapidly and at reasonable expense, are therefore highly attractive. Excellent recent texts and useful reviews are available that deal with the simulation of microporous materials.[1–4] Here I summarise the most widely used methods and the information they give.

The initial challenge to simulation was to model known structures correctly. Once this was successfully achieved, the more difficult issues of the modelling of disorder and defects were addressed. Most recently, efforts are being made to understand on the molecular scale the processes of nucleation and growth in synthesis, the molecular motion of adsorbates, and reaction states and reactivity in catalysis. The impact of modelling extends far beyond the detailed atomistic simulation of known structures or of adsorption and reaction, however. The computer offers scope for the prediction of hypothetical structure types, their evaluation in terms of likely thermodynamic stability and even possible structure directing agents for their synthesis. In more specific, targeted studies, modelling methods have been demonstrated to aid in the solution of structures based on well-defined building units that were intractable using other approaches. In addition, modelling in the broader sense, where structural building units and adsorbed molecules can be suitably simplified, can be used to understand and predict the morphology of growing crystals, the structure of mesoporous solids and the shape of their adsorption isotherms.

The theoretical details of the computational approaches and the encoding of their associated mathematics is largely beyond the scope of this book, but relevant codes that are frequently quoted in the literature and that are publicly or commercially available will be referred to where appropriate. The most important simulation methods and their applications to porous solids are described here: Pair Potential or Force Field methods (sometimes referred to, with more or less specificity, by the term Molecular Mechanics), Monte Carlo methods and Simulated Annealing, Molecular Dynamics simulations and Quantum Mechanical Modelling *via* the Hartree–Fock combination of atomic orbitals or Density Functional Theory. These require ever-increasing levels of computing power and expertise and the computational hardware is a practical limitation on what may be achieved. More importantly, great care should be taken in interpreting the results from calculations. Only detailed successful comparison with available experimental evidence gives confidence in the reliability of the methods. Once simulations accurately reproduce observed results they can be extrapolated to unmeasured or inaccessible systems.

At the very simplest level of simulation, computer graphics is ideally suited to the representation of these well-ordered, delicate architectures, and provides clear insights into structural features. Whatever simulation technique is applied, the results are most sensibly interpreted when related to the structural model, and this is best represented graphically.

Some of the better known computer codes for structure simulation and modelling described in this chapter are given in Table 4.1.

4.2 Structure Simulation Using Interatomic Potentials: Molecular Mechanics

One of the first challenges in modelling molecular sieves is to simulate the structure of the framework itself and, where appropriate, the position of extra-framework cations, including organic cations taken up during synthesis. If the effects of thermal vibrations are ignored the problem reduces to one of finding the lowest free energy. If only enthalpic (potential energy) terms are considered, the result of an energy minimisation becomes the calculation of a zero Kelvin free energy minimum. Simulating structures at elevated temperatures requires an estimate of the contributions of entropic terms. The entropy of molecular sieves will contain contributions from the configurational entropy, as extra-framework species distribute themselves between different energy levels according to the Boltzmann distribution, and from vibrational modes. Very often, to a first approximation, the potential energy terms from atom–atom interactions dominate and can be used to simulate structures. Furthermore, since the parameterisation of these atom–atom potentials is performed using experimental data measured over a range of temperatures, the estimated structures implicitly include contributions from temperature-dependent terms. Once general expressions for the energy are derived as a function of the position of all atoms in the unit cell, determination of the energy-minimised structure becomes

Table 4.1 Types of computer structure simulations and programs.

Type of computer simulation	*Details of simulation approach*	*Code*
Pair potential methods for energy minimisation to simulate structural details	Inclusion of short- and long-range forces Shell model for polarisabilities Three-body terms included as required Ewald summation of energies	METAPOCS[5] THBREL[6] GULP[7]
Surface energy minimisation	Bulk + Surface: Two region calculation	MARVIN[8]
Molecular mechanics	Simulation based on harmonic forcefields, using classical mechanics to predict structure and dynamics. Forcefields include Dreiding, UVFF, CVFF, *etc.*	
Quantum mechanical (QM) approaches to simulation of structures, chemisorbed species and transition states		
Orbital-based cluster and periodic lattice calculations	Non-periodic and periodic Hartree–Fock calculations (using basis sets, *e.g.* B3LYP, 6-31G)	GAUSSIAN[9] CRYSTAL[10, 11]
Density Functional approach (energy a function of the electron density) including electron correlation approximations	Local basis sets implementation possible Plane wave approach also adopted Pseudopotentials applied to core electrons LDA or GGA approximations to account for electron correlation effects	SIESTA[12] DMOL[13] CASTEP[14a] VASP[14b] DGAUSS[15]
Hybrid DFT/H–F QM approach	*e.g.* DFT/6-31G combinations	
Embedded QM calculations	Embedded QM cluster calculations within bulk region minimised by pair potential methods	
Monte Carlo methods of sampling combined with energy calculation and minimization for structural models, template-framework configurations, adsorption simulation, slow diffusion, *etc.*	Random sampling of starting structures or assemblages, combined with energy minimisation *via* pair potential (and QM) methods	

Table 4.1 (*Continued*).

Type of computer simulation	*Details of simulation approach*	*Code*
	Combination with simulated annealing routines permit full exploration of configuration and avoidance of local minima (MC-SA) Grand Canonical Monte Carlo (GCMC) methods to achieve equilibrium configurations, for example in the simulation of adsorption. Configurational bias GCMC to allow study of larger adsorbate molecules Kinetic Monte Carlo for simulation of diffusion mechanisms with high energy barriers	
Molecular Dynamics methods for the study of motion	Newtonian equations of motion applied to follow motion of frameworks and adsorbates at defined temperatures, *etc.*	DL-POLY[16] DISCOVER in INSIGHT[17]

a well-defined mathematical problem. Calculations of this sort, for example when applied to give minimum energy configurations and dynamic properties, are sometimes described under the term 'Molecular Mechanics' (as opposed to Quantum Mechanics).

4.2.1 Structural Simulation Using Pair Potentials: Energy Calculation

Once a starting structure is available, in terms of atomic positions within a unit cell, the total energy of the system can be calculated by summing the atom–atom potential energies. An expression for the pairwise, two-body interaction that has been used successfully for a range of ionic and semi-ionic oxides is used for modelling atom–atom interactions in microporous solids. Examples of parameters used in interatomic potential models of zeolites and AlPOs are those of Sanders *et al.*[18] and Gale and Henson,[19] respectively.

For two atoms i and j, a distance r apart, the potential energy is taken to be that of a Coulombic (electrostatic) term supplemented by a Lennard-Jones potential:

$$V_{ij}(r) = \frac{kq_iq_j}{r} + \frac{B_{ij}}{r^{12}} - \frac{C_{ij}}{r^6}$$

where C_{ij} relates to the van der Waals interaction and can be estimated from atomic or ionic polarisability data. A similar expression:

$$V_{ij}(r) = \frac{kq_iq_j}{r} + A_{ij}\exp\left(\frac{-r_{ij}}{\rho_{ij}}\right) - \frac{C_{ij}}{r^6}$$

may also be used where the dispersion forces are modelled using a Buckingham potential.

The Coulombic potential may be calculated either by attributing formal charges to the ions or, as is chemically more reasonable for most framework materials, by partial charges. For example, for SiO_2 frameworks, the partial charges on Si and O are close to +2 and −1, respectively. The advantages of using formal charges include transferability of potentials between systems with different chemical compositions, because the size of the partial charge will depend on the electronegativity of the different atoms. Furthermore, it becomes much more straightforward to deal with the substitution of ions of different charge. In any case, subsequent parameterisation of formal charge models tends to take deviations from formal charges into account implicitly.

In order to model the effects of ionic polarisability, particularly for the oxide ions, which are more polarisable than the smaller cations, the 'shell model'[20] is adopted. In this approach, a positive core and a negative shell, with the net charge of the ion (−2 for the oxide ion), are coupled by a harmonic spring and allow polarisation in this way. The constant for the 'shell model' spring, along with constants for the short range forces, may be derived either theoretically (for example by *ab initio* Hartree–Fock methods) or by deriving best fits to experimentally measured data on well-characterised samples of similar chemistry. For example, for zeolites, potential parameters can be derived by obtaining a best fit to structural and physical (dielectric, elastic) properties of α-SiO_2 and α-Al_2O_3. Examples of parameters used in interatomic potential modelling of zeolites and AlPOs are those of Sanders *et al.*[21] and Gale and Henson,[22] respectively.

Finally, because of the covalent nature of the bonding within most molecular sieves (and zeolites in particular) it is necessary to model the directionality of the bonds. This empirical three-body term takes the form:

$$V_{ijk} = \frac{1}{2}k(\theta - \theta_0)^2$$

where θ_0 is the equilibrium three-atom bond angle, θ the angle in the structure and k the factor derived from model compounds.

Atom–atom two- or three-body interactions of these sorts are incorporated within available 'forcefields' that have been derived to fit empirical data. Commonly used forcefields include the universal force field (UFF)[23] and the constant valence force field (CVFF).[24]

The total energy of the starting structure may then be calculated by summing up all interaction terms for atoms within the unit cell. For periodic structures, in contrast with the modelling of discrete molecules, it is not reasonable to neglect the effects of atoms located outside the unit cell. This is particularly true of the long-range Coulombic terms. In order to get round this difficulty, and to reduce in size what would otherwise be very large and unwieldy computations, the Ewald technique is adopted. This is a mathematical approach that enables all long-range interactions throughout a periodically repeating crystal structure to be determined efficiently. Complication arises from inherent disorder within the structures, such as that of the location of framework and extra-framework cations with occupancies less than unity. The distribution of these charges can, in some cases, reasonably be described as randomly disordered, and the most likely distributions and associated energies can be derived from Monte Carlo methods, as described in Section 4.2.3.

4.2.2 Energy Minimisation and Simulated Annealing Techniques

Once the energy of a starting model has been calculated by summation of the two- and three-body potentials described above, the equilibrium structure can be simulated by energy minimisation. This is performed by allowing the atomic positions of all atoms or ions to vary in directions that result in a lowering of the overall potential energy. If the unit cell is kept constant, this is a constant volume free energy minimisation. Where appropriate, it may be possible to vary the unit cell parameters to reduce strain energy within the crystal, and the unit cell may be predicted in this way *via* energy minimisation at constant pressure.

The mathematical algorithm that is used within most modelling programs is the Newton–Raphson method. This involves calculation of the second derivatives of the potential energy with respect to atomic displacements and the use of these values to relax the structure to a lower energy configuration. The method does require that the model be close to the correct structure to begin with, however, and for systems where this may not be the case, it may be necessary to apply the technique of Simulated Annealing. Whereas the Newton–Raphson method will enable an energy-minimised structure to be arrived at by continuously varying atomic parameters, Simulated Annealing methods permit displacements that result in an increase in overall energy, with a probability that decreases with increasing overall energy. This allows a wider range of possible structural configurations to be sampled and therefore increases the probability of achieving the true energy minimised structure. In this sense the process is analogous to the real thermal annealing of condensed oxides, by which equilibrium structures free of defects can be prepared.

The simulation of structures using pair potential methods gives important information, including unit cell dimensions, atomic positions and details of atomic motion including lattice vibrations (phonon modes). Further analysis permits the calculation of heat capacities, the dependence of volume with temperature and the prediction of vibrational spectra, such as IR and neutron spectroscopies. Codes that perform such periodic structure energy minimisation using pair potential models include METAPOCS,[5] THBREL[6] and GULP[7] (Table 4.1). All have been used successfully to model framework structures.

Such calculations also successfully predict the observed variation of unit cell volume with temperature for zeolitic structures, including the effect of negative thermal expansivity.[25] This latter effect results from the motion of essentially rigid tetrahedra that can result in the decrease of average distances between tetrahedral cations. Other applications of energy minimisation studies are given below.

4.2.3 Application of Monte Carlo Methods to Structure Simulation

For cases where disorder arises, such as the distribution of aluminium within the tetrahedral framework sites of zeolites, or of extra-framework cations over different cation sites, the simulation is more complicated. For aluminium distribution, the simplest approach is to assume an average charge per cation, calculated from the framework composition, effectively smearing out the overall negative framework charge. Alternatively, it is possible to determine the lowest energy configuration by sampling the energetics of a very wide range of different distributions. In order to estimate the minimum energy state, Monte Carlo methods are applied. These generate, in a random way, a very large number of different configurations within a fixed framework and calculate their associated energies. A similar approach can also be adopted for extra-framework, charge-balancing cations. If the most favourable distributions are then taken, the whole structure can then be allowed to relax to an energy minimum by minimisation of the free energy *via* Newton–Raphson methods using the potential energy expressions described in Section 4.2.1.

Use of Monte Carlo methods, together with simulated annealing algorithms, enables a very large number of starting possibilities to be assessed, and increases the reliability of finding the true minimum energy configuration. It should be noted, however, that the aluminium distribution within the framework may be influenced by factors other than their energetics within the framework. For example, in frameworks templated by organic cations, the position of the charge on the organic species can influence the location of aliovalent substituting cations through its varying proximity to different sites. Nevertheless, in the absence of such information, the Monte Carlo approach is a powerful one for simulating cationic zeolites. It also finds widespread application in simulating templating, adsorption and diffusion, as we will see below.

4.2.4 Application of Pair-potential Methods to the Study of Surfaces

Modelling the external surface structure of open framework solids is an area of much current interest.[26] The surface structure is highly relevant in processes of crystal growth and in adsorption and catalysis. In the first process, the surface represents the site of attachment of species from solution, whereas in catalysis there is evidence that appreciable catalysis can take place at the external pore mouths, particularly in the case of materials of small particle size or delaminated zeolites that are made up of individual zeolitic layers, each covered in terminal silanols.[27]

Whereas the bulk crystalline framework is periodic in three dimensions, surfaces are only periodic in two. A number of simulation codes using interatomic potentials have been written to take this into account, including the MARVIN program of Gay and Rohl.[8] The general approach is to define a region including the surface attached to another representing the bulk. In principle, this follows the Mott–Littleton approach of modelling the structure of other, localised defects. Once the surface to be studied has been chosen, it is modelled by considering a block that is made of two regions, each with the same two-dimensional periodicity. A block of finite thickness is chosen, because the surface structure will be influenced by the arrangement of underlying layers, and significant relaxations could occur below the surface layer. The first region is close to the interface and includes the surface atoms. The termination of the surface (defined by the position within the crystallographic unit cell of the plane that cuts through the framework) must also be chosen. In practice, terminations that sever fewer bonds will be energetically favoured, but there may be several different possibilities for a given structure and surface. The exposed surface is typically terminated in a physically reasonable manner. For silicates, surface silanol groups (SiOH) are expected to be the most likely termination, since these are observed experimentally by IR spectroscopy. The second region contains atoms that are deeper within the crystal, placed at their bulk equilibrium positions, and with a close fit to the atoms at the interface with the first region. The minimum energy configuration is then determined by allowing the atoms in the first region to relax in a manner analogous to that used for energy minimisation on the bulk, keeping the atoms in the second region fixed at their bulk equilibrium positions. From such calculations the relaxed positions of the surface atoms can be calculated, as well as the surface energy per unit area, defined as the difference in energy from that of the same number of atoms in the bulk structure. All surfaces have a positive surface energy (being obtained by breaking bonds): the most stable surfaces have small positive surface energies.

Studies of this type are relatively recent, but already promise to give important information on surface structures, particularly when taken in combination with experimental studies using surface electron microscopy and scanning microscopies (AFM and STM). Recent modelling studies, for example, suggest that the most stable termination of the (001) surface of zeolite L consists of complete, relatively undistorted double six-membered rings, which is supported by HRTEM observations and has important consequences for understanding growth of this zeolite.[28]

In fact, surfaces are only one kind of defect that can occur in the structures of microporous frameworks. Others include twin planes and stacking faults, dislocations and isolated defects such as substituting cations or hydroxyl nests (where an absent framework silicon is replaced by four protons, giving a cluster of four SiOH silanols). For many structures of this kind, defect energies and structures can be simulated by methods based on pair potentials, if suitable large periodic repeat units can be constructed. Where electronic properties of such defects are desired, quantum mechanical modelling is required.

4.3 Structure Simulation Using Quantum Mechanical Methods

Quantum mechanical approaches to structure modelling aim to simulate structures *ab initio*, that is beginning from a description of the component atoms in terms of their atomic orbitals. The resulting models include a full description of the electronic distribution in the solid and therefore genuine insights into bonding and, for microporous solids, such important details as the structural origin of acidity and the electronic and catalytic properties of transition metal cations either in the framework (such as titanium) or as extra-framework cations and clusters. Furthermore, quantum mechanical methods can be used to investigate the stability of species present in synthesis gels and, of importance in catalysis, chemisorbed species and catalytic intermediates, including transition states.

Quantum mechanical (QM) approaches divide into two broad categories, those based on the Hartree–Fock approach, where a so-called basis set is used to describe each atomic orbital, and Density Functional methods, where the distribution of electrons is described by localised basis sets or as the summation of series of plane waves of different amplitudes and frequency. Hybrid methods, incorporating attractive features of each method, have also been developed. These higher-level calculations are much more computationally demanding than modelling using pair potentials, so that approaches have been sought to reduce the size of the calculations without severely compromising their accuracy (for example by cluster calculations) and also to improve the efficiency of the computer simulations. Details are beyond the scope of this monograph. Accessible general discussions of QM methods based on the Hartree-Fock approach and their applications in solid state chemistry are given by Catlow *et al.*;[4] a description of density functional methods is given by Gale in a recent text.[29] Table 4.1 lists available codes that are frequently quoted in the literature in applications in microporous solids.

4.3.1 Quantum Mechanical Methods

Historically, most early QM approaches adopted the Hartree–Fock (HF)-based approach, because codes (such as GAUSSIAN[9]) were readily available and widely used for calculations on molecules. Early studies of this kind concentrated on selected clusters taken to represent the region of chemical interest, and the clusters were terminated by chemically reasonable bonds, such as silanol groups. The accuracy of such approaches is improved by including electron correlation effects.

Long-range Coulombic interactions, which only decrease according to the inverse of distance, are known to be important for crystalline materials, and periodic HF calculations have been possible for some time *via* the CRYSTAL program.[10,11] However, because of the very considerable computational cost of the HF approach, only calculations including relatively few atoms have been possible (for example where high symmetry greatly simplifies the calculation). For calculations of defects, or chemisorbed species or transition state intermediates, the ensembles of interest have low symmetry and fully periodic calculations assume impractical dimensions. To get round this problem, while still including important long-range interactions, the approach of 'embedded' calculations has been developed, in which a central core that is treated by Hartree–Fock theory is literally embedded within a region treated by pair potentials. Although this method improves the accuracy of the simulation, there remains uncertainty in the choice of the cluster size and of the three-dimensional boundary between the cluster and the crystalline continuum. A fully periodic, HF approach remains beyond current computational capabilities for most systems of interest.

4.3.2 Density Functional Theory

Density functional theory (DFT), developed within solid state physics, is based on the theorem of Hohenberg and Kohn[30] that the ground state energy of a system depends on the electron density. It can be applied to calculations performed either with localised basis sets or by combination of plane waves. Both approaches have been applied to microporous solids,[31] although the plane wave methods have been used more commonly. The SIESTA code, for example,[12] permits DFT calculations using localised basis sets as does GAUSSIAN.

The complex wavefunction of the periodic system can be represented by the Fourier summation of sinusoidal plane waves. Most DFT calculations use Kohn–Sham theory,[32] so that the DFT basis sets introduce an orbital-type nature to the simulation. The method was improved by including a consideration of electron density gradients in calculations, as well as magnitudes of the electron density (rather than just taking the electron density into account). This is the generalised gradient approximation (GGA), rather than the local density approximation (LDA) in the terminology of DFT. In addition, the method has been rendered more applicable by using a 'pseudopotential' to account for the effect of nuclei and core electrons,[33] reducing the problem to one of considering the distribution of valence electrons, which are responsible for the chemical behaviour of the system. An excellent discussion of the plane wave pseudopotential method is given by Gale.[29]

DFT methods have become increasingly useful as their implementation within computer codes has benefited from improvements in computational techniques. These include the important steps introduced by Car and Parrinello[34] and other more recent developments. Such changes have enabled the method to utilise fully increases in computing speed available through the implementation of parallel computing. As a result, DFT calculations have become an attractive method for the study of fully periodic microporous solids and their chemical interactions with adsorbates. Examples of codes that enable plane wave DFT calculations to be performed include DMOL,[13] CASTEP,[14a] VASP,[14b] and DGAUSS[15] (Table 4.1).

It should be noted, however, that DFT simulations tend to simulate dispersive interactions poorly, so that in many cases interactions of non-polar adsorbates within the pores of microporous solids are better treated using well-parametrised forcefield methods that use established atom–atom pair potentials.

4.4 Applications of Modelling to Structure Simulation

Both molecular mechanics (using interatomic potentials, two- and three-body terms, *etc.*) and quantum mechanical modelling techniques are widely used. Pair potential methods are favoured for both energy minimisations where rapid structural determinations and energy calculations are required and also in the study of physisorption. Computer intensive *ab initio* QM-based methods are used where details of the electronic structure or of chemical interactions are required, including studies of bonding and reactivity. Such applications are referred to in the later chapters as appropriate, with some illustrative examples being given below.

4.4.1 Structural Studies Through Energy Minimisation

Energy minimisation methods provide a useful means of examining and refining model structures, particularly those arrived at from powder diffraction data. QM methods give details on the electron distribution, giving direct information on bonding type (see next section). Originally applied to zeolites and related inorganic structures, such methods have also been successfully applied to hybrid solids.

Modelling using interatomic potentials can often assist in refining structural models which are not unambiguously determined from diffraction data, for instance when the observed symmetry is lower than that of the idealised symmetry. For example, the structural model of the zeolite NU-87 was only confirmed and refined when the minimum energy structure was predicted by energy minimisation. The minimised structure was found to account for features of the powder pattern which the original model could not.[35] The aluminophosphate MgAPO-36, which energy minimisation showed to possess triclinic symmetry rather than the maximum possible topological symmetry, which is orthorhombic, is another example (Figure 4.1).[36]

The actual values of lattice energies can also be calculated. Comparison of such values for pure silica polymorphs of known zeolite structure types[37] or of aluminophosphates[38] reveal clear trends when normalised to framework cation content and plotted against framework density (Figure 4.2). The framework stability of the observed structures with respect to dense phases such as quartz (SiO_2) or berlinite ($AlPO_4$) decreases linearly as the framework density

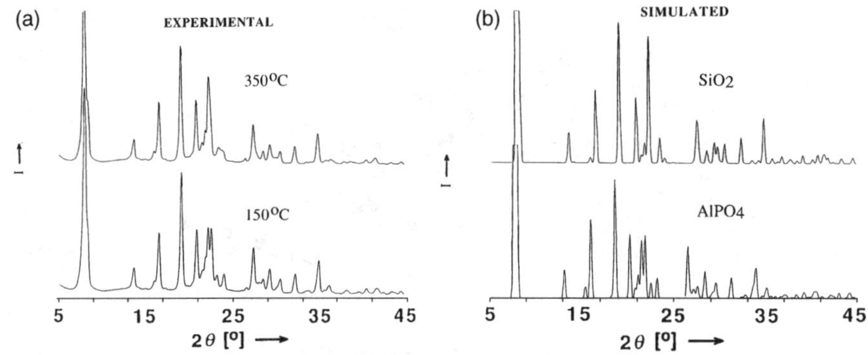

Figure 4.1 Comparison of the observed X-ray powder diffraction pattern of calcined MgAPO-36 (run at 150 and 350 °C) with that calculated from constant pressure atomistic simulation using the program GULP. [Reproduced from reference 36 with permission. Copyright 1992 Wiley-VCH Verlag GmbH & Co. KGaA.] Whereas the minimum energy configuration of a pure silica composition with the postulated framework structure is orthorhombic, alternating Al and P atoms in framework sites of the structure resulted in the monoclinic distortion observed at 150°C (and the consequent splittings seen in some of the peaks).

decreases.[39] The calculated trend is in close agreement with experimental calorimetric measurements.[40–42] It is likely that any hypothetical structure plotting well above this trend will involve strained bonds, and is likely to be unstable as the silica or aluminophosphates composition. The trend also highlights the thermodynamic role of structure-directing agents in stabilising high silica zeolites and $AlPO_4$s with low framework densities with respect to dense phases.

4.4.2 Bonding in Microporous Solids: Substitutional Behaviour

Although the atomic charges on aluminium, silicon and phosphorus in tetrahedral coordination in silicates and phosphates are formally +3, +4 and +5, analysis of quantum mechanical calculations of the valence electron distribution indicate that the bonding has a strong covalent character. Work by van Santen on silicates suggests a partial charge on silicon in silicates of close to +2.[43,44] Cora, in a detailed computational study of silicates and aluminophosphates by a range of HF, DFT and hybrid HF/DFT calculations, concluded that the true net atomic charges on Si and P are less than one half of their formal values.[39] T–O bonds of the TO_4 tetrahedra therefore show increasing polarisation in the sequence Al–O < Si–O < P–O, as expected. Whereas a microporous silica can be considered to possess a continuous semicovalent network, an aluminophosphate is better considered as a network consisting of phosphate molecular ions and aluminium cations. The greater ionic character

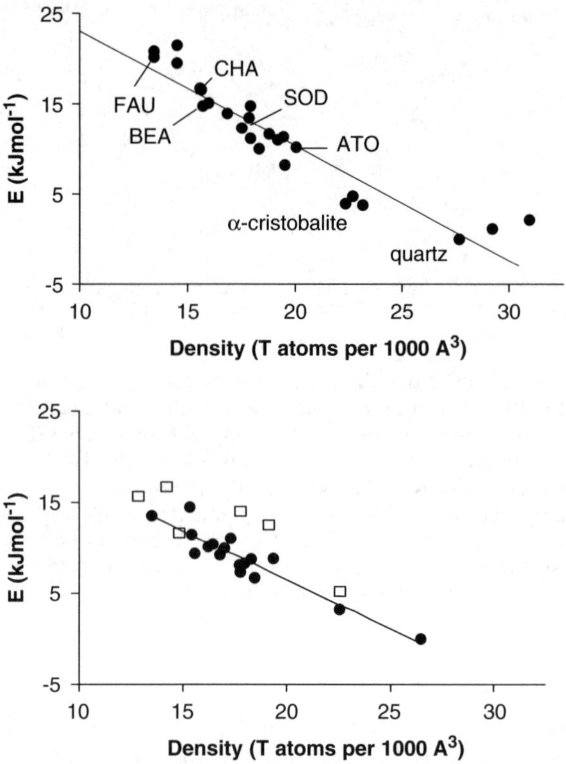

Figure 4.2 (Above) Lattice energies of different zeolitic framework topologies with hypothetical pure silica compositions, calculated using the GULP program, expressed per mole of SiO_2 relative to quartz and plotted against framework density. [Reproduced from the data of reference 37.] (Below) Lattice energies determined experimentally by solution calorimetry for pure silica zeolite polymorphs (solid symbols) and aluminophosphate polymorphs (open symbols).[40–42]

of framework aluminium in aluminophosphates than in zeolites explains its increased tendency to increase its coordination to 5 or 6 upon adsorption of H_2O, while remaining in the framework. This is commonly observed in $AlPO_4$s, but not in zeolites (a more ionic cation will tend to form more non-directional bonds).

These insights into the nature of the chemical bonding also help us understand patterns of framework substitution. In particular, strongly ionic metals readily replace aluminium in phosphates, but less readily in silicates. Once substitution has occurred, QM-based methods can be applied to understand the Lewis acid behaviour of such dopants as they coordinate to adsorbed molecules. The activation of hydrogen peroxide by tetrahedral Ti is of particular importance for selective oxidation and is discussed further in Chapters 7 and 9.

4.4.3 Organic and Inorganic Cation Sites: Monte Carlo/Simulated Annealing Approaches

Modifications of the Monte Carlo (MC) techniques described above can be applied directly to the study of templating in microporous solids. The method of Freeman,[45] for example, is based on MC and energy minimisation methods, in which molecular dynamics is used to generate a library of guest conformations, which can be docked into the porous framework and energy minimised. The method has been developed by Cox *et al.*[46] by adding a simulated annealing procedure to avoid local energy minima. In particular, the calculations give a measure of the goodness of fit of the amines or alkylammonium cations within the cages in the as-prepared solids, and the specificity or otherwise of the 'templating' effect (discussed in more detail in Chapter 5). For a charged template, the interaction energy consists of a long-range Coulombic term and a short-range 'non-bonding' energy resulting from the dispersive forces and short-range repulsion. This latter term, although lower in magnitude, really describes the steric factors, and dominates the specific geometry of the interaction. The match between predicted and experimentally observed template location, where that is known from experiment, is commonly very close (Figure 4.3)[47] and leads to confidence in predicting novel examples.

In many cases, where interactions of an organic cation with two frameworks are being compared, the Coulombic interaction may be ignored. Calculations often reveal a very close, rather than a unique, match between organic and framework, and readily explain observed product specificities. Attempts are ongoing to develop such methods together with computer-aided molecular design, to screen or predict organic cations as potential structure directing agents for targeted framework types. The ZEBEDDE (ZEolites By Evolutionary De-novo DEsign) program,[48] in which an organic template is computationally 'built' in the pore structure of a target structure to give an optimum fit, epitomises this approach. This is an attractive prospect, but the complexity of variables in the synthesis and process and the similarities between pore geometries of many open frameworks means that it remains a difficult challenge.

The predicted location of extra-framework cation sites in zeolites can also be studied by MC methods, although the problem is further complicated by the disorder in both framework aluminium location (and associated charge) and partial occupancy of cation sites. In this case a large number of possible extra-framework cation site distributions has to be considered, and a model assumed for the location of framework charge.

4.4.4 Structure Simulation as an Aid to Structure Solution: Hypothetical Structures

Model building has played an important role in crystal structure solution since its earliest days, and particularly before the advent of direct methods of structure solution from single crystal diffraction data. In the field of

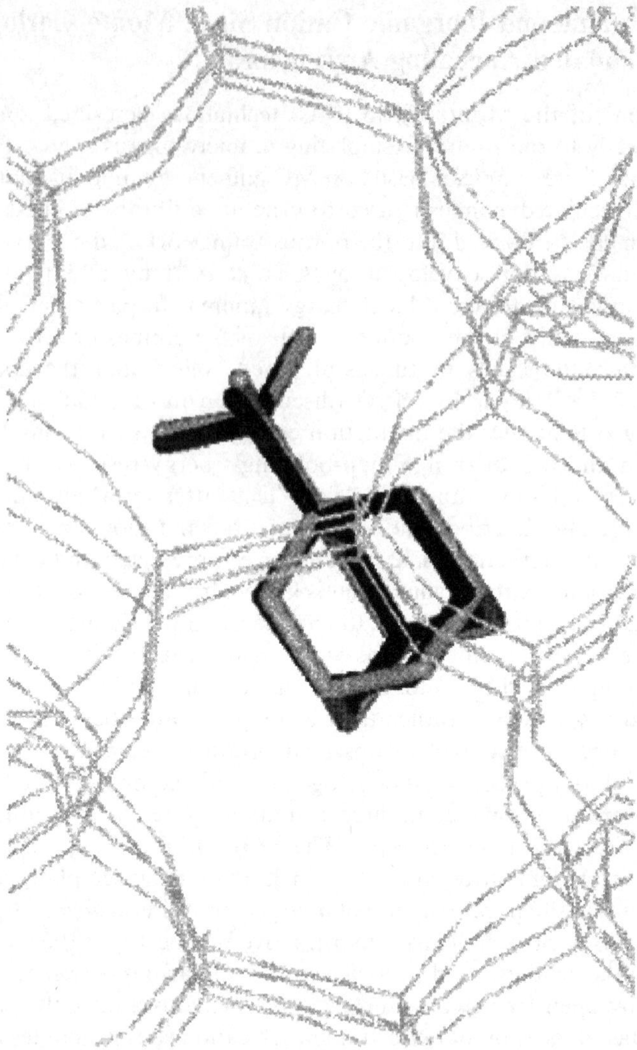

Figure 4.3 Comparison of the measured and simulated positions of the trimethylammonium adamantane cation within cages of the high silica zeolite SSZ-23, the framework of which is represented by a line diagram. [Reproduced from reference 47 with permission. Copyright 1999 American Chemical Society.]

microporous solids, where single crystals are not usually available, or where structural disorder is present, it continues to be a fruitful approach, particularly in combination with high-resolution electron microscopy, adsorption data and X-ray powder diffraction. The usual approach entails the examination of unit cell dimensions or symmetry for relationships with related, known structures, or of direct interpretation of electron micrographs in terms of specific building units. Important structure solutions of this kind include those of zeolite Beta, ETS-10 and MCM-22.

A computational approach to structure solution was developed for zeolites by Deem and Newsam.[49] In this approach, the input data is the X-ray powder diffraction data, together with the unit cell size and symmetry, and the number and multiplicity of crystallographically distinct tetrahedral sites. Trial structures that are consistent with the cell geometry and contents are generated by the program and can be modified by Monte Carlo and simulated annealing methods. The goodness of fit of X-ray diffraction patterns calculated from the models is taken as an index of the closeness of the model to the actual structure. Once reasonable trial structures are obtained, they can be refined against experimental data in the usual way. Examples of successful structure solutions of this kind include aluminophosphate UIO-7 of the group of Lillerud.[50]

As interest in mixed coordination and hybrid, MOF-type microporous structures has increased, so computer-modelling methods of the same general type have been developed that permit the inclusion of more complex secondary building units, rather than tetrahedra. This is particularly relevant for MOF-type materials, where the metallocentric clusters and the ligands are well defined. This AASBU approach (automated assembly of secondary building units), pioneered by Mellot-Draznieks *et al.*[51] for metal organic frameworks, has recently met with spectacular success in the structure solution of the chromium carboxylate structures MIL-100[52] and MIL-101,[53] based on trimers of chromium octahedra, that are described in Chapter 2. In this general approach, the input to the program is the type of secondary building units (inorganic and organic units expected on the basis of the inorganic solution chemistry and the identity of the organic ligands) and their ratio. Using a Monte Carlo-simulated annealing approach these are allowed to link *via* connecting points, for example by sharing oxygen atoms. This gives a series of hypothetical structures which can be energy minimised using molecular mechanics approaches outlined above, and ranked in terms of their energetic feasibility. The method has two attractive features: firstly, it can facilitate the structure solution of experimental phases, and secondly it can predict possible new materials. In the example of MIL-100, a consideration of the chromium trimesate chemistry suggested chromium trimers could link *via* trimesate units to give a supertetrahedron, the apices of which were the chromium trimers and the faces of which would be occupied by the trimesate ions. Use of the AASBU approach to the assembly of this type of unit predicts three structures of similar energies, the predicted experimental diffraction pattern of one of which matched the experimental one. MIL-101 is based on a similar arrangement of supertetrahedral SBUs based on the same metal trimers, linked across the edges of the SBUs *via* the terephthalate units.

A very large number of different frameworks can be generated by the manipulation of secondary building units, for example by changing the way in which particular cages or layers are connected. Extensive enumerations of hypothetical structures have been produced.[54–56] The approach lends itself naturally to computer-assisted modelling, using different mathematical approaches. Libraries of hypothetical structures can be prepared, with associated minimised energies and simulated diffraction patterns. A database of some such structures already exists on the IZA website for tetrahedrally connected networks. In this way it may be possible to match up a newly synthesised, unknown material of known symmetry and diffraction pattern with a hypothetical structure or to predict feasible frameworks for synthesis.[57] The very rapid increase of possible structure types with decreasing symmetry and increasing numbers of distinct tetrahedral sites, however, remains a daunting challenge, particularly in the light of the structural complexity observed in some of the more recently discovered high silica zeolite structures, such as TNU-9 (Chapter 3).

The modelling of mesoporous solids such as MCM-41 should be mentioned here, for comparison. For these materials the amorphous nature of the walls means there is no unique structure. Instead, a suitable model must be consistent with the experimental observations of pore size, wall thickness and density, low angle diffraction and NMR-derived values for framework cation connectivities (for Si, this corresponds to the distribution of Q^n silicon environments). One approach for mesoporous silica is to take as a starting point an amorphous silicate network with the structure of silica glass (as derived, for example, from wide angle diffraction, NMR and reverse Monte Carlo modelling) and to cut pores of appropriate size and symmetry from it to represent the porous structure, allowing bond relaxation after the termination step.[58] Another is to simulate the synthesis process, by allowing silicate coated micelles to pack, and to allow silicate units to move and condense, according to so-called Kinetic Monte Carlo methods that assign realistic rates to the processes of bond formation and motion and then allow the processes to occur in a random way, with the probability of occurrence determined by the energy of the process.[59] Both routes allow the generation of physically reasonable atomistic models of the structure, which can then be used, for example, in realistic simulations of adsorption isotherms by the GCMC methods described in Section 4.5. Figure 4.4 illustrates an atomistically modelled structure for MCM-41 achieved by the second approach.

4.5 Simulating Physisorption in Porous Solids

The physisorption of non-polar molecules within microporous solids, where no electron transfer is involved, determines adsorptive and diffusional characteristics. These are also important in their catalytic performance, which depends at least in part on molecular access to and from the active catalytic sites. Simulations that account for atom–atom interactions through pair potential methods can be used to determine accurately the energy of a physisorbed molecule, but they must include routines that enable a search through the available space to find the distribution of adsorption energies and also the global energy minimum.

Such searches use *Monte Carlo* techniques, in which adsorbate molecules are inserted with random positions and orientations into the porous host structure.[60] For complex molecular guests, a series of low-energy configurations of

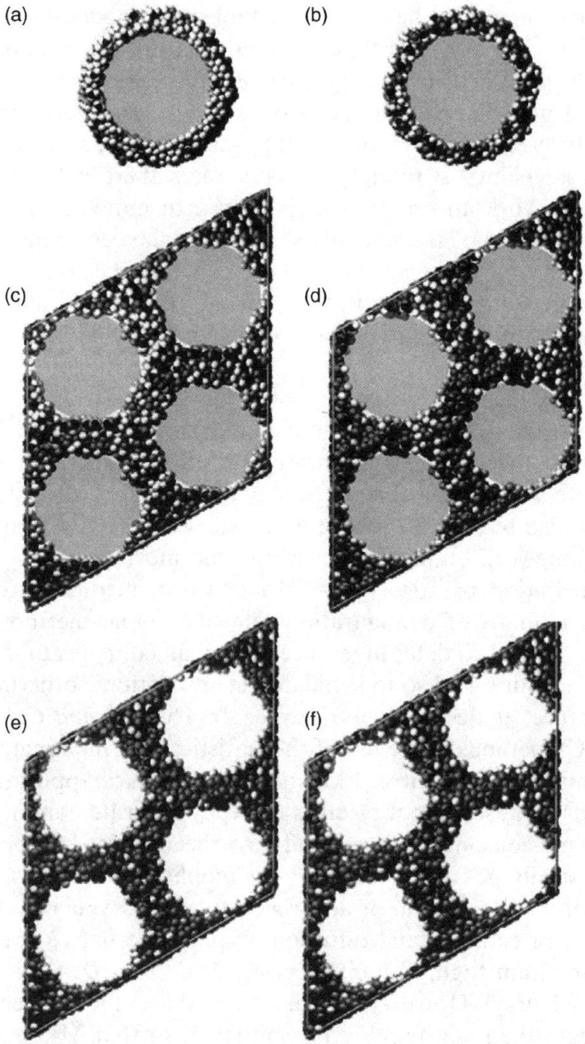

Figure 4.4 An atomistic model of the mesoporous silica MCM-41, viewed along the channels, built up computationally by a kinetic Monte Carlo approach described in the text in which silica-coated micelles aggregate and condense to give a periodic array with amorphous silica walls. Snapshots of the different stages of the KMC simulation. The micelles are represented as areas shaded in grey; silicon, bridging oxygen, and non-bridging oxygen are represented as black, grey and white spheres, respectively. Grey parallelograms indicate periodic boundaries. The snapshots are (a) the initial configuration of a micelle surrounded by silicic acid monomers, (b) the micelle with a partly condensed layer of silica, (c) two–by-two unit cells of the periodic silica structure after aggregation of multiple images of the micellar rod in a hexagonal array and (d), (e), (f) the silica structure at the beginning of the calcination, after removal of the micelle, and after the calcination. [Reproduced from reference 59 with permission. Copyright 2006 American Chemical Society.]

the guest species may first be derived by molecular modelling and molecular dynamics in the gas phase, and these configurations inserted randomly into the host. For each trial, the total energy is calculated, and compared with the sum of energies calculated for the guest species in the gas phase and the empty porous solid to give the interaction energy (or enthalpy, since it is strictly the energy at zero Kelvin). Although in many cases there will be some chance overlap of framework and molecule, giving rise to unfavourable interactions, the very large number of trials involved results in the generation of a series of candidate low-energy arrangements. These can then be used as the starting points for further energy minimisation, for example by simulated annealing. The final minimisation gives a selection of low-energy locations, as well as their interaction energies.

These routines identify low-energy sites for adsorbates, and are a crucial element of statistical mechanical routines that permit the study of processes that are strongly affected by temperature. The effect of increasing temperature is to distribute adsorbates over a range of sites according to Boltzmann statistics, and also results in the motion of adsorbates and thermal vibrations within the framework solid. To determine the most thermodynamically favourable distribution of adsorbates, Monte Carlo methods are required to sample many millions of configurations and determine the most stable. The approach can be used to determine the equilibrium configuration as a function of adsorbate pressure, and so to simulate the adsorption isotherm (uptake at a fixed temperature, as described in Chapter 7). These *Grand Canonical Monte Carlo* (GCMC) methods make use of the statistical thermodynamic concept of the Grand Canonical Ensemble. In fact, straightforward application of Monte Carlo methods to adsorbate molecules of even moderate complexity becomes unworkably time-consuming, so methods have been devised to spend more time sampling physically possible configurations (configuration bias GCMC).

Investigation of the motion of adsorbed molecules, which give mechanisms and rates of re-orientation and diffusion, require alternative approaches. For systems that contain highly mobile species, *Molecular Dynamics* (MD) techniques are widely used. However, for many adsorbates the timescales of motion are much longer than can feasibly be simulated, so that MD is only relevant either for small molecules or at high temperatures. In order to simulate slower diffusion, the process must be considered in terms of rare events with significant activations. The activated processes are then usefully treated by transition state theory, and the associated processes treated over extended timescales and volumes by, for example, *Kinetic Monte Carlo* (KMC) techniques.

4.5.1 Monte Carlo Methods: Grand Canonical Monte Carlo (GCMC)

Monte Carlo methods can be used to simulate the adsorption of molecules as a function of temperature and the pressure of the gas in contact with the porous solid. In this way adsorption isotherms of single or multiple component gas mixtures can be simulated, which saves experimental expense and time. In essence, the temperature and pressure are specified and therefore the chemical potential of the adsorbate in equilibrium with the gas phase is known. Addition of molecules to the empty solid using the Monte Carlo method is then performed, and after each step the total energy is calculated using pair potential models of adsorbate–solid and adsorbate–adsorbate interactions. Typically these are modelled using Lennard-Jones potentials, and CH_2, CH_3 and CH_4 groups are usually modelled as 'united atoms', a simplifying assumption that is found to work well for the simulation of liquid hydrocarbons. Random moves to add, subtract and move molecules are continued until the Grand Canonical ensemble of adsorbed molecules has the same chemical potential as the adsorbing gas, and is therefore at equilibrium. The uptake at this point (the number of adsorbed molecules N) represents a single point on the adsorption isotherm. The process is repeated with different gas pressures (and chemical potentials) until a full isotherm has been generated. The results of the process can subsequently be analysed to give snapshots of the adsorption, so that an atomistic picture of the process can be obtained.

A recent application of this method is illustrated in Figure 4.5. Experimental measurements of the uptake of dinitrogen, N_2, within the aluminium methylphosphonate-α reveal a distinct step in the isotherm at low pressure at a loading of 2 molecules per unit cell, before complete filling at 6 molecules per unit cell at high pressures. GCMC calculations of the process show that the simulated isotherms were very sensitive to the Lennard-Jones radius, σ, taken for the methyl groups lining the pores. Using a σ value of 0.37 nm reproduced the isotherm step remarkably accurately. Furthermore, snapshots of the process suggest that the step results from an ordering process of the N_2 within the pores. Below the step, a single molecule is located at any height within the channels, whereas at higher pressures the molecules begin to order in the pore so that up to three can pack at any height within the channels. A fully ordered and efficiently packed arrangement is achieved at the higher pressure corresponding to pore filling.[61]

4.5.2 Configurational Bias GCMC Methods

For molecules more complex than the smallest hydrocarbons, it becomes impossible for GCMC calculations to reach equilibrium within feasible simulation times. This is because the probability of locating a single atom at a low energy position (without overlap with framework or other adsorbates) is already small, particularly as pore filling increases, so the chance that two or more additional, attached atoms will not overlap rapidly becomes very small indeed, even when space exists. To get round this problem, a modified approach is adopted. Once one atom of the complex molecule has successfully been positioned by MC, the other atoms of the molecule are 'grown' sequentially, searching for the lowest configuration during the process. This configurational bias GCMC greatly improves the success rate of finding possible adsorption

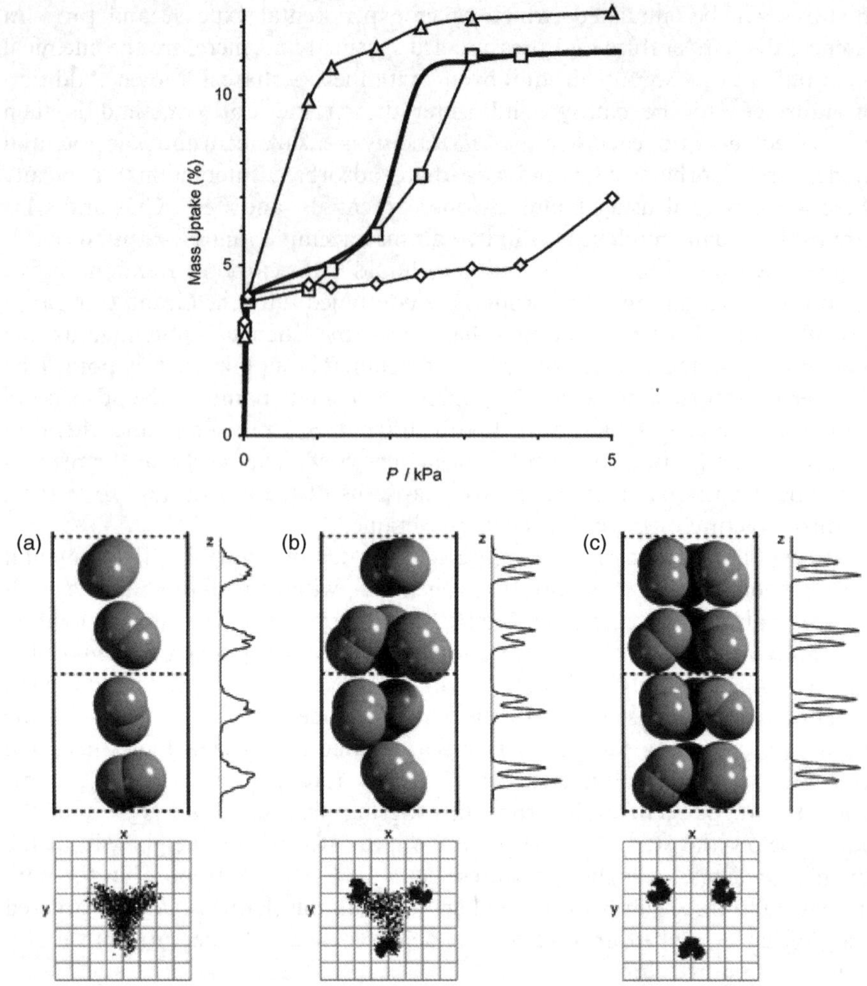

Figure 4.5 Simulations of the adsorption of nitrogen at low pressures and at 77 K within the triangular channels of AlMePO-α (Figure 2.27) using Grand Canonical Monte Carlo methods, taking Lennard-Jones diameters for the rotating -CH_3 groups of (△) 0.367 nm, (□) 0.371 nm and (◇) 0.375 nm. The experimental isotherm, which displays a low pressure step at 2 molecules per channel in the unit cell, is fitted most closely when the -CH_3 group is modelled by a sphere of diameter 0.371 nm. Snapshots of the molecular simulation at 2, 3 and 6 molecules per channel in the unit cell (below, from left to right) indicate that the step corresponds to a change of the type of packing in the channels from 2 molecules to more than 2, with full ordering and pore filling occurring at 6. Projections of the centres of mass of the adsorbate molecules down the channel axes are shown below, indicating the ordering. [Reproduced from reference 61 with permission. Copyright 2005 Royal Society of Chemistry.]

sites,[62] and can be used to determine the adsorption behaviour of more complex molecules, once the bias that has been introduced is taken into account.

Studies of this type have been used by Smit and co-workers[63,64] to explain the so-called 'inverse shape selectivity' observed in the conversion of long chain *n*-alkanes over acid zeolites. In such reactions, product distributions are found to depend on the pore structure, particularly for medium-pore zeolites such as ZSM-5. In some cases branched alkanes are favoured over linear alkanes in the products of medium-pore zeolites compared to the reaction selectivities of large-pore zeolites such as zeolite Y. For example, doubly branched isomers are favoured over ZSM-5. This is in contrast with what would be expected from diffusion rates and is attributed to the enhanced thermodynamic stability of some branched intermediates in the medium-pore zeolites that is predicted by configurational bias GCMC.

4.5.3 Molecular Dynamics

Whereas Monte Carlo methods investigate the energies of many different configurations of adsorbed molecules, without taking into account real processes by which they adopt those configurations, Molecular Dynamics simulations aim to predict the real time-dependent motion. They do this by using classical Newtonian equations of motion for a many-body system to determine the trajectories of molecules and atoms of the lattice. Where translational motion is sufficiently rapid, it is possible to calculate rates and coefficients of diffusion (where these are *ca.* $5 \times 10^{-10}\,m^2\,s^{-1}$ and above). Typically MD simulations sample motion over time periods from 10^{-12} to 10^{-9} seconds, a timescale that is relevant to experimental techniques such as solid state NMR, pulsed field gradient NMR, and neutron and IR spectroscopies. The review of Demontis and Suffritti remains an important description of the method applied to zeolites.[65]

In a typical MD run, the input parameters are the initial positions and velocities of the particles, together with the dependence of the potential energy of the system on the atomic positions, which is usually defined by pair potential models of the kind described in Section 4.1. The simulation is composed of two parts, an initial run in which the ensemble (N atoms, Volume V, Energy E) is allowed to reach thermal equilibrium, and a second, longer run in which the atom trajectories can be followed and used to derive the diffusivities. These trajectories can be further analysed by calculating their energy profile. For example, for pore geometries that contain constricted openings, it is possible to calculate activation energies for migration through these. The timesteps for these calculations have to be sufficiently short to observe the type of motion with the greatest frequency, so intervals of 10^{-15} s are typically taken. Even fast computers and long MD runs only sample time intervals of nanoseconds. This means that motions of larger molecules in neutral microporous solids, or smaller molecules through cationic zeolites, where activation energies are high, cannot be studied by this method.

There are many examples where the approach has been fruitfully applied, however, particularly for small molecules through pure silica forms of zeolites. The diffusion of methane and other non-polar gases such as xenon and SF_6 through silicalite, for example, readily fulfils these conditions[66] (studied later by transition state methods, as well[67]). The migration of xylene isomers through silicalite, of relevance in understanding product selectivities in the xylene isomerisation reactions (Chapter 8) has been modelled, as has, more recently, the motion of these isomers through the zeolite CIT-1, which contains both 10MR and 12MR apertures.[4]

A recent example of the comparison of MD simulation and experiment is observed in the simulation of the re-orientation of benzene in the AlMePO-α and -β polymorphs.[68] MD predicts that benzene rotates freely within the pores of AlMePO-β but is constrained to take up one of three symmetry related positions in the more triangular channels of AlMePO-α, with the plane of the aromatic ring approximately parallel to the channel axis (Figure 4.6). Deuterium wideline (static) NMR spectra (shown in Figure 7.7), which are highly sensitive to molecular motion, support these predictions, with C_6D_6 in AlMePO-β giving a sharp resonance typical of free tumbling, whereas in AlMePO-β the ^{2}H NMR spectra are as predicted for the more constrained form of motion.

4.5.4 Transition State Theory and Related Methods

For systems where time periods of *ca.* 10^{-9} seconds do not encompass all the important types of motion, the rates of which may be limited by significant energy barriers (such as hopping between separate adsorption sites on extra-framework cations), MD cannot be applied. As a result, other approaches have been developed which make use of transition state theory.

In these approaches the free energy along a proposed trajectory between sites can be calculated and the rate coefficient derived using kinetic theory. The review of Auerbach in reference 69 is particularly strong in the description of such methods. The group of Auerbach has published an important series of papers in which the motion of benzene through cationic faujasitic zeolites X and Y is discussed in detail, from both a simulational and experimental standpoint, and the rate coefficients predicted. Once the kinetics for a particular step have been determined, theoretically or experimentally, simulations of motion within the crystalline pore structure are required in order that mass transport can be described in terms of diffusion coefficients. These coefficients relate to either Fick's law of diffusion, where the diffusion flux is proportional to the concentration gradient, or the Stefan law, where diffusion is proportional to the gradient in chemical potential (See Chapter 7).

Such lattice models require details of site connectivity to be established, and usually assume that migration proceeds over well-defined energy barriers that do not change with the number of sites that are occupied and that subsequent jump probabilities are not correlated. Kinetic Monte Carlo methods can then be applied, where the probability of a successful random jump is related to the

Figure 4.6 The trajectories of a benzene molecule in the AlMePO-α (upper) and -β (lower) polymorphs obtained from a molecular dynamics (MD) simulation indicate that whereas benzene readily re-orients isotropically (freely) in the more cylindrical channels of the β polymorph, it undergoes more restricted motion in the more markedly triangular channels of the α polymorph. This is confirmed by ^{2}H wideline NMR of C_6D_6 (see Section 7.2.3 for details).

rate coefficient, rather than to the energy (*via* the Boltzmann constant) as in conventional Monte Carlo algorithms.

Kinetic Monte Carlo and related methods, in combination with transition state theory, have successfully modelled experimental observations of features such as the concentration dependence of benzene diffusion through zeolite Na-X. NMR-derived diffusivities decrease monotonically with increased loading for benzene in Na-X, for example, and Auerbach *et al.*[69] show that KMC simulations can explain this. Such calculations are becoming increasingly important for the chemical engineer modelling the performance of porous sorbents and catalysts under operating conditions.

4.6 Modelling Chemical Bonding and Reactivity

4.6.1 Nucleation and Crystal Growth

Recent computational approaches to nucleation and growth have made use of a range of modelling techniques, ranging from cluster modelling of silicate species present in a synthesis gel to atomistic modelling of the growing surfaces of crystals and numerical modelling of the step-by-step layer growth of crystalline solids.[70]

Recently, quantum mechanical cluster calculations of the oligomerisation and cyclisation reactions of hydrated silicate species have been carried out using density functional theory and taking into account the effects of solvation.[71] Figure 4.7 indicates some of the reaction steps that have been investigated, and the calculations emphasise that cyclic species, especially 4-membered rings, are thermodynamically favoured over linear oligomers. Such 4MRs are of course important components of many zeolite structures. Experimental NMR studies have long shown a very wide and pH-dependent distribution of such species in silicate solutions. Although these initial attempts do not yet take into account the roles of metal cations or organic templates, they do represent important developments in understanding the mechanism of hydrothermal crystallisation, particularly when combined with modelling of the energetics of growing surfaces.[72] Surface calculations of the 2-region kind described in Section 4.2.4 can indicate which surface planes are stable, their likely termination surfaces and the activation energy for growth. Although in their early stages, such studies may ultimately predict the crystal morphology as a function of synthesis conditions.

In the absence of *ab initio* calculated rate constants for crystal growth, the group of Anderson has shown that growth features observed by atomic force microscopy can be simulated by mathematical modelling. This requires correct assignment of the relative rates of attachments of secondary building units at different types of growth sites at the surface, namely edge, kink and terrace locations. Description and further references to studies that model crystallisation and growth are given in Chapter 5.

4.6.2 Chemisorption

Chemisorption, which involves the formation of a chemical bond between adsorbate and adsorbent, can only adequately be described using quantum

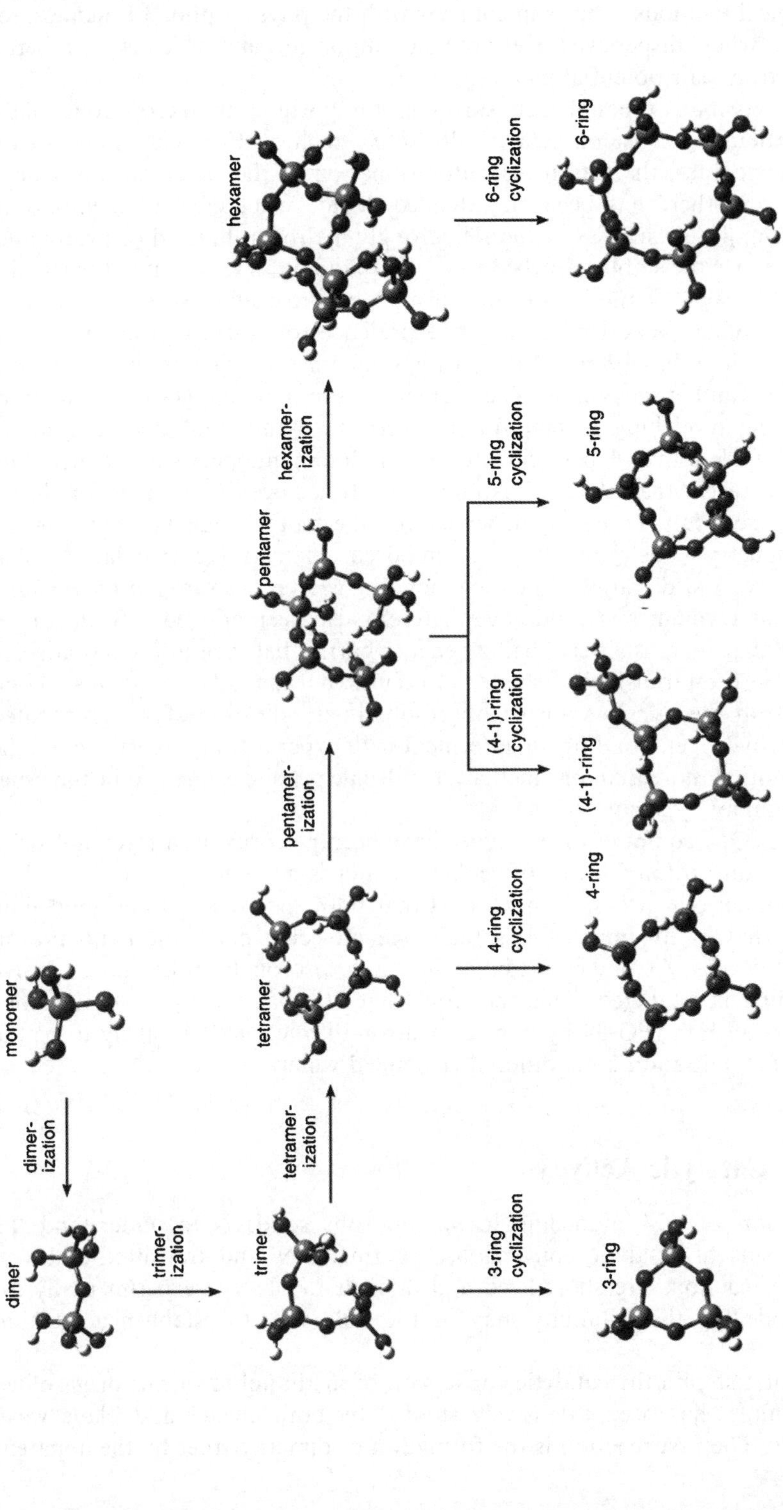

Figure 4.7 Silicate condensation reactions simulated computationally by Lewis *et al.*[71] Free energy calculations, taking the effect of solvent into account, underline the favourability of the formation of cyclic tetramers, or 4MRs. [Reproduced from reference 71 with permission. Copyright 2005 Wiley-VCH Verlag GmbH & Co. KGaA.]

mechanical methods. This is in contrast with the physisorption of alkanes, for example, where dispersive forces are more important, and which is often better modelled by pair potential models.

There are many types of chemisorption of importance in microporous solids, and particularly those on catalytically active zeolites. For acid catalysis over protonic zeolites, the action of proton donation to the adsorbed molecule is crucial, and there has been an extended debate whether, for an important reactant molecule such as methanol, strongly hydrogen-bonded or protonated molecules are the stable adsorbed form. Methanol is an adsorbate of particular interest, because of the importance of methanol-to-olefin (MTO) and methanol-to-gasoline (MTG) processes. The presence (or not) of protonated methoxonium ions has been debated, with sometimes conflicting evidence from simulation and spectroscopy. The current picture that emerges from quantum mechanical modelling is that for adsorbed methanol molecules present at loadings of less than 1 per acid site, the predominant species is the methanol molecule, rather than the methoxonium ion. It has been found that for cluster models, the result is critically dependent on the zeolite fragment taken and on the symmetry constraints that are imposed. Plane wave calculations that sample a very wide range of possible bonding geometries through the *ab initio* Molecular Dynamics method have therefore also been applied.[73] Such studies indicate that there is a very shallow energy barrier between molecular adsorption and proton transfer for isolated molecules, but that adsorption of a second molecule at the same site invariably results in stabilisation of the protonated form by hydrogen bonding, in agreement with experimental observations. The adsorption of more than one methanol molecule per site is likely to be the usual state in many applications.

Similar QM computational studies have been performed on H_2O molecules adsorbed on protonic forms of zeolites. Water is a slightly weaker base than methanol, so that it is not unexpected that both spectroscopy and modelling suggest that at loadings of less than one molecule per proton, no proton transfer occurs. At higher loadings, however, proton transfer and extensive H-bonding is predicted. This therefore suggests that the hydroxonium ions observed on H-SAPO-34 by powder neutron diffraction[74] are likely to be the result of stabilisation by additional H-bonded water.

4.6.3 Catalytic Activity

A key aim of QM modelling in microporous solids is to understand the mechanisms of catalytic conversions. Intermediates and transition states in catalytic reactions are short lived and difficult to observe experimentally, so that modelling their stability may be the only way to establish a reaction pathway.

The first step in the catalytic conversion of methanol to olefins or gasoline, for example, has been extensively studied by both cluster and plane wave methods. The first reaction is the formation of dimethylether by the apparent dehydration of two molecules of methanol:

$$2\,CH_3OH \rightarrow CH_3OCH_3 + H_2O$$

Two mechanisms have been proposed for this reaction. The first includes a step in which the first methanol is protonated and dehydrated to give a reactive methoxy group attached to the framework, whereas the second involves two methanol molecules undergoing an S_N2 reaction inside the zeolite cage, catalysed by protonation of one of the methanols and with the transition state solvated by the zeolite cage. The conclusion from QM studies (using both cluster[75,76] and plane wave[77,78] approaches) is that the reaction proceeds through the latter pathway, in which the zeolite stabilises the transition state.

The next step in the methanol-to-hydrocarbons reaction, and in fact the crucial one for the generation of hydrocarbon products is C–C bond formation. Very many proposed mechanisms exist for potential routes at the acid sites of the zeolites, but recent evidence suggests that the reaction instead proceeds *via* a reactive hydrocarbon pool (See Chapter 8). In fact, an extensive series of high-level theoretical calculations[79] suggests that no single combination of direct reaction steps can link methanol to ethene, and so provides strong indirect evidence that the hydrocarbon pool mechanism is the correct one.

Use has also been made of QM methods in determining the transition state configurations and energies for hydrocarbon transformations over the protonic forms of zeolites. The review of Rozanska and van Santen gives several examples of this approach.[80] For example, in the alkylation of benzene with propene, the first transition state is taken to be the protonation of propene, and the second transition state is the protonated propylbenzene structure. In such studies, it becomes apparent that the zeolite stabilises the charged transition state and that the degree of stabilisation of the transition state depends strongly on the geometry of the zeolite framework. This then is the theoretical basis of the often-proposed transition state selectivity, in which the product selectivity is strongly affected by the zeolite structure type. These ideas have been examined further in other recent QM modelling studies, for example for xylene disproportionation.[81]

4.7 Summary

Computer simulation is now an established technique in the study of porous solids, and improves our understanding of their synthesis and structure and their activity as adsorbents and catalysts. Modelling techniques assist in structure solution, and act to inspire designed synthesis. They are capable of the accurate prediction of adsorption and diffusion behaviour, so that the need for expensive measurements can be reduced. A range of such techniques, from mathematical modelling to quantum mechanical modelling, gives insights into processes as diverse as crystallisation, chemisorption and catalysis. Further specific examples of the use of modelling of these processes are given in later chapters.

References

1. C. R. A. Catlow, *'Computer Modelling in Inorganic Crystallography'*, Acad. Press Ltd., London, 1992.
2. *'Computer Modelling of Microporous Materials'*, Eds. C. R. A. Catlow, R. A. van Santen and B. E. Smit, Elsevier, Amsterdam, 2004.
3. R. Millini, *Catal. Today*, 1998, **41**, 41.
4. C. R. A. Catlow, L. Ackermann, R. G. Bell, F. Cora, D. H. Gay, M. A. Nygren, J. C. Pereira, G. Sastre, B. Slater and P. E. Sinclair, *Faraday Discuss.*, 1997, **106**, 1.
5. C. R. A. Catlow, A. N. Cormack and F. Theobald, *Acta Cryst. Sect. B*, 1984, **40**, 195.
6. M. Leslie, *Daresbury Laboratory, Warrington, Cheshire, UK WA4 4AD.*
7. J. D. Gale and A. L. Rohl, *Mol. Simulat.*, 2003, **29**, 291.
8. D. H. Gay and A. L. Rohl, *J. Chem. Soc. Farad. Trans.*, 1995, **91**, 925.
9. Gaussian 03, Revision C.02, M. J. Frisch *et al.*, Gaussian, Inc., Wallingford CT, 2004.
10. R. Dovesi, C. Roetti, C. Freyria-Fava, E. Apra, V. R. Saunders and N. M. Harrison, *Phil. Trans. A*, 1992, **341**, 203.
11. V. R. Saunders, R. Dovesi, C. Roetti, M. Causa, N. M. Harrison, R. Orlando, C. M. Zicovich-Wilson, K. Doll and B. Civalleri, *CRYSTAL2002 User's Manual, University of Torino*, 2002.
12. J. M. Soler, E. Artacho, J. D. Gale, A. Garcia, J. Junquera, P. Ordejon and D. Sanchez-Portal, *J. Phys. Cond. Matter*, 2002, **14**, 2745.
13. Molecular Simulations, *DMOL4.2.*
14. (a) M. C. Payne, M. P. Teter, D. C. Allan, T. A. Arias and J. D. Joannopoulos, *Rev. Mod. Phys.*, 1992, **64**, 1045; (b) G. Kresse and J. Furthmuller, *Phys. Rev. B*, 1996, **54**, 11169.
15. Molecular Simulations, *Dgauss4.1.*
16. See W. Smith, Molecular Simulation, 2006, 32, 933.
17. Insight II, Accelrys Inc., Cambridge 2003.
18. M. J. Sanders, M. Leslie and C. R. A. Catlow, *J. Chem. Soc., Chem. Commun.*, 1984, 1271.
19. J. D. Gale and N. J. Henson, *J. Chem. Soc. Farad. Trans.*, 1994, **90**, 3175.
20. B. G. Dick and A. W. Overhauser, *Phys. Rev. B*, 1958, **112**, 90.
21. J. M. Sanders, M. Leslie, and C. R. A. Catlow, *J. Chem. Soc., Chem. Commun.*, 1984, 1271.
22. J. D. Gale and N. J. Henson, *J. Chem. Soc. Farad. Trans.*, 1997, **93**, 629.
23. A. K. Rappe, C. J. Casewit, K. S. Colwell, W. A. Goddard and W. M. Skiff, *J. Am. Chem. Soc.*, 1992, **114**, 10024.
24. P. Dauber-Osguthorpe, V. A. Roberts, D. J. Osguthrope, J. Wolff, M. Genest and A. T. Hagler, Proteins: Structure, Function and Genetics, 1988, 4, 31: Implemented in the *Discover* program: *Discover*, version 99.1, Accelrys, San Diego, USA.
25. J. W. Couves, R. H. Jones, S. C. Parker, P. Tschaufeser and C. R. A. Catlow, *J. Phys. Condens. Mat.*, 1993, **5**, 329.

26. B. Slater, J. O. Titiloye, F. M. Higgins and S. C. Parker, *Curr. Opin. Solid State Mater. Sci.*, 2001, **5**, 417.
27. A. Corma, V. Fornes, S. B. Pergher, T. L. M. Maesen and J. G. Buglass, *Nature*, 1998, **396**, 353.
28. T. Ohsuna, B. Slater, F. Gao, J. Yu, Y. Sakamoto, G. Zhu, O. Terasaki, D. E. W. Vaughan, S. Qiu and C. R. A. Catlow, *Chem.-Eur. J.*, 2004, **10**, 5031.
29. J. D. Gale, in *'Computer Modelling of Microporous Materials'*, Ed. C. R. A. Catlow, R. A. van Santen and B. E. Smit, Elsevier, Amsterdam, 2004, 129.
30. P. Hohenberg and W. Kohn, *Phys. Rev. B*, 1964, **136**, 864.
31. G. Poulet, P. Sautet and E. Artacho, *Phys. Rev. B*, 2003, **68**.
32. W. Kohn and L. J. Sham, *Phys. Rev. A*, 1965, **140**, 1133.
33. L. Kleinman and D. M. Bylander, *Phys. Rev. Lett.*, 1982, **48**, 1425.
34. R. Car and M. Parrinello, *Phys. Rev. Lett.*, 1985, **55**, 2471.
35. M. D. Shannon, J. L. Casci, P. A. Cox and S. J. Andrews, *Nature*, 1991, **353**, 417.
36. P. A. Wright, S. Natarajan, J. M. Thomas, R. G. Bell, P. L. Gai-Boyes, R. H. Jones and J. Chen, *Angew. Chem. Int. Ed. Engl.*, 1992, **31**, 1472.
37. N. J. Henson, A. K. Cheetham and J. D. Gale, *Chem. Mater.*, 1994, **6**, 1647.
38. N. J. Henson, A. K. Cheetham and J. D. Gale, *Chem. Mater.*, 1996, **8**, 664.
39. F. Cora, M. Alfredsson, C. M. Barker, R. G. Bell, M. D. Foster, I. Saadoune, A. Simperler and C. R. A. Catlow, *J. Solid State Chem.*, 2003, **176**, 496.
40. I. Petrovic, A. Navrotsky, M. E. Davis and S. I. Zones, *Chem. Mater.*, 1993, **5**, 1805.
41. N. Hu, A. Navrotsky, C.-Y. Chen and M. E. Davis, *Chem. Mater.*, 1995, **7**, 1816.
42. P. M. Piccione, C. Laberty, S. Yang, M. A. Camblor, A. Navrotsky and M. E. Davis, *J. Phys. Chem. B*, 2000, **104**, 10001.
43. A. J. M. de Man and R. A. van Santen, *Zeolites*, 1992, **12**, 269.
44. P. -P. E. A. de Moor, T. P. M. Beelen and R. A. van Santen, *J. Phys. Chem. B*, 1999, **103**, 1639.
45. C. M. Freeman, C. R. A. Catlow, J. M. Thomas and S. Brode, *Chem. Phys. Lett.*, 1991, **186**, 137.
46. A. P. Stevens, A. M. Gorman, C. M. Freeman and P. A. Cox, *J. Chem. Soc. Farad. Trans.*, 1996, **92**, 2065.
47. M. A. Camblor, M. J. Diaz-Cabanas, P. A. Cox, I. J. Shannon, P. A. Wright and R. E. Morris, *Chem. Mater.*, 1999, **11**, 2878.
48. D. W. Lewis, D. J. Willock, C. R. A. Catlow, G. J. Hutchings and J. M. Thomas, *Nature*, 1996, **382**, 604.
49. M. W. Deem and J. M. Newsam, *Nature*, 1989, **342**, 260.
50. D. E. Akporiaye, H. Fjellvag, E. N. Halvorsen, T. Haug, A. Karlsson and K. P. Lillerud, *Chem. Commun.*, 1996, 1553.
51. C. Mellot-Draznieks, J. M. Newsam, A. M. Gorman, C. M. Freeman and G. Férey, *Angew. Chem. Int. Ed.*, 2000, **39**, 2270.

52. G. Férey, C. Serre, C. Mellot-Draznieks, F. Millange, S. Surble, J. Dutour and I. Margiolaki, *Angew. Chem. Int. Ed.*, 2004, **43**, 6296.
53. G. Férey, C. Mellot-Draznieks, C. Serre and F. Millange, *Acc. Chem. Res.*, 2005, **38**, 217.
54. D. E. Akporiaye and G. D. Price, *Zeolites*, 1989, **9**, 23.
55. M. M. J. Treacy, K. H. Randall, S. Rao, J. A. Perry and D. J. Chadi, *Z. Kristallogr.*, 1997, **212**, 768.
56. O. D. Friedrichs, A. W. M. Dress, D. H. Huson, J. Klinowski and A. L. Mackay, *Nature*, 1999, **400**, 644.
57. M. D. Foster, A. Simperler, R. G. Bell, O. D. Friedrichs, F. A. A. Paz and J. Klinowski, *Nature Mater.*, 2004, **3**, 234.
58. B. P. Feuston and J. B. Higgins, *J. Phys. Chem.*, 1994, **98**, 4459.
59. C. Schumacher, J. Gonzalez, P. A. Wright and N. A. Seaton, *J. Phys. Chem. B*, 2006, **110**, 319.
60. S. Yashonath, J. M. Thomas, A. K. Nowak and A. K. Cheetham, *Nature*, 1988, **331**, 601.
61. C. Schumacher, J. Gonzalez, P. A. Wright and N. A. Seaton, *Phys. Chem. Chem. Phys.*, 2005, **7**, 2351.
62. B. Smit and J. I. Siepmann, *Science*, 1994, **264**, 1118.
63. M. Schenk, B. Smit, T. J. H. Vlugt and T. L. M. Maesen, *Angew. Chem. Int. Ed.*, 2001, **40**, 736.
64. M. Schenk, S. Calero, T. L. M. Maesen, T. J. H. Vlugt, L. L. van Benthem, M. G. Verbeek, B. Schnell and B. Smit, *J. Catal.*, 2003, **214**, 88.
65. P. Demontis and G. B. Suffritti, *Chem. Rev.*, 1997, **97**, 2845.
66. A. I. Skoulidas and D. S. Sholl, *J. Phys. Chem. B*, 2002, **106**, 5058.
67. R. L. June, A. T. Bell and D. N. Theodorou, *J. Phys. Chem.*, 1991, **95**, 8866.
68. J. Gonzalez, R. N. Devi, P. A. Wright, D. P. Tunstall and P. A. Cox, *J. Phys. Chem. B*, 2005, **109**, 21700.
69. S. M. Auerbach, F. Jousse and D. Vercauteren, in *'Computer Modelling of Microporous Materials'*, Ed. C. R. A. Catlow, R. A. van Santen and B. E. Smit, Elsevier, Amsterdam, 2004, 49.
70. J. R. Agger, N. Hanif and M. W. Anderson, *Angew. Chem. Int. Ed.*, 2001, **40**, 4065.
71. M. J. Mora-Fonz, C. R. A. Catlow and D. W. Lewis, *Angew. Chem. Int. Ed.*, 2005, **44**, 3082.
72. M. E. Chiu, B. Slater and J. D. Gale, *Angew. Chem. Int. Ed.*, 2005, **44**, 1213.
73. I. Stich, J. D. Gale, K. Terakura and M. C. Payne, *J. Am. Chem. Soc.*, 1999, **121**, 3292.
74. L. Smith, A. K. Cheetham, R. E. Morris, L. Marchese, J. M. Thomas, P. A. Wright and J. Chen, *Science*, 1996, **271**, 799.
75. S. R. Blaszkowski and R. A. van Santen, *J. Am. Chem. Soc.*, 1996, **118**, 5152.
76. S. R. Blaszkowski and R. A. van Santen, *J. Phys. Chem. B.*, 1997, **101**, 2292.

77. R. Shah, J. D. Gale and M. C. Payne, *J. Phys. Chem. B*, 1997, **101**, 4787.
78. E. Sandre, M. C. Payne and J. D. Gale, *Chem. Commun.*, 1998, 2445.
79. D. Lesthaeghe, V. Van Speybroeck, G. B. Marin and M. Waroquier, *Angew. Chem. Int. Ed.*, 2006, **45**, 1714.
80. X. Rozanska and R. A. van Santen, in *'Computer Modelling of Microporous Materials'*, Ed. C. R. A. Catlow, R. A. van Santen and B. E. Smit, Elsevier, Amsterdam, 2004, 165.
81. M. Clark, M. Sierka and J. Sauer, *J. Am. Chem. Soc.*, 2004, **126**, 936.

CHAPTER 5

Synthesis

5.1 Introduction

Microporous framework solids are synthesised *via* solvent-mediated crystallisations from mixtures of reactive precursors. The reaction pathway is controlled by kinetic as well as thermodynamic considerations so that equilibrium phase diagrams, so relevant in the high-temperature preparation of ceramics, are not useful here. Rather, synthetic routes have been developed empirically *via* a major synthetic effort that continues today. The continuing industrial and academic interest in these materials provides a powerful incentive to understand the principles underlying their formation through the processes of gel formation and evolution, nucleation and crystal growth.

These fundamental processes are influenced by a large number of synthetic parameters. For a hydrothermal synthesis these are the gel composition, in terms of the inorganic and organic components and water; the reaction pH; temperature and time; the presence of additives such as mineralising agents and additional modifications such as seeding, aging and stirring. This scope for variation, coupled with the interdependence of variables, gives zeolite synthesis the appearance of something of a 'black art', with literature recipes that should be followed exactly. The compilation of verified syntheses, as prepared by Robson,[1] is certainly of great use, giving syntheses of zeolites and aluminophosphates that have been checked independently, but it is ultimately of more value to understand the effect of variables so that preparations can be developed and optimised. Enough is now known for this to be possible. However, it remains a major challenge to predict the products of reaction of a new gel composition or to devise a route to a novel structure type with desired properties.

This chapter concentrates on principles of the hydrothermal synthesis of zeolites. Silicate-based systems are the most important industrially and have been studied in the most detail and over the longest period. There is therefore a great deal of experimental (and theoretical) evidence on which to base reaction schemes, and a series of thorough reviews (see particularly those of Cundy and Cox,[2,3] and references within) and at least one monograph[4] have appeared on the subject. Phosphate syntheses are less thoroughly studied, but follow the same principles. Hybrid solids are such recent additions, and so varied in type, that studies have so far concentrated on the discovery of new phases, rather than the synthesis mechanism.

The synthesis of mesoporous solids is a fundamentally different process, one that involves the formation of a composite inorganic-organic solid by the organisation of inorganic units around a liquid crystalline arrangement of surfactant micelles. These arrangements are controlled by interfacial energies and no true crystallisation step is involved. Nevertheless, gel chemistry, pH, temperature and time are also important variables in determining the structure type, pore size and degree of order of the product, which can in some cases show crystal-like morphology. This is a new and fast-moving field, so I will summarise the current concepts in this area.

5.2 Principles of Hydrothermal Synthesis

In most syntheses, a gel is initially prepared that contains framework-building inorganic species, typically aluminates, silicates, phosphates or metal-oxyanions, available in a reactive form. There are also syntheses that result in crystallisation from clear solution, and others that proceed under essentially 'dry gel' conditions (without added water). The sources of the reagents can determine the products that form: in general more reactive sources are favoured. In addition, for zeolites and related solids, structure directing species must be present in the gel that can stabilise the open framework structures relative to denser phases of the same composition. Hydrated metal cations and/or organic cations perform this function in zeolite preparations.

The gel is made up in a solvent medium which enables the framework-forming species to come into solution, but should dissolve the crystalline product only sparingly. By far the most widely used solvent is water, which possesses a suitably high dipole moment ($\varepsilon_r = 78$) to permit dissolution of ionic and partially ionic solids, as well as acceptably low vapour pressures at the temperatures at which synthesis occurs at convenient rates (usually 100–200 °C). Under these conditions the syntheses are described as hydrothermal. A more limited set of syntheses has been performed in other, non-aqueous, solvents such as alcohols, pyridine-based solvents and ionic liquids, and these are generally described as solvothermal and ionothermal.[5,6]

The solubility and speciation of the gel components must be optimised by working at the pH that permits their transfer from the gel to a growing crystal face. For example, the solubility of silica in pure water is too low to permit this process, but increases as the pH (and the temperature) increases.[4,7] Furthermore, the speciation in solution is strongly pH dependent. The pH should not be such that a component of the gel will precipitate to an inert compound that plays no further part in the synthesis, as is the case for certain transition metals (such as Fe(III)) at high pH. The solubility of the reactants can also be controlled by adding mineralising agents, such as the fluoride ion, which increases solubility by forming soluble complexes and which catalyses the formation of Si–O–Si bonds.

A homogeneous gel made up of these components is prepared and heated in an inert container at temperatures above 100 °C and consequently under pressures that are above atmospheric. Typically PTFE, 'Teflon'-lined stainless steel autoclaves are used with recommended maximum fill levels of around 75% to avoid problems associated with thermal expansion and excessive pressure build up. The upper limit on operation temperature is set at around 200 °C by the PTFE, which softens and flows at temperatures a little above this. A typical commercial design such as that of the Parr Instrument Co. is given in Figure 5.1. Under these conditions equilibrium is established between the gel and the species present in solution. For silicate-based systems these include a very wide range of oligomers, which have been identified on the basis of ^{29}Si NMR, mass spectrometric and crystallisation studies. Historically, two synthetic routes have been proposed, in which crystallisation occurs either from species in solution, or by solid-to-solid reaction in the sol-gel. Although the second route is difficult to disprove in those syntheses that possess a gel phase, the mediating role of a solvent phase is more widely accepted.

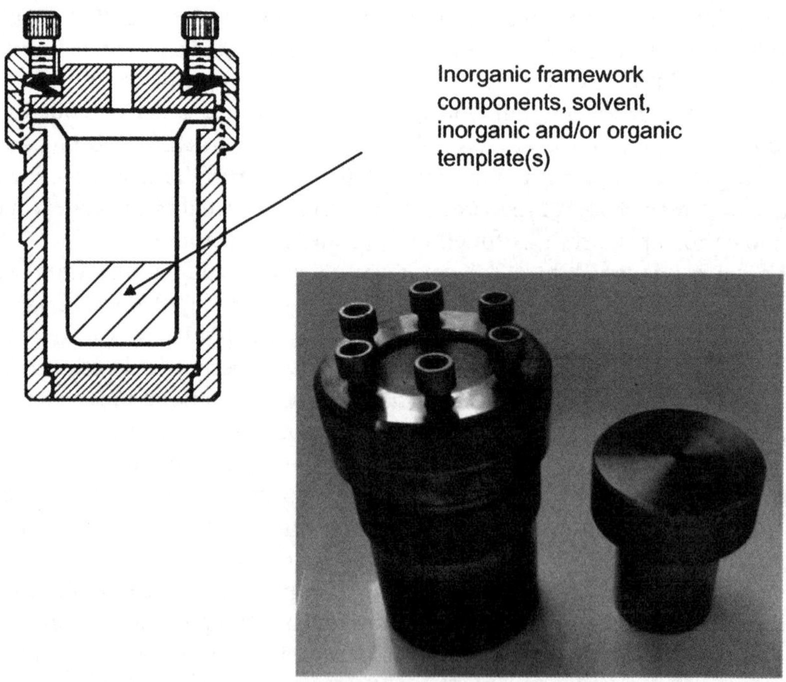

Figure 5.1 Most hydrothermal syntheses between 100 °C and 200 °C are carried out in Teflon containers within stainless steel jackets. Commercial autoclave assemblies are designed to prevent leakage from the Teflon liner and to burst safely if the pressure increases above expected limits (due to over-temperature, for example). (Reproduced by permission of the Parr Instrument Co.)

A solution containing silicate or other framework-forming species therefore results at elevated temperature and reaches supersaturation. There is then a thermodynamic driving force for crystallisation, but nuclei of a critical size must first be produced. The general model of nucleation suggests that the formation of nuclei from a supersaturated solution occurs as particles agglomerate from solution to form a part of the crystalline structure. These are very small, and consequently have a high surface area and therefore a positive free energy of formation. These initial nuclei will be unstable, and most will redissolve, but under conditions of supersaturation some may persist long enough to grow. At a certain critical size the lowering in free energy as a consequence of the contribution of the lattice energy (an r^3 term) outweighs the increase in free energy due to increasing surface area (an r^2 term). Addition of further species from solution then results in a net reduction in the free energy and crystal growth occurs.

The number of nuclei that form will depend on the degree of supersaturation of the solution, and can be augmented by the addition of seed crystals of the same structure, or even other structures. Studies suggest that addition of seed crystals is usually accompanied by the addition of amorphous nuclei present on the surface of the seed crystals. These 'initial bred' nuclei also accelerate crystallisation.[8] Heterogeneous nucleation is also observed to occur at the gel surface. In the process of aging, which is often reported in the literature, the gel is kept at a lower temperature prior to the high temperature at which crystal growth occurs. During this aging period nuclei stable at the lower temperature may be formed that subsequently grow during the second step. The nature of the very first nuclei to form is difficult to measure, since they will have a very similar composition to species present in the gel and solution, many of which will not be incorporated intact into the growing solid. Once the nuclei have reached a few nanometres in size it is possible to detect them *in situ* as discrete entities by X-ray or light scattering or Raman spectroscopy, or to observe them directly in quenched samples by high resolution transmission electron microscopy. It is also possible to stop the crystallisation once very small nanoparticles have been formed.

Nuclei that have achieved their critical radius grow by adding species from solution to certain preferred crystal faces. Under the conditions of crystallisation from solution, the slower growing faces will be the best developed. For example, in crystals that adopt a needle-like (acicular) form, growth along the needle axis is the fastest and the area of the end face is small. The relative growth rates on different planes will therefore determine the crystal morphology. In general, those planes with the lowest free energies, usually corresponding in the case of zeolites to those with the lowest surface density of dangling hydroxyls, are those that are exhibited in the final crystal form. In addition, the crystal morphology may be modified strongly by the presence of additives adsorbed at the growing surface that alter the relative growth rates. Crystal morphologies, often beautifully expressed for microporous solids (see Figure 3.16), often exhibit the true point group symmetry of a structure. The ultimate size of the product crystals will depend on the growth rate and the number of nuclei. High temperature and high dilution favour large crystal size, whereas high supersaturation and seeding will produce very small crystallites. Crystal growth ends when the species from the solution have been exhausted. Measuring the rate of crystal growth is possible experimentally by X-ray diffraction, because the crystals are in the domain (> *ca.* 10 nm) where sharp Bragg diffraction maxima are obtained. Following the intensity increase of these maxima as crystallisation proceeds gives the crystallisation curve. This is possible both on multiple sets of quenched data, or *in situ* at synchrotron X-ray sources using 'white' X-radiation beam-lines. Whereas the former is more widely available, the latter offers more possibilities of observing intermediates during synthesis. A typical sigmoid crystallisation curve is shown in Figure 5.2, exhibiting periods of induction and rapid growth that tails off as the reagents become exhausted.

In some cases crystallisation sequences are observed where a first formed phase transforms after extended reaction times to other phases. This kind of behaviour is in accord with Ostwald's law, which states that under kinetically controlled conditions, the first phases to form will be those with higher entropy, and these may transform towards the thermodynamically most stable phase *via* phases with progressively lower entropy and lower free energy. The sequential synthesis of zeolite Na-Y and then the denser zeolite Na-P from the same

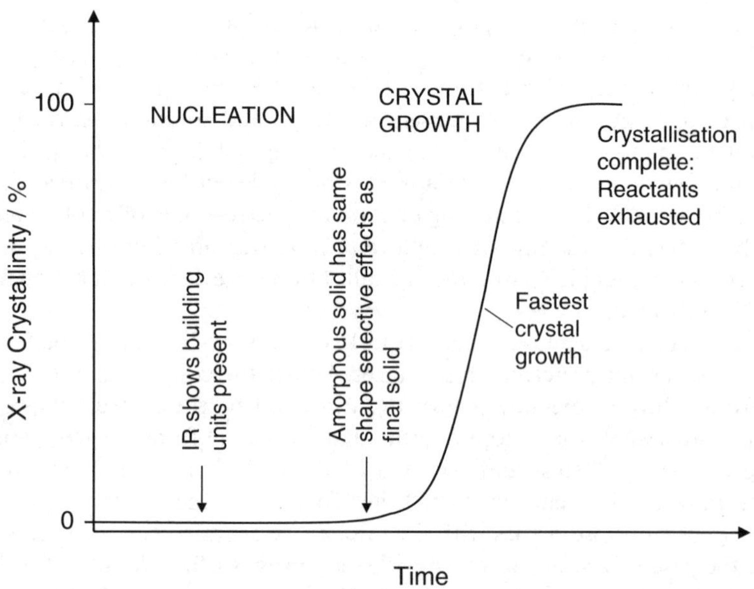

Figure 5.2 A typical hydrothermal crystallisation proceeds via a sigmoidal crystallisation curve, with an initial nucleation period followed by crystal growth. Even when there is little long-range crystallinity, the products can exhibit 'zeolitic' features due to the presence of structural building units (shown by characteristic IR absorption bands) or microporosity similar to that of the final crystalline solid.

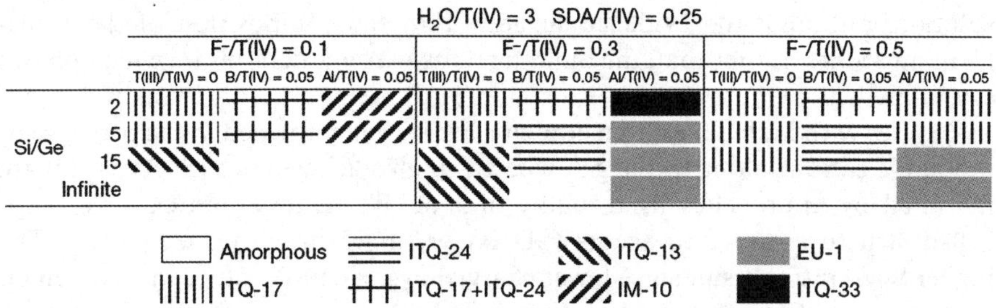

Figure 5.3 The results of exploratory synthesis in terms of solid products can helpfully be displayed in tabular form. In the table shown, from reference 10, the phase selectivity to different zeolites in the germanosilicate system with just one template, hexamethonium $((CH_3)_3N^+(CH_2)_6N^+(CH_3)_3)$, is given in terms of the content of fluoride, boron and aluminium. [Figure reproduced from reference 10 with permission. Copyright 2006 Nature Publishing Group.]

synthesis mixture is one example; the successive crystallisation of zeolite Y and then ZSM-4 (MAZ) in the presence of tetramethylammonium ions is another.[9] Commonly two or more phases may co-crystallise, and under these conditions the desired phase may be obtained pure by changing the gel composition and/or temperature or by seeding the mixture. The results of a large number of crystallisations at different compositions are conveniently plotted on 'composition space-phase' diagrams, such as those illustrated for syntheses in the SiO_2-GeO_2-HF-H_2O system using the hexamethonium $[((H_3C)_3N(CH_2)_6N(CH_3)_3)]^{2+}$ template (Figure 5.3).[10] These were performed under a high throughput regime (see Section 5.6).

These considerations are general to the synthesis of open framework solids. Details specific to particular chemical compositions are given below. By far the largest amount of synthetic work has been devoted to the aluminosilicates and other metallosilicates and silicas, which are synthesised hydrothermally at alkaline pHs (or pH 7 for fluoride syntheses). Phosphate-based solids have also received much attention and there is much current interest in inorganic-organic hybrid solids, such as carboxylates and phosphonates, which tend to have higher solubilities than silicates, and crystallise at more acidic pHs.

5.3 Synthesis of Zeolites

5.3.1 Gel Formation

Aluminosilicate gels are prepared by the mixing of reactive sources of alumina and silica. These include fumed silica, silica sols and alkoxides of silica, and aluminium, aluminates, aluminium salts and alkoxides of aluminium. Other synthetic routes have been explored that use structured aluminosilicate precursors, such as zeolites themselves,[11] to supply aluminate and silicate species to solution, although the evidence suggests that the composition of the transforming zeolite, rather than the building units present, determines which phases crystallise.

Zeolites with high aluminium contents are typically formed in the presence of inorganic cations alone in the gel, whereas high silica zeolites (Si/Al > 20) are favoured by approaches using bulky organic alkylammonium cations as so-called structure directing agents (SDAs) or, more broadly, 'templates'. The higher Si/Al ratios result since fewer charged organic molecules can be taken up by the available pore space, so that fewer negative charges associated with aluminium atoms substituting for silicon atoms are required. Higher aluminium contents impart higher framework charge density, more cation exchange capacity and enhanced gas adsorption properties but lower hydrothermal stability and lower acid site strength.

Alkali and alkaline earth metal cations (such as Li^+, Na^+, K^+, Rb^+, Ca^{2+}, Sr^{2+} and Ba^{2+}) have been used to facilitate zeolite syntheses. Some characteristic syntheses in which inorganic cations control zeolite crystallisation are given in Table 5.1. They are commonly used in the soluble hydroxide form to ensure pH values of 12–13 in the reaction mix, which are ideal for crystallisation to occur. For cations such as calcium, the lower solubility in alkaline solution requires higher reaction temperatures and longer crystallisation times. Once in solution, the inorganic cations exist as hydrated 'structure-making' species, which are surrounded by extensively H-bonded shells of water molecules.

Since the recognition by Barrer and Denny,[12] and Kerr and Kokotailo,[13] that *organic cations* such as tetramethylammonium (TMA) could be used as structure-directing agents in zeolite syntheses, a very wide range of quaternary ammonium ions, amines and cationic complexes have been investigated. Early successes were achieved with the use of tetraethylammonium (TEA) and tetrapropylammonium (TPA) cations, which resulted in the zeolites β[14] and ZSM-5,[15] both of which are used commercially. Other examples where commercially available organics have been shown to give new zeolite structures are given in Table 5.2. Very often, inorganic cations must be added as well as the organic cations for crystallisation to occur. Examples of other structures prepared with commercially available structure-directing agents include synthetic ferrierite (ZSM-35), EU-1 and NU-87, theta-1 (ZSM-22), ZSM-23, ZSM-12 and MCM-22. More recently, template design and synthesis has further extended the range of high silica zeolites. This is discussed in some detail in Section 5.4.2.

The function of these metal and organic cations has been the subject of prolonged debate. Their primary role is to stabilise open silicate frameworks with respect to dense silicates. This has been demonstrated by calorimetric data, for example by Petrovic *et al.*,[20] who showed that all zeolitic silicas are only 7–14 kJ$(mol SiO_2)^{-1}$ less stable than quartz, the stability decreasing with decreasing density (Figure 4.3). They concluded that this relatively small difference would be made up by the non-bonding interactions

Table 5.1 Important zeolite types synthesised with inorganic cations as structure directing agents using verified synthesis procedures[1] and also those reported by Barrer.[4]

Cations added	*Gel composition[1] (formally as metal oxides)*	*Zeolite (topbotology type)*	*Composition*
Li, Na	8.6 LiCl:Na_2O:Al_2O_3:2 SiO_2	Li-ABW (ABW)	$Li_4Al_4Si_4O_{16}$
Na	3.165 Na_2O:Al_2O_3:1.926 SiO_2	Na-A (LTA)	$Na_{12}Al_{12}Si_{12}O_{48}$
Na	17 Na_2O:Al_2O_3:8 SiO_2	Na-X (FAU)	$Na_{86}Al_{86}Si_{106}O_{384}$
Na	4.62 Na_2O:Al_2O_3:10 SiO_2 (seeded)	Na-Y (FAU)	$Na_{56}Al_{56}Si_{136}O_{384}$
Na	6 Na_2O:Al_2O_3:30 SiO_2	Na-Mordenite (MOR)	$Na_5Al_5Si_{43}O_{96}$
Na	3.25 Na_2O:Al_2O_3:30 SiO_2 (seeded)	Na-ZSM-5 (MFI)	$Na_7Al_7Si_{89}O_{192}$
Na, K	5.5 Na_2O:1.65 K_2O:Al_2O_3:2.2 SiO_2 $Na_{73}K_{22}Al_{95}Si_{97}O_{384}$	Na,K-X (FAU)	
K	2.35 K_2O:Al_2O_3:10 SiO_2	K-L (LTL)	$K_9Al_9Si_{27}O_{72}$
K	20 K_2O:Al_2O_3:2 SiO_2	K-F (EDI)	$K_{10}Al_{10}Si_{10}O_{40}$
K, Na	0.17 Na_2O:2.0 K_2O:Al_2O_3:5.18 SiO_2	K-chabazite (CHA)	$K_{11}Al_{11}Si_{25}O_{72}$
K, Na	4.2 Na_2O:1.3 K_2O:Al_2O_3:16.5 SiO_2 $Na_{2.9}K_{5.4}Al_8Si_{28}O_{72}$	K,Na-offretite (OFF)	
K, Sr	2.3 K_2O:0.1 SrO:Al_2O_3:10 SiO_2	K-ZK-5 (KFI)	$K_{22}Al_{22}Si_{74}O_{192}$

System[4]	*Zeolites reported*
CaO-Al_2O_3-SiO_2-H_2O	analcime (ANA), thomsonite (THO), epistilbite (EPI), harmotome (HAR), mordenite (MOR)
SrO-Al_2O_3-SiO_2-H_2O	analcime (ANA), ferrierite (FER), cancrinite (CAN), mordenite (MOR), yugawaralite (YUG), gmelinite (GME), chabazite (CHA), heulandite (HEU)
BaO-Al_2O_3-SiO_2-H_2O	zeolite L (LTL)

between the organic and the framework. The closeness of fit of the organic cation (which determines the strength of the non-bonding interaction) therefore determines the magnitude of the enthalpic driving force for the templated silicas.

The selectivity with which frameworks can be crystallised by using different cations (or combination of cations) is a crucial feature of these syntheses. The relative size and charge of the metal or organic cations clearly play an important role in deciding which structure forms. Crystallography shows many examples where after crystallisation the cation remains included within a cage from which it cannot easily escape (K^+ in the cancrinite cage of zeolite L, TMA^+ in the sodalite cage of ZK-4 (LTA)). Nevertheless, it is relatively rare to find strict one-to-one relationships of cations to structures. One frequently cited example of a very close fit of organic cation to pore space is for the zeolite

Table 5.2 Important zeolite types synthesised with readily available organic compounds as structure-directing agents or co-structure-directing agents according to verified synthesis procedures (v)[1] and also those reported elsewhere.

Organics/Cations added	*Gel composition (content formally as oxides)*	*Zeolite (topology)*	*Composition*
TMA, Na	1.55 Na_2O:4.1 $(TMA)_2O$:Al_2O_3:3.9 SiO_2:320 H_2O (v)	ZK-4 (LTA)	$Na_{9.2}(TMA)_{0.8}Al_{10}Si_{14}O_{48}$
TMA, Na	3.5 Na_2O:1.24 TMA Br:Al_2O_3:9.2 SiO_2:0.7 Na_2SO_4:125 H_2O (v)	Mazzite (MAZ)	$Na_{7.3}(TMA)_{0.7}Al_8Si_{28}O_{72}$
TEA	2 Na_2O:K_2O:12.5 $(TEA)_2O$:Al_2O_3:50 SiO_2:750 H_2O:2.9 HCl (v)	Beta (BEA)	$Na_{0.9}K_{0.6}(TEA)_{7.6}Al_{4.5}Si_{59.5}O_{128}$
MTEA methyl-triethylammonium Na	10 Na_2O:20 MTEA Br:Al_2O_3:100 SiO_2:2000 H_2O (v)	ZSM-12 (MTW)	$Na_{0.5}(MTEA)_{1.3}Al_{0.8}Si_{27.2}O_{56}$
TPA	0.04 NH_4F:0.08 TPABr:SiO_2:20 H_2O (v)	ZSM-5 (MFI)	$TPA_4Si_{96}O_{192}F_4$
TBA tetrabutyl-ammonium	60 Na_2O:9.4 TBABr:$Al_2(SO_4)_3$:90 SiO_2:2500 H_2O[16]	ZSM-11 (MEL)	$Na_{0.6}TBA_{2.6}Al_{2.2}Si_{93.8}O_{192}$
Hexamethonium	10 Na_2O:6 RBr_2:Al_2O_3:60 SiO_2:3000 H_2O[17]	EU-1 (EUO)	$(Na,R)_xAl_4Si_{108}O_{224}$
Decamethonium	6 Na_2O:NaBr:5 RBr_2:Al_2O_3:40 SiO_2:2000 H_2O[18]	NU-87 (NES)	$(Na,R)_xAl_4Si_{64}O_{136}$
Hexamethylene-imine (HMI), Triethanolamine (TEOA), Na	0.7 Na_2O:20 TEOA:4 HMI:B_2O_3:12 SiO_2:80 H_2O[19]	MCM-22 (MWW)	$Na_3HMI_7B_6Si_{66}O_{144}$
Pyrrolidine, Na	21 Na_2O:46 C_4H_9N:Al_2O_3:100 SiO_2 (v)	ZSM-23 (MTT)	$NaAlSi_{23}O_{48}.wH_2O$ (no organic)
18-crown-6, Na	2.2 Na_2O:0.87 (18-crown-6):Al_2O_3:10 SiO_2:140 H_2O (v)	EMC-2 (EMT)	$Na_{20}(18\text{-crown-6})_4Al_{20}Si_{76}O_{192}$

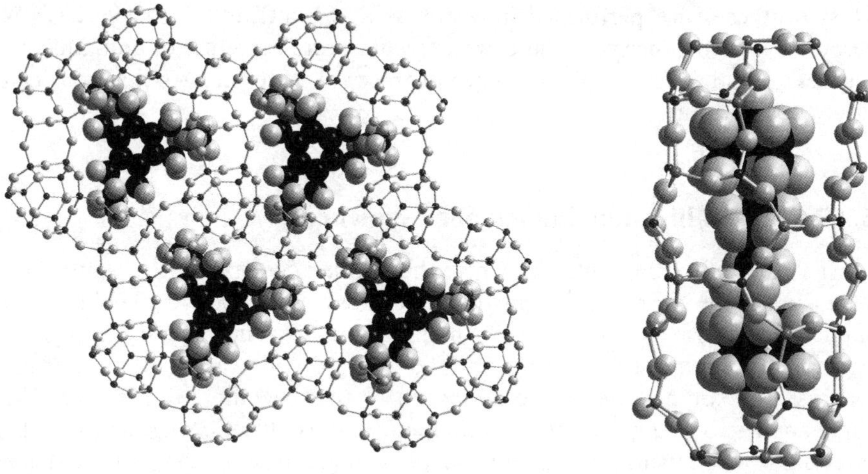

Figure 5.4 The close fits of (left) the triquaternary alkylammonium cation, 2,3,4,5,6,7,8,9-octahydro-2,2,5,5,8,8-hexamethyl-1H-benzo[1.2-c:3,4-c′:6-c″]tripyrrolium 'triquat', $[(C_4H_4N(CH_3)_2)_3]^{3+}$ within the pores of the zeolite ZSM-18[21] (Coordinates courtesy P.A. Cox) and (right) of the diquaternary cation $(C_7H_{13}N\text{-}(CH_2)_4\text{-}NC_7H_{13})^{2+}$ within cages in the magnesioaluminophosphate STA-2 are good examples of the templating effect of organics in the synthesis of zeolites.

ZSM-18,[21] templated by the triquaternary octahydrohexamethyl benzotripyrrolium cation, which modelling shows fits the available pore space very closely (Figure 5.4).[2] More frequently it is found that more than one organic cation can direct crystallisation to a framework type, or that a single organic cation can direct synthesis to more than one framework type, depending on the other synthesis variables. Notably, a combined experimental and modelling study by Harris and Zones showed that, among a series of organic templates that direct the crystallisation of SSZ-13 (a synthetic chabazite), the fastest crystallisation resulted from the use of the templates that fitted the cage structure with the most favourable energetics (arising mainly from van der Waals interactions).[22] Whether the organic cation plays a void-filling or a templating role, it is certainly true that the use of different organic cations in the synthesis is the most fruitful route to novel structure types and this is discussed further in Section 5.4.1.

Typical zeolite synthesis conditions are strongly alkaline, enabling silica and alumina to be dissolved at concentrations that permit a sufficient supply of species to nuclei and growing crystals. At these high pHs the alumina exists as tetrahedral aluminate $Al(OH)_4^-$ species, whereas silica exists as a variety of oligomeric species in solution (Section 5.3.3). The hydroxide ion is also acting as a mineraliser under these conditions, *i.e.* it increases solubility and catalyses the subsequent formation of Si–O–Si bonds. Fluoride ions can also perform this function, even at neutral pH, and this is described further in Section 5.4.3.

Most syntheses are performed in water as a solvent, and their synthesis is described as hydrothermal. There are a few literature examples of non-aqueous syntheses of silicates, but the presence of traces of water in such reactions can rarely be ruled out.

5.3.2 Crystallisation Curves and Sequences

Most studies of zeolite synthesis, where the degree of crystallisation is based on a measure of the X-ray crystallinity per unit mass of the product, give sigmoidal curves of the form shown schematically in Figure 5.2. The general features are an induction period, followed by a period of rapidly increasing growth, which reaches a maximum rate before tailing off as the reactants are used up. The curve applies to examples where no competitive crystallisation takes place. The rates of growth depend on temperature, pH and gel composition, but the shape remains the same.

In general, increasing the temperature decreases both induction and growth times, and the magnitude of apparent activation energies measured for the process (*via* treatment of the crystallisation data *via* the Avrami equation, for example)[23] indicates crystallisation is governed by thermally activated reactions and not by rates of diffusion. A high temperature limit is often reached where dense silicates rather than open frameworks are produced. This is thought to be associated with loss of waters of crystallisation from metal cations in inorganic-only systems, or with decomposition of organic cations where these are the structure-directing agents.

Crystallisation is found to proceed over a relatively narrow pH range: for zeolite mordenite, for example, this is between 11.5 and 13.0.[24] According to this study, higher pH values in this range lead to faster crystallisation, until at the high pH limit fast growth is followed by dissolution and reaction to dense phases. Lowering the pH below 11.5 reduces the growth rate below values of practical significance.

5.3.3 The Induction Period and Nucleation

During the induction period, during which no particles with long range order are formed, experimental evidence indicates that equilibrium is established between the gel and the species in solution. Particles a few nanometres in size form and agglomerate into particles that in turn grow into colloidal zeolites. Perhaps the most convincing evidence that crystallisation can occur from a mobile, solvent phase is that these solids can be prepared from clear solutions, as demonstrated, for example, by Mintova *et al.* for zeolite A.[25] They found that, starting from a clear solution and at room temperature, colloidal gel particles formed, in each of which nucleated a single zeolite A nanocrystal that replaced the gel particle (Figure 5.5). Heating the resultant colloidal suspension gave large crystals *via* solution transport as the nanocrystals dissolved.

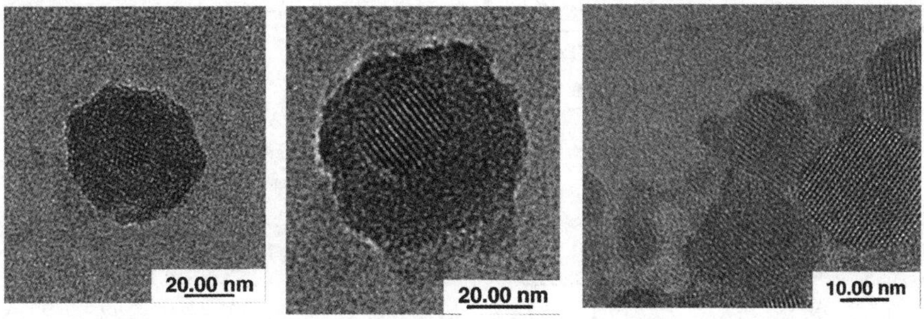

Figure 5.5 High resolution of nanocrystals of zeolite A formed by the crystallisation of colloidal amorphous silicate particles in clear suspensions at room temperature. Initial nucleation is followed by full crystallisation of the particles (from left to right). [Figure reproduced from reference 25 with permission. Copyright 1999 Science.]

The silicate species present in alkaline silicate solutions have been analysed by different methods. These include quenching, followed by surface passivation by trimethylsilylation and analysis by chromatography and mass spectrometry, the structural analysis of products crystallised from the solutions and by solution NMR. All these methods indicate that the solutions contain a wide range of silicate oligomers, distribution of which depends on the cationic species that is present. For example, early studies of tetramethyl ammonium (TMA)-silicate solutions indicated that when $TMA/SiO_2 = 1$, the predominant species was the double 4MR unit found in zeolite A, whereas for solutions containing tetrabutylammonium cations (TBA) a tetraalkylammonium silicate was crystallised containing the $Si_{10}O_{25}{}^{10-}$ anion.[26] Such solutions are the source of species from which nucleation occurs.

One of the most thoroughly studied zeolite crystallisations is that of the synthesis of the pure silica form of ZSM-5 from a clear solution using TPA as the structure-directing agent. A series of low-angle X-ray scattering experiments by de Moor *et al.*[27–29] indicate the presence of populations of scattering particles 2.8 nm and 10 nm in size. The 2.8 nm particles are thought to be primary units of silicate assembling around TPA cations, whereas the 10 nm species are aggregates of these. Diffuse light scattering studies by Schoeman,[30] also of the TPA-silicalite system, also showed two populations of particles during the initial stages of synthesis: one at around 3 nm in size, which is present at a constant concentration and a second population of a mean size that is first distinguished when it is 10 nm and which grows as the synthesis proceeds (Figure 5.6).

Chang and Bell showed that major changes occur in the TPA-silica gel, with units being formed that contained the TPA cation, and that these units rearranged with time.[31] Spectroscopic ^{1}H–^{29}Si CP MAS NMR evidence from Burkett and Davis[32] on the same system indicates that the TPA cations become included in organised inorganic-organic composites before crystallinity is apparent, and that their conformation is similar to that which they adopt in the

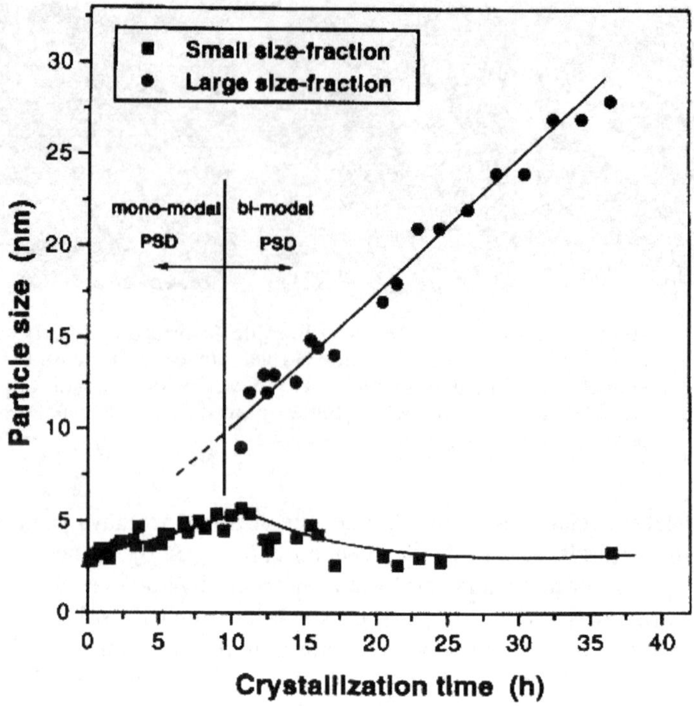

Figure 5.6 Light scattering experiments, interpreted in terms of the sizes of colloidal particles in heated gels crystallising to silicalite, show that one population of particles *ca.* 3–5 nm in size is present throughout the synthesis, whereas a second population of particles forms and grows into zeolite crystals upon heating. [Figure reproduced from reference 30 with permission. Copyright 1997 Elsevier.]

zeolite product (which has also been determined by single crystal crystallography). Recent HRTEM/small angle scattering studies of room temperature crystallisation of templated silicalite/ZSM-5 suggest that zeolite particles form by aggregation of structured precursor nanoparticles that only make up a part of the total population of colloidal particles.[33] Furthermore, studies[34] have demonstrated that, if separated before crystallinity appears, the amorphous solids that result show shape selective adsorption and catalytic properties in the calcined form that resemble those of the zeolites into which the gels transform if left longer in the reaction mixture. This supports the synthesis mechanism in which structured colloidal units form prior to crystallisation.

The picture that emerges is therefore one in which amorphous precursors assemble from solution and then reorganise to give a colloidal suspension of zeolite crystals similar to that observed by Mintova *et al.* at room temperature for zeolite A.[25] The mechanism of structure direction in the TPA-ZSM-5 system, as proposed by Burkett and Davis on the basis of the experimental

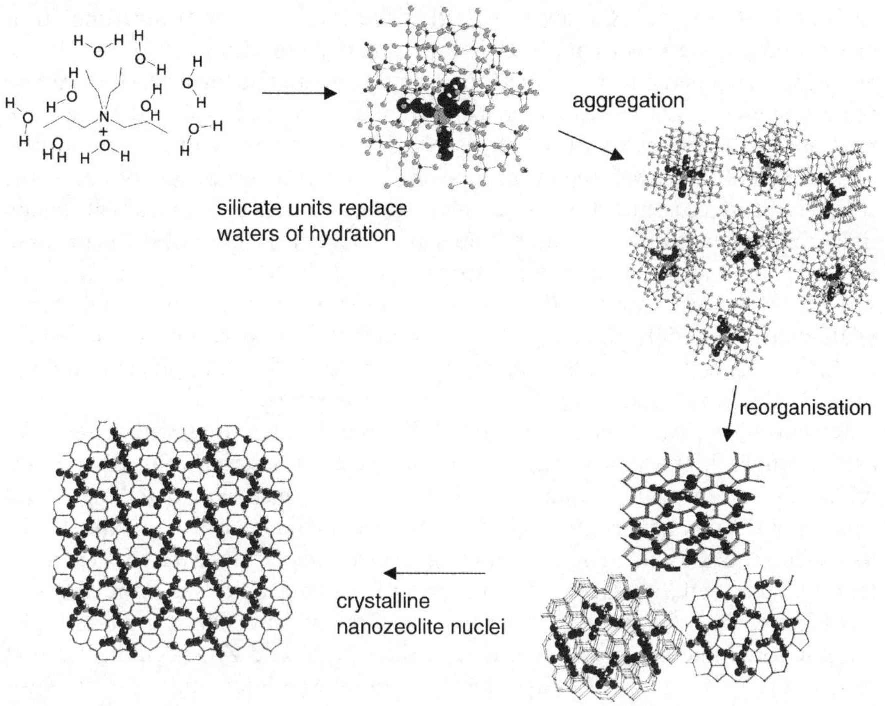

Figure 5.7 A schematic representation of the crystallisation of silicalite, templated by tetrapropyl ammonium ions.

information,[35,36] proceeds as follows (Figure 5.7 – the position of TPA is that taken from crystallographic studies of TPA-silicalite[37]):

Step 1. A coordinating shell of water molecules arranges around the TPA cation. This process of cation hydration is known to occur from other studies of quaternary ammonium cations.
Step 2. Hydrated silicate species interact with hydrated organic cations, resulting in replacement of water molecules in the coordination sphere of the TPA by silicate anions. The hydrophobicity/hydrophilicity of the organic is known to be critical in this stage.[38] (Replacement of TPA with ethanoltripropylammonium ions results in much slower crystallisation of the pure silica ZSM-5, whereas diethanoldipropylammonium does not nucleate the structure.)
Step 3. Agglomeration of these TPA-silica units results in the formation of nuclei which reorganise by bond breaking and making and grow by addition of species from solution to give ordered colloidal zeolites. Further addition of silicate units takes place in the crystal growth regime.

5.3.4 Crystal Growth

The next stage in the crystallisation curve after the induction period is the initial period of rapid increase in growth rate. Pioneering studies by Zhdanov on zeolites Na-A and Na-X showed that all of the crystals grow at the same linear growth rate.[39] He was then able to model the crystallisation curve by considering the combined effects of nucleation and growth. The linear growth rate in these systems increases with increasing temperature and is found to be proportional to the concentration of dissolved silica. At least one essential precursor to zeolite growth therefore exists in concentrations proportional to the soluble silicate concentration. The concentration of dissolved silica will remain in equilibrium with the gel and remains at a constant level until the gel is used up and the concentration of the nutrient species decreases. The linear crystal growth then slows down and the growth curve achieves its sigmoidal shape. Mathematical models that predict the crystallisation curve can be developed for zeolite crystallisations (see Section V.B in reference 2, for example) and find use in predicting large-scale industrial zeolite preparations.

Recent advances in scanning probe microscopies, together with a series of astonishingly high-quality high-resolution electron micrographs of growth-related defects or of zeolite surfaces, taken 'edge-on', have elucidated important aspects of the mechanism of zeolite growth. A crystal face is usually labelled by its Miller Index (hkl). The structure of a face can be described in terms of features such as flat terraces, edges of terraces which drop down from one layer to the next, kink sites where the edge is not straight and islands upon terraces. Atomic force microscopy of synthetic zeolites A, Y and ZSM-5 indicates that growth occurs by a layer-by-layer mechanism by which low index crystal faces grow outwards by the nucleation and subsequent expansion of new layers.[40–42] This growth mechanism has become clear upon the observation of steps that define the edges of growing layers. These steps are usually related to the lattice repeats perpendicular to the face, for example for zeolite A for (100) crystal faces the observed edge steps are 1.20(15) nm, compared to 1.22 nm for the lattice repeat – which comprises one sodalite cage and one D4R unit) (see Figure 5.8). Growth occurs by nucleation on terrace sites and the outward expansion of islands to give layers by addition of species at both edge and kink sites. The cross sections of the growing surfaces are parabolic in cross section, implying that the area of a terrace is proportional to elapsed growth time, as predicted by the model. Furthermore, by modelling of the observed features of zeolite growth on Na-A, it has been possible to estimate for the first time the relative rates of addition at each of the possible sites: kink sites are found to be the most favoured.[40]

In AFM studies of the surfaces of silicalite (Figure 5.9),[41] 'oversized' terrace steps are observed (up to 110 nm on the (100) surface and 20 nm on the (010) face) in addition to edges that are closely related to the crystallographic repeats. These bear no relation to any simple structural element and extension of the layer-by-layer mechanism necessitates that they must be due to a linear defect resulting in the terrace growth being impeded. This is explained in terms of stacking defects within the framework that hinder the growth along a surface. ZSM-5 is known from HRTEM to contain faults due to the arrangement of structural layers through mirror planes rather than inversion centres, and such defects expressed at a surface could lead to the observed defects.

Figure 5.8 Atomic force microscopy (AFM) of the (100) surface of zeolite A supports a model of layer by layer growth, with the events of nucleation on a flat terrace and growth at edge sites and kink sites (the latter are where two straight edges meet). Note the curved terrace edges that result from a combination of the last two growth mechanisms. [Figure from reference 40 with permission. Copyright 2001 Wiley-VCH Verlag GmbH & Co. KGaA.]

Whereas AFM studies such as these give important information on zeolite crystal growth in terms of features a few nanometres in size, individual HRTEM micrographs give important clues to the nature of layer-by-layer growth on an atomic level. Features within crystals can highlight the growth history, whereas 'side-on' projections of the surfaces can indicate how the faces are terminated. There are many examples of defect structures in zeolites, and HRTEM is the method of choice for their study. Zeolite Y, for example, is known to show stacking defects on (111) faces, particularly when caesium cations or mixtures of 15-crown-5 and 18-crown-6 cyclic ethers are added to the synthesis gel. These stacking faults arise from different modes of stacking of layers of sodalite cages as a result of layer-by-layer growth with different orientations in adjacent layers[43] (see Section 5.4.1). Similarly, zeolite Beta grows by the addition of silicate layers to the surface of the structure, and these can stack in any of four possible positions relative to the underlying plane. In two dimensions, TEM indicates the stacking sequence that occurs as the growth proceeds. A recent micrograph (Figure 3.11) shows that if two growing regions on the same layer are nucleated separately but with different stacking orientations in the same layer, they cannot join up properly for two subsequent layers, so that 'double pore' defects are formed. Similar features are observed for the

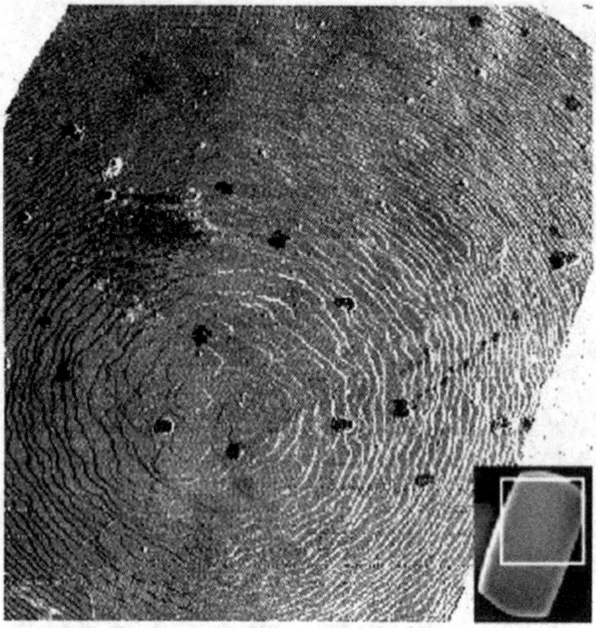

Figure 5.9 AFM of the (010) surface of silicalite, from which relative rates of terrace growth in different directions of the surface have been determined. [Figure reproduced from reference 41 with permission. Copyright 2003 American Chemical Society.]

structurally related titanosilicate ETS-10,[44] strongly supporting the layer-by-layer growth mechanism.

'Edge-on' surface micrographs have been recorded that reveal details of the surface termination structure in zeolites such as Y, Beta and L. For zeolite L, for example, the surfaces appear to show (remembering that they are a projection) that all surfaces terminate in a D6R.[45] This indicates that either D6R-containing units are attached directly from solution, or that once their surface growth has begun, attachment of silicate species to them is very much faster than subsequent outward growth from the complete unit. A similar situation of D6R-terminated surfaces exists for the surfaces of zeolite Y. HRTEM of polymorph C of zeolite Beta,[46] by contrast, indicates that D4R units are the favoured surface termination units for this structure (Figure 5.10), which is consistent with indirect evidence for the importance of this unit in F- and Ge-containing silicate syntheses.

Using such information from AFM and TEM, in collaboration with computer simulation of the distribution of silicate monomers and oligomers in solution and the stability of different faces described in Sections 4.2.4 and 4.6.1, it should become possible to establish details of the growth mechanism on the atomic scale. This could be important in preparing particles of desired sizes and shapes for specific applications.

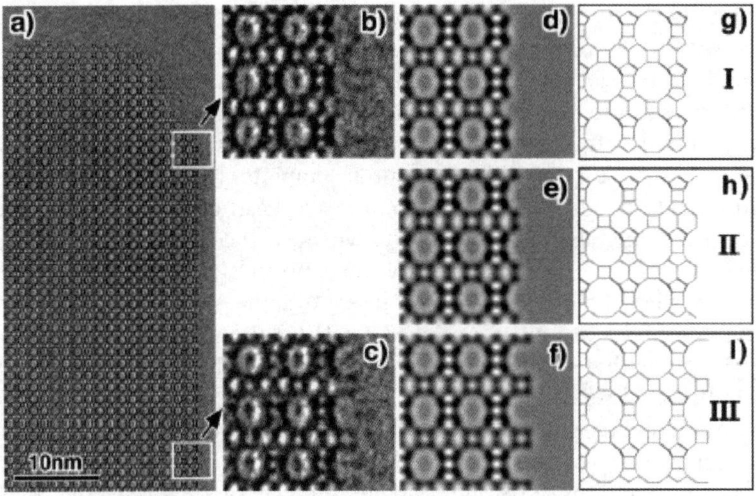

Figure 5.10 HRTEM of a microcrystal of a silicate with the Beta-C structure, showing a projection of the (100) surface termination (a, b, c) and simulated images (d–f). The differences in observed surfaces at (b) and (c) suggest that the surface could be built up (g–i) by the addition of D4Rs (I → III) or via the rapid sequential addition of pairs of 4MRs (I → II → III). [Figure reproduced from reference 46 with permission. Copyright 2002 Wiley-VCH Verlag GmbH & Co. KGaA.]

5.4 Issues in Zeolite Synthesis

5.4.1 Role of the Structure-directing Agent: Designed 'Templates'

The most successful monocationic organic templates have C/N ratios of between 10 and 14.[47] The criteria appear to be that the molecules possess sufficient charge to make them soluble, but a hydrophobicity that matches that of the high silica framework. The organic cations must form strong non-bonding interactions with the framework oxygens as well as electrostatic interactions with charges on the framework. From a practical point of view they must also be stable under the high temperature, alkaline conditions that favour the Hofmann elimination reaction of alkylammonium ions.

The strength of non-bonding interactions will depend in part on the geometric fit and this is well modelled by computer simulation using forcefield calculations (Section 4.4.3). In addition, though, the hydrophobicity of the organic cation is a frequently used concept for understanding the efficacy of the molecule as a template, particularly for pure or high silica zeolites.[38] For example, whereas TPA is known to be an excellent template for silicalite, TEA is too hydrophilic to form high silica zeolites under similar conditions and tetrapentylammonium is too hydrophobic. It is thought that the H_2O hydration sphere around the TEA cation is too strongly held to be dislodged by silicate oligomers. For pure silica zeolite polymorphs, therefore, tetraalkylammonium cations with C/N ratios of *ca.* 12 are favoured for synthesis. The more hydrophilic TEA is an effective template for aluminium-containing zeolite Beta, but does not template pure silica Beta under similar conditions, indicating the need to match the hydrophobicity of the template to that of the zeolite.

When commercially available potential templates had been thoroughly examined, attention shifted to the design and synthesis of molecules that possess features of charge, shape and stability, which, if incorporated intact into the growing framework, must necessarily yield novel structures with interesting pore geometries. Examples of generalised organic synthetic routes to such potential templates are given in Scheme 5.1: (a) the alkylation of amines, for example with methyl iodide, to give the quaternary ammonium cations; (b) the Menschutkin and related substitution reactions of primary organohalides with tertiary or secondary amines, including the use of dihalides to give diquaternary or spiro-quaternary cations; (c) cyclisation, for example by Diels–Alder condensations or reactions of secondary amines to give rigid bicyclic and polycyclic molecules, including 'propellane'-type molecules[48–50]; (d) fused bicyclo organonitrogen class of compounds, described as having ring construction [l.m.n], where n is not 0, prepared from starting cyclic ketones that are converted to imines *via* a Beckman rearrangement reaction[51] and the conversion of (e) ketones and (f) nitriles to trimethylammonium derivatives. Ketones are converted by the Leuckhart reaction,[52] whereas nitriles are converted to the corresponding amines, and then methylated.[53,54]

Very many high silica zeolites have been prepared using organic molecules prepared in these and similar ways. Synthesis of the Mobil zeolite ZSM-18 via the triquat cation shown in Figure 5.4, is often quoted as 'real' templating behaviour – the shape of the triquaternary template is a very close fit to that of the pore system. More recently, Zones and co-workers at Chevron[47,55–57] and Corma and co-workers at the ITQ, Valencia,[58–64] have independently developed this template design approach and many new high silica zeolites have been prepared within the SSZ and ITQ series. Some successful examples are shown in Figure 5.11, along with the organic molecules that directed their crystallisation.

Metal complexes can also be used as structure-directing agents for zeolites. Addition of the crown ether 18-crown-6 to zeolite Y syntheses templates the formation of the hexagonal EMT structure described in Section 2.2.2 because the crown ether complexes the sodium cations and the complex fits closely to one set of cavities unique to the EMT structure.[43,70] This cavity results from the way the layers of sodalite cages are stacked (Figure 2.5). In addition, the extra-large-pore zeolite UTD-1 was prepared using the organometallic permethylated cobaltacinium ion as a template, which stacks within the 14MR channels.[71] In another approach, the pure silica version of zeolite A (ITQ-29) has been prepared by using molecules that 'self-assemble' in dimers *via* π-stacking effects to template the large alpha cages in the structure.[72]

Hypothetical modelling studies of the kind described in Chapter 4 suggest that there is a very large number of possible structure types accessible through

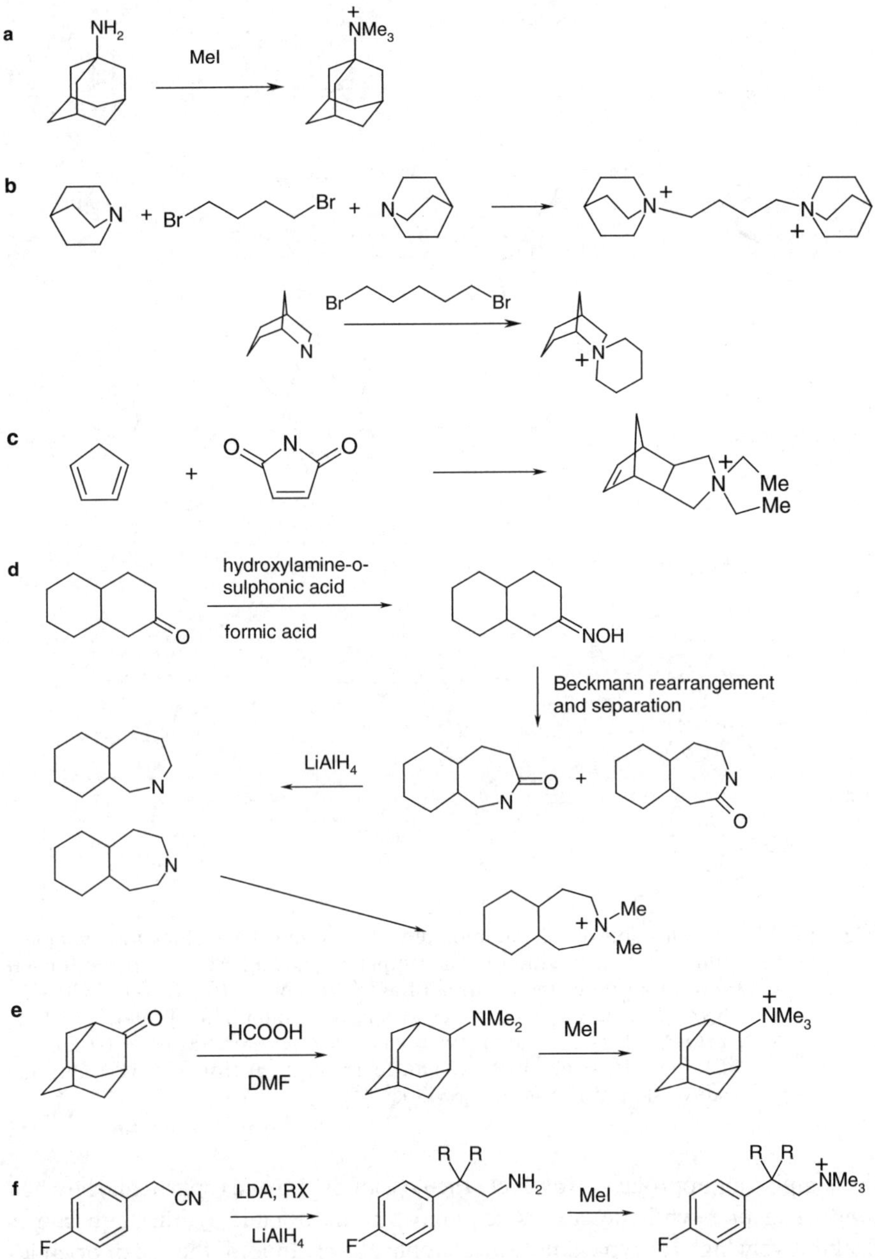

Scheme 5.1 Synthetic routes to novel alkylammonium ions as structure directing agents.

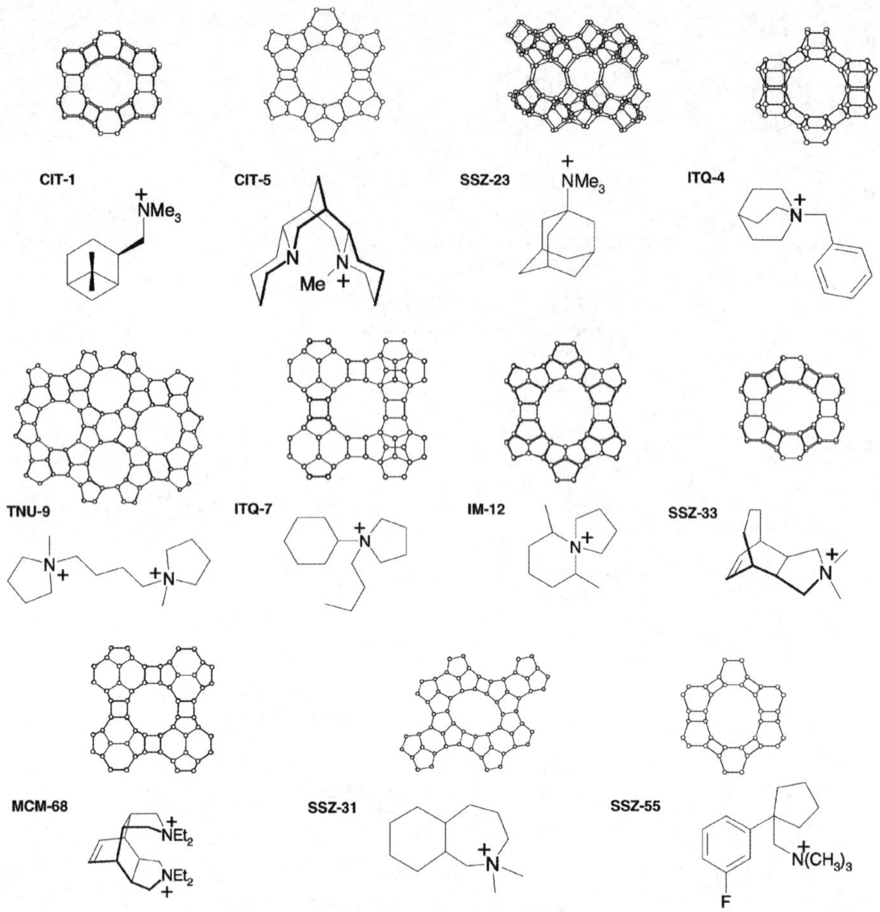

Figure 5.11 A collection of alkylammonium cations and the zeolites they template under certain conditions. The templates are synthesised by (from top left to bottom right) methylation of available amines (for CIT-1,[65] CIT-5,[66] SSZ-23), reaction of bromides with tertiary amines (for ITQ-4,[58] TNU-9, ITQ-7,[67] IM-12[68]), aldol type condensations (for SSZ-33[57] (related to CIT-1), MCM-68[69]), Beckmann rearrangement (for SSZ-31[51]) and the conversion of nitriles (for SSZ-55[54,69]).

the templating approach. However, the cost of such solids, particularly when more than one synthetic step is required for the organic synthesis, becomes prohibitively high for typical industrial applications. Indeed, the use of organics as templates is usually restricted to those produced cheaply for other processes (such as TMA, TEA and TPA), especially since their ultimate fate is to be thermally decomposed to leave the pores. An ingenious modification of the synthesis of ZSM-5 has been suggested by Davis and Zones to avoid this problem, who propose the use of acetals as templates.[73] These are stable under the alkaline conditions of synthesis, but can be hydrolysed by acids after crystallisation and the components reassembled and recycled. The method is of particular interest for zeolites with three-dimensional pore systems, where extraction of the organic is more readily achieved than for one-dimensional structures.

5.4.2 The Fluoride Route

In silicate synthesis performed at high pH, the hydroxide ion acts as a mineraliser, increasing the solubility of silica and catalysing condensation reactions. Fluoride ions can also perform this function, particularly in the synthesis of pure silica polymorphs of zeolites. The use of fluoride was recognised by Flanigen and Patton in 1978,[74] and was developed further by Guth and Kessler in the early 1990s for silicates[75] and phosphates.

The most spectacular successes with the method were achieved by Camblor and co-workers in the mid 1990s with the synthesis of a series of silica polymorphs in the ITQ series of solids using the following general composition as a starting point:

$$SiO_2 : 0.5\ HF : 0.5\ ROH : x\ H_2O$$

They found[76,77] that for syntheses of this kind, by using high concentrations of quaternary ion structure directing agents, low H_2O/SiO_2 ratios ($x < 15$) and working within a pH range of 7–9 and at temperatures of 135–150 °C resulted in the crystallisation of a series of very open framework silicas. Higher concentrations of the SDAs in the reaction mix give more open products. As well as promoting the crystallisations by catalysing hydrolysis of the silicate precursors, the fluoride ions show structure directing action, and tend to favour structures containing small cages. In the open structure ITQ-7,[76] for example, the fluoride appears to favour the formation of [4^6] cages and thereby directs crystallisation to the structure in the presence of an appropriate template (1,3,3-trimethyl-6-azonia tricycle[3.2.1.$4^{6,6}$]dodecane[76] or more recently *N*-butyl-*N*-cyclohexyl-pyrrolidinium – see Figure 5.11). The crystallisation of new phases is therefore a result of combined structure direction between the organic cation and the fluoride anion.

The fluoride ions remain in the structure upon crystallisation so that the product solids have the general formula:

$$SiO_2F_x^- \cdot R_{x/n}^{n+} \cdot yH_2O$$

^{29}Si and ^{19}F MAS NMR and single crystal diffraction show that the fluoride ions are bound to framework silicon atoms, often within small cages, such as those found in octadecasil (AST in a [4^6] cage), nonasil (NON, [$4^15^46^2$]), EU-1 (EUO, [$4^15^46^2$]), silicalite (MFI, [$4^15^26^2$]), ITQ-4, (IFR, [4^26^4]), SSZ-23 (STT, [4^35^4]), chabazite (CHA, [4^66^2]) and ITQ-9 (STF, [$4^15^26^2$]) where they expand the silicon coordination to five-fold (SiO_4F). The silicate-fluoride version of EU-1, for example, in which fluoride is bound to silicon atoms from within the [$4^15^46^2$] cages, is shown in Figure 3.1.[78] The ^{19}F NMR chemical shifts (δ) are very sensitive to the location of the fluoride ion, where, for example, F^- in the $[4^6]$ cage of AST has δ −38 ppm, whereas for nonasil F^- in the $[4^1 5^4 6^2]$ cage, δ is −76 ppm.

In one synthetic germanosilicate-fluoride system, it is possible to prepare mixtures of the A-, B- and C-polymorphs of zeolite Beta.[79] In this case, the fluoride MAS NMR provides a powerful tool to establish the relative amounts of the C-polytype, in which the fluoride occupies D4Rs that are in the structure, compared to the A- and B-polymorphs, which have no D4Rs and in which the fluoride resides in $[4^3 5^4]$ cages, with a clearly different chemical shift.

The fluoride ions balance the charge introduced by the organic cation without the need for negatively charged defects that are observed in the preparation of high silica zeolites from alkaline solution. If the organic and the fluoride ion are removed by calcination, these low-defect pure silica polymorphs have higher hydrophobicity than other solids as a direct result of the conditions employed in their synthesis. Zeolitic aluminosilicates can also be prepared in fluoride media, where the charge balancing is performed by a mixture of the fluoride ions and the aluminium cations substituting for silicon.

5.4.3 Incorporation of Aluminium and Other Heteroatoms

Aluminosilicate zeolites can exhibit Si/Al ratios that vary widely within the same structure type. The amount of aluminium that is incorporated within a structure depends on both the structure and the conditions of its synthesis. For example, zeolites with the LTA topology can be prepared directly over the maximum possible range that obeys the Loewenstein aluminium avoidance principle (Si/Al from 1 to infinity) by different synthetic routes. The most aluminium-rich solid is the usual material prepared using sodium cations, Na-A, whereas the pure silica polymorph has recently been prepared by the group of Corma using the 'supramolecular' templating approach for the alpha-cage previously described.[72] Zeolite ZK-4 (LTA) has been prepared with intermediate silica/alumina ratios by the use of TMA cations in the synthesis in addition to sodium cations. The TMA ions template the sodalite cages. This achievement is far from academic, since whereas Na-A and ZK-4 cannot be prepared in the acidic form, the high silica material can. In addition the high silica form has higher hydrothermal stability.

Zeolites with the FAU structure type can be prepared by direct synthesis with Si/Al ratios over a much narrower range (from 1 to *ca.* 5). The higher values refer to syntheses that use the crown ether 15-crown-5 as an additive.[80] In these cases the sodium complex of 15-crown-5 is included in the structure. The lower charge density of the included complex compared to hydrated sodium cations requires less framework charge for charge balancing. Zeolite Y with much higher Si/Al ratios, required for catalytic applications, must be prepared by post-synthetic treatment (Chapter 6).

It is notable that many high aluminium content zeolites contain 4-membered rings, which may be due to the ease with which such even-numbered ring structures take up higher concentrations of aluminium without violating Loewenstein's principle. At the other end of the compositional range, many of the high silica zeolites prepared with organic cations contain 5MRs, although this is not exclusively the case. Small changes in the aluminium and alkali metal contents in the synthesis can control which phase crystallises, for reasons that are incompletely understood.[81] Table 5.3 illustrates this for syntheses using the template 1,4-bis(*N*-methylpyrrolidinium) butane (see Figure 3.23),[81] where XRD patterns confirm the series of different zeolites prepared with changing aluminium contents in the gel.

It is also the case that some of the high silica zeolites prepared in the laboratory using organic cations are found as rare natural minerals with much higher aluminium content, cation contents rich in calcium and no organic cations: examples include gottardiite (natural equivalent to NU-87), mutinatite (ZSM-5) and tschernickite (Beta). These natural minerals provide both inspiration and a challenge for zeolite synthetic chemists to prepare novel material compositions at reasonable cost that would have applications in adsorption or ion exchange.

During the synthesis of aluminosilicate zeolites, the aluminium tends to remain in the solid phase, whether in the gel or in the crystalline product. It does not necessarily end up evenly distributed throughout the crystals,

Table 5.3 Representative products obtained using 1,4-bis(*N*-methylpyrrolidinium) butane (1,4-MPB) as an organic SDA.[a]

Gel composition		*Time (days)*	*Product*[b]
SiO_2/Al_2O_3	$NaOH/SiO_2$		
60	1.00	14	TNU-10
60	1.00	14[c]	mordenite
15	1.00	14	analcime
30	1.00	7	TNU-10
120	1.00	14	TNU-10
240	1.00	14	analcime + TNU-10
60	1.13	14	analcime
60	0.87	14	IM-5 + TNU-9
60	0.73	14	TNU-9
40	0.73	7	MCM-22
∞	0.73	7	MCM-47
60	0.60	7	quartz + ZSM-12
60	0.47	7	ZSM-12
∞	0.47	7	MCM-47
60	0.33	7	ZSM-12

[a] The oxide composition of the synthesis mixture is $4.5R \cdot xNa_2O \cdot yAl_2O_3 \cdot 30SiO_2 \cdot 1200H_2O$, where R is 1,4-MPB, and x and y are varied between $5.0 \leq x \leq 17.0$ and $0.0 \leq y \leq 2.0$, respectively. All the syntheses were carried out under rotation (100 rpm) at 160 °C, unless otherwise stated.
[b] The product appearing first is the major phase, and the product obtained in a trace amount is given in parentheses.
[c] Run performed under static conditions. [Reproduced from reference 81, with permission. Copyright 2004 American Chemical Society.]

however. ZSM-5 crystals prepared in the presence of TPA and sodium cations showed clear evidence of zoning, with silica-rich cores (resulting from nucleation around hydrophobic TPA) surrounded by rims richer in aluminium.[82]

The ordering of aluminium on the atomic scale within the crystal structure of zeolites is also of interest, since it will affect the location of framework charge and can affect the strength of any Brønsted acid sites that are generated in high silica structures. In a remarkable study of the ^{29}Si NMR of a range of fifteen as-prepared zeolites X and Y with different Si/Al ratios (from 1.26 to 5.31), Melchior *et al.*[83] addressed the question of aluminium ordering in the faujasite structure. Although at first sight there are only five peaks, corresponding to silicon in one crystallographic site with from 0 to 4 aluminium nearest neighbours, the authors were able to deconvolute the peaks to resolve resonances from environments with different numbers of aluminium atoms in the second nearest tetrahedral neighbour shell. Careful modelling indicates an ordering scheme in which Al–O–Si–O–Al linkages are more strongly avoided within the D6R sub-units of the FAU structure than are the same linkages between different D6R units, an observation that implies that the D6R is the tertiary building unit in FAU crystallisation, as suggested by TEM and AFM. Furthermore, the results are consistent with 4MRs being the secondary building unit, indicating the crystallisation pathway:

$$4MR \rightarrow D6R \rightarrow FAU$$

For high silica zeolites templated by organic cations, there is growing evidence that the position of the aluminium in the framework is correlated with the location of the positive charge on the organic template. For example, careful CP MAS and REDOR NMR studies on as-made Al-ZSM-12 templated by benzyltrimethylammonium ions ($C_6H_5CH_2N^+(CH_3)_3$) show that the methylene protons are preferentially located near the silicon atoms adjacent to the included aluminium, so that the framework aluminium is closely associated with the template's positive charge.[84]

The synthesis of silicates that contain metal cations other than aluminium in framework positions results in solids with modified adsorptive and catalytic properties. The criteria for successful incorporation of cations into tetrahedrally coordinated silicate frameworks are that they should exhibit solubility under synthesis conditions without the formation of an insoluble oxide or oxyhydroxide and that the substituting species should be able to adopt tetrahedral coordination. Preparation of a phase that is shown to contain other metals (by selected area chemical analysis in an electron microscope, for example) is no guarantee that the metal has adopted a framework site. Physicochemical methods must be used to determine whether this is the case. This is complicated when the metal adopts more than one site or is present at very low levels. Confirmatory evidence for framework substitution may be available from unit cell size determination (substitution of a cation larger than Si^{4+} will in general result in an increase in unit cell dimension), and determination of coordination geometry by NMR, UV-visible and EXAFS spectroscopies (Chapter 3), which are able to distinguish whether the metal is within the silicate or in nanoparticles of its own oxide, and whether it is tetrahedrally coordinated or not. IR and Raman spectroscopies are also sensitive to new vibrational modes that result from framework substitution.

Incorporation of the elements Li, Be, B, P, Ti, Fe, Zn, Ga, Ge and Sn into fully tetrahedral silicate or aluminosilicate frameworks has been demonstrated convincingly (Chapters 2 and 3). The evidence for incorporation of other transition metals is more ambiguous and in general for transition metals particular care must be taken in preparation to ensure the metal cations are not precipitated as unreactive hydroxides or oxides. In some cases incorporation of the heteroatom results in novel structures – the zincosilicates VPI-7[85] and -9,[86] germanosilicates of the ITQ series described recently by the group of Corma[10,87] and some gallosilicates[88] are good examples. It is thought that the different sizes and electronegativities of the substituting atoms affect the stability of the metallosilicate building units during the synthesis. For example, incorporation of beryllium is found to favour 3MRs and inclusion of germanium favours 4MRs and D4Rs even at relatively low levels of incorporation. The novel 18MR germanosilicate ITQ-33,[10] for example, which is prepared in the presence of fluoride and germanium, possesses both 3MRs and D4Rs.

Inclusion of these cations does impart new catalytic activities, but in many cases the active site results from a metal ion that has left the framework and entered the pore space upon heating, especially in the presence of water vapour. This is thought to be the case for zinc- and gallium-containing solids used in the dehydrocyclisation of butane and propane to aromatics in the Cyclar process (Chapter 9). Boron, iron, chromium and vanadium all appear to leave the framework under harsh conditions. The incorporation of titanium and more recently tin into framework sites within silicates have become very important substitutions, because both titanosilicates and stannosilicates have been shown to contain stable Lewis acid sites of importance in selective oxidation catalysis. The metal atom can coordinate additional water molecules in the as-prepared material, but these can be removed by heating. In the synthesis of titanosilicates, titanium is usually added to the gel as the alkoxide, and synthesis performed in the absence of sodium hydroxide to avoid precipitation of sodium titanate or nanoparticulate titanium oxides.

The incorporation of phosphorus into aluminosilicates was already noticed by Kuhl in 1971,[89] but remained the subject of discussion as to whether it was included in framework sites until this was confirmed by re-examination in 1990. By changing the pH of synthesis, Wilson and Flanigen were able to form the aluminophosphate family of solids, and open the way to a multitude of new materials.

5.4.4 Modifying Crystallite Size: Nano- and Giant Zeolite Crystals

Typical syntheses of microporous silicates result in the preparation of fine powders with particle sizes of a few microns, which may be agglomerates of many crystallites. For phosphates and hybrids, crystals at least an order of magnitude larger are more usual, as a result of the different reactant and product solubilities. Whereas particle sizes of the order of microns are acceptable for most applications of zeolites, the preparation of crystals that are either larger or smaller can be very useful.

5.4.4.1 Nanozeolites

The synthesis of zeolite nanocrystals can take place from clear solutions as single colloidal gel particles crystallise to give crystals 40–80 nm in size. Mintova and Bein have developed this method and have prepared a number of zeolites (and aluminophosphates) by this route.[25] Investigation of nanozeolites has given much information on the first steps of zeolite crystallisation and the nanozeolites themselves have many potential applications.

The synthesis of colloidal zeolites is best achieved at relatively low temperature from clear solution under conditions of high supersaturation of the silicate species. This favours nucleation over growth. High concentrations of organic structure-directing agents (rather than inorganic cations) act to keep the silicate species in solution and limit the aggregation of the proto-zeolites. The review of Tosheva and Valtchev[90] documents many of the successful synthetic routes to nanozeolites. Colloidal zeolites can be collected from their synthesis mixture by centrifugation or by using semi-permeable membranes (rather than filtration) and care must be taken to avoid their post-synthetic aggregation. Once redispersed, they can be used as coatings, for example by making use of their negatively charged surfaces to bind them to surfaces that are made positively charged. Calcination usually results in irreversible aggregation, so nanozeolites prepared using organic templates must be calcined once 'in position'. The application of nanozeolites is discussed further in Chapter 10.

5.4.4.2 Large Single Crystals

The synthesis of crystals of *ca.* 30 micron or more in size enables structure solution *via* laboratory single crystal diffraction. The size limit is still smaller given access to synchrotron-based diffractometers. Although structure solution from powder data is possible, it is not routine, and single crystal structures give bond lengths and angles of the framework and the positions and occupancies of extra-framework species much more accurately than is possible by refinement of powder diffraction data. Large single crystals are also of use in studies of diffusion by PFGNMR, frequency response spectroscopy, interference and IR microscopy and other techniques that benefit from the longer intra-crystalline diffusion pathways they present (see Chapter 7). Recent studies of the mechanism of growth by AFM have also benefited greatly from the availability of large single crystals with well-defined crystal faces.

Lethbridge *et al.*[91] have comprehensively reviewed the reported syntheses of large silicate crystals, and have identified routes to larger silicate crystals. These

Table 5.4 Summary of routes to large zeolite crystals (after Lethbridge *et al.*[91]).

Route	*Comments*	*Structure type examples*
Charnell method	Addition of aluminium chelating agent ethanolamine or related compounds	LTA, FAU
Fluoride route	Addition of fluoride ions, as HF or complexes with amines. Operates at pHs down to 7.	EUO, FER, IFR MFI, MEL, MTT, MTN, TON, STT
High P and T	Hydrothermal, temperatures up to 700 °C	ANA, CAN, FAU, GIS, HEU, MFI, NAT, OFF, SOD, STI, THO
Bulk silica dissolution	Silica source is bulk (unreactive) silica	ANA, MFI, MOR, SOD
Gel diffusion	Slow counter-diffusion of reagents	KFI, LTA, FAU
Zero gravity	Crystallisation in space in absence of temperature fluctuations or vibrations	FAU, MFI

are summarised in Table 5.4. In all cases the underlying principle is to reduce the nucleation rate relative to the growth rate, which may itself be slow. The classic Charnell syntheses of zeolites A and X, for example, involves the addition of triethanolamine to an otherwise inorganic synthesis, resulting in single crystals up to 125 μm in dimension. Studies suggest that the triethanolamine suppresses nucleation and chelates aluminium in solution, reducing the rate at which species can be delivered to the growing crystal surface.[92] More recently, Shimizu *et al.* have described a 'bulk silica dissolution' route, where bulk silica acts as the source and releases nutrients slowly, resulting in crystals of silicalite that are several mm in dimension.[93] The use of pieces of silicate glass of different composition under very high P and T conditions have been also shown to give large crystals of a number of zeolites with natural analogues. Interestingly, these syntheses are without organic templates, and the silicate product depends on the inorganic composition of the glass. Of all the methods, the fluoride route described earlier is the most generally accessible and applicable, particularly for high silica zeolites. Large crystals of several framework types suitable for single crystal diffraction have already been prepared in this way.[78,94–96]

5.5 Synthesis of Other Microporous Solids

5.5.1 Aluminium and Other Metal Phosphates

Many of the same general principles of gel preparation, nucleation and crystal growth apply to the synthesis of metal phosphates. One major difference is that the syntheses are best performed in aqueous solution at pH values close to neutral. This results in a different chemistry of metal ion speciation in solution and it is found that metal cations may readily be incorporated directly during

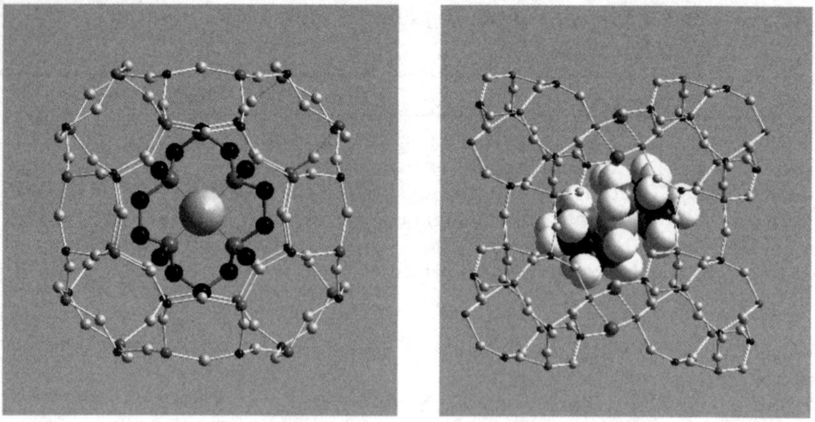

Figure 5.12 Metal amine complexes can act as templates for the crystallisation of framework aluminophosphates: $[Ni(tetramethylcyclam)]^{2+}$ templates the larger B cage of the STA-7 structure (the figure, left, shows the experimentally determined position of the tetramethylcyclam complex, shown without hydrogen atoms); $[Ni(diethylenetriamine)_2]^{2+}$ directs the synthesis of the aluminophosphate fluoride, UT-6 (right).

synthesis. For the aluminophosphate ($AlPO_4$) composition, charge considerations result in strict alternation of the aluminium and phosphorus in adjacent cation sites, so that Al/P ratios in tetrahedral three-dimensional framework solids of this type are close to unity. This restricts the possible variation of synthesis parameters. At the lower pH values of these preparations, amines are likely to be protonated, and they have found to be as effective as structure directing agents as alkylammonium cations. Tertiary amines have been found to be useful, whereas secondary and especially primary amines more typically result in the formation of layered or chain aluminophosphate structures, where the nitrogen atoms of the amines are hydrogen-bonded to phosphate oxygens that are not shared with framework cations. The organic amine or hydroxide has a dual function. In addition to acting to promote the crystallisation of a porous solid by structure direction, it must also be present at sufficient levels to bring the pH close to 7 (suitable for open framework solids to crystallise) because of the presence of phosphoric acid. The addition of alkali metal hydroxides to arrive at a similar pH favours the preferential formation of less open frameworks. Under the nearly neutral conditions of aluminophosphate synthesis, species other than alkylammonium ions may also be used as structure directing agents. These include azamacrocycles or linear polyamines containing complexed metals such as copper(II) and nickel(II).[97,98] At pH values close to 7, these retain the metal cation during synthesis and become included in the crystals. Figure 5.12 shows the configuration of the nickel tetramethylcyclam complex within a cage of STA-7, as determined by single crystal diffraction and modelling, and the location of a nickel(diethylenetriamine)$_2$ complex in $AlPO_4$(F)-34, as determined from modelling and powder diffraction.

Table 5.5 gives template-structure relationships for some of the better-known aluminophosphate based solids.

Incorporation of charged amines or alkylammonium ions into a neutral aluminophosphate framework necessitates charge balance by the incorporation of negative ions, such as hydroxide or fluoride. The fluoride synthesis route is very fruitful for the synthesis of aluminophosphate fluorides, where the fluoride ion often occupies well-defined sites. The ability of framework aluminium to expand its coordination sphere to become five- or six-fold coordinated is a key feature of these materials, and results from the greater ionicity of aluminosilicates compared to aluminophosphates. Although aqueous synthesis is usually the preferred route, aluminophosphates can also be prepared from reaction gels made up in non-aqueous solutions, such as pyridine:HF, ethylene glycol:HF and imidazolium:HF.[5,6] Under these conditions fluoride is a very effective mineraliser, and structures prepared in this way can contain aluminium with coordination numbers greater than 4.

Substituted aluminophosphates are of interest as acid and oxidation catalysts. Typical substitutions include M^{2+} for Al^{3+} (M = Mg, Mn, Fe, Co, Zn), Si^{4+} for P^{5+} and $2Si^{4+}$ for $Al^{3+} + P^{5+}$ (described in Chapters 2 and 3). In cases where the degree of substitution is at a few percent, similar methods to those for substitutional metallosilicates can be used for confirmation (measurement of unit cell dimensions, ^{31}P MAS NMR for measuring the substitutions of cations for Al and ^{29}Si NMR for measuring the mode of silicon substitution). For some of the metals, much higher metal substitutions are possible than for silicates, and complete replacement of aluminium by cations such as cobalt in tetrahedrally connected frameworks has been reported (see Chapter 2). A number of other metals have been claimed to have been substituted into the aluminophosphate framework, including Ti^{4+} and Cr^{3+}.[103,104]

In many of the cases described above the substitution is aliovalent, that is the substitution of a cation by another of different, and usually a lower, charge. This results in the framework having a negative charge. This is usually balanced by positive charge(s) on the organic molecule used as a template. For some metalloaluminophosphates the degree of negative charge on the framework, and therefore the degree of cation substitution, is determined by the sum of the positive charge on the templates that can be taken into the channel system. As is the case with the aluminosilicates, a range of metalloaluminophosphates has been prepared *via* the use of rational template design. Mono-, di- and triquaternary cations shown in Table 5.5 have successfully templated novel MAPO and SAPO materials (see also Figure 5.4): in a number of these cases no unsubstituted $AlPO_4$ versions of the structure have been obtained.

The crystallisation of aluminophosphate-based solids is particularly well suited to *in situ* study by X-ray diffraction[105] and NMR, because the syntheses are in general much faster than those of zeolites and because of the high abundance of NMR active nuclei in these gels (^{27}Al, ^{31}P, ^{29}Si, ^{13}C, ^{19}F). The faster growth rates arise because the solubility of aluminophosphate species are much higher and growth rates are consequently faster than those of zeolites. The incorporation of metal ions into the reaction mixture tends to accelerate

Table 5.5 Organic species used in the synthesis of some aluminophosphate-based microporous solids. Synthetic conditions to be found in Robson's Verified Syntheses[1] and references within the Atlas[99] and in the review of Patarin *et al.*[100]

Template (+ co-template)	*$AlPO_4$-number (Actual materials formed)*	*Topology*
tri-*n*-propylamine, TPA, *N*-methyl, dicyclohexylamine	$AlPO_4$-5 ($AlPO_4$, MAPO, SAPO)	AFI
di-*n*-butylamine	$AlPO_4$-8	AET
di-*n*-propylamine	$AlPO_4$-11 ($AlPO_4$, MAPO, SAPO)	AEL
hexamethonium	$AlPO_4$-17 ($AlPO_4$, MAPO, SAPO)	ERI
TEA	$AlPO_4$-18 ($AlPO_4$, SAPO)	AEI
di-*n*-propylamine 1,4-diazabicyclo[2.2.2]octane-based polymeric electrolytes	$AlPO_4$-31 ($AlPO_4$, MAPO)	ATO
TEA	$AlPO_4$-34 ($AlPO_4$(F), MAPO, SAPO)	CHA
tri-*n*-propylamine, ethyl,dicyclohexylamine	$AlPO_4$-36 ($AlPO_4$, MAPO)	ATS
TPA + TMA	$AlPO_4$-37 ($AlPO_4$, SAPO)	FAU
TPA	$AlPO_4$-40 ($AlPO_4$)	AFR
di-*n*-propylamine	$AlPO_4$-41 ($AlPO_4$, MAPO, SAPO)	AFO
TMA (+ diethanolamine) Kryptofix K222	$AlPO_4$-42 ($AlPO_4$, MAPO, SAPO)	LTA
di-*n*-propylamine	$AlPO_4$-46 (MAPO)	AFS
di-*n*-propylamine	$AlPO_4$-50 (MAPO)	AFY
TEA + tri-*n*-propylamine	$AlPO_4$-52 ($AlPO_4$)	AFT
N, N, N', N'-tetramethylhexane-1,6-diamine tetramethylenediquinuclidinium	$AlPO_4$-56 (MAPO, SAPO)	AFX
di-*n*-propylamine	VPI-5 ($AlPO_4$)	VFI
decamethonium	DAF-1 (MAPO)	DFO
heptamethylenediquinuclidinium	STA-1 (MgAPO)	SAO
tetramethylenediquinuclidinium	STA-2 (MgAPO, SAPO, $AlPO_4$)	SAT
1,3,5-tris(triethylammoniomethyl)benzene	STA-5 (MgAPO)	BPH
tetramethylcyclam	STA-6 (MAPO)	SAS
tetramethylcyclam (+ TEA) cyclam (+TEA)	STA-7 (MAPO, SAPO)	SAV
TMA	UIO-7 ($AlPO_4$, MAPO)	ZON
Hexamethonium	EMM-3[101,102]	

the crystallisation rate with respect to the pure aluminophosphate, presumably by reducing the energy of formation of nuclei interacting with positively charged organic cations. *In situ* XRD and Raman studies of the formation of aluminophosphates-5 and -34 using tetraethylammonium as a structure-directing agent by the group of Weckhuysen[106,107] indicate that the role of added divalent metal (Zn^{2+}) in the crystallisation is to act as a nucleating agent and

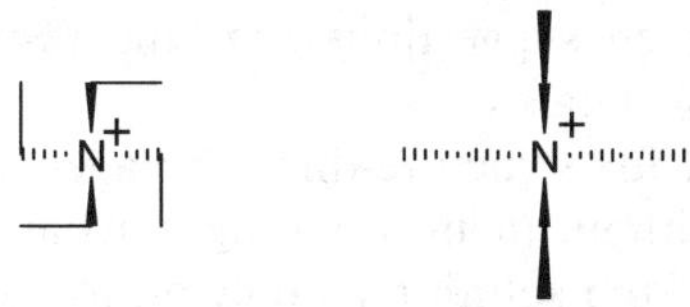

Scheme 5.2 The tg.tg (left) and tt.tt (right) conformations of tetraethylammonium ions.

favour the formation of the $AlPO_4$-34 structure, and also that inclusion of the zinc favours the tetraethylammonium ion adopting the tg.tg conformation in preference to the tt.tt conformation (see Scheme 5.2, below), which fits neatly into the CHA cage. The tt.tt conformation is adopted in AlPO-5, which has a channel structure that encloses the cation less effectively.

In situ studies have given important insights into the synthesis of phosphate solids, where there appears more possibility of observing building blocks in solution (Pre-Nucleation Building Units or PNBUs in the terminology of Taulelle[108]) than in silicate systems. *In situ* studies of the synthesis of aluminophosphate $AlPO_4$-CJ2, for example, which contains both five- and six-fold coordinated aluminium in SBUs of the as-prepared crystal form, reveal a well-defined peak in the ^{27}Al that is attributed to five-fold aluminium. PNBUs of this type are thought to 'clip' together to form the $AlPO_4$-CJ2 crystals. Furthermore, recent study of the synthesis of $AlPO_4$-34 (framework type CHA) and related phases by both XRD and NMR has shown that a layered precursor is involved in the synthesis, and that 4MRs are present in different structural units throughout the crystallisation.[109,110]

5.5.2 Metal-organic Frameworks

Very many microporous organic-inorganic hybrids have been prepared, through the reaction of amines, phosphonic acids and particularly carboxylic acids with metal cations. The syntheses have been variously conducted at temperatures from room temperature to 220 °C, and in water and different organic solvents (diethylformamide, for example) and even in the absence of solvent. The main difference between these syntheses and those of silicates and phosphates is that the geometry of at least part of the framework (the organic part) is predetermined. Furthermore, a desired geometry of metal complex, such as the paddle wheel unit of HKUST-1, the Zn_4O_{13} cluster of MOF-5 or the M_3O_{16} trimers of MIL-88, MIL -100 and MIL-101 can also be achieved by conducting the synthesis in a controlled manner. This has been demonstrated experimentally by Surblé *et al.*,[111] who have performed EXAFS measurements at different stages of the synthesis of Fe-MIL-89 (closely related to MIL-88), and shown that trimeric iron oxide secondary building units remain intact from starting acetate, *via* an amorphous phase through to crystallisation into the final solid. There are even suggestions that the resultant products are the thermodynamically stable phases.[112]

There is no real need for structure-directing agents in these hybrids – the components assemble without them. Building units are expected to add to the growing crystal surface from solution as they do for zeolitic frameworks.

5.6 High-throughput Synthesis

The last decade has witnessed an explosion in the structural and compositional variety of solids prepared *via* the hydrothermal and solvothermal methods outlined above. The number of experimental variables suggests the need for many exploratory syntheses and so these systems are suitable candidates for combinatorial, or at least high-throughput, approaches, in which trays of different gels can be prepared automatically, heated hydro- or solvothermally, filtered and characterised automatically by X-ray diffraction. Such approaches have been developed and implemented by a number of research groups, including those within industry at SINTEF[113] and UOP,[114] and in academia at ITQ[115] and Munich.[116] These have served both to discover new phases and to comprehensively delineate crystallisation fields within triangular 'composition space' diagrams. The discovery of the extra-large-pore germanosilicate ITQ-33[10] using high-throughput methods under conditions that are easily accessed but not 'typical' (as shown in Figure 5.3) is a remarkable product of this approach – there will be many more.

5.7 Synthesis of Ordered Mesoporous Solids

5.7.1 A General Synthesis Pathway to Mesoporous Solids

The synthesis of mesoporous silicas proceeds by a different route from that of the crystalline materials described above, and involves the assembly of aggregates of surfactant micelles and silicate species into regular, ordered arrays, followed by condensation of the silicate to give the solid inorganic component. Figure 5.13 shows the scheme originally proposed by workers at Mobil for this process[117] and discussed in more detail by Monnier *et al.*[118] Subsequent *in situ* NMR and X-ray diffraction studies of the first stages of the assembly support this general model.

The steps in a typical synthesis are therefore:

(1) *Formation of surfactant micelles or aggregates, in which the ionic head-group is solvated by water and ions in the solution.*

Mesostructured organic-silica structures are formed with surfactant concentrations above the critical micelle concentration (CMC), indicating the importance of this first self-assembly step. The thermodynamically favoured shape of the micelles in this dilute regime (without inter-micelle interactions or presence of silicate ions) has been shown by

Israelachvili[119] to depend on the value of a dimensionless packing parameter, g,

$$g = V/a_o l_c$$

where V is the effective volume of the hydrophobic chain of the surfactant, a_o is the mean micelle surface area per hydrophilic head group and l_c is the critical length of the hydrophobic chain. Different amphiphile/liquid crystal geometries correlate with different ranges of g (Table 5.6).

For example, whereas low area headgroup surfactants (high g values, $g = 1$) favour lamellar liquid crystal type phases or cubic phases with low curvature, such as the Ia-3d bicontinuous gyroidal structure, surfactants with higher area headgroups with lower g values favour strongly curved micellar shapes, such as cylindrical rods (MCM-41) or spheres (*e.g.* SBA-1, SBA-2).

(2) *Hydrolysis of silicate precursors, typically alkoxides,* $Si(OR)_4$, *to give silicate species in solution.*

The rate of this hydrolysis step, which is important in determining the degree of order in some systems, depends on the precursor, pH and the presence of mineralisers such as fluoride ions that catalyse the hydrolysis. For example, tetramethoxysilane hydrolyses more rapidly than tetraethoxysilane or higher homologues. Hydrolysis rates are of particular importance where mixtures of siloxanes are present (for example including functionalised ones) and are to be co-condensed to give homogeneous solids. In this case their hydrolysis must be at similar rates.

(3) *Association of the silicate species with the micellar arrays, as they replace water and anions in the coordination sphere of the micelle headgroups, and subsequent clustering of the micelles.*

This has the effect of reducing repulsive interactions between the micelles, while the long-range attractive van der Waals interactions remain and result in assembly of the micelles. That mesoporous silicas only form above the CMC and that silicate species must be added to induce their precipitation confirms the involvement of silica-covered micelles. Electron microscope images of the quenched early stages of MCM-41 formation (Sadasivan *et al.*,[121] Figure 5.13(b)), suggest how this may occur. At this stage the material is a soft solid, so that its

Table 5.6 Effect of *g*-parameter on sequence of stable surfactant-water phases.[120]

g	*Symmetry of stable surfactant-water mesophase*
1/3	Cubic Pm3n
1/2	Hexagonal p6m
1/2–2/3	Cubic Ia3d (gyroid surface)
1	Lamellar

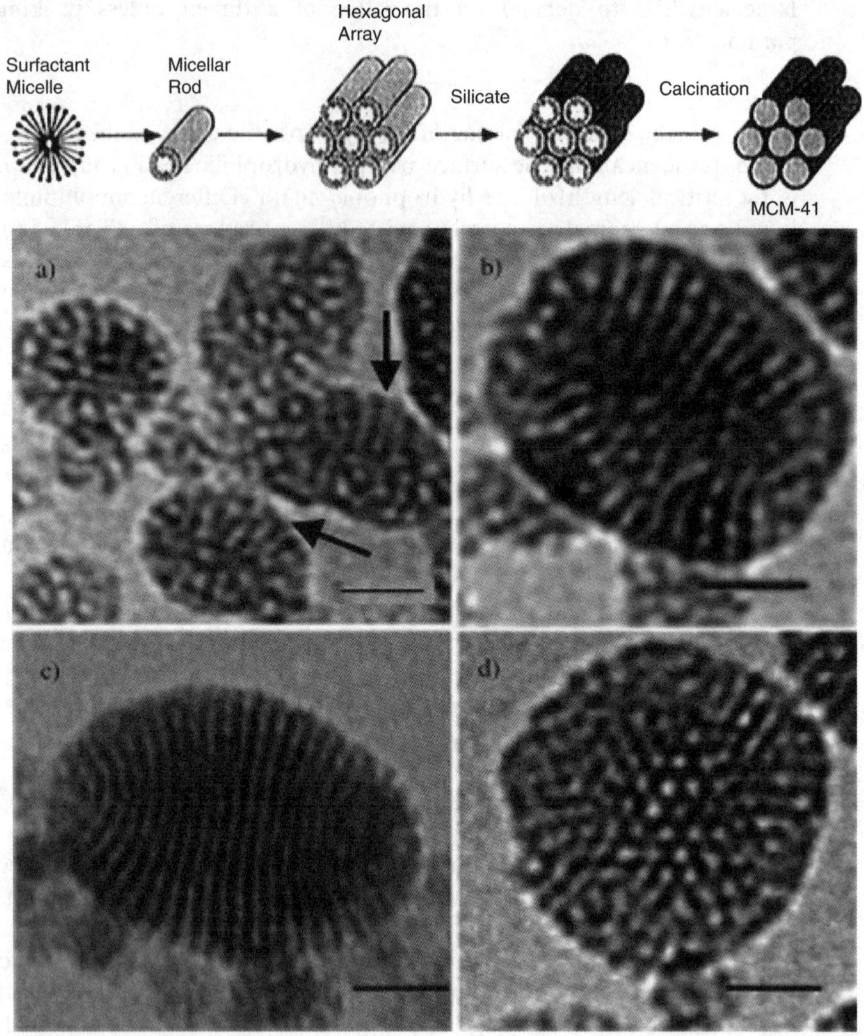

Figure 5.13 (Above) The aggregation of silica-coated cylindrical micelles, followed by condensation, is the commonly accepted model for the synthesis of MCM-41. [Reproduced from reference 117 with permission. Copyright 1992 American Chemical Society.] (Below) TEM images of nanoparticles prepared by quenching the synthesis of MCM-41 (a–d) suggests that under these conditions a soft disordered aggregate of silica-coated micelles reorganises to a low-energy hexagonally arranged mesophase before full condensation occurs. [Reproduced from reference 121 with permission. Copyright 2002 Wiley-VCH Verlag GmbH & Co. KGaA.]

microstructure can adjust to an energetically favoured one. Although the *g* parameter is strictly only for the dilute regime, the relative trends in stability of liquid crystalline phases remain valid, so that lower values of *g* favour composites with highly curved interfaces.

(4) *To form a mesoporous solid, the silicate species surrounding and separating the surfactant regions must undergo condensation.*

This bond-forming event may overlap with the hydrolysis step or may occur in a second, separate step. If the silicate-surfactant assembly is able to form under equilibrium conditions before the onset of crystallisation, very well-ordered solids can result, even showing crystalline morphologies. If the two steps overlap, the products are less likely to show long-range order. During condensation, the amount of Q^4 silica increases and the silica becomes solid.

(5) *To render the solid porous, the template can be removed by solvent extraction or calcinations at high temperature. The latter results in further condensation, with an associated reduction in unit cell and pore size.*

5.7.2 Mechanisms of Silicate Mesophase Formation from Aqueous Solution

There are several important variations of this general process. Mesoporous solids can be prepared under alkaline, acid or neutral pHs, with cationic, anionic or neutral surfactants of different types.[122] Furthermore, the synthesis can be influenced by the concentration and type of anions in the solution, by organic additives that concentrate within the core of the micelles and by the inclusion of other framework-forming cations. Aluminium, boron and titanium are among the many cations that can be introduced into mesoporous silicas, depending on whether acidic or alkaline synthesis conditions are adopted.

A list of different types of mechanism is given in Table 5.7, based on the species present in the final composite mesophase. The isoelectric point of silica is around 2, so that syntheses at pH values above this will involve anionic silicates, whereas under strongly acidic conditions the silica will carry a positive charge. The inorganic matrix, I, is represented as I^- or I^+, respectively, and the surfactant as S^+ (cationic), S^- (anionic) or S^o (uncharged). X^- represents anions within the solution.

Syntheses of MCM-41 and MCM-48 both proceed *via* the $S^+ I^-$ mechanism, where under alkaline conditions anionic silicates assemble around hexagonal p6mm and cubic Ia-3d arrays of the surfactant cetyltrimethylammonium ($CH_3(CH_2)_{19}N^+(CH_3)_3$). Formation of the cubic structure, which has a lower surface curvature, is favoured by conditions where the effective surfactant headgroup area is larger (due to solvation effects, for example). Using the dicationic surfactant $CH_3(CH_2)_{15}N^+(CH_3)_2CH_2CH_2CH_2N^+(CH_3)_3$, (C16-3-1), which has a very large headgroup (giving a low characteristic *g* value) and favours spherical micelles, the mesocage structure SBA-2 is formed.[123] As shown in Figure 3.14, the structure of SBA-2 is based on a close packing of spheres, which are arranged in both cubic and hexagonal close packed arrays. It is also possible to prepare mesoporous solids with hybrid organic-inorganic frameworks *via* this mechanism by using organically bridged siloxanes (of the

Table 5.7 Mechanisms for the synthesis of mesostructured solids.

Mechanism	*Surfactant type*	*Synthesis pH*	*Mesophase structures and symmetries*
$S^+ I^-$	Cationic	Basic	MCM-41 (p6mm), SBA-2 ($P6_3$/mmc/Fm3m)
$S^+ X^- I^+$	Cationic	Acidic	SBA-1 (Pm3n), SBA-16 (Im3m)
$S^- I^+$	Anionic	5–10	AMS-n
$S^o H^+ X^- I^+$	Neutral (block copolymers)	Acidic	SBA-15 (p6mm), FDU-5 (Ia-3d), FDU-12 (Fm3m), SBA-16 (Im3m)

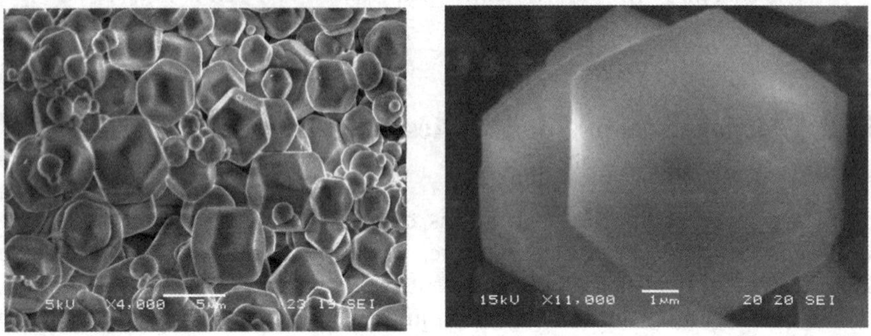

Figure 5.14 Crystalline shapes of mesoporous silicas such as SBA-1 (left) and FDU-12 (right) formed under acidic conditions indicate that the mesophase organises into regular shapes as it forms as a soft solid during the hydrolysis stage, becoming lithified during the later condensation stage.

form $(CH_3O)_3Si$-R-$Si(OCH_3)_3$, R = CH_2, C_2H_4, C_6H_4, *etc.*) as the silicate precursor.

Mesoporous silicas can also be prepared using cationic templates under strongly acidic conditions. The silica is then also positively charged, so that the necessary negative charge for electrical neutrality is derived from anions of the added acid (*e.g.* Cl^- from HCl). SBA-1 is an example of a product from such a synthesis (Figure 2.41).[124,125] It is based on the packing of two kinds of globular micelles of surfactant molecules with large headgroups and low '*g*' values such as cetyltriethylammonium or C16-3-1 cations. The hexagonal close-packed structure related to SBA-2 can also be prepared in acidic media.[126] Anderson suggests that for cage structures of this sort, which contain windows, the micelles must be solvated by aqueous solution, and the micelles themselves must be separated at the window region by this aqueous layer.[127] The crystalline shapes commonly obtained in these acidic syntheses[125,128] (Figure 5.14) indicate that the two steps of hydrolysis and condensation are well separated under these conditions.

The work of Che *et al.*[129] indicates that highly ordered solids can also be prepared using anionic surfactants, including those with amino acid headgroups at pHs either side of neutral, but including siloxanes functionalised with positively charged groups such as trimethylammonium. There is then well-defined charge matching of the anionic polar headgroup of the surfactant and the surface of the micelle during the organisation process, leading to the family of well-ordered AMS-n materials, some of which have novel topologies (Table 2.3).[130]

Non-ionic surfactants, in particular those based on polyethylene oxide – polypropylene oxide – polyethylene oxide (PEO-PPO-PEO) block copolymers such as P123 (molecular formula EO_{20}-PO_{70}-EO_{20}) and F127 (EO_{106}-PO_{70}-EO_{106}), also act as surfactant templates under acidic conditions. The PEO chains are the hydrophilic region, whereas the PPO chain is the hydrophobic core. These give rise to highly ordered, extra-large-pore mesoporous solids that are similar in geometry to some of the solids prepared with molecular cationic and anionic surfactants. SBA-15, SBA-16, FDU-5 and FDU-12 (Table 2.3) are examples. These materials form by a S° H^{+} X^{-} I^{+} mechanism, where both the surfactant and the silica pick up positive charge, and this is balanced by anions from added acids. The calcined solids of this type also possess microporosity, indicating that chains of the polymer become included in the inorganic walls during synthesis (*ca.* 20% of the surfactant cannot be removed by extraction).

These mechanisms account for the overall synthetic process. In fact, many changes can be made to the conditions to modify pore size, degree of order and structure type by their effect on the relative interfacial energies. Changing the molecular structure of the surfactant determines both the pore size and the product structure, by determining the shape and size of the micellar aggregates. MCM-41 of different pore sizes can be prepared by using alkyltrimethylammonium surfactants with different numbers of methylene groups in the alkyl chain, for example, whereas surfactants such as C16-3-1 with larger, doubly charged headgroups tend to give mesocage structures. Added co-solvents, such as 1,3,5-trimethylbenzene (mesitylene), can preferentially partition within the cores of micelles, swelling them and ultimately resulting in larger pores, whereas adding large concentrations of salts affects the synthesis by modifying the ionic strength of the solution, and 'salting out' the mesophase. This addition of salt reduces the size of the electrostatic double layer surrounding the micelles and changes the balance of attractive and repulsive forces between micellar aggregates in favour of attraction. A mineralising agent such as fluoride catalyses the hydrolysis step and can also improve the long-range order of phases formed at intermediate pHs by separating the hydrolysis and condensation steps, so that the soft solid can achieve an ordered, low-energy geometry before condensation (solidification) occurs. If the two steps overlap, poorly ordered 'worm-hole' microstructures tend to result.

It is also possible to add partially formed zeolite nanoparticles as a form of silicate precursor. If these are sufficiently reactive to cluster around the micelles and condense to form mesoporous solids, the resulting phase, once calcined, has both mesoporosity (between the walls) and microporosity (within the walls). This is discussed further in Chapter 10.

5.7.3 True Liquid Crystal Templating and Evaporation-induced Mesostructure Formation: Non-silica Mesoporous Solids

Mesoporous solids can also be prepared by mechanisms other than those described above in Section 5.6.2, and this enables mesoporous solids of a very wide compositional range to be prepared.

(1) The 'true liquid crystal templating' (TLCT) route was originally suggested by Beck *et al.* as a possible alternative mechanism for MCM-41 mechanism,[117] but micelle formation/agglomeration is now thought to be the operative under those conditions. The TLCT route has been used successfully to make mesoporous silicas[131–133] and Attard has also successfully used it to prepare mesoporous solids with a wide range of chemical compositions. In this route a much higher concentration of surfactant is used, with the aim to prepare a liquid crystalline phase composed of only surfactant and water. The phase diagrams for such systems have been extensively studied; for example that of CTAB contains lamellar, cubic Ia-3d and hexagonal liquid crystalline phases. The next step is then to introduce precursor species into one of the two regions of the liquid crystalline assembly that slowly react to give a mesoporous solid. Conceptually very straightforward, this is a successful route to prepare mesoporous solids with a wide range of chemistries, including mesoporous metals.[134]

(2) Liquid crystalline mesophases can also be prepared in non-aqueous solution. Using ethanolic solutions of non-ionic block copolymers as a medium, Zhao *et al.*,[135] for example, have developed a route by which mesoporous metal phosphates and borates can readily be prepared. A mixture of the metal alkoxide and the acidic chloride is used to give a complex in ethanolic solution. This acid–base pair reacts in one component of the liquid crystalline assembly to give a mesophase solid of uniform composition. Evaporation of the ethanol leaves the mesostructure. This appears to be one of the most promising routes to the formation of non-silica mesoporous solids. Many reports of these solids have appeared, particularly of metal oxides such as titanium dioxide, but loss of ordering on the mesoscale frequently occurs upon template removal. The route of Sanchez,[136] which involves the use of titanate precursor species in the sol-gel, is a promising approach to stable mesostructured titania that retains its porosity.

5.8 Summary

Much is now known about the general mechanism of zeolite synthesis, and recent microscopic evidence has shed light on both nucleation (TEM of nanoparticles) and growth. For individual cases, however, the product of crystallisation remains a complex process that is difficult to predict *ab initio*.

The use of designed templates has been largely responsible for the increase in the number of new zeolites. In fact, modelling shows there to be very many attractive target structures. The number of these that will be made will depend to a large extent on the ingenuity of the chemists responsible for making new template molecules. Against this background, a degree of planning in research of this kind is required. In general, the most efficient templates have been those with a degree of rigidity and well-defined shapes, such as the polycyclic molecules of Nagakawa and Zones. The cost of preparing ever-more-complex molecules is likely to prevent their widespread industrial application, even for the synthesis of fine chemicals of high value but there is still a major opportunity for this approach in the synthesis of extra-large-pore zeolites (pore sizes > 8 Å). The syntheses of UTD-1, CIT-5 and ITQ-33 encourage this. The approaches of self-assembling molecules by hydrogen bonding to act as supramolecular templates, or of using molecules that can be reversibly assembled and taken apart after synthesis are also attractive and further work of this kind is expected. The optimal approach would be to generate a hypothetical structure with promising features of pore size and connectivity by computational methods and then to design and synthesise an organic molecule to direct the crystallisation of that solid. This is the basis of the approach adopted by Lewis *et al.*, using the ZEBEDDE code, to generate templates for the known aluminophosphate $AlPO_4$-34.[137] The utility of this method is likely to be greatest for predicting templates for structures with cavities in the form of cages, where the space to be occupied by the molecule is very well defined.

The synthesis of zeolites without organic templating agents has advantages of cost, safety and ease of subsequent activation for use as catalysts and sorbents: for this reason zeolites A, Y and mordenite are all of commercial importance. In this light it is of interest to note that the purely inorganic natural mineral tchernichite has an identical structure to zeolite Beta which is prepared synthetically in the presence of tetraethylammonium ions. Similarly the minerals terranovaite and boggsite are all found in nature without organic components. In principle, then, template-free syntheses of these and other such silicates and phosphates should be possible in the presence of inorganic cations. Organic-free synthesis is already observed for mixed coordination microporous solids such as ETS-4 and ETS-10.

The principles governing the crystallisation of microporous solids of novel inorganic compositions such as germanates will be similar to those for zeolites and aluminophosphates, the major differences being in the solubilities and speciation in solution. For example, the remarkable new crystalline mesoporous germanate is characterised by specific $Ge_{10}O_{24}(OH)_3$ building units. The degree to which the synthesis mechanisms of such solids are investigated will depend on whether these materials find applications. In the synthesis of organic-inorganic hybrids, the organic component (the carboxylate, phosphonate or amine) is pre-determined, and the freedom is in the organisation of the metal and organic units. The synthetic design lies in the choice of organic ligands of required geometry and in finding conditions suitable for the generation of the required metal clusters and for achieving suitable solubilities for the ligands.

By contrast, the mechanisms of the synthesis of ordered mesoporous solids are very different from those of silicate and phosphate frameworks. For control over the synthesis of mesoporous solids, the key is to understand the interactions of micellar surfactants with condensable inorganic framework-building units. Synthetic routes are also being developed to prepare mesoporous silicates made up of nanoparticles of zeolites, with the aim of combining advantages of microporous and mesoporous solids.

References

1. H. Robson, *'Verified Syntheses of Zeolitic Materials'*, Elsevier, Amsterdam, 2001.
2. C. S. Cundy and P. A. Cox, *Chem. Rev.*, 2003, **103**, 663.
3. C. S. Cundy and P. A. Cox, *Micropor. Mesopor. Mater.*, 2005, **82**, 1.
4. R. M. Barrer, *'Hydrothermal Chemistry of Zeolites'*, Academic Press, London, 1982.
5. R. E. Morris and S. J. Weigel, *Chem. Soc. Rev.*, 1997, **26**, 309.
6. E. R. Cooper, C. D. Andrews, P. S. Wheatley, P. B. Webb, P. Wormald and R. E. Morris, *Nature*, 2004, **430**, 1012.
7. D. W. Breck, *'Zeolite Molecular Sieves'*, Wiley, New York, 1974.
8. L.-Y. Hou and R. W. Thompson, *Zeolites*, 1989, **9**, 526.
9. F. G. Dwyer and P. Chu, *J. Catal.*, 1979, **50**, 263.
10. A. Corma, M. J. Diaz-Cabanas, J. L. Jorda, C. Martinez and M. Moliner, *Nature*, 2006, **443**, 842.
11. S. Khodabandeh and M. E. Davis, *Micropor. Mater.*, 1997, **12**, 347.
12. R. M. Barrer and P. J. Denny, *J. Chem. Soc.*, 1961, 971.
13. G. T. Kerr and G. T. Kokotailo, *J. Am. Chem. Soc.*, 1961, **83**, 4675.
14. R. L. Wadlinger, G. T. Kerr and E. J. Rosinski, 1967, US Patent.
15. R. J. Argauer and G. R. Landolt, 1972, US Patent, 3, 702, 886.
16. J. B. Nagy, Z. Gabelica and E. G. Derouane, *Zeolites*, 1983, **3**, 43.
17. J. L. Casci, B. M. Lowe and T. V. Whittam, 1985, US Patent 4, 537,754.
18. M. D. Shannon, J. L. Casci, P. A. Cox and S. J. Andrews, *Nature*, 1991, **353**, 417.
19. M. E. Leonowicz, J. A. Lawton, S. L. Lawton and M. K. Rubin, *Science*, 1994, **264**, 1910.
20. I. Petrovic, A. Navrotsky, M. E. Davis and S. I. Zones, *Chem. Mater.*, 1993, **5**, 1805.
21. S. L. Lawton and W. J. Rohrbaugh, *Science*, 1990, **247**, 1319.
22. T. V. Harris and S. I. Zones, *Stud. Surf. Sci. Catal.*, 1994, **84**, 29.
23. A. N. Christensen, T. R. Jensen, P. Norby and J. C. Hanson, *Chem. Mater.*, 1998, **10**, 1688.

24. D. Domine and J. Quobex, *'Molecular Sieves'*, Society of Chemical Industry, London, 1968, 78.
25. S. Mintova, N. H. Olson, V. Valtchev and T. Bein, *Science*, 1999, **283**, 958.
26. D. Hoebbel, W. Wieker, P. Francke and A. Otto, *Z. Anorg. Allg. Chem.*, 1975, **418**, 35.
27. P.-P. E. A. de Moor, T. P. M. Beelen, B. U. Komanshek, O. Diat and R. A. van Santen, *J. Phys. Chem. B*, 1997, **101**, 11077.
28. P.-P. E. A. de Moor, T. P. M. Beelen and R. A. van Santen, *J. Phys. Chem. B*, 1999, **103**, 1639.
29. P.-P. E. A. De Moor, T. P. M. Beelen, R. A. van Santen, K. Tsuji and M. E. Davis, *Chem. Mater.*, 1999, **11**, 36.
30. B. J. Schoeman, *Zeolites*, 1997, **18**, 97.
31. C. D. Chang and A. T. Bell, *Catal. Lett.*, 1991, **8**, 305.
32. S. L. Burkett and M. E. Davis, *Chem. Mater.*, 1995, **7**, 920.
33. T. M. Davis, T. O. Drews, H. Ramanan, C. He, J. S. Dong, H. Schnablegger, M. A. Katsoulakis, E. Kokkoli, A. V. McCormick, R. L. Penn and M. Tsapatsis, *Nature Materials*, 2006, **5**, 400.
34. A. Corma and M. J. Diaz-Cabanas, *Micropor. Mesopor. Mater.*, 2006, **89**, 39.
35. S. L. Burkett and M. E. Davis, *Chem. Mater.*, 1995, **7**, 920.
36. S. L. Burkett and M. E. Davis, *Chem. Mater.*, 1995, **7**, 1453.
37. G. D. Price, J. J. Pluth, J. V. Smith, J. M. Bennett and R. L. Patton, *J. Am. Chem. Soc.*, 1982, **104**, 5971.
38. A. V. Goretsky, L. W. Beck, S. I. Zones and M. E. Davis, *Micropor. Mesopor. Mater.*, 1999, **28**, 387.
39. S. P. Zhdanov and N. N. Samulevich, *'Proceedings of the 5th International Conference on Zeolites'*, Heyden, London, 1980, 75.
40. J. R. Agger, N. Hanif and M. W. Anderson, *Angew. Chem. Int. Ed.*, 2001, **40**, 4065.
41. J. R. Agger, N. Hanif, C. S. Cundy, A. P. Wade, S. Dennison, P. A. Rawlinson and M. W. Anderson, *J. Am. Chem. Soc.*, 2003, **125**, 830.
42. M. W. Anderson, J. R. Agger, J. T. Thornton and N. Forsyth, *Angew. Chem. Int. Ed.*, 1996, **35**, 1210.
43. T. Ohsuna, O. Terasaki, V. Alfredsson, J. O. Bovin, I. D. Watanabe, S. W. Carr and M. W. Anderson, *Proc. Roy. Soc. Lond. A*, 1996, **452**, 715.
44. M. W. Anderson, J. R. Agger, N. Hanif and O. Terasaki, *Micropor. Mesopor. Mater.*, 2001, **48**, 1.
45. T. Ohsuna, B. Slater, F. Gao, J. Yu, Y. Sakamoto, G. Zhu, O. Terasaki, D. E. W. Vaughan, S. Qiu and C. R. A. Catlow, *Chem. Eur. J.*, 2004, **10**, 5031.
46. B. Slater, C. R. A. Catlow, Z. Liu, T. Ohsuna, O. Terasaki and M. A. Camblor, *Angew. Chem. Int. Ed.*, 2002, **41**, 1235.
47. S. I. Zones, Y. Nakagawa, G. S. Lee, C. Y. Chen and L. T. Yuen, *Micropor. Mesopor. Mater.*, 1998, **21**, 199.

48. Y. Nakagawa and S. I. Zones, in *'Molecular Sieves'*, Ed. M. L. Occelli and H. E. Robson, Van Nostrand Reinhold, New York, 1992.
49. S. I. Zones, Y. Nakagawa, L. T. Yuen and T. V. Harris, *J. Am. Chem. Soc.*, 1996, **118**, 7558.
50. P. Wagner, Y. Nakagawa, G. S. Lee, M. E. Davis, S. Elomari, R. C. Medrud and S. I. Zones, *J. Am. Chem. Soc.*, 2000, **122**, 263.
51. G. S. Lee, Y. Nakagawa, S. J. Hwang, M. E. Davis, P. Wagner, L. Beck and S. I. Zones, *J. Am. Chem. Soc.*, 2002, **124**, 7024.
52. S. I. Zones, S. J. Hwang and M. E. Davis, *Chem. Eur. J.*, 2001, **7**, 1990.
53. S. Elomari and S. I. Zones, *'Zeolites and Mesoporous Materials at the Dawn of the 21st Century: Proc. 13th International Zeolite Conference'*, Elsevier, New York, 2001, 167.
54. M. G. G. Wu, M. W. Deem, S. A. Elomari, R. C. Medrud, S. I. Zones, T. Maesen, C. Kibby, C. Y. Chen and I. Y. Chan, *J. Phys. Chem. B*, 2002, **106**, 264.
55. P. Wagner, Y. Nakagawa, G. S. Lee, M. E. Davis, S. Elomari, R. C. Medrud and S. I. Zones, *J. Am. Chem. Soc.*, 2000, **122**, 263.
56. S. I. Zones, Y. Nakagawa, L. T. Yuen and T. V. Harris, *J. Am. Chem. Soc.*, 1996, **118**, 7558.
57. S. I. Zones, S. J. Hwang, S. Elomari, I. Ogino, M. E. Davis and A. W. Burton, *Compt. Rend. Chim.*, 2005, **8**, 267.
58. P. A. Barrett, M. A. Camblor, A. Corma, R. H. Jones and L. A. Villaescusa, *J. Phys. Chem. B.*, 1998, **102**, 4147.
59. A. Corma, M. Puche, F. Rey, G. Sankar and S. J. Teat, *Angew. Chem. Int. Ed.*, 2003, **42**, 1156.
60. R. Castaneda, A. Corma, V. Fornes, F. Rey and J. Rius, *J. Am. Chem. Soc.*, 2003, **125**, 7820.
61. A. Corma, F. Rey, S. Valencia, J. L. Jorda and J. Rius, *Nature Mater.*, 2003, **2**, 493.
62. T. Blasco, A. Corma, M. J. Diaz-Cabanas, F. Rey, J. Rius, G. Sastre and J. A. Vidal-Moya, *J. Am. Chem. Soc.*, 2004, **126**, 13414.
63. A. Cantin, A. Corma, S. Leiva, F. Rey, J. Rius and S. Valencia, *J. Am. Chem. Soc.*, 2005, **127**, 11560.
64. D. L. Dorset, G. J. Kennedy, K. G. Strohmaier, M. J. Diaz-Cabanas, F. Rey and A. Corma, *J. Am. Chem. Soc.*, 2006, **128**, 8862.
65. R. F. Lobo and M. E. Davis, *J. Am. Chem. Soc.*, 1995, **117**, 3764.
66. K. Tsuji, P. Wagner and M. E. Davis, *Micropor. Mesopor. Mater*, 1999, **28**, 461.
67. S. Leiva, M. J. Sabater, S. Valencia, G. Sastre, V. Fornes, F. Rey and A. Corma, *Compt. Rend. Chim.*, 2005, **8**, 369.
68. J. L. Paillaud, B. Harbuzaru, J. Patarin and N. Bats, *Science*, 2004, **304**, 990.
69. D. L. Dorset, S. C. Weston and S. S. Dhingra, *J. Phys. Chem. B*, 2006, **110**, 2045.
70. C. Baerlocher, L. B. McCusker and R. Chiappetta, *Micropor. Mater.*, 1994, **2**, 269.

71. C. C. Freyhardt, M. Tsapatsis, R. F. Lobo, K. J. Balkus and M. E. Davis, *Nature*, 1996, **381**, 295.
72. A. Corma, F. Rey, J. Rius, M. J. Sabater and S. Valencia, *Nature*, 2004, **431**, 287.
73. H. Lee, S. I. Zones and M. E. Davis, *Nature*, 2003, **425**, 385.
74. E. M. Flanigen and R. L. Patton, 1978, US Patent 4, 073, 865.
75. J. L. Guth, H. Kessler, J. M. Higel, J. M. Lamblin, J. Patarin, A. Seive, J. M. Chezeau and R. Wey, in '*Zeolite Synthesis*', Ed. M. L. Occelli and H. E. Robson, ACS Symposium Series 398, Washington, 1989, 176.
76. M. A. Camblor, L. A. Villaescusa and M. J. Diaz-Cabanas, *Top. Catal.*, 1999, **9**, 59.
77. L. A. Villaescusa and M. A. Camblor, *Rec. Res. Devel. Chem.*, 2003, **1**, 93.
78. M. Arranz, J. Pérez-Pariente, P. A. Wright, A. M. Z. Slawin, T. Blasco, L. Gomez-Hortiguela and F. Cora, *Chem. Mater.*, 2005, **17**, 4374.
79. J. A. Vidal-Moya, T. Blasco, A. Corma, M. T. Navarro and F. Rey, *Stud. Surf. Sci. Catal.*, 2004, **154**, 1289.
80. F. Delprato, L. Delmotte, J. L. Guth and L. Huve, *Zeolites*, 1990, **10**, 546.
81. S. B. Hong, E. G. Lear, P. A. Wright, W. Z. Zhou, P. A. Cox, C. H. Shin, J. H. Park and I. S. Nam, *J. Am. Chem. Soc.*, 2004, **126**, 5817.
82. R. M. Dessau, E. W. Valyocsik and N. H. Goeke, *Zeolites*, 1992, **12**, 776.
83. M. T. Melchior, D. E. W. Vaughan and C. F. Pictroski, *J. Phys. Chem.*, 1995, **99**, 6128.
84. D. F. Shantz, C. Fild, H. Koller and R. F. Lobo, *J. Phys. Chem. B*, 1999, **103**, 10858.
85. M. J. Annen, M. E. Davis, J. B. Higgins and J. L. Schlenker, *J. Chem. Soc., Chem. Commun.*, 1991, 1175.
86. L. B. McCusker, R. W. Grosse-Kunstleve, C. Baerlocher, M. Yoshikawa and M. E. Davis, *Micropor. Mater.*, 1996, **6**, 295.
87. A. Corma, M. Diaz-Cabanas, J. Martinez-Triguero, F. Rey and J. Rius, *Nature*, 2002, **418**, 514.
88. K. G. Strohmaier and D. E. W. Vaughan, *J. Am. Chem. Soc.*, 2003, **125**, 16035.
89. G. H. Kuhl, *Inorg. Chem.*, 1971, **10**, 2488.
90. L. Tosheva and V. P. Valtchev, *Chem. Mater.*, 2005, **17**, 2494.
91. Z. A. D. Lethbridge, J. J. Williams, R. I. Walton, K. E. Evans and C. W. Smith, *Micropor. Mesopor. Mater.*, 2005, **79**, 339.
92. G. Scott, R. W. Thompson, A. G. Dixon and A. Sacco Jr, *Zeolites*, 1990, **10**, 44.
93. S. Shimizu and H. Hamada, *Micropor. Mesopor. Mater.*, 2001, **48**, 39.
94. I. Bull, L. A. Villaescusa, S. J. Teat, M. A. Camblor, P. A. Wright, P. Lightfoot and R. E. Morris, *J. Am. Chem. Soc.*, 2000, **122**, 7128.
95. C. A. Fyfe, D. H. Brouwer, A. R. Lewis, L. A. Villaescusa and R. E. Morris, *J. Am. Chem. Soc.*, 2002, **124**, 7770.
96. L. A. Villaescusa, P. S. Wheatley, I. Bull, P. Lightfoot and R. E. Morris, *J. Am. Chem. Soc.*, 2001, **123**, 8797.

97. R. Garcia, E. F. Philp, A. M. Z. Slawin, P. A. Wright and P. A. Cox, *J. Mater. Chem.*, 2001, **11**, 1421.
98. R. Garcia, I. J. Shannon, A. M. Z. Slawin, W. Z. Zhou, P. A. Cox and P. A. Wright, *Micropor. Mesopor. Mater.*, 2003, **58**, 91.
99. C. Baerlocher, W. M. Meier and D. H. Olson, *'Atlas of Zeolite Framework Types'*, Elsevier, 2001.
100. J. Patarin, J. L. Paillaud and H. Kessler, in *'Handbook of Porous Solids'*, Ed. F. Schuth, K. S. W. Sing and J. Weitkamp, Wiley-VCH, New York, 2002, 815.
101. M. Afeworki, D. L. Dorset, G. J. Kennedy and K. G. Strohmaier, *Chem. Mater.*, 2006, **18**, 1697.
102. M. Afeworki, G. J. Kennedy, D. L. Dorset and K. G. Strohmaier, *Chem. Mater.*, 2006, **18**, 1705.
103. D. B. Akolekar and R. Ryoo, *J. Chem. Soc. Farad. Trans.*, 1996, **92**, 4617.
104. J. D. Chen, J. Dakka, E. Neeleman and R. A. Sheldon, *J. Chem. Soc., Chem. Commun.*, 1993, 1379.
105. F. Rey, G. Sankar, J. M. Thomas, P. A. Barrett, D. W. Lewis, C. R. A. Catlow, S. M. Clark and G. N. Greaves, *Chem. Mater.*, 1995, **7**, 1435.
106. A. M. Beale, A. M. J. van der Eerden, S. D. M. Jacques, O. Leynaud, M. G. O'Brien, F. Meneau, S. Nikitenko, W. Bras and B. M. Weckhuysen, *J. Am. Chem. Soc.*, 2006, **128**, 12386.
107. M. G. O'Brien, A. M. Beale, C. R. A. Catlow and B. M. Weckhuysen, *J. Am. Chem. Soc.*, 2006, **128**, 11744.
108. F. Taulelle, *Curr. Opin. Solid State Mater. Sci.*, 2001, **5**, 397.
109. O. B. Vistad, D. E. Akporiaye and K. P. Lillerud, *J. Phys. Chem. B*, 2001, **105**, 12437.
110. O. B. Vistad, D. E. Akporiaye, F. Taulelle and K. P. Lillerud, *Chem. Mater.*, 2003, **15**, 1639.
111. S. Surblé, F. Millange, C. Serre, G. Férey and R. I. Walton, *Chem. Commun.*, 2006, 1518.
112. C. Lee, C. Mellot-Draznieks, B. Slater, G. Wu, W. T. A. Harrison, C. N. R. Rao and A. K. Cheetham, *Chem. Commun.*, 2006, 2687.
113. D. E. Akporiaye, I. M. Dahl, A. Karlsson and R. Wendelbo, *Angew. Chem. Int. Ed.*, 1998, **37**, 609.
114. L. M. Knight and G. J. Lewis, *Stud. Surf. Sci. Catal.*, 2004, **154**, 171.
115. A. Cantin, A. Corma, M. J. Diaz-Cabanas, J. L. Jorda and M. Moliner, *J. Am. Chem. Soc.*, 2006, **128**, 4216.
116. S. Bauer, T. Bein and N. Stock, *Inorg. Chem.*, 2005, **44**, 5882.
117. J. S. Beck, J. C. Vartuli, W. J. Roth, M. E. Leonowicz, C. T. Kresge, K. D. Schmitt, C. T. W. Chu, D. H. Olson, E. W. Sheppard, S. B. McCullen, J. B. Higgins and J. L. Schlenker, *J. Am. Chem. Soc.*, 1992, **114**, 10834.
118. A. Monnier, F. Schuth, Q. Huo, D. Kumar, D. Margolese, R. S. Maxwell, G. D. Stucky, M. Krishnamurty, P. Petroff, A. Firouzi, M. Janicke and B. F. Chmelka, *Science*, 1993, **261**, 1299.

119. J. N. Israelachvili, *'Intermolecular and Surface Forces'*, Academic Press, 1992.
120. U. Henriksson, E. S. Blackmore, G. J. T. Tiddy and O. Soderman, *J. Phys. Chem.*, 1992, **96**, 3894.
121. S. Sadasivan, C. E. Fowler, D. Khushalani and S. Mann, *Angew. Chem. Int. Ed.*, 2002, **41**, 2151.
122. Q. S. Huo, D. I. Margolese, U. Ciesla, P. Y. Feng, T. E. Gier, P. Sieger, R. Leon, P. M. Petroff, F. Schuth and G. D. Stucky, *Nature*, 1994, **368**, 317.
123. Q. S. Huo, R. Leon, P. M. Petroff and G. D. Stucky, *Science*, 1995, **268**, 1324.
124. Q. S. Huo, D. I. Margolese and G. D. Stucky, *Chem. Mater.*, 1996, **8**, 1147.
125. A. E. Garcia-Bennett, S. Williamson, P. A. Wright and I. J. Shannon, *J. Mater. Chem.*, 2002, **12**, 3533.
126. S. N. Che, S. Y. Lim, M. Kaneda, H. Yoshitake, O. Terasaki and T. Tatsumi, *J. Am. Chem. Soc.*, 2002, **124**, 13962.
127. M. W. Anderson, C. C. Egger, G. J. T. Tiddy, J. L. Casci and K. A. Brakke, *Angew. Chem. Int. Ed.*, 2005, **44**, 3243.
128. S. Che, Y. Sakamoto, O. Terasaki and T. Tatsumi, *Chem. Mater.*, 2001, **13**, 2237.
129. S. Che, A. E. Garcia-Bennett, T. Yokoi, K. Sakamoto, H. Kunieda, O. Terasaki and T. Tatsumi, *Nature Mater.*, 2003, **2**, 801.
130. A. E. Garcia-Bennett, O. Terasaki, S. Che and T. Tatsumi, *Chem. Mater.*, 2004, **16**, 813.
131. E. Kramer, S. Forster, C. Goltner and M. Antonietti, *Langmuir*, 1998, **14**, 2027.
132. N. R. B. Coleman and G. S. Attard, *Micropor. Mesopor. Mater.*, 2001, **44**, 73.
133. G. S. Attard, J. C. Glyde and C. G. Goltner, *Nature*, 1995, **378**, 366.
134. G. S. Attard, P. N. Bartlett, N. R. B. Coleman, J. M. Elliott, J. R. Owen and J. H. Wang, *Science*, 1997, **278**, 838.
135. B. Z. Tian, X. Y. Liu, B. Tu, C. Z. Yu, J. Fan, L. M. Wang, S. H. Xie, G. D. Stucky and D. Y. Zhao, *Nature Mater.*, 2003, **2**, 159.
136. G. Soler-Illia, A. Louis and C. Sanchez, *Chem. Mater.*, 2002, **14**, 750.
137. D. W. Lewis, D. J. Willock, C. R. A. Catlow, J. M. Thomas and G. J. Hutchings, *Nature*, 1996, **382**, 604.

CHAPTER 6

The Chemistry of Microporous Framework Solids

6.1 Introduction

Microporous solids display a wide range of chemical behaviour: strong solid acidity; cation exchange; adsorption and inclusion; redox activity. All of these reactions can take place within a framework that is relatively inflexible and usually chemically inert, the main role of which is to provide a well-defined host with tunable pore geometry and framework charge. The high thermal and chemical stability of the aluminosilicate zeolites, which can be enhanced by framework dealumination, and of related silicas is the key to their widespread applicability. As a consequence, they have been used as hosts for most of the chemistry described in this chapter. Many of the newer families of microporous solids have lower chemical stabilities, so that reactions within their pores remains largely unexplored. Mesoporous solids are thermally robust and present reactive internal surfaces and much larger molecules may enter their pores than is possible for zeolites. They are the subjects of intensive current study, which will be dealt with only briefly here.

6.2 Stability and Post-synthetic Modification

6.2.1 Thermal Stability of the Framework

The thermal stability of aluminosilicate zeolites is generally very high, and in the absence of water vapour many cation-exchanged zeolites can be heated to temperatures in excess of 800 °C without loss of crystallinity or recrystallisation to denser phases. Aluminophosphates are usually less stable. Physisorbed water is lost reversibly from all these solids at temperatures up to *ca.* 300 °C and where hydroxyl groups are present, either as structural defects or as bridging hydroxyls, these are removed as water molecules at temperatures of 400 °C and above, giving local structural defects. Where microporous solids are prepared containing organic templates, calcination in oxygen at temperatures between 450 and 600 °C removes the organic species. Template removal usually proceeds by degradation of the organic amine (for example by the Hofmann degradation to give an amine and an alkene) followed by combustion of residual hydrocarbons remaining trapped within the pores. As a result it is sometimes preferable to preheat in nitrogen to remove most of the organic before heating in oxygen – this is safer and results in less generation of water within the pores, which can result in framework hydrolysis, particularly of $AlPO_4$s. For most high-silica zeolites, the organic molecule is readily removed leaving the framework intact, but for many new open framework solids, structural collapse occurs. Figure 6.1 shows thermogravimetric analyses (TGA) of templated aluminophosphates STA-6 and STA-7 heated in flowing air, and combined TGA and differential scanning calorimetry (DSC) of templated stilbite, TNU-10.

It is not easy to predict which frameworks will remain intact upon template removal, and which will collapse, but it should be borne in mind that all open frameworks are less stable (when empty) than dense crystalline forms, and barriers to recrystallisation are kinetic rather than thermodynamic. High-temperature treatment of microporous solids eventually results in recrystallisation to dense ceramics rather than other microporous solids. For example, the magnesium form of zeolite P transforms to magnesian cordierite,[1] and aluminophosphates transform to dense $AlPO_4$ polymorphs.

A few thermally induced structural rearrangements are observed between different microporous frameworks that result from the breaking and forming of chemical bonds. Typically these are topotactic, that is, the crystallographic axes of the initial and final structures are related, and the transformations arise from the reaction of a relatively small fraction of the bonds. Examples include the transformation of chabazite to sodalite,[2] of VPI-5 to $AlPO_4$-8[3] and of AlMePO-β to AlMePO-α (Figure 6.2).[4] All these reactions are catalysed by the presence of water vapour, and result in the formation of the thermodynamically favoured product. Rarely, non-topotactic, reconstructive transformations (involving bond-breaking) of one microporous solid to another can occur, presumably *via* a poorly crystalline intermediate. The thermal transformation of $AlPO_4$(F)-34 containing a nickel complex to $AlPO_4$-5 is one such example.[5]

Other thermally induced phase changes in microporous solids tend to be changes in symmetry as a result of the increased thermal motion of the framework atoms, cation migration or desorption of adsorbed species. The changes in unit cell size and symmetry of the cationic forms of zeolite Rho upon heating is an example – the symmetry changes from Im3m to I-43m as the framework becomes distorted away from maximum symmetry upon cation dehydration and motion. The distortion away from maximum symmetry increases at lower temperatures as the thermal motion is reduced.[6,7] For aluminophosphates the presence of adsorbed species on the framework aluminium atoms frequently results in their adopting five- or six-fold coordination and gives rise to symmetry changes that are reversed upon calcinations or dehydration. Much greater changes are observed upon dehydration of flexible MOFs such as MIL-53 and MIL-88, and upon their uptake of adsorbate molecules (see Chapters 2 and 7).

An interesting example of formation of a zeolite structure by the thermal treatment of a precursor outside of the hydrothermal autoclave is the observed

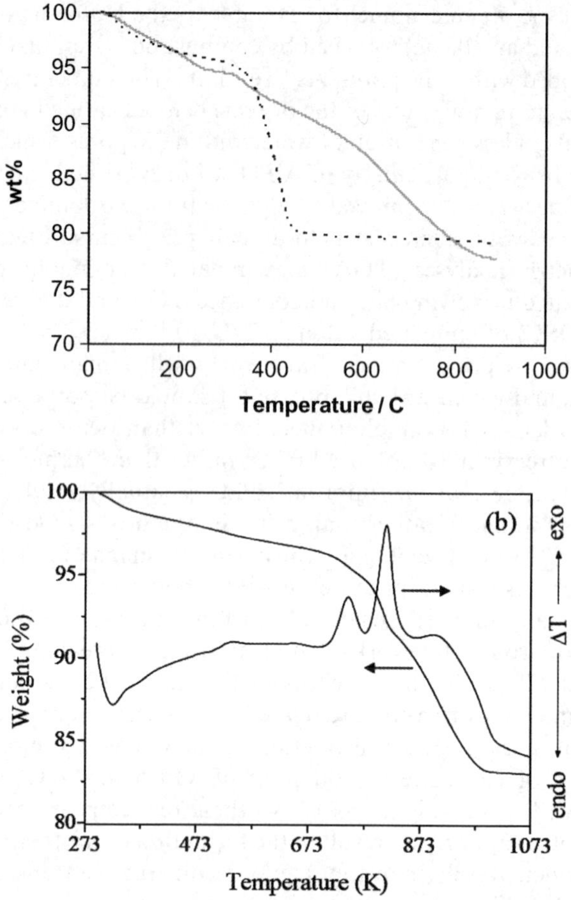

Figure 6.1 (Above) Thermogravimetric analyses of aluminophosphates STA-6 and STA-7 in flowing air. The template is removed more easily from the STA-7 (dotted line), in which the cages are connected via a 3D network of 8MRs, than from STA-6 (grey line), where the cages are connected by 8MR windows along one axis only (see Figure 2.18 for structures of STA-6 and -7). (Below) Combined TGA and DTA plots of the calcinations in air of the synthetic stilbite TNU-10, which show that the template removal is strongly exothermic.
(Courtesy S. B. Hong).

condensation reaction of the layered precursor, MCM-22(P), to give zeolite MCM-22.[8] A zeolite with the same tetrahedral connectivity as MCM-22, known as MCM-49, can also be prepared directly.[9,10] In the precursor material, each layer of the precursor is covered on both sides by terminal hydroxyl groups. Upon heating, these condense with those of the adjacent layer to give the fully tetrahedrally connected microporous framework structure, topology type MWW (Figure 6.3). If pillaring agents are introduced into the precursor before calcination, a pillared zeolite is prepared (MCM-36).[11] A similar

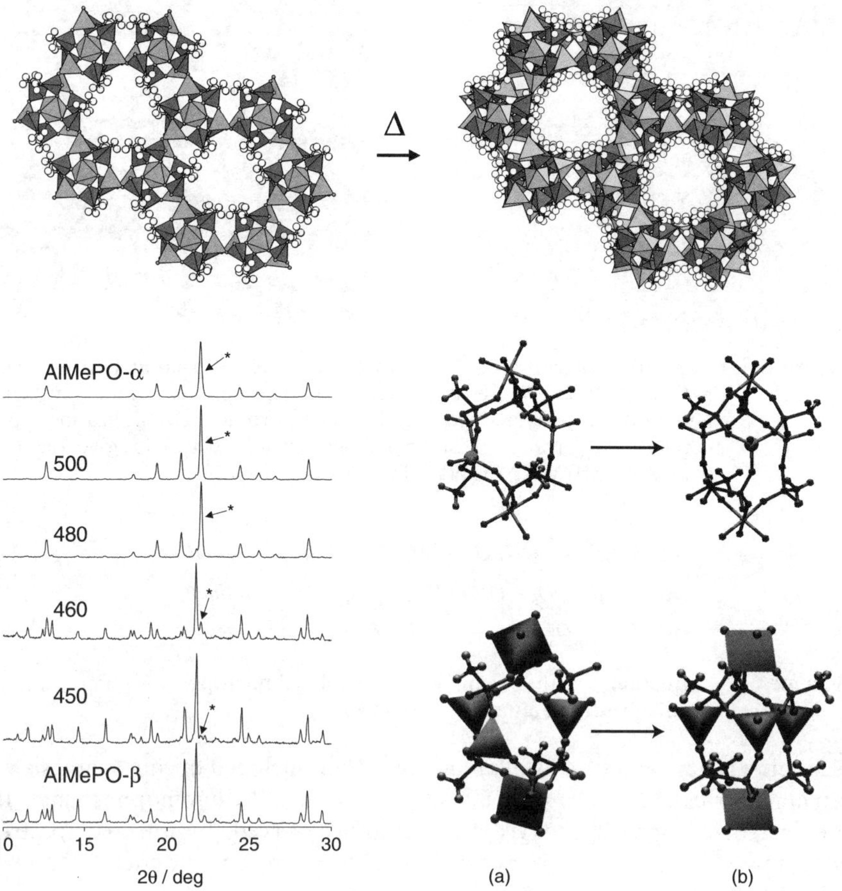

Figure 6.2 AlMePO-β transforms topotactically to AlMePO-α upon heating in the presence of water vapour at temperatures of 450 °C and higher (above) as shown by X-ray diffraction (below, left). The change results from the rotation of a single type of AlO_4 tetrahedron in the structure resulting from a bond breaking/bond making event (below right). [Reproduced from reference 4 with permission. Copyright 1997 Royal Society of Chemistry.]

transformation is observed upon heating the layered precursor 'PREFER', when the layers condense to give the ferrierite structure[12] and with other layered silicates that give zeolites upon calcination.[13,14]

6.2.2 Chemical Conversions during Calcination

Calcination of the as-prepared forms of organically templated zeolites and substituted aluminophosphates[15] in oxygen at *ca.* 500 °C removes the organics to leave the protonic forms, where the protons are in the form of bridging hydroxyls (Scheme 6.1). These balance the negative framework charge

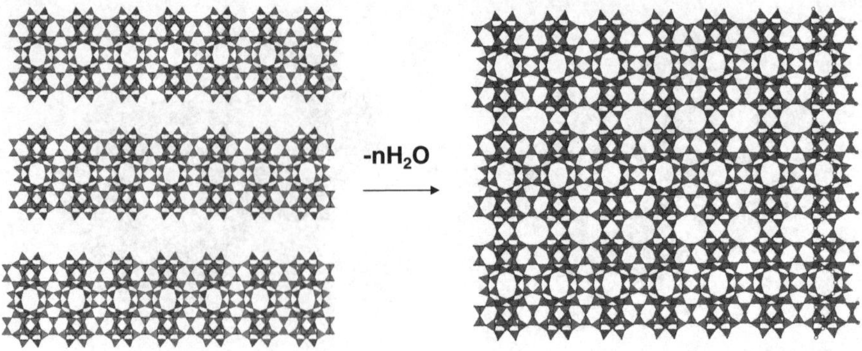

Figure 6.3 Formation of MCM-22 by thermal treatment of a laminated precursor MCM-22(P). In the precursor the silicate sheets are terminated on the top and bottom surfaces by silanol (SiOH) groups. Upon heating, these condense to give Si–O–Si bonds and liberate water to give the fully connected MCM-22 (MWW) framework.

$$M_y^+ R_{x-y}^+ Al_x Si_{1-x} O_2 + O_2 \rightarrow M_y^+ H_{x-y}^+ Al_x Si_{1-x} O_2 + products$$
$$R_x^+ M_x^{2+} Al_{1-x} PO_4 + O_2 \rightarrow H_x^+ M_x^{2+} Al_{1-x} PO_4 + products$$
$$R_x^+ AlSi_x P_{1-x} O_4 + O_2 \rightarrow H_x^+ AlSi_x P_{1-x} O_4 + products$$

Scheme 6.1 Equations for the calcination of templated microporous aluminosilicates and substituted aluminophosphates.

(see below). For zeolites, any alkali metal cations included during synthesis will remain after calcination. For some substituted metalloaluminophosphates the presence of bridging hydroxyls is not clearly observed (by infrared spectroscopy, for example) and in these cases it is likely a more complex defect structure exists around the substituting metal cation.[16]

Where the frameworks are neutral and the organic is either neutral or charge-balanced by ionic species such as hydroxyl or fluoride ions, coordinated to framework cations, calcination gives defect-free, neutral solids (Scheme 6.2). Examples include silica polymorphs prepared *via* the fluoride route, aluminophosphates and the neutral AlMePO-β, which is prepared using 1,4-dioxane as a structure directing agent. For the last solid, the calcination is best performed in nitrogen to avoid oxidation of the organic groups.[4]

If the solids are prepared using metal complexes as templates (Section 5.4.1), calcination removes the ligand, leaving metal cations as extra-framework charge-balancing species if the framework is negatively charged, or as oxides if the final framework is neutral (Scheme 6.3). The nature of the metal species left in the pores is readily determined by a combination of selected area analysis by EDX and X-ray absorption spectroscopy, as shown for the thermally induced transformations of the metal complexes of azamacrocycles and linear polyamines in aluminophosphates such as STA-6 (SAS), STA-7 (SAV) and $AlPO_4$-34 (CHA).[5,17] The change in EXAFS signal associated with the liberation of nickel from an azamacrocyle in SAPO STA-6 to give extra-framework

$$R_x^+ SiO_2F_x^- \rightarrow SiO_2 + products$$

$$R_x^+ AlPO_4F_x^- \rightarrow AlPO_4 + products$$

$$(C_4H_8O_2)_x Al_2(CH_3PO_3)_3 \rightarrow Al_2(CH_3PO_3)_3 + xC_4H_8O_2$$

Scheme 6.2 Equations for the thermal generation of neutral frameworks from templated microporous solids.

$$[Ni(cyclam)]_{x/2}^{2+}.AlP_{1-x}Si_xO_4(SAS) + O_2 \rightarrow Ni_{x/2}^{2+}AlP_{1-x}Si_xO_4(SAS) + products$$

$$[Ni((H_2N(CH_2)_2)_2NH)_2]_{y/2}^{2+} AlPO_4(F)_y(CHA) + O_2 \rightarrow (NiO)_{y/2} AlPO_4(CHA) + products$$

Scheme 6.3 Equations for the calcination of as-prepared aluminophosphates containing nickel complex templates.

cations is shown in Figure 3.30, whereas the change in the EXAFS associated with the formation of small nickel oxide particles upon calcining the aluminophosphate fluoride UT-6 templated by a nickel complex of the linear amine diethylenetriamine is shown in Figure 6.4.

As-prepared metal organic frameworks are rendered porous by heating, the main purpose of which is to remove neutral species such as solvent molecules or unreacted carboxylic acid left over from the synthesis. The thermal stability of MOFs varies widely, often depending on the chemical make up of the metal 'nodes'. Among the terephthalates, MIL-100 and MOF-5, for example, are stable to at least 300 °C, and the microporous scandium terephthalate $Sc_2(O_2CC_6H_4CO_2)_3$ to more than 400 °C. The nickel phosphonate $Ni_2(O_3PCH_2N(C_2H_4)_2NCH_2PO_3) \cdot xH_2O$ is also thermally stable to around 400 °C, losing water in two steps, including loosely bound water from the pores and water molecules coordinated to the nickel. Among the MOFs based on amines, the metal imidazolate ZIFs described in Section 2.4.2 are the most thermally stable, staying intact up to 550 °C.[18]

The as-prepared forms of mesoporous solids can also be made porous by heating. This should be performed in two steps, first under nitrogen to remove most of the organic, followed by heating under oxygen to remove residual organic material. This is important because direct heating under oxygen results in violent combustion of the large content of surfactant template. In fact, removal of most (or all) of the surfactant is usually readily achieved by solvent extraction, which is the only acceptable method when the mesoporous framework is either organically functionalised or an organic-inorganic hybrid. In addition, extraction enables recycling of the surfactant.

6.2.3 Dealumination and the Preparation of 'Ultrastable' Zeolite Y

The largest use of a zeolite as a catalyst is the acid form of zeolite Y in catalytic cracking of the heavier fraction of crude oil to more valuable hydrocarbon

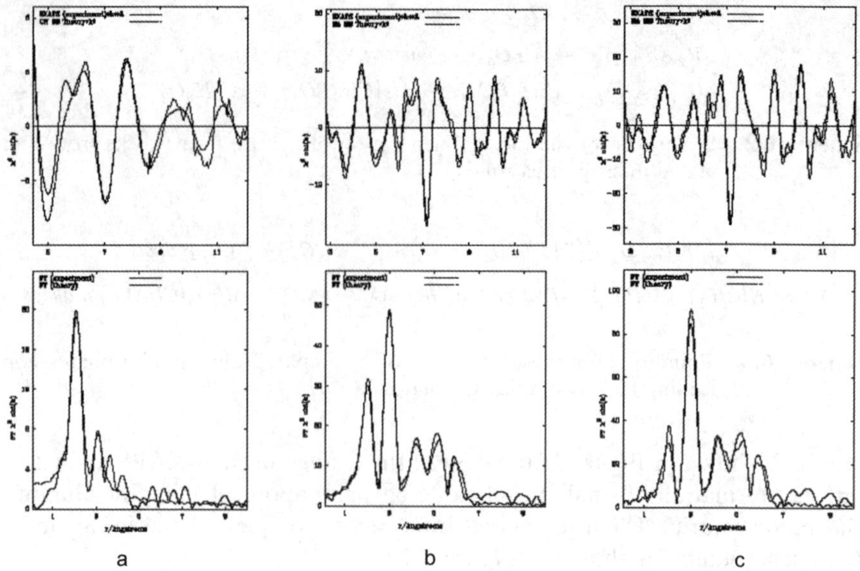

Figure 6.4 Nickel K-edge k^3-weighted EXAFS spectra, with fitted curves according to chemical model (top row) and the associated Fourier transforms (bottom row) for: (a) Ni(diethylenetriamine)$_2$-$Al_6P_6O_{24}F_2$ (Ni(deta)$_2$-UT-6); (b) Ni(deta)$_2$-UT-6 calcined in oxygen at 600 °C; (c) bulk NiO, which has the rocksalt structure. The similarity between (b) and (c) indicates that NiO particles are produced upon calcination of the as-prepared complex-templated AlPO.

products. The reaction requires a catalyst that can be regenerated in air at temperatures of 700 °C or more. A typical zeolite Y, synthesised with a Si/Al ratio of 2.5 and prepared in the protonic form by ion exchange with ammonium and thermal removal of ammonia, rapidly loses crystallinity under these conditions.

One way to increase the stability would be to synthesise directly a zeolite Y with high silica content. However, the high positive charge associated with the alkali metal cations that act to 'template' the structure requires a high negative framework charge and therefore a high aluminium content to balance it. Incorporation of the ether 15-crown-5 in the synthesis results in Y zeolites with Si/Al ratios higher than from typical syntheses (5 *cf.* 2.5), because the effective templates are then the coordinated cations, which have a low charge-to-volume ratio, and therefore tend to give high silica frameworks.[19] However, crown ethers are expensive and the use is largely of academic interest. Fortunately, post-synthetic routes can also give high silica Y. The reaction of the ammonium-form of zeolite Y with silicon tetrachloride vapour at elevated temperatures,[20,21] for example, results in direct substitution of aluminium by silicon to give a defect-free pure silica Y that is of great interest for academic studies, and possibly for niche applications, and a related approach involves treatment of the zeolite with a solution of ammonium hexafluorosilicate at 100 °C[22] (see equations below) (Scheme 6.4).

$$(NH_4)_x Al_x Si_{1-x} O_2 + xSiCl_4 \rightarrow (N_2, 400\,°C) \rightarrow SiO_2 + xNH_4Cl + xAlCl_3$$

$$M_x^+ Al_x Si_{1-x} O_2 + x(NH_4)_2 SiF_6 \rightarrow (aq, 100\,°C)\ SiO_2 + xMF + x(NH_4)_2 AlF_5$$

Scheme 6.4 Routes for chemical dealumination.

These methods are impracticable for large-scale preparation of cracking catalysts, however. One of the great achievements in zeolite technology was therefore the development of a straightforward and cheap route to an acid form of zeolite Y that is stable under the extreme conditions of catalytic cracking. This route is known as ultrastabilisation. It was found that if the ammonium form of Y is deammoniated in a shallow bed at temperatures of around 450 °C, rather unstable H-Y results, whereas slower heating and deammoniation of a more deeply packed sample of the zeolite (deep-bed calcination) results in a highly stable and crystalline solid. Characterisation of the 'ultrastable' Y[23] revealed that the unit cell parameter had decreased, and the silicon to aluminium ratio within the framework (as revealed by solid state ^{29}Si MAS NMR) had increased, while the bulk silicon to aluminium ratio had remained constant. Finally, solid state ^{27}Al MAS NMR showed that a major part of the aluminium had left the tetrahedral cation sites to move to extra-framework positions. Figure 6.5 illustrates the same process occurring in zeolite Rho, which is an important acid catalyst for the synthesis of methylamines (Chapter 8). A starting Cs,K-Rho zeolite is first exchanged to the ammonium form and steamed under deep-bed conditions. The sample retains crystallinity. Although the overall Si/Al ratio remains the same (as shown by EDX), the framework Si/Al increases and the extra-framework aluminium (in five- and six-fold coordination) is observed in addition to tetrahedrally coordinated framework aluminium.

In the steaming process, there is a loss of some potential Brønsted acidity associated with tetrahedral aluminium, and the generation of significant Lewis acidity associated with the extra-framework species. The solid can be further modified by further ammonium ion exchange, removal of extra-framework aluminium (by acid washing or use of chelating agents such as ethylenediaminetetraacetic acid, EDTA) and further deammoniation under deep-bed conditions to give highly stable solids. These are shown to contain secondary mesoporosity by nitrogen adsorption measurements and, recently, by 3D TEM (Figure 3.15).[24] The key to the ultrastabilisation reaction appears to be the retention of water vapour and ammonia within the deep zeolite bed during calcination. This, combined with the inherent acidity of protonated zeolites, promotes framework dealumination and silicate migration at comparable rates and enables most of the framework to remain intact. The silicon atoms migrate (from regions that become the mesopores) to replace sites vacated by the aluminium atoms. This secondary mesoporosity in turn provides enhanced molecular transport routes through these treated solids. The process may also be performed under conditions of elevated water vapour pressure, under conditions known as steaming, and very high framework silicon contents can be achieved in a number of zeolites in this way. For zeolites prepared with

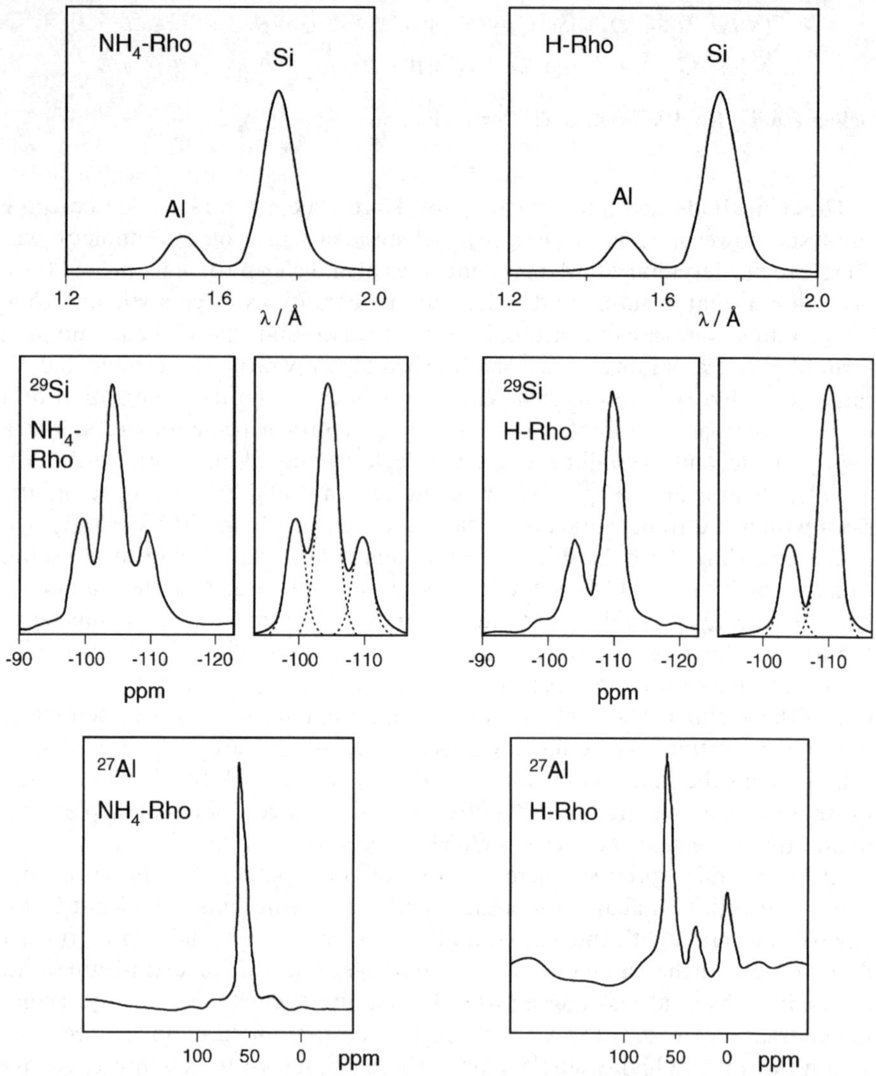

Figure 6.5 Chemical changes upon the ultrastabilisation by 'steaming' of zeolite Rho by heating NH_4-Rho in N_2 saturated with water vapour at 450 °C. XRD (not shown) indicates that crystallinity is retained. Measurements on NH_4-Rho are shown in the left hand column of spectra, whereas the corresponding spectra of ultrastabilised H-Rho are shown in the right hand column. Whereas Energy Dispersive analysis in the electron microscope (EDX analysis) (top) shows no significant change in the bulk composition (the K_α X-ray emission lines of Al and Si have equivalent ratios before and after treatment, giving $(Si/Al)_{bulk} = 3.4$), the ^{29}Si (middle, observed and deconvoluted) and the ^{27}Al (bottom) MAS NMR show major changes. The Si/Al ratio of the framework (obtained from the expression in Section 3.4.1.2) increases from 3.5 to 14.4, and the aluminium atoms lost from the framework take up five-fold and octahedral coordination. [P. A. Wright, PhD Thesis, Cambridge University, 1986.]

aluminium contents in restricted ranges, where no direct routes to high silica equivalents are known, this may be the only route to high silica materials.

Microporous silicates synthesised with isomorphous substitution of elements such as boron and gallium for silicon show similar demetallation behaviour, where the heteroatoms leave the structure more readily than aluminium atoms. In particular, the behaviour upon calcination of boron-containing solids has been examined by ^{11}B and ^{29}Si MAS NMR.[25] Boron is observed to move from tetrahedral to trigonal coordination upon the formation of the protonic borosilicate form, and studies on the protonated form of zeolite B-Beta have shown that the boron may be removed from the framework stepwise by hydrolysis of Si–O–B bonds, ultimately giving boric acid.[26] This is lost from the structure if put into contact with aqueous solution.

6.2.4 Stability in Aqueous Solution

Whereas many aluminosilicate zeolites are almost indefinitely stable at solution pH values close to 7, and are consequently widely used in ion exchange reactions, they do show dissolution at low- and in particular high-pH values. They are typically prepared from strongly alkaline solutions, where alumina and silica have appreciable solubility, so it is possible to re-dissolve them in these media. Indeed, zeolites can re-dissolve and recrystallise during hydrothermal synthesis and can act as sources of aluminates and silicates for further zeolite synthesis. Their behaviour in acid solutions is more variable: frameworks rich in aluminium collapse quickly due to hydrolysis of Al–O bonds and removal of aluminium, whereas high silica zeolites are much more stable. Zeolites with high aluminium contents, including zeolites A, X and L, as well as the titanosilicate ETS-10, are so unstable to acids that they cannot be prepared in the protonic form. In contrast to these very acid-sensitive high alumina zeolites, high silica zeolites such as ferrierite and mordenite are remarkably resistant to mineral acids, even in refluxing concentrated solutions. Loss of aluminium from the framework occurs at a similar rate to migration of silicate anions from elsewhere in the framework to fill the vacancies left by aluminium loss. These vacancies are thought to be in the form of local 'hydroxyl nests', where the tetrahedral cation is replaced by terminal hydroxyl groups. The eventual result of this kind of dealumination is the production of high silica zeolite containing secondary mesopores distributed throughout the bulk. In general, the solubility of silicates in typical mineral acids is low, although hydrofluoric acid will dissolve them, *via* the formation of soluble fluorosilicate species.

Other families of microporous solids, such as the aluminophosphates, are much more reactive in water, even in vapour form. Aluminium in these structures readily coordinates water to adopt five- or six-fold coordination, and exposure of many calcined AlPOs to moisture results in hydrolysis of framework bonds and loss of crystallinity. One way to avoid this, and thereby to enable handling, is to adsorb volatile organic compounds upon cooling after calcination. This protects the most reactive parts of the framework against moisture, and the volatile organics can be removed readily under vacuum to leave the porous structure intact. Metal organic frameworks show a wide range of stability in aqueous solvents, commensurate with their variable chemical composition. Whereas MOF-5, once fully desolvated, transforms rapidly in contact with moist air, the metal imidazolate ZIFs are stable in boiling water (and a range of other solvents). This enhanced stability is attributed to the metal centres in the ZIFs being sterically highly protected against coordinating solvent molecules.[18]

6.2.5 Post-synthetic Modification: Metallation and Pore-size Modification

The process by which vacancies and associated hydroxyl nests can be produced within zeolite frameworks and then filled by using reactive silicon sources can be modified to incorporate a range of other metals into silicate frameworks. Indeed, the strong susceptibility of boron-containing silicates to losing the boron and leaving vacancies makes these solids particularly useful starting materials for such an approach. The parent borosilicates should have high Si/B ratios, to allow crystallinity to be retained upon boron loss. Aluminium can be included by such an approach, for example *via* solutions of aluminium nitrate, which gives the strong acid form of solids not synthesised directly with aluminium in the framework.[27] This approach has been adapted to include titanium and vanadium, among other metal cations.

Although the true pore size of microporous solids is determined by the framework structure and the location of any extra-framework cations that may partially block the pore windows, it can also be modified by post-synthetic treatments. In catalytic processes the deposition of carbonaceous residue ('coke') has this effect, and generally leads to deactivation. In some cases, though, it has been found that the deposition of a thin organic coating on the external surface can improve the selectivity of a zeolite in shape selective processes, presumably by passivating the unselective surface sites and acting as an additional product 'filter' by modifying the external surface pore size. For example, Kaeding[28] demonstrated that deposition of a carborane-silicone polymer onto ZSM-5 passivated the surface and led to enhanced *para*-selectivity in xylenes prepared by toluene disproportionation, and work at DuPont showed that the selectivity to dimethylamine in the synthesis of methylamines over zeolite Rho could be enhanced by the surface deposition and heat treatment of phosphates and alkoxides.[29]

6.3 Cation Exchange

6.3.1 Cation Exchange in Aqueous Solution

The cation exchange properties of zeolites give rise to their largest single application in terms of volume, as detergent additives. They are used for the softening of water (*i.e.* the replacement of calcium ions by sodium ions) in domestic washing. Their widespread use in this application stems from the environmental need to replace phosphates as cation sequestering agents. The presence of effluent phosphate detergent builders in natural water promotes algal blooms and depletion in oxygen, resulting in the death of aquatic fauna. Using zeolites instead of the phosphates removes this problem. Typically, zeolites with a high cation exchange capacity are required for this application and zeolite A (Si/Al = 1) has been the preferred zeolite, although other forms are also widely used (for example, a high aluminium content zeolite P known as MAP (maximum aluminium P), which possesses the gismondine GIS structure type[30]).

The uptake of ammonium into zeolites from aqueous solution is important for the removal of waste ammonia in the farming of animals and fish. It is also used for the preparation of slow-release fertilisers and as the first step in the preparation of acidic catalysts. For the latter application, heating the ammonium form removes ammonia leaving the protonic, acidic zeolite. The selective uptake of radioactive caesium ions from waste streams of nuclear power plants by zeolites and other microporous solids, such as silicotitanates, is another important application. The ready availability of natural clinoptilite makes it particularly attractive in low-level applications, whereas high-grade waste remediation is more demanding, since it involves ion exchange from caustic solutions, and silicotitanates may be more useful in this case.[31] For catalytic purposes, metal cations such as caesium, lanthanum, nickel, palladium and platinum are readily included by ion exchange, the first three as solutions of simple salts, the latter commonly as cationic tetra-ammine complexes, such as $[Pt(NH_3)_4]^{2+}$, which can subsequently be decomposed to leave the metal cation and reduced to give the metal as finely dispersed clusters.

The subject of ion exchange is an important and specialised one, and several detailed reviews are available.[32] The experimental ion exchange behaviour can be described in terms of temperature-dependent equilibria, although in practice the kinetics of the process are also important, for example to ensure rapid water softening in washing. Where high degrees of cation exchange are required, these can most easily be achieved by repeated contact with fresh solutions of an excess of the exchanging ion, although complete exchange may not be possible if a fraction of the cations reside in poorly accessible sites. The pH of the aqueous solution is also important, for highly acidic or alkaline solutions can cause structural degradation. This is likely to be of particular importance for zeolites with high aluminium contents or for aluminophosphate based solids. For di- and trivalent ions, hydrolysis of the aqua-ions may also be important, and for this reason ion exchange is commonly performed at slightly acidic pH to reduce the concentration of hydroxymetal species. One consequence of ion exchange with metalhydroxy cations is the phenomenon of over-exchange, where there are apparently more extra-framework cations than are required to balance the framework charge.[33,34]

In discussing the equilibrium between solution and zeolite, the exchange of cation P^{p+} in the solution for Q^{q+} in the zeolite can be described by the equation:

$$qP_S^{p+} + pQ_Z^{q+} \Leftrightarrow qP_Z^{p+} + pQ_S^{q+} \qquad (6.1)$$

so that the equilibrium constant K_a is of the form

$$K_a = \frac{a_{P_Z}^q a_{Q_S}^p}{a_{P_S}^q a_{Q_Z}^p} = \frac{f_P^q Z_P^q}{f_Q^p Z_Q^p} \cdot \frac{\gamma_Q^p m_Q^p}{\gamma_P^q m_P^q} \tag{6.2}$$

where a_P and a_Q are activities of P and Q at equilibrium, f_P and f_Q are the ionic activity coefficients in the zeolite, γ_P and γ_Q are the ionic activity coefficients in solution, m_P and m_Q are the molarities in solution and Z_P and Z_Q are the equivalent fractions of the two cations in the zeolite, weighted according to their charge,

$$Z_P = \frac{pn_{P_z}}{pn_{P_z} + qn_{Q_z}}; \quad Z_Q = \frac{qn_Q}{pn_{P_z} + qn_{Q_z}} \tag{6.3}$$

$$S_P = \frac{pm_{P_S}}{pm_{P_S} + qm_{Q_S}}; \quad S_Q = \frac{qm_{Q_S}}{pm_{P_S} + qm_{Q_S}} \tag{6.4}$$

Plotting these values of Z against the analogous equivalent fractions of ions in solution, S_P and S_Q, enables the selectivities of binary cation exchange experiments to be displayed, whereas a thermodynamic treatment of K_a can be used to obtain the values of ΔG°, ΔH° and ΔS° for the exchange.

Typical isotherm types for binary ion exchange of a general system are shown schematically in Figure 6.6, showing cases where there is selectivity for the cation M from solution, selectivity to cation M in the zeolite and incomplete ion exchange by M. In addition, it is sometimes observed that the selectivity for an ion can reverse as the loading changes, giving a sigmoidal isotherm. Every zeolite shows its own characteristic cation exchange behaviour, which depends on its composition and structure, and the starting location of cations within the structure. The behaviour can also show strong temperature dependence, particularly when the ion exchange reaction involves strong entropy changes. The ion exchange of a cation is likely to result in a loss of water of hydration as the cation coordinates to framework oxygen atoms and may even require loss of water of hydration for the hydrated ion to enter small-pore zeolites.

One additional effect is observed when zeolites are suspended in concentrated salt solutions (>0.5 M). Under these conditions it becomes thermodynamically favourable for additional inclusion of both cationic and anionic species, in addition to cationic species required for charge balance. This is known as salt imbibition or occlusion.

6.3.1.1 Examples of Ion Exchange Behaviour in Microporous Solids: Zeolites and Silicotitanates

A great deal is known about ion exchange in zeolites, because of their importance as detergent builders – Zeolite A is the most widely used. Most of the cation sites in zeolite A are accessible in or from the large α-cages (Figure 6.7)

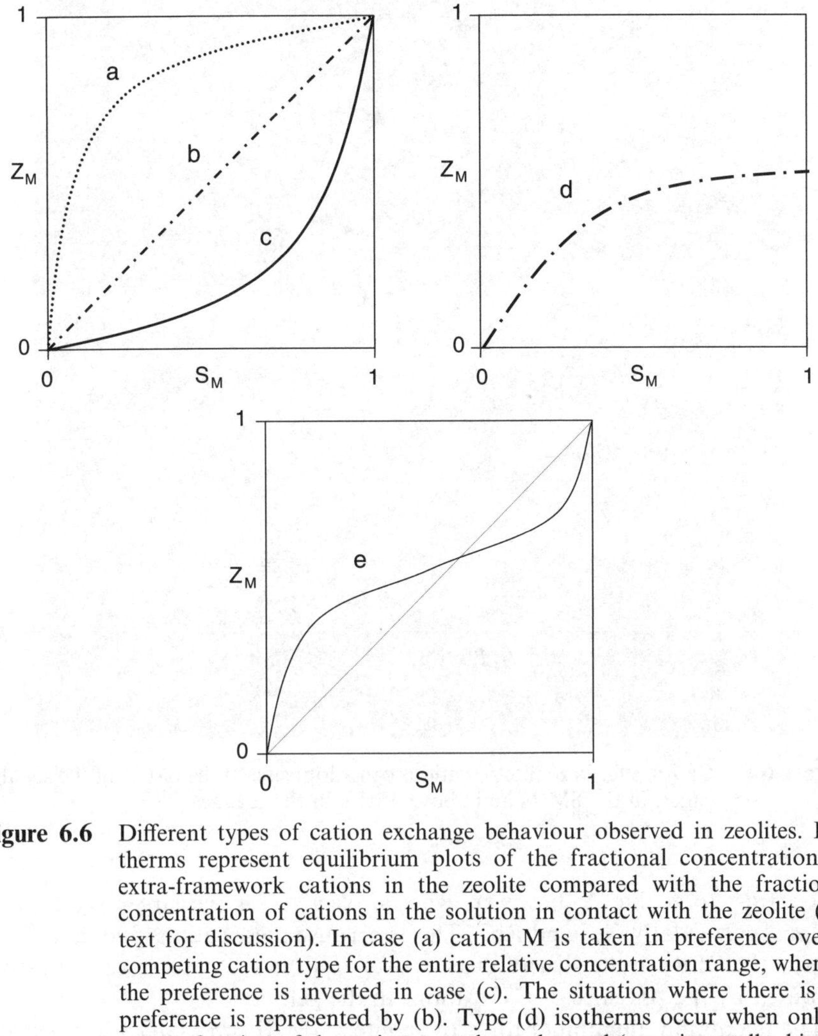

Figure 6.6 Different types of cation exchange behaviour observed in zeolites. Isotherms represent equilibrium plots of the fractional concentration of extra-framework cations in the zeolite compared with the fractional concentration of cations in the solution in contact with the zeolite (see text for discussion). In case (a) cation M is taken in preference over a competing cation type for the entire relative concentration range, whereas the preference is inverted in case (c). The situation where there is no preference is represented by (b). Type (d) isotherms occur when only a certain fraction of the cations may be exchanged (experimentally, kinetic barriers may also result in this behaviour). Finally, isotherms of type (e) indicate that the selectivity changes as the relative concentration of the cations in solution changes.

and most cations can fully exchange for sodium. The selectivity for alkali metal cations decreases in the order $Na^+ > K^+ > Rb^+ > Cs^+, Li^+$.[35] Zeolite A shows a strong preference for the univalent cations Ag^+ and Tl^+ and all the alkali earth metal cations other than Mg^{2+} over Na^+.[36] Mg^{2+} can be exchanged for Na^+, but requires a higher excess of Mg^{2+} in solution. This is because (as is the case for Li^+) the waters of hydration are strongly held, and must be removed for the Mg^{2+} ions to reach the cation sites. Similar complete ion exchange is observed for monovalent and divalent cations in zeolite Rho,[37] in which, like zeolite A, all cation sites are accessible through 8MRs (Figure 6.7).

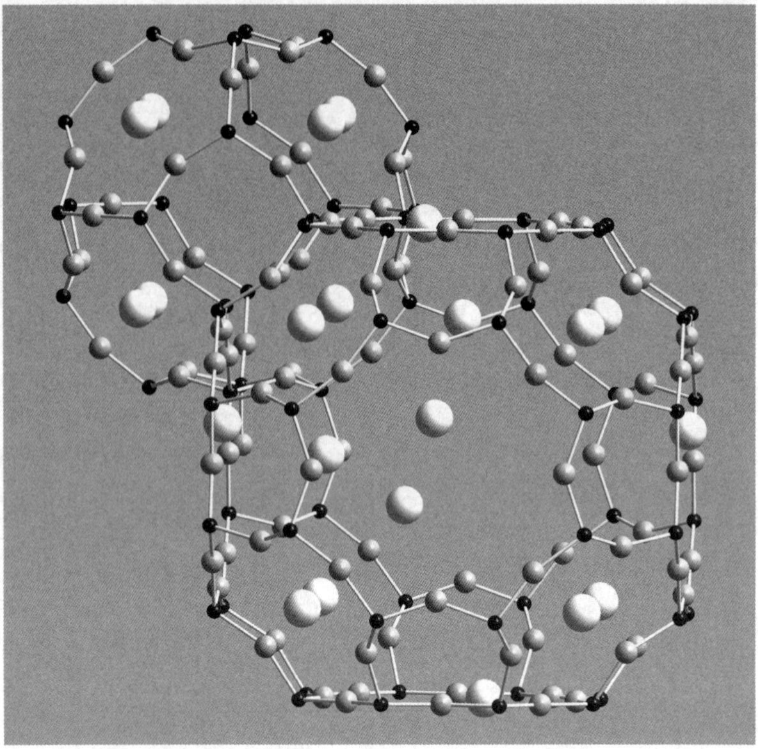

Figure 6.7 Cation sites in zeolite A: cations can adopt sites at the 6MRs of the sodalite cages, in the 8MRs and above 4MRs in the α-cages.

In contrast to zeolite A, however, Rho is selective for the larger alkali metal cations, particularly Cs^+ and Rb^+. This is because these cations find favourable sites in the D8Rs in the structure: as a corollary, the synthesis of zeolite Rho is favoured by the presence of Cs^+ cations in the gel.

The maximum aluminium form of zeolite P (MAP, $MAlSiO_4$) has recently achieved widespread use as an alternative detergent builder to zeolite A.[30] Also a small-pore (8MR) zeolite, the sodium form of its flexible framework structure[38] shows a high selectivity for alkali earth and alkali metal cations (other than Li^+) over the whole compositional range. Its ion exchange performance in washing is enhanced by its small particle size, requiring short ion exchange distances and delivering high rates.

For zeolites such as X, Y and L, the ion exchange behaviour is complicated.[32] Some cations in the starting forms are present in closed sites, which can only be accessed *via* narrow 6MR openings. (For the cation locations in Y, see Figure 2.6.) For this reason, there is incomplete exchange of large cations such as Cs^+ in zeolite Y, because they cannot access sites in the sodalite cages from the main channel. Similarly, alkaline earths such as Ca^{2+}, Sr^{2+} and Ba^{2+} are unable to exchange fully for Na^+ in Y. In zeolite K-L, cations are located within cages, intercage regions or in the large-pore channels. At room temperature, only those present in the large channels can be ion exchanged.

Cation exchange into the high silica zeolite ZSM-5 is important for the preparation of catalysts rather than for ion exchange applications, because of its low cation exchange capacity. All of the cation sites are available *via* 10MR openings, so that Na-ZSM-5 shows complete and selective exchange by other alkali metal cations.[39] The observed behaviour is similar for Zn^{2+} and Cu^{2+}, but for alkaline earth metals and La^{3+} only partial exchange is possible. One explanation for the lower efficiency of ion exchange of divalent alkaline earths and lanthanum is that the low charge density of the high silica framework is unable to 'solvate' the exchanging cations sufficiently well to strip away their waters of hydration, so that they cannot gain access to the favoured sites.

The selective uptake from nuclear waste of radioactive cations which emit a high intensity of penetrating radiation is an important target. Once separated and concentrated, the waste could be encapsulated in borosilicate glass and buried in deep salt mines. The γ-emitters ^{137}Cs and ^{90}Sr are two products of nuclear power generation that are targets for such treatment. Although zeolites can be used for their separation under some conditions, in high-level waste 'cooling ponds' the radioactive cations are typically present in solutions of very high ionic strength (up to 10 M in Na^+) that may have pHs over a very wide range (including highly alkaline values that would dissolve zeolites). Any ion exchanger useful for their removal must therefore be both highly selective and chemically very stable. The silicotitanate CST and the silicotitanate version of pharmacosiderite (see Section 2.5.1) show both properties.[31] Of these, CST is reportedly the most selective material known for this purpose and is currently under investigation for use in these radioactive waste treatments in the USA. Caesium exchange for sodium or protons is strongly favoured, proceeding *via* a site-by-site mechanism, in which first one site and then another is filled. Starting with the H-form, two Cs^+ sites are filled, resulting in a change of symmetry and reaching a Cs^+ exchange capacity of 40% of the maximum extra-framework cation content.[40] The very high selectivity for caesium results from the close fit of the cation within the tunnels of CST and permits use of the material for Cs^+ removal even when the Na^+ content is up to 9 orders of magnitude greater than that of Cs^+. This degree of selectivity is reminiscent of the very selective complexation of alkali metal cations by crown ethers, and it can be fine-tuned by the cation substitutions within the framework which modify the size and geometry of extra-framework cation sites.

Selective ion exchange is also reported for the Sandia octahedral molecular sieves (SOMS), such as those based on niobates.[41] Furthermore, once ion exchange has been achieved, high-temperature treatment gives dense perovskitic phases. Such a separation/encapsulation route may be a viable strategy for radioactive waste removal and storage. There have been relatively few reports of cation exchange behaviour in MOFs, because many of the materials have neutral frameworks. Nevertheless, examples are known, and we have recently observed cation exchange behaviour in lanthanide N, N′-piperazine bismethylene phosphonates (Na La (O_3P-CH_2N $C_4H_8NCH_2PO_3$)).[42]

6.3.2 Solid State Ion Exchange and Intra-zeolite Cation Migration

6.3.2.1 *Solid State Ion Exchange*

As well as readily exchanging cations in aqueous solution, it has been found that ion exchange also occurs upon bringing a cationic zeolite in contact with a metal salt, typically a chloride, and heating the intimate mixture. In particular, Beyer and co-workers have followed the reaction of protonated or ammonium forms of zeolites with metal halides (particularly volatile halides) at elevated temperatures (Scheme 6.5).[43–45] This results in hydrogen chloride and ammonium chloride removal and proton exchange by the metal cations. Many such examples have been reported, including zinc, iron(II) and lanthanum into zeolites Y, mordenite and ZSM-5.

In a variation of the method, it is possible to incorporate indium(I) into high silica zeolites (such as zeolite Beta) by heating a mixture of NH_4-zeolite and In_2O_3 in hydrogen, in a process described as reductive solid state ion exchange (RSSIE)(Scheme 6.6).[46,47] This provides a route to the In(I)-zeolite that could not readily be achieved *via* solution exchange.

In fact, the reaction also occurs by heating the mixture in a vacuum (autoreductive SSIE).[47,48] The In^+ species are readily oxidised to InO^+ species upon exposure to air, as demonstrated by infrared spectroscopy using pyridine as a molecular probe of the metal species present. Similar chemistry is observed upon heating Ga_2O_3/H-zeolite mixtures.

6.3.2.2 *Intra-zeolite Cation Migration*

Solid state ion exchange must involve the migration of cations through the zeolite structure in the absence of water of hydration. Cation migration also occurs in cation-exchanged zeolites at high temperatures. One class of such behaviour occurs when aqueous ion exchange leaves hydrated di- or trivalent cations in the large pores of zeolites such as Y or L that have cages that are only accessible from the large pores *via* narrow windows. Once dehydrated, these cations can find thermodynamically favoured sites of improved coordination within 'closed' cage sites, but there is a kinetic barrier to entering the sites. High temperature (and in some cases the presence of traces of water vapour) can provide the activation energy for internal cation exchange to occur. For example, aqueous ion exchange of large cations such as lanthanum into zeolite

$$\textit{e.g.}\ FeCl_2 + 2NH_4\text{-Zeo} \rightarrow Fe(II)\text{-Zeo} + 2NH_4Cl$$

Scheme 6.5 Solid state ion exchange.

$$H^+\text{-Zeo} + \tfrac{1}{2}In_2O_3 + H_2 \rightarrow In(I)\text{-Zeo} + 1\tfrac{1}{2}H_2O\ (500\ ^\circ C)$$

Scheme 6.6 Reductive solid state ion exchange RSSIE.

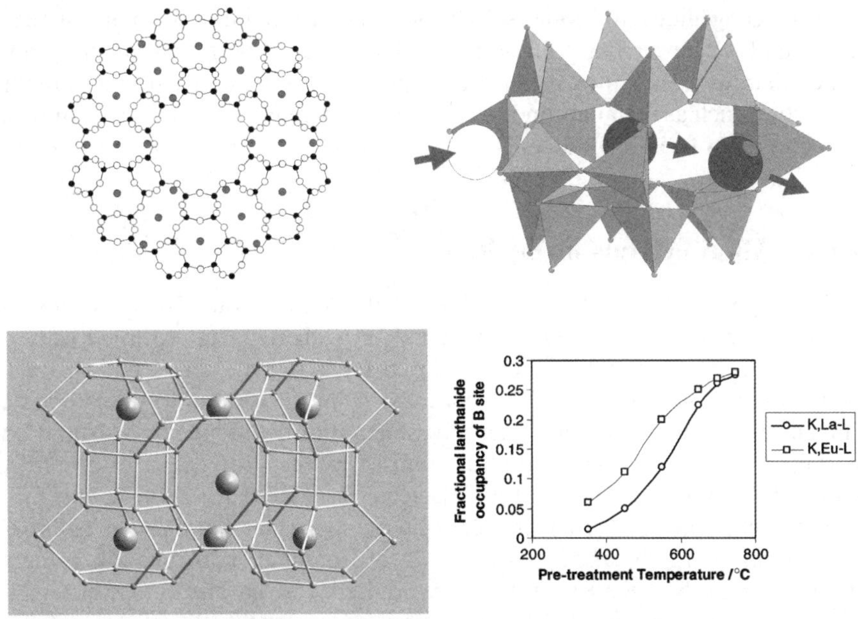

Figure 6.8 Cation migration in zeolite L. When K-L zeolite is stirred at 70 °C with aqueous solutions of lanthanide salts, only potassium ions with direct access to the large channels (left, above and below) are exchanged, giving *ca.* 1.2 Ln^{3+} cations per unit cell. These lanthanide cations move readily into sites between the cancrinite cages (B sites) upon heating to 300 °C, but motion into the cancrinite cages is highly thermally activated. Motion of Ln^{3+} cations into and K^+ cations out of the cancrinite cage occurs only above 400 °C, as shown by the results of Rietveld refinement of the X-ray powder diffraction data of heated samples (right, below).
[P. A. Wright, PhD Thesis, Cambridge University, 1986; P. A. Wright and C. J. Satterley, unpublished results.]

K-L results in potassium cations in the main channel being replaced. Subsequent heating at 400 °C results in dehydration of the La^{3+} cations and their migration into sites away from the main channel and between the columns of cancrinite cages and D6Rs (Figure 6.8).[49] This involves migration through narrow 8MRs. Remarkably, at temperatures > 600 °C, the lanthanum cations are able to migrate from these sites and replace potassium cations within the cancrinite cages. This requires strongly thermally activated migration of both La^{3+} and K^+ cations through highly non-planar 6MRs.

6.4 Inclusion Chemistry

In addition to extra-framework charge-balancing cations, a variety of other elements and compounds can be included within the pores, usually from the vapour. These include volatile organic and inorganic molecules, which are described in the next chapter, but also species such as metals and non-metals, metal chalcogenides and ionic salts of low melting point. For many of these compounds, their preparation as isolated wires or clusters within the pores, rather than as bulk solids in their own phase, imparts unusual and valuable properties, such as metallic conductivity, strong pigmentary colour and shifts in the band gap of semiconductors.

6.4.1 Alkali Metals in Zeolites

As long ago as 1966, Rabo found that alkali metal vapours, such as those of sodium or potassium, could be included into alkali metal cation forms of zeolites.[50] Subsequent studies indicated that the Na_4^{3+} cluster could be formed by inclusion of sodium metal into zeolite Na-Y or the sodium form of sodalite. This intra-zeolitic process has been investigated in detail by Edwards and co-workers.[51] The alkali metal atoms are ionised by the strong electrostatic fields within the zeolite and the resulting electrons become delocalised within multinuclear clusters of alkali metal cations or, at higher loadings, over larger numbers of charge balancing cations. At low loadings, adsorbed alkali metal atoms such as Na, K, Rb and Cs are ionised and an electron transfers to multinuclear clusters of alkali metal cations of the host zeolite (Scheme 6.7):

These clusters are readily identified by ESR spectroscopy, where the unpaired electron couples equally to each of the metal nuclei of the cluster. Figure 6.9 shows the characteristic 13-line spectrum of a Na_4^{3+} cluster formed within the sodalite cage of zeolite Na-Y upon inclusion of sodium or potassium vapour. This cluster has been confirmed to form in the sodalite cages of sodium loaded Na-Y by structural study.[52] Similar clusters are observed upon alkali metal loading of K-exchanged zeolites. The clusters that are formed are made up of cations of the host zeolite, rather than by cations of the included metal. Furthermore, mixed clusters (containing metal cations of different metals) are not observed. Notably, lithium cations do not accept electrons to form clusters. This is thought to be because, being small, they strongly polarise the aluminosilicate framework and have lower associated positive charge than the other alkali metal cations. As a result, they do not trap the electrons. Instead the electrons are delocalised elsewhere in the structure.

At higher alkali metal loadings, metal atom ionisation results in electrons that are more widely delocalised, or in clusters that interact strongly with each other. The exact behaviour depends on the structure of the zeolite. For example, in zeolite Y the first effect is to fill each sodalite cage with a cluster, whereupon the electron spins interact and give a singlet in the ESR.[53] In zeolite K-L, incoming potassium atoms are ionised and the resultant excess electron is delocalised among the K^+ cations that line the large one-dimensional channels of this structure.[54]

$$M_{(g)} + (Na^+)_n - zeolite \rightarrow (M^+ + [nNa]^{(n-1)+}) - zeolite$$

Scheme 6.7 Loading alkali metals into cationic zeolites.

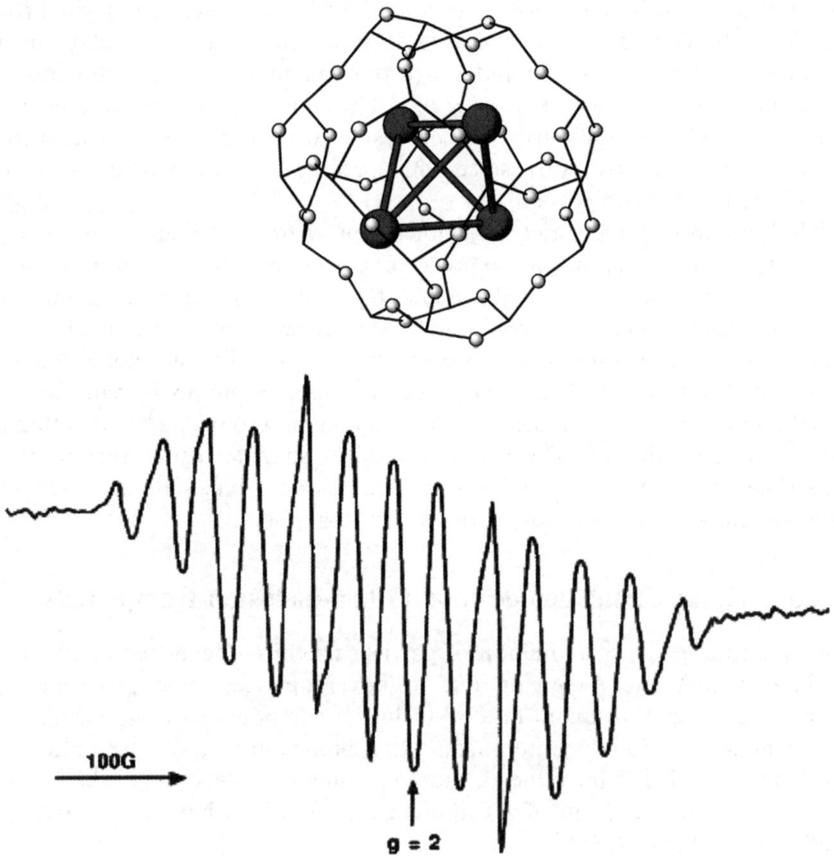

Figure 6.9 Na_4^{3+} cluster and 13-line ESR spectrum observed in zeolite Na-A treated with sodium vapour.
[Courtesy P. A. Anderson, University of Birmingham.]

Alkali metal inclusion gives rise to very strongly basic zeolites, with strong electron donating power (Lewis bases). These are very active as basic catalysts, but their reactivity is too high to permit useful application. Where reducible cations are present in the framework such as in the titanosilicate, ETS-10, inclusion of sodium metal results in the reversible reduction of the framework titanium cation from Ti^{4+} to Ti^{3+}, with concomitant formation of Na^+ cations in the pores.[55]

6.4.2 Sulfide Chromophores: The Ultramarine Family

The beautiful blue and turquoise colours observed in natural forms of sodalite (examples of which are known as *lapis lazuli* and ultramarine) are known to result from the inclusion of sulfide S_3^- (blue) and smaller concentrations of S_2^- (yellow) chromophores within the sodalite cages. Similar materials based on synthetic sodalite have been synthesised artificially since the original route developed by Guimet.[56] In a typical large-scale process the china clay mineral kaolinite is mixed with sulfur and anhydrous sodium sulfate or carbonate, and additional carbon added as a reductant. The sulfide species become included within the sodalite cage during the process, and have been investigated thoroughly by ESR spectroscopy, since both species are paramagnetic. The resultant pigments are widely used.

There is considerable variety in synthetic routes to blue pigments of this kind, including various aluminosilicate precursors, sodium salts and sulfur. Typically the sodalite structure is prepared in a first step, and a subsequent calcination step gives the vivid colour. The royal blue colour appears to require three components: alkali metal, high sulfur content and the sodalite structure. A recent report indicates that the blue pigment can be prepared simply by calcination of sodalites prepared with sulfate and alkylammonium groups included within the pores during synthesis.[57] The importance of the sodalite cage in this process is underlined by the reported synthesis of thermochromic blue samples of zeolite A by high temperature inclusion of the S_3^- chromophore.[58]

6.4.3 Metal Chalcogenides and Other Inclusion Compounds

The quantum effect of reduction of particle size in reducing the band gap of semiconductors and so giving rise to novel optoelectronic properties has stimulated interest in 'quantum dot' inclusion compounds of nanoparticles of semiconductors within zeolite pores.[59] In a pioneering study, Herron and coworkers succeeded in introducing cadmium sulfide clusters within the pores of zeolite Y *via* the reaction of a cadmium-exchanged zeolite Y with hydrogen sulfide gas (Scheme 6.8).[60]

Careful analysis of both powder X-ray diffraction and EXAFS spectroscopic data located the cadmium sulfide as $(CdS)_4$ cubes occupying the space within sodalite cages, with the Cd^{2+} ions coordinated to framework oxygen atoms (Fig. 6.10). Furthermore, the clusters were observed to order between adjacent sodalite cages, to give 'superclusters' or a superlattice structure. In subsequent work, a variety of compounds and elements have been prepared as well-defined clusters within zeolite frameworks, including metal oxides, selenides and phosphides, and these have been studied mainly with the view of determining the effects of cluster size on optical and electronic properties.

6.4.4 Inclusion of Complexes by MOCVD

There have been many successful attempts to include organometallic complexes into zeolites, for example for use as heterogenised catalysts. An interesting variation on this approach is the inclusion of organometallic complexes such as

$$Cd^{2+}\text{-zeol} + H_2S \leftrightarrow CdS + 2H^+ + \text{zeol}^{2-}$$

Scheme 6.8 Formation of CdS clusters within a zeolite.

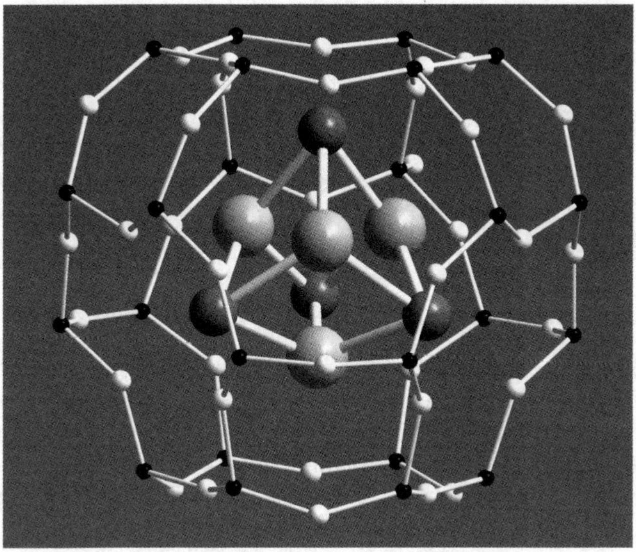

Figure 6.10 $Cd_4(SO)_4$ clusters formed in the sodalite cage of zeolite Cd-Y upon treatment with H_2S.

$[(\eta^5\text{-}C_5H_5)Pd(\eta^3\text{-}C_3H_5)]$, $[(\eta^5\text{-}C_5H_5)Cu(PMe_3)]$ and $[(CH_3)Pd(PMe_3)]$ into MOF-5 by Metal Organic Chemical Vapour Deposition (MOCVD).[61] The non-polar nature of the molecular sieve allows ready access of the metal complexes into the pores, and careful UV photolysis results in their conversion to finely divided (1 or 2 nm) metal particles.

6.5 Reduction and Oxidation Chemistry

It is possible to oxidise and reduce atoms in the framework and also those within the pores of microporous (and mesoporous) solids of appropriate chemical compositions. Although pure aluminosilicate, silicate and aluminophosphate frameworks cannot be oxidised or reduced, transition metal and some lanthanide cations within the framework can exist in different oxidation states. Also, although typical alkali, alkali metal and most lanthanide cations in extra-framework positions possess no redox chemistry, transition metal cations such as nickel, copper and platinum do. In the case of the transition metals, this enables them to become important catalysts. The included sulfide species in ultramarine-related pigments described above are also prepared through the reduction of sulfate species.

6.5.1 Reduction of Extra-framework Transition Metal Cations

Transition metal cations such as Ni^{2+}, Pd^{2+} and Pt^{2+} present as charge balancing cations as the result of cation exchange (for Pd and Pt as the ammine complexes, $[M(NH_3)_4]^{2+}$) are readily reduced at elevated temperatures in hydrogen gas. This gives metallic particles (*via* the univalent cation in the case of nickel) and associated protons (to maintain charge balance). The combined metal/proton forms of the zeolites give bifunctional catalysts that find use in the reforming of hydrocarbons from crude oils. These reductions are usually irreversible because prolonged heating leads to progressive sintering of the metal particles and their exit from the pore space.

6.5.2 Redox Behaviour of Framework Transition Metal Cations

Inclusion of transition metals in tetrahedral framework sites in zeolites is only possible for a few metals in the first row, because second and third row elements have ionic sizes that are too large. Even for first row transition elements relatively few examples have been unambiguously synthesised or characterised. For manganese, cobalt, nickel and copper the highly alkaline conditions of zeolite synthesis tend to result in the precipitation of oxyhydroxides and hydroxides. Titanium is a notable exception, and titanosilicates have been shown to exhibit a range of activities in catalysing oxidation reactions, but there is no evidence that the titanium is oxidised or reduced. Aluminophosphates display a much greater ability to take up first row transition metals into their frameworks, as a result of the suitable synthesis pH and the greater ionic character of the framework, and both doped and stoichiometric transition metal phosphates are readily crystallised. It is not possible to remove the organic molecules from organically templated open framework transition metal phosphates without collapse of the structures but aluminophosphate solids in which up to 20 atomic percent of the framework aluminium in tetrahedral positions is replaced by Mn^{2+}, Fe^{2+} or Co^{2+} are stable to removal of the template by calcination in oxygen. As well as rendering the solids porous, this treatment oxidises the transition metals to the trivalent state, whilst leaving the cations tetrahedrally coordinated. There are strong changes in the UV-visible spectra, and the cobalt aluminophosphates change from blue (tetrahedral Co(II)) to green (tetrahedral Co(III)) whereas manganese aluminophosphates change from almost colourless to mauve. The cations may subsequently be reduced upon exposure to hydrogen, hydrocarbons or even water. These changes are determined quantitatively by UV-visible and X-ray absorption spectroscopy (Chapter 3). Such solids have recently been found to be shape selective oxidation catalysts.

In the examples discussed above, the transition metals are heteroatoms that occupy tetrahedral sites within the framework. There are also examples where transition metals occupy octahedral framework sites within a porous silicate framework. In the much studied titanosilicate ETS-10, in which the structure contains a mixture of octahedral titanium and tetrahedral silicon, a small fraction of the octahedral titanium may be reduced to Ti(III) by hydrogen. In addition, it is possible to include the lanthanide cerium into the mixed coordination microporous silicate AV-5.[62] As described in Section 2.5.1, this compound is composed of microporous silicate layers and octahedral sheets containing cerium and sodium cations. The cerium in this solid is incorporated as Ce(III) during the hydrothermal synthesis, but upon heating in oxygen at 300 °C it is thought to undergo oxidation to Ce(IV) (on the basis of indirect NMR evidence).

There is a growing class of manganese oxide-based octahedral molecular sieves, based entirely on edge-sharing frameworks, related to the hollandite structure (Section 2.6). These manganese oxides show redox behaviour[63] and potential as cathode materials for rechargeable lithium batteries through their ability to take up and release lithium cations, with attendant reduction of the framework manganese cations.[63]

6.6 Intra-zeolitic Chemistry

The cages and channels within microporous solids provide unique, crystallographically well-defined, nanometre-scale environments in which chemical reactions may be performed. Molecules can be introduced into sites where they are separated from other molecules. Ingenious 'ship-in-a-bottle' type syntheses of complexes and the study of photochemical reactions on isolated molecules are examples of this kind where the spatial separation of molecules is important in the chemical behaviour. The ability of zeolites to control the position and orientation of included molecules can introduce potentially interesting optical or electronic properties to the resultant organic-inorganic arrays.

6.6.1 Isolated Molecules and Arrays of Molecules

6.6.1.1 Encapsulated Metal Complexes: Ships-in-a-bottle

One of the most elegant synthetic approaches to using the internal cavities of zeolites as nanometre size reactors is that of the 'ship-in-a-bottle' synthesis of metal complexes within zeolite cages, which are then too large to escape through the cage windows. The term was initially coined by Herron to describe metal complexes, such as those with salen-(bis(salicylidene)ethylendiamine-)[64] or phthalocyanine[65] (Scheme 6.9) that were formed in the supercages of faujasitic zeolites. Zeolites X and Y are most commonly used, but the fully

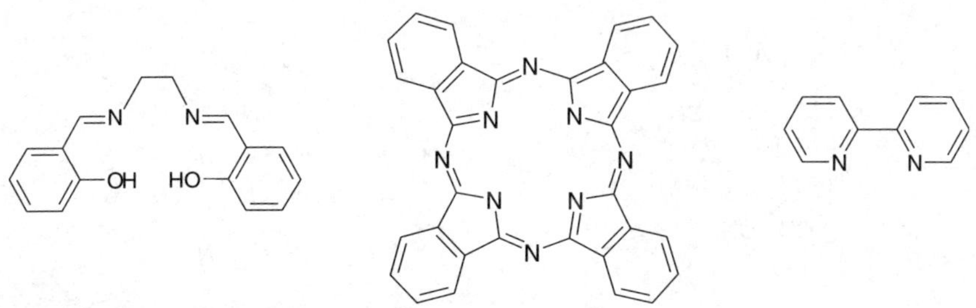

Scheme 6.9 Salen (left), phthalocyanine (centre) and bipyridyl (right) ligands.

hexagonal variant of faujasite (EMT) also possesses large enough cavities to be of interest.

There are two conceptually distinct routes to encapsulated complexes. The encapsulation of salen-complexes is achieved by adsorption of the salen ligand into, for example, a cobalt-exchanged zeolite Y. Whereas the uncomplexed ligand is flexible and able to migrate through the 12MR windows of Y, the cobalt salen complex of the so-called Schiff base (this refers to the imine group that complexes to the metal) is rigid, due to its coordination of the transition metal cation, and consequently is unable to move through the window. Many other metal salen complexes have been prepared in zeolites in this way, for example for $M{\equiv}VO^{2+}$, Pd^{2+} and with modified and even chiral salen ligands. As well as catalytic properties discussed in Chapter 9, the cobalt salen complex retains its ability, observed in solution, to reversibly bind O_2 in a 1:1 complex. Other metal complexes can be assembled in this way, and encapsulated within the pores: among them, tris(bipyridyl) metal complexes, for example of ruthenium(II), cobalt(II), iron(II) and vanadium(IV) as VO^{2+}. The ruthenium complex is of particular interest because of its property of photo-induced electron transfer.

The preparation of metallocyanines proceeds by chemical assembly of the 'ship', rather than by complexation. Catalytic properties of encapsulated phthalocyanine complexes are reported in Chapter 9. Herron prepared iron phthalocyanine within zeolite Y by templating the tetramerisation of o-phthalodinitrile around iron(II) cations within the zeolite (Scheme 6.10). The same process may be performed around Cu^{2+} as a template. The phthalocyanine complex is encapsulated within the supercage, and cannot escape. One drawback to this route is that by-products are formed in the pores, which may be difficult to remove and characterise. An alternative route, achieved by Balkus, is to crystallise the zeolite around the phthalocyanine, in a 'bottle-around-the-ship' approach.[66] This has been successful, for example, in preparing Ru(II)-perchloro- and perfluoro-phthalocyanines in zeolite X. The generality of this crystallisation route is limited by the solubility of the complex and its stability in solution. Furthermore, it relies on fortuitous uptake of the complex during crystallisation into large pores, rather than true templating. In these cases the close fit of molecule to framework usually leaves little space for any desired catalysis.

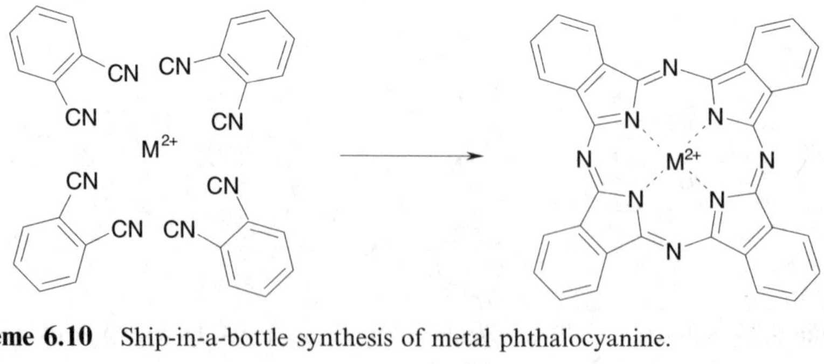

Scheme 6.10 Ship-in-a-bottle synthesis of metal phthalocyanine.

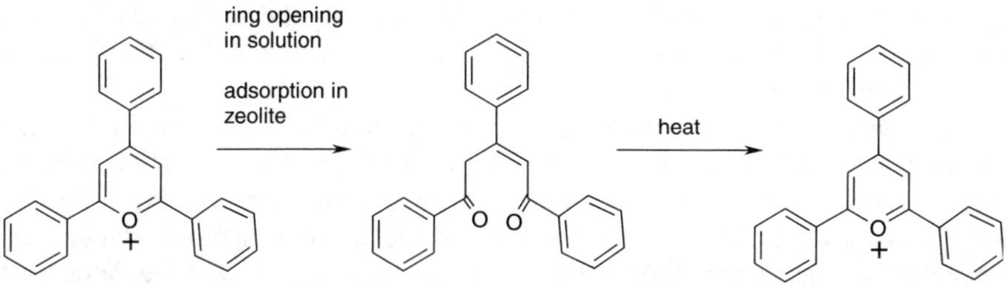

Scheme 6.11 Inclusion of the organic photosensitive tripyrilium ion in a zeolite.

6.6.1.2 Encapsulation of Functional Organic Molecules

Most initial attention on ship-in-a-bottle complexes was paid to the encapsulation of metal complexes, as described above. More recently, however, the technique has been adapted for fully organic syntheses of large functional organic molecules within zeolite cavities. The molecules synthesised by C–C bond-forming reactions within zeolite pores include pyrilium and trityl cations, which have photocatalytic and electrochemical properties. The encapsulation of 2,4,6-triphenylpyrilium, for example, occurs *via* a ring opening of the cation in solution to give a flexible dione that can adsorb into the supercages of the zeolite. Heating the composite solid reforms the pyrilium heterocyclic ion for applications in photocatalysis (Scheme 6.11).[67] Other examples are given in the review of Corma and Garcia.[68]

Conducting organic polymers such as polyacetylenes, polyanilines and polypyrroles are of interest in electronic devices. One of the difficulties associated with their application is that they are degraded upon exposure to the atmosphere. For this reason, attempts have been made to prepare them encapsulated within zeolites, for example by polymerising acetylene over metal-exchanged zeolites. A recent report shows that polyacetylenes can also be prepared inside functionalised MOFs (see Section 10.3.3). It remains a challenge to prepare materials of acceptable properties for applications.

Just as some metal complexes can be incorporated during synthesis, functional organic molecules can also be entrained in this way. In one such example, Mintova *et al.* incorporated the photosensitive 2-(2′-hydroxyphenol)benzothiazole in nanocrystals of zeolite Y.[69] Such approaches aim toward the development of nanoscale devices for sensing, UV filtering and molecular switching.

6.7 Chemistry of Mesoporous Solids

6.7.1 Hydrothermal Stability and Post-synthetic Functionalisation

Typical mesoporous silicas are more weakly acidic and less hydrothermally stable than zeolites, particularly at elevated temperatures, and for these reasons cannot be used as high-temperature cracking catalysts. Their chemistry is much closer to amorphous high surface area silicas (and silica-aluminas, if some aluminium is included) and therefore do not show the well-defined cation exchange properties observed in zeolites.

Chemical modifications of mesoporous silicas, and the type of chemistry that can be performed within their pores, is different from that of zeolites because their pores are larger and can be lined with reactive silanol (SiOH) and other groups. Silanol and SiOEt groups are left on the internal surfaces after solvent extraction of surfactant templates by acidified ethanol, for example. When the solids are calcined, surface hydroxyls are removed by condensation, but refluxing in water hydrolyses surface Si–O–Si bonds and regenerates surface hydroxyls. These silanol groups can react as nucleophiles with metal chlorides and alkoxides. In particular, they can act as points of attachment for chlorosilanes or ethoxysilanes, such as those of the form $(EtO)_3$-Si-R-X (where X is a functional organic group). This modification of the surface can directly attach functional groups such as alkylammonium, carboxylic acid or thiol. An additional versatile Grignard route uses lithiated organics to add functional groups to the hydroxylated silica surface that are not available as siloxanes.[70,71] In addition to these post-synthetic routes to surface functionalisation, functional groups can be included during synthesis by co-condensation of functionalised siloxanes with tetraethoxysilane. Removal of the surfactant template by extraction leaves the functional groups accessible to adsorbed molecules.

Functionalised mesoporous silicas can be used for subsequent immobilisation of catalytic complexes.[72] The large available pore space enables a very wide range of functional molecules and complexes to be immobilised in this way. Mesoporous silicas functionalised with alkylammonium ions can be used for anion exchange. This can be used to prepare the solids as the hydroxide form, which demonstrates basic character,[73] or to ion exchange in anionic metal complexes, such as $[AuCl_3]^-$ or $[PtCl_4]^{2-}$. These species can subsequently be reduced to prepare finely dispersed metal particles throughout the pores.[74]

Mesoporous solids can be prepared in acidic form, either inorganically, by incorporation of aluminium in the framework, or organically, by oxidising thiol groups to sulfonic acid groups. As well as imparting catalytic activity for molecular transformations, this has been shown to be useful for promoting the formation of carbon within the pores, for example by the dehydration of sugars that fill the pores. Very high surface area carbons with regular pore geometries can be prepared by such routes:[75] in these the silica is effectively used as a cast, and subsequently removed by dissolution, leaving a regular inverse replica of the original structure.

6.8 Summary

Although a common view of microporous solids is as fixed frameworks which are neutral or have a negative charge that permits cation exchange behaviour, they are rarely completely inert. To be of use, they should not lose their regular order upon template removal and should be able to withstand changes in framework composition without structural collapse. The most important and best studied of these result from dealumination (or more generally demetallation) of silicate frameworks, when metal cations are lost from framework positions and the subsequent vacancies replaced by migrating silicon atoms.

The frameworks of some inorganic microporous solids show changes in dimensions and symmetry, for example as a result of dehydration or cation relocation: much greater flexibility is observed in some MOFs where flexible organic linkers can change configuration and adapt to adsorption. In addition, the frameworks of aluminophosphates can tolerate redox behaviour of included transition metal cations and both AlPOs and silicates show variation in coordination around framework cations (*e.g.* aluminium atoms in AlPOs, titanium atoms in zeotypic titanosilicates, and both aluminium and silicon atoms in fluoride-containing, as-prepared aluminophosphates or silicas, respectively).

Zeolites were used as well-defined hosts with well-defined nanospaces that can be filled or used as reactors long before nanotechnology achieved its current popularity. 'Ship-in-a-bottle' complexes, laser dyes and chromophore species, nanoparticulate oxides and sulfides and conducting wires and polymers have all been assembled in regular arrays within the pores. At present, catalytically active nanoparticles of supported metals remain the most widely applied of such composites.

Mesoporous solids provide well-ordered scaffolds for chemistry based largely on surface functionalisation. This is readily achieved at the internal surfaces because of the large numbers of silanol hydroxyls that are present there, and which can be used to attach ligands *via* established synthetic procedures.

References

1. G. Sankar, P. A. Wright, S. Natarajan, J. M. Thomas, G. N. Greaves, A. J. Dent, B. R. Dobson, C. A. Ramsdale and R. H. Jones, *J. Phys. Chem.*, 1993, **97**, 9550.
2. S. Cartlidge and W. M. Meier, *Zeolites*, 1984, **4**, 218.
3. E. T. C. Vogt and J. W. Richardson, *J. Solid State Chem.*, 1990, **87**, 469.
4. V. J. Carter, P. A. Wright, J. D. Gale, R. E. Morris, E. Sastre and J. Pérez-Pariente, *J. Mater. Chem.*, 1997, **7**, 2287.
5. R. Garcia, I. J. Shannon, A. M. Z. Slawin, W. Z. Zhou, P. A. Cox and P. A. Wright, *Micropor. Mesopor. Mater.*, 2003, **58**, 91.
6. J. B. Parise, T. E. Gier, D. R. Corbin and D. E. Cox, *J. Phys. Chem.*, 1984, **88**, 1635.
7. J. B. Parise, L. Abrams, T. E. Gier, D. R. Corbin, J. D. Jorgensen and E. Prince, *J. Phys. Chem.*, 1984, **88**, 2303.
8. M. E. Leonowicz, J. A. Lawton, S. L. Lawton and M. K. Rubin, *Science*, 1994, **264**, 1910.
9. S. L. Lawton, A. S. Fung, G. J. Kennedy, L. B. Alemany, C. D. Chang, G. H. Hatzikos, D. N. Lissy, M. K. Rubin, H. K. C. Timken, S. Steuernagel and D. E. Woessner, *J. Phys. Chem.*, 1996, **100**, 3788.

10. G. J. Kennedy, S. L. Lawton, A. S. Fung, M. K. Rubin and S. Steuernagel, *Catal. Today*, 1999, **49**, 385.
11. A. Corma, V. Fornes, J. Martinez-Triguero and S. B. Pergher, *J. Catal.*, 1999, **186**, 57.
12. L. Schreyeck, P. Caullet, J. C. Mougenel, J. L. Guth and B. Marler, *Micropor. Mater.*, 1996, **6**, 259.
13. B. Marler, N. Stroter and H. Gies, *Micropor. Mesopor. Mater.*, 2005, **83**, 201.
14. B. Marler, M. A. Camblor and H. Gies, *Micropor. Mesopor. Mater.*, 2006, **90**, 87.
15. G. Sankar, D. Gleeson, C. R. A. Catlow, J. M. Thomas and A. D. Smith, *J. Synch. Rad.*, 2001, **8**, 625.
16. M. P. J. Peeters, J. H. C. van Hoof, R. A. Sheldon, V. L. Zholobenko, L. M. Kustov and V. B. Kazansky, in '*Proceedings of the 9th IZC*', Washington DC, 1993, 651.
17. R. Garcia, T. D. Coombs, M. J. Maple, W. Zhou, I. J. Shannon, P. A. Cox and P. A. Wright, *Stud. Surf. Sci. Catal.*, 2004, **154**, 993.
18. K. S. Park, Z. Ni, A. P. Cote, J. Y. Choi, R. D. Huang, F. J. Uribe-Romo, H. K. Chae, M. O'Keeffe and O. M. Yaghi, *Proc. Natl. Acad. Sci. USA*, 2006, **103**, 10186.
19. F. Dougnier, J. Patarin, J. L. Guth and D. Anglerot, *Zeolites*, 1992, **12**, 160.
20. H. K. Beyer, I. M. Belenykaja, F. Hange, M. Tielen, P. J. Grobet and P. A. Jacobs, *J. Chem. Soc. Farad. Trans. I*, 1985, **81**, 2889.
21. M. W. Anderson and J. Klinowski, *J. Chem. Soc. Farad. Trans. I*, 1986, **82**, 1449.
22. G. W. Skeels and D. W. Breck, in '*Proc. 6th International Zeolite Conference*', 1984, 87.
23. C. V. McDaniel and P. K. Maher, in '*Molecular Sieves*', Society of Chemical Industry, London, 1968, 186.
24. A. H. Janssen, A. J. Koster and K. P. de Jong, *Angew. Chem. Int. Ed.*, 2001, **40**, 1102.
25. C. Fild, D. F. Shantz, R. F. Lobo and H. Koller, *Phys. Chem. Chem. Phys.*, 2000, **2**, 3091.
26. H. Koller, C. Fild and R. F. Lobo, *Micropor. Mesopor. Mater.*, 2005, **79**, 215.
27. R. F. Lobo and M. E. Davis, *J. Am. Chem. Soc.*, 1995, **117**, 3764.
28. W. W. Kaeding, C. Chu, L. B. Young, B. Weinstein and S. A. Butler, *J. Catal.*, 1981, **67**, 159.
29. D. R. Corbin, S. Schwarz and G. C. Sonnichsen, *Catal. Today*, 1997, **37**, 71.
30. C. J. Adams, A. Araya, K. J. Cunningham, K. R. Franklin and I. F. White, *J. Chem. Soc. Farad. Trans.*, 1997, **93**, 499.
31. A. Clearfield, *Solid State Sci.*, 2001, **3**, 103.
32. W. Schmidt, in '*Handbook of Porous Solids*', Ed. F. Schuth, K. S. W. Sing and J. Weitkamp, Wiley-VCH, New York, 2002, 1058.
33. J. Valyon and W. K. Hall, *Catal. Lett.*, 1993, **19**, 109.

34. H. S. Park and K. Seff, *J. Phys. Chem. B*, 2000, **104**, 2224.
35. R. M. Barrer, L. V. C. Rees and D. J. Ward, *Proc. Roy. Soc. A*, 1963, **273**, 180.
36. H. S. Sherry and H. F. Walton, *J. Phys. Chem.*, 1967, **71**, 1457.
37. R. M. Barrer, S. Barri and J. Klinowski, *J. Chem. Soc. Farad. Trans. I*, 1980, **76**, 1038.
38. B. R. Albert, A. K. Cheetham, J. A. Stuart and C. J. Adams, *Micropor. Mesopor. Mater.*, 1998, **21**, 133.
39. D. P. Matthews and L. V. C. Rees, *Chem. Age India*, 1986, **37**, 353.
40. A. J. Celestian, D. G. Medvedev, A. Tripathi, J. B. Parise and A. Clearfield, *Nucl. Instr. Meth. Phys. Res. B*, 2005, **238**, 61.
41. M. Nyman, A. Tripathi, J. B. Parise, R. S. Maxwell, W. T. A. Harrison and T. M. Nenoff, *J. Am. Chem. Soc.*, 2001, **123**, 1529.
42. J. A. Groves, P. Lightfoot and P. A. Wright, 2007 (unpublished).
43. H. K. Beyer, H. G. Karge and G. Borbely, *Zeolites*, 1988, **8**, 79.
44. H. G. Karge, B. Wichterlova and H. K. Beyer, *J. Chem. Soc. Farad. Trans.*, 1992, **88**, 1345.
45. H. G. Karge, G. Palborbely and H. K. Beyer, *Zeolites*, 1994, **14**, 512.
46. R. M. Mihalyi, H. K. Beyer, V. Mavrodinova, C. Minchev and Y. Neinska, *Micropor. Mesopor. Mater.*, 1998, **24**, 143.
47. M. R. Mihalyi and H. K. Beyer, *Chem. Commun.*, 2001, 2242.
48. Y. Neinska, R. M. Mihalyi, V. Mavrodinova, C. Minchev and H. K. Beyer, *Phys. Chem. Chem. Phys.*, 1999, **1**, 5761.
49. P. A. Wright, '*Structure of zeolites and the zeolite-sorbate complex*', PhD Thesis, Cambridge, 1986.
50. J. A. Rabo, *Discuss. Farad. Soc.*, 1966, **41**, 328.
51. P. P. Edwards, P. A. Anderson and J. M. Thomas, *Acc. Chem. Res.*, 1996, **29**, 23.
52. A. R. Armstrong, P. A. Anderson and L. J. Woodall, *J. Am. Chem. Soc.*, 1995, **117**, 9087.
53. P. A. Anderson and P. P. Edwards, *J. Am. Chem. Soc.*, 1992, **114**, 10608.
54. P. A. Anderson, A. R. Armstrong and P. P. Edwards, *Angew. Chem. Int. Ed.*, 1994, **33**, 641.
55. S. Bordiga, G. T. Palomino, A. Zecchina, G. Ranghino, E. Giamello and C. Lamberti, *J. Chem. Phys.*, 2000, **112**, 3859.
56. R. J. D. Tilley, *'Colour and Optical Properties of Materials'*, John Wiley and Sons, 2000.
57. D. Arieli, D. E. W. Vaughan and D. Goldfarb, *J. Am. Chem. Soc.*, 2004, **126**, 5776.
58. S. Kowalak and M. Strozyk, *J. Chem. Soc. Farad. Trans.*, 1996, **92**, 1639.
59. G. A. Ozin, S. Kirkby, M. Meszaros, S. Ozkar, A. Stein and G. D. Stucky, *ACS Symp. Ser.*, 1991, **455**, 554.
60. N. Herron, Y. Wang, M. M. Eddy, G. D. Stucky, D. E. Cox, K. Moller and T. Bein, *J. Am. Chem. Soc.*, 1989, **111**, 530.
61. S. Hermes, M. K. Schroter, R. Schmid, L. Khodeir, M. Muhler, A. Tissler, R. W. Fischer and R. A. Fischer, *Angew. Chem. Int. Ed.*, 2005, **44**, 6237.

62. J. Rocha, P. Ferreira, L. D. Carlos and A. Ferreira, *Angew. Chem. Int. Ed.*, 2000, **39**, 3276.
63. R. N. Deguzman, Y. F. Shen, B. R. Shaw, S. L. Suib and C. L. O'Young, *Chem. Mater.*, 1993, **5**, 1395.
64. N. Herron, *Inorg. Chem.*, 1986, **25**, 4714.
65. N. Herron, *J. Coord. Chem.*, 1988, **19**, 25.
66. K. J. Balkus Jr., A. Khanmamedova and M. Eissa, *Stud. Surf. Sci. Catal.*, 1995, **97**, 189.
67. A. M. Amat, A. Arques, S. H. Bossmann, A. M. Braun, S. Gob and M. A. Miranda, *Angew. Chem. Int. Ed.*, 2003, **42**, 1653.
68. A. Corma and H. Garcia, *Eur. J. Inorg. Chem.*, 2004, 1143.
69. S. Mintova, V. de Waele, M. Holzl, U. Schmidhammer, B. Mihailova, E. Riedle and T. Bein, *J. Phys. Chem. A*, 2004, **108**, 10640.
70. J. E. Lim, C. B. Shim, J. M. Kim, B. Y. Lee and J. E. Yie, *Angew. Chem. Int. Ed.*, 2004, **43**, 3839.
71. S. Angloher and T. Bein, *J. Mater. Chem.*, 2006, **16**, 3629.
72. A. Taguchi and F. Schuth, *Micropor. Mesopor. Mater.*, 2005, **77**, 1.
73. I. Rodriguez, S. Iborra, A. Corma, F. Rey and J. L. Jorda, *Chem. Commun.*, 1999, 593.
74. C. M. Yang, M. Kalwei, F. Schuth and K. J. Chao, *Appl. Catal. Al.*, 2003, **254**, 289.
75. S. N. Che, A. E. Garcia-Bennett, X. Y. Liu, R. P. Hodgkins, P. A. Wright, D. Y. Zhao, O. Terasaki and T. Tatsumi, *Angew. Chem. Int. Ed.*, 2003, **42**, 3930.

CHAPTER 7

Adsorption and Diffusion

7.1 Introduction and Definitions

7.1.1 Introduction

The ability of zeolites and related solids to take up molecules from the liquid or gas phase is the basis for their application in gas purification and separation and is important in catalysis. Molecules small enough to diffuse through the pore windows have access to an internal surface area that is very many times greater than that of the external surface. Once at the internal surface they can bind to particular sites. The process involved is known as adsorption, since the adsorbate molecules are bound to an inorganic surface (ad = to (Latin)). The strength of the adsorption process varies from weak physisorption, for example in gas separation, to much stronger chemisorption, where molecules form chemical bonds with the internal surface adsorption sites. These sites are framework and extra-framework cations and framework oxygen atoms in the case of inorganic microporous solids and include organic groups in hybrid solids. In some cases the structure may relax as molecules are adsorbed, to enable a more favourable interaction to occur. For example, framework cations are able to adjust their coordination to bind to adsorbates and many hybrid solids are found to be highly flexible and responsive to adsorption. Once adsorbed within the pores, the molecules may diffuse between adsorption sites in different cages and along channels. Diffusion rates are determined by the strength of adsorption and the proximity to other adsorption sites, the relationship of molecular shape to channels, the concentration of co-adsorbed species and the temperature.

Experimental studies of adsorption may be grouped into three main categories: measurement of adsorption uptakes and associated thermodynamic parameters, spectroscopic analysis of adsorbate molecules and NMR and diffraction-based analysis of the location of molecules in the pores. Measurement of adsorption uptakes and thermodynamic parameters gives information on the strength of adsorption and the heterogeneity of adsorption sites. Spectroscopic methods probe changes in the vibrational and electronic energy levels of both adsorbate and molecular sieve, and indicate the strength and chemical type of specific interactions. In ideal circumstances, diffraction and NMR spectroscopy can determine the precise geometry of the adsorbate-solid complex. The complementary processes of re-orientational motion and diffusion are studied experimentally through measurements of the kinetics of adsorption and desorption, by optical and IR microscopy and by NMR and neutron scattering. In addition to these experimental methods, computational simulation of both adsorption and diffusion is possible to an increasing degree, as the structures and physical parameters are more accurately established and as computing power increases. Examples of this approach were given in Chapter 4, including combined pair-potential, Molecular Mechanics and Monte Carlo methods for modelling physisorption, quantum mechanical modelling of chemisorption and molecular dynamics and transition state theory for simulating diffusion.

In the first part of this chapter I outline the theory and practice of adsorption studies, before going on to examine the nature of adsorption of different molecules at specific types of sites on the internal surface and the diffusion of adsorbates through channels and cages in microporous solids. In the light of this, I discuss the role of adsorption in particular applications.

7.1.2 Definitions

7.1.2.1 Physisorption and Chemisorption

In general terms, physical adsorption, or physisorption, refers to weak bonding of molecules to surfaces through the interactions of induced or permanent dipoles and/or quadrupoles, whereas chemisorption describes adsorption where transfer of chemical charge between adsorbate and surface takes place. Physisorption is characteristically observed at low temperatures, is not an activated process and is completely reversible. Chemisorption, by contrast, involves the formation of bonds, persists to elevated temperatures and can lead to chemical changes. For the adsorption of molecules on microporous solids, important physisorption interactions include the uptake of simple non-polar molecules such as dinitrogen and dioxygen on cationic forms of zeolites whereas the adsorption of molecules onto acid sites is the most important type of chemisorption, because of its importance in catalysis.

7.1.2.2 Molecular Sieving

The ability of crystalline microporous solids to adsorb molecules selectively on the basis of their size is a consequence of the well-defined pore windows that limit access of molecules to the internal surface area. The size and shape of these windows are determined mainly by the structure of the framework, which is specific to each type of material. For low silica zeolites it can often be modified by ion exchange, where the cations can adopt sites that partially block pore windows. The *Atlas of Zeolite Framework Types*[1] gives the dimensions of the free diameters of these pores for each structure type, taking into account the radii of the framework oxygen atoms (taken as 1.35 Å) and neglecting the effect of extra-framework cations.

In general, the pore window sizes of zeolites are described in broad categories, that is small, medium, large and most recently extra large, which broadly correspond to free diameters around 4 Å, 5.5 Å, 7 Å and > 8 Å, as described in Chapter 3. The pore size depends mainly on the number of atoms within the ring of tetrahedral cations and oxygen atoms that make up the pore, and also on the ring's configuration. In many zeolites the rings are far from regular or planar, decreasing their effective free diameter. For low silica zeolites, with high cation exchange capacity, the presence of extra-framework cations can also reduce the effective pore size. Zeolite A in the sodium form is known commercially as zeolite 4A (the pores are roughly 4 Å in size), whereas in the calcium form, where fewer cations are present to balance the framework charge, the effective free diameter is greater and the material is known commercially as zeolite 5A. Figure 7.1, which is based on the crystallographic structures of these materials,[2,3] illustrates this. Exchange instead with the larger monovalent K^+ results in a reduction in pore size, and the material is known as 3A. The pore sizes and shapes of other, non-zeolitic, microporous solids show more variety because of their greater structural diversity, and as a result they are likely to find important applications in sorption technology. Table 7.1 compares the pore sizes of selected structural types.

The free diameters of some typical adsorbate molecules are given in Table 7.2, which usually refer to the Lennard-Jones σ, or distance of closest approach. These should be taken as a guide, because the relevant measure of effective molecular size will depend on the temperature, the chemical interaction between adsorbate and solid and the shape of the molecular opening. Indeed, for molecular size alone, there are many approaches to deriving σ, including fitting it by molecular simulation to the second virial coefficient, the gas viscosity, vapour-liquid equilibria or, for liquids, the molar volumes. For example, the kinetic diameters of O_2 and N_2 are usually quoted, from Breck,[34] as 3.46 and 3.64 Å, which correspond to Lennard-Jones models of these molecules as spheres. However, it is their minimum molecular widths that are likely to be more important in diffusion through narrow pore openings, and for this they are better considered as two site quadrupoles, with a cross-sectional diameter, σ, and a length between the two sites, L. These two-centre Lennard-Jones Quadrupole (2CLJQ) parameters of Vrabec and Stoll,[35,36] fitted against thermodynamic data, are given in the table: the diameters of O_2 and N_2 are 3.11 and 3.32 Å, respectively. Furthermore, since the oxygen radius of 1.35 Å usually taken to determine the free diameter of the molecular sieve is the ionic radius, this is not consistent with Lennard-Jones derived molecular sizes. For example, ZSM-5 adsorbs cyclohexane (the second dimension of which is 6.2 Å), but the pore size derived from the crystal structure is 5.6 Å, significantly smaller than this (Table 7.1). As a result, Cook and Conner[37] suggest that the effective pore size of zeolites is consistently 0.7 Å more than that usually quoted on the basis of the hard sphere model.

Temperature is also an important factor in molecular sieving behaviour, because the framework of a microporous solid is not fixed at the crystallographic positions given by the atomic parameters, but is able to

Figure 7.1 The 'Sentinel Effect' in zeolite A. Crystal structures of (below) dehydrated Ca-A and (above) dehydrated Na-A, showing the positions of extra-framework cations in relation to the α-cage. Note that in Ca-A all the 8MR intercage windows are free of cations.

Table 7.1 Pore window dimensions and pore volumes of selected microporous and mesoporous solids.

Material	*Composition*	*Pore window size (A°)*[1]	*Pore Vol. (cm³ g⁻¹)*[2]	*Ref.*
Zeolite 5A	$CaAl_2Si_2O_8$	4.1 × 4.1 × 4.1	0.22	4
ITQ-29(LTA)	SiO_2	4.1 × 4.1 × 4.1	0.24	4
ZSM-5	$H_xAl_xSi_{1-x}O_2$ (Si/Al >20)	(5.5 × 5.1)–(5.3 × 5.6)	0.18	5
NaX	$Na_x(Al_xSi_{1-x})O_2$	7.4 × 7.4 × 7.4	0.28	6
Beta	$H_xAl_xSi_{1-x}O_2$ (Si/Al >10)	(6.6 × 7.7) × (6.6 × 7.7)× (5.6)	0.22	7
ITQ-21	$(Si,Ge)O_2$ (Si/Ge >1.2)	7.4 × 7.4 × 7.4	0.24	8,9
CIT-5	$H_xAl_xSi_{1-x}O_2$ (Si/Al >10)	(10.0 × 9.9)	0.13	10
IM-12	$Si_{0.82}Ge_{0.18}O_2$	(9.5 × 7.1)–(8.5 × 5.5)	0.26	11
ITQ-33	$Si_{0.66}Al_{0.04}Ge_{0.3}O_2$	12.2 × (6.1 × 4.0) × (6.1 × 4.0)	0.30	12
$AlPO_4$-34	$AlPO_4$	3.8 × 3.8 × 3.8	0.27	13
STA-7	$AlP_{1-x}Si_xO_4$	3.9 × 3.8 × 3.8	0.30	14
$AlPO_4$-11	$AlPO_4$	(6.5 × 4.0)	0.13	6
$AlPO_4$-5	$AlPO_4$	7.3	0.18	6
MgAPO-36	$Mg_xAl_{1-x}PO_4$	(7.5 × 6.5)	0.21	13
VPI-5	$AlPO_4$	12.7	0.25	6
SU-M	$Ge_{10}O_{20.5}(OH)_3$	12	0.08	15
ETS-10	$Na_2Si_5TiO_{13}$	7.6 × 7.6 × 7.6	0.15	16,17
VSB-5	$Ni_{20}[OH]_{12}(H_2O)_6(HPO_4)_8(PO_4)_4$	10.2	0.18	18
AlMePO-β	$Al_2(CH_3PO_3)_3$	5.8	0.12	19
Scandium	$Sc_2(O_2CC_6H_4CO_2)_3$	3	0.26	20

(Continued)

Table 7.1 (*Continued*).

Material	*Composition*	*Pore window size* $(A^{\circ})^{1}$	*Pore Vol.* $(cm^{3}\,g^{-1})^{2}$	*Ref.*
terephthalate				
MOF-5 (IRMOF-1)	$Zn_4O(O_2CC_6H_4CO_2)_3$	8	0.59	21
MOF-177	$Zn_4O(C_6H_3(CO_2)_3)_2$	10.8	1.59	22
MOF-500	$Fe_{12}S_{12}C_{156}H_{96}N_{12}O_{87}$	4 cage types: windows 3.4–9.5	0.92	23
MIL-53	$Cr(OH)(O_2CC_6H_4CO_2)$	8.5	0.55	24
MIL-68	$V(OH)(O_2CC_6H_4CO_2)$	16	0.31	25
MIL-100	$Cr_3O\ (H_2O)_2(OH,F)(C_6H_3\ (CO_2)_3)_2$	See Table 2.2	1.15	26
MIL-101	$Cr_3O(H_2O)_2(OH,F)(O_2C\ C_6H_4CO_2)_3$	See Table 2.2	1.96	27
HKUST-1	$Cu_3(C_6H_3(CO_2)_3)_2(H_2O)_3$	$9.5 \times 9.5 \times 9.5$	0.68	28
Pillared zinc terephthalate	$Zn_2(O_2CC_6H_4CO_2)_2(NC_6H_{12}N)$	$7.5 \times 4 \times 4$	0.72	29
ZIF-8	$Zn(N_2C_3H_2CH_3)_2$	$3.4 \times 3.4 \times 3.4$	0.64	30
ZIF-11	$Zn(N_2C_3H_2C_6H_5)_2$	$3.0 \times 3.0 \times 3.0$	0.58 (H_2)	30
Nickel bisphosphonate	$Ni_2(O_3PCH_2NC_4H_8NCH_2PO_3)$	10.0	0.21	31
SBA-1	SiO_2	Variable (pore diameters 15–30)	0.50	32
MCM-41	SiO_2	>30	0.65	33

[1] From crystal structure. Brackets indicate an elliptical pore window.
[2] Measurement using N_2 or Ar adsorption.

Table 7.2 Molecular size parameters (Å).

Molecule	*Formula*	*σ (LJ sphere)*[34]	*σ (LJ sphere)*[35–37]	*2CLJQ model*[35–37]		*Courtauld space-filling model*[38]
				σ	*L*	
Neon	Ne		2.80			
Argon	Ar		3.40			
Krypton	Kr		3.63			
Xenon	Xe		3.90			
Hydrogen	H_2	2.89				
Oxygen	O_2	3.46		3.11	0.97	
Nitrogen	N_2	3.64		3.32	1.05	
Carbon dioxide	CO_2	3.3		2.98	2.42	
Carbon monoxide	CO	3.76		3.30	1.14	
Water	H_2O	2.46				
Sulfur hexafluoride	SF_6	5.5	3.96			
Methane	CH_4	3.8	3.73			
Ethene	C_2H_4			3.76	1.27	
Ethane	C_2H_6			3.49	2.38	
Iso-butane	$CH(CH_3)_3$	5.0				
n-Hexane	C_6H_{14}					$3.9 \times 4.3 \times 9.1$
3-Methylpentane	C_6H_{14}					$4.6 \times 5.8 \times 8.6$
Cyclohexane	C_6H_{12}	6.0				$4.7 \times 6.2 \times 6.9$
Benzene	C_6H_6	5.85				$3.4 \times 6.2 \times 6.9$
Toluene	C_7H_8					$3.4 \times 6.2 \times 8.6$
para-Xylene	C_8H_{10}					$3.7 \times 6.2 \times 8.6$
ortho-Xylene	C_8H_{10}					$4.1 \times 6.9 \times 7.5$
Mesitylene	C_9H_{12}					$3.7 \times 7.8 \times 8.5$
Triisopropylbenzene	$C_{15}H_{24}$	8.5				

vibrate. The vibrational amplitude increases with temperature, which results in changes in the effective free diameters of the pores. At the same time, the kinetic diameters of the molecular adsorbates will change as a function of temperature. As an example of temperature-dependent molecular sieving, zeolite Na-A does not take up N_2 at 77 K, but does take up O_2 (which is slightly smaller), whereas both N_2 and O_2 are taken up at room temperature. In fact, more nitrogen than oxygen is adsorbed at room temperature, but this is because of stronger cation-quadrupole and cation-induced dipole interactions between N_2 and Na^+ than between O_2 and Na^+, rather than differences in molecular size.

Molecular sieving by controlling access of molecules to the internal surfaces and by restricting molecular diffusivity is particularly important in processes that require the separation of branched from linear alkanes or in resolution of the different isomers of xylene: relevant dimensions of some important hydrocarbons of these types are given in Table 7.2. Small-pore zeolites such as Na-A are particularly important for the separation of *n*-alkanes and *n*-alkanols from their branched isomers, whereas medium-pore zeolites such as ZSM-5 show adsorption of *p*-xylene but very slow (or no) adsorption of *o*-xylene. Molecular sieving is also important in restricting the size of molecules that leave the pores of zeolites after catalytic reaction within them. This product diffusivity selectivity is described in detail for specific examples in the next chapter, but the intra-zeolitic isomerisation of xylenes in ZSM-5 to give predominantly *p*-xylene product is an excellent example.

For microporous solids of unknown structure, simple measurement of the uptake of a series of hydrocarbons of different sizes immediately gives an estimate of the pore window sizes. At the upper end of zeolitic pore sizes, large probe molecules such as 1,3,5-trimethylbenzene (mesitylene) ($3.7 \times 7.8 \times 8.5$ Å) and perfluorotri-*n*-butylamine (*ca.* 10 Å) are able to determine the pore size of large- or extra-large-pore zeolites. For solids with the largest pores (VPI-5, ECR-34[39]), the larger of these molecules adsorb. On a practical note, care should be taken in adsorption experiments using such large adsorbates: their vapour pressures are low, so there is a need to avoid condensation between particles, and their purity must be assured, since even minor amounts of more volatile impurities can compromise the measurements.

7.2 Theory and Methods for the Study of Adsorption

7.2.1 Adsorption Isotherms

An adsorption isotherm is the equilibrium uptake of a sorbate (for example as moles of adsorbate per gram of sorbent) measured at a constant temperature as a function of the concentration of the sorbate. For adsorption from the gas phase the adsorption is therefore measured as a function of pressure.

There are two main ways to measure the amount of adsorption – gravimetric and volumetric. In gravimetric methods the uptake is measured as the increase in weight of a sample as the adsorbate pressure is varied. This requires great accuracy in weighing, which is usually performed using an electrobalance. In volumetric methods the changes in pressure upon dosing known amounts of gas into a volume containing the sample are measured and the uptake calculated at the resultant equilibrium pressures. Each method has its advantages and disadvantages. The adsorption of many vapours is most reliably performed gravimetrically, where condensation away from the sample has a negligible effect on the measurements. For low-temperature, low-pressure studies, however, the volumetric method is preferred, because the sample is in contact with the wall of the sample container, itself immersed in the cryogenic bath. Heat transfer from the sample is therefore faster in the volumetric apparatus, and thermal equilibrium is reached more quickly. Adsorption isotherms are typically classified according to their shape within the Brunauer classification (Figure 7.2).[40,41]

The physical adsorption of nitrogen at 77 K is the most widespread general method to measure surface areas and pore size distributions of solids. Special care is needed to apply the technique to crystalline microporous solids. For such solids the isotherm shape is typically Type I, with high uptake at low pressures and a sharply defined maximum level of uptake as the internal pores are filled (Figure 7.2). The maximum uptake can most usefully be converted to a pore volume per unit mass by assuming a density for the adsorbed nitrogen equal to that of liquid nitrogen at that temperature (0.807 g cm^{-3}). Surface areas of microporous solids are also commonly quoted in the literature, to give a comparison with open surface solids, even though the Langmuir or BET models (see below) used to obtain the surface area values are not strictly applicable. Typical pore volumes of microporous solids are given in Table 7.1 for a range of microporous (and some mesoporous) solids. It is possible to measure adsorption below maximum uptake accurately, even for nitrogen or argon adsorption at

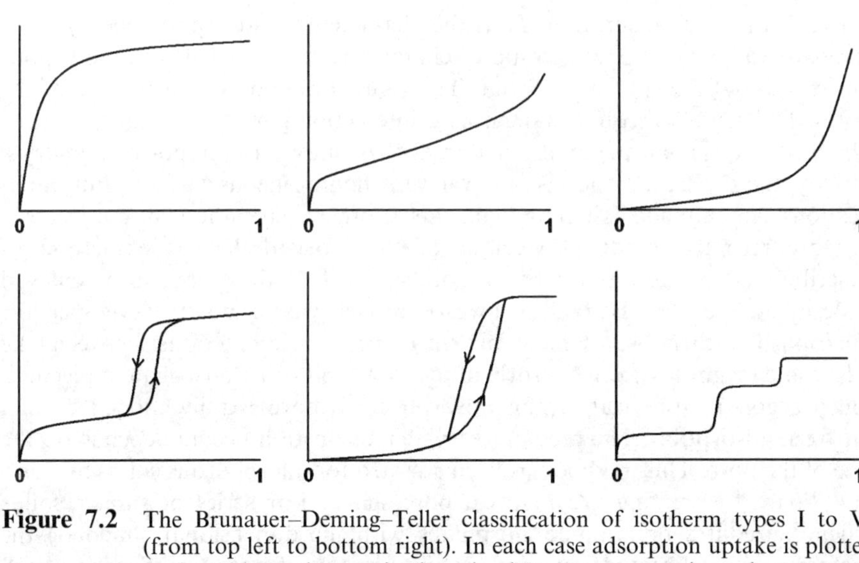

Figure 7.2 The Brunauer–Deming–Teller classification of isotherm types I to VI (from top left to bottom right). In each case adsorption uptake is plotted against p/p_o, where p is the adsorbate pressure and p_o the saturated vapour pressure of the pure liquid adsorbate at the isotherm temperature.

77 K, if accurate measurement and control of low adsorbate pressures is possible. It should be noted, however, that for the adsorption of nitrogen or argon at 77 K on microporous solids the times to reach equilibrium at such low temperatures can be very long and it may be preferable to use adsorbates that show appreciable physisorption at room temperature or above (such as CO_2). All such methods effectively probe the environment experienced by the adsorbates. For a series of chemically similar solids the strength of interaction at low coverage will depend on the number and proximity of adsorption sites. In high silica zeolites, therefore, where the initial interaction at low coverage is between the adsorbate molecule and the silica walls, smaller pore solids should show adsorption at lower adsorbate pressure than those with larger pores.

The adsorption of gases and vapours onto microporous solids, while nominally of type I, is not well described by a simple Langmuir model, because the assumptions made for its derivations are not valid. For example, there is no open surface for adsorption and desorption, there is significant interaction between adsorbates and the surface is heterogeneous in terms of energies of adsorption at different possible sites. Similarly, models of weak multilayer adsorption that are applicable to Type II isotherms characteristic of open surfaces (such as that of Brunauer, Emmett and Teller[42]) are not applicable, because there is no available space for multilayer formation. Alternative empirical equations, such as the Dubinin–Radushkevitch (D–R) or Dubinin–Astakov (D–A) models have long been used for microporous solids, particularly to describe micropore filling in carbons but also for zeolites:[43,44]

$$W = W_0 \exp\left[-RT \ln\left(p_o/p\right)\Big/\beta E_0\right]^n$$

where W_0 is total adsorption, E_0 is the characteristic adsorption energy for a standard vapour and β is the ratio of characteristic energy between the vapour under test and the standard vapour. The value of n depends on the size of the pores. The methods require standardised interaction parameters that have been obtained on similar materials, and make assumptions on pore geometries. Strictly, the D–R expression is only valid for homogeneous surfaces, but modifications can be made assuming Gaussian or other distributions of pore sizes.

More recently, modified versions of the Horvath–Kawazoe method of describing adsorption and deriving pore size distributions have also achieved widespread use. The Horvath–Kawazoe model was originally developed for microporous carbons,[45] and took into account interactions between the adsorbate (argon) with the adsorbent (carbon) and other adsorbate molecules. Since argon is spherical, argon adsorption is more readily modelled than nitrogen adsorption. The pressure at which adsorption occurs depends on the size of the pore. This method has been adapted for microporous solids by Saito and Foley,[46] assuming a cylindrical pore shape. For series of similar solids (such as zeolites or aluminophosphates with uni-dimensional channels) the method, once calibrated, can give useful estimates of pore size. Sorbents with different pore geometries, or surface chemistries, cannot be directly compared, although Horvath–Kawazoe pore diameters are commonly quoted and give a first estimate of the pore size.

Such methods of analysing and describing adsorption data have considerable merit in describing microporosity in porous carbons, which are not crystalline, or for microporous solids of unknown structure, but for zeolites of known structure they add little to our understanding. In such cases, the form of the adsorption isotherms can be modelled by computer simulation using Grand Canonical Monte Carlo methods. In this approach all the parameters are known or can be measured or calculated (see Section 4.5.1) so that the adsorption isotherm can be simulated using a physically well-characterised model.

In some cases, adsorption isotherms are observed that demonstrate 'Type I-like' adsorption behaviour for adsorbates, but with stepped uptake in the low-pressure region. There are a number of such examples, including nitrogen in AlMePO-α, as described in Section 4.5.1, CO in $AlPO_4$-11[47] and benzene in silicalite.[48] These are thought to derive from phase changes in the adsorbate structure. In AlMePO-α, a change in the packing of nitrogen molecules in the channels results in the step, whereas for benzene on silicalite, the first step corresponds to filling of channel intersections. In each case the ordering scheme above the step evolves as more molecules are adsorbed. Yaghi recently reports argon adsorption data on the hierarchically assembled MOF-500,[23] which displays a different type of multistep adsorption extending into the mesoporous regime. The adsorption steps correspond to pore filling of three types of pores present in the structure, with diameters of 10.4 Å, 13.0 Å and 18 Å.

In addition, behaviour that is strongly divergent from Type I may occur, particularly in neutral frameworks. In pure silica zeolites Y, for example, polar molecules such as water and methanol are adsorbed with Type V isotherms, with low initial uptake due to weak interactions with the silica, followed by an increase in adsorption at higher pressures resulting from adsorbate-adsorbate interactions of increasing importance. Similarly, the adsorption of water on the hydrophobic organic-inorganic porous solid AlMePO-β shows low adsorption at low partial pressures, because the initial sorbate-sorbent interaction is weak (lower than the heat of liquefaction) but at higher water vapour pressures ($p/p_o = 0.8$) sorbate-sorbate interactions enhance the total adsorption energy as clusters form, giving rise to a sharp increase in uptake.[19]

Finally, at very high adsorbate pressures, additional structural effects can be seen in some zeolites as molecules are able to access small cages in the structure. This can lead to reversible and irreversible changes in the structure, including, for the zeolite natrolite, expansion at high (*ca.* 1.2 GPa) fluid pressures as ordering of water molecules and extra-framework cations occurs.[49,50]

7.2.1.1 Adsorption on Mesoporous Solids

Adsorption onto mesoporous solids, *i.e.* those with pores larger than *ca.* 20 Å, continues to increase at partial pressures well above those at which uptake on microporous solids is complete. Multilayer uptake is possible and the BET treatment is applicable, particularly until any sharp increase of uptake due to capillary condensation might arise. The pressure p_c at which this is observed is derived from the Kelvin equation for the pressure drop across a liquid meniscus:

$$p = p_o e^{-2\gamma V_m/rRT}$$

The corresponding pore radius r_p is then given by the following equation:

$$r_p = \frac{2\gamma V_m}{RT \ln p_o/p} + t = \frac{9.53}{\ln p_o/p} + t \quad for \quad N_2$$

where γ is the surface tension of nitrogen adsorbate, V_m the molar volume and t is the thickness of the multilayer at the pressure of condensation and depends on the substrate chemistry. For silica the variation of multilayer thickness with pressure for N_2 at 77 K is given in the literature.[51] The value of t can also be estimated using the methods of de Boer,[52]

$$t(Angstroms) = \left[{}^{13.99}\!/\!_{\log(p_o/p)} + 0.034\right]^{\frac{1}{2}}$$

or Halsey,[53]

$$t(Angstroms) = 3.54\left[{}^{5}\!/\!_{2.03\log(p_o/p)}\right]^{\frac{1}{3}}$$

Such equations are the key to calculations of pore size distributions of mesoporous solids, *via* the Barrett, Joyner and Halenda (BJH) method,[54] as explained by Thomas and Thomas.[51]

The prototype mesoporous solid MCM-41 gives N_2 adsorption isotherms of type IV, as shown in Figure 7.3. Initial monolayer coverage is followed by multilayer adsorption until at a higher pressure a steep increase in uptake occurs due to capillary condensation of liquid nitrogen within the mesopores. Standard analysis enables calculation of a surface area, a pore size distribution (corresponding to the pressure range at which the condensation occurs) and a total pore volume (from the maximum uptake before liquid nitrogen condenses in the voids between particles). The calculation algorithms for such pore size analyses are usually included with commercial porosimeters.

A detailed discussion of adsorption onto mesoporous solids is beyond the scope of this text, but certain features relevant to microporous solids should be described. Firstly, microporous solids can themselves contain mesoporosity. The most important example of this is observed in zeolites such as Y or mordenite that have been treated after synthesis to remove aluminium from the framework (Section 6.2.3). The migration of silica leaves mesopores that are evident from nitrogen adsorption isotherms and directly visible by electron microscopy.[55,56] The presence of secondary mesopores enhances diffusion and catalytic properties. Conversely, mesoporous solids that are well ordered on the mesoscale can contain disordered micropores in their walls. The mesoporous channels of calcined SBA-15, for example, are connected by micropores that result from removal of block copolymer chains that run between the large channels in the as-synthesised material. This is observed from nitrogen

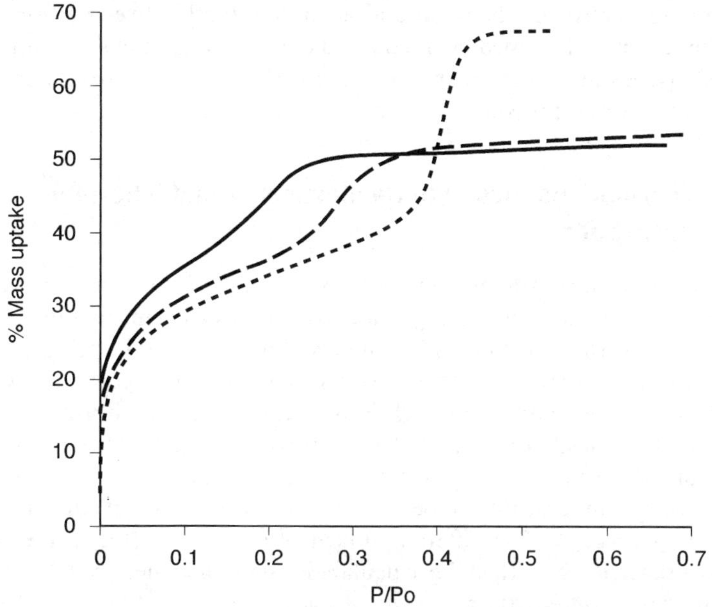

Figure 7.3 Experimental adsorption isotherms of N_2 at 77 K on mesoporous solids: MCM-41 (dotted line), which has cylindrical channels, SBA-2 (dashed line), which has one set of large spherical cages, linked via windows, and SBA-1 (solid line), which contains two sorts of cages, with diameters in the 10–20 Å range.

adsorption data, where the micropore volume increases sharply between extracted samples, in which polymer remains within the walls, and calcined samples, in which the template is fully removed from these sites.[57] Direct evidence for the microporosity is also seen by high resolution electron microscope images of the silicas[58] and of platinum 'casts'[59] synthesised within the calcined silicas that show narrow connecting bridges (formed in the micropores) between the larger rods that form in the mesopores. In solids such as these, where distinct micro- and mesoporosity are present, methods to determine the proportions of micropore and mesopore contributions have been used, such as the α_s plot method.[41]

Finally, the 'mesocage' silicas (*e.g.* SBA-1, -2, -6 and -12) and metal organic frameworks such as MIL-100 and MIL-101 possess porosity on a length scale that bridges the microporous and mesoporous regimes. As a result there is no clearly defined capillary condensation step in the isotherm of SBA-1 (Figure 7.3). Whereas this results from the crystallographically defined structure for the hybrids, the precise details of the pore geometry and dimensions of the mesocage silicas are not known. The general structural arrangement can be determined by electron microscopy, however, and it is then possible to model the adsorption isotherms by taking a starting structural model, calculating the adsorption isotherm by density functional or GCMC methods and varying the structural parameters to optimise the fit to the data. Density functional methods[60] sum up the interaction between adsorbent and adsorbate (and between adsorbate and adsorbate) given an assumed geometry and equate the chemical potential of the adsorbed species at a particular uptake with that of the adsorbate in the gas phase at any desired pressure.

7.2.2 Thermodynamics: Microcalorimetry and Thermal Desorption

Measurement of the enthalpies and entropies of adsorption and their variation as the pore space is filled is important to establish chemical models of the interactions and to enable chemical engineers to model and optimise catalytic reactors and large-scale separations. For physisorption of gases and vapours, the most widely available approach is to measure the equilibrium uptake as a function of adsorbate pressure and at a series of constant temperatures, giving a series of isotherms. This then permits the dependence of the equilibrium pressure with temperature to be derived for any value of the uptake, or fractional coverage, θ. The differential heat of adsorption (the heat at a particular coverage, θ_i) can then be calculated using the linearised form of the Clausius–Clapeyron equation:

$$\ln p_{\theta_i} = -\frac{\Delta H_{ads,\theta_i}}{R} \cdot \frac{1}{T} + c$$

where p is the pressure at which a given uptake is achieved as the temperature T is varied. The differential free energy and entropy of adsorption can also be calculated from measurements of this type *via* thermodynamic relationships.

For separation applications, it is important to measure adsorption in two- or multi-component mixtures. In these cases the compositions of both the adsorbed and the gas phase are needed to obtain a full description of the system. These can be obtained by measurement, or, if data on the single component adsorptions are available, the adsorption behaviour of the mixture can be simulated if the interaction parameters between the two adsorbates are known.

For strongly chemisorbed species it can be difficult to obtain equilibrium uptakes below saturation at usual temperatures and measurable pressures (the adsorption isotherms rise very steeply with pressure to the monolayer coverage). Furthermore, chemisorption may be thermally activated, resulting in very long equilibration times. For these reasons other approaches are required to measure the thermodynamics of chemisorption. The two main ways are microcalorimetry and thermal desorption.

Calorimetry is a highly accurate method to measure the heat of adsorption, be it physisorption or chemisorption.[61] It is typically performed using microcalorimeters of the Tian–Calvet type, in which known volumes of the adsorbate are sequentially dosed onto the solid from the gas phase at the required temperature and the liberated heat is determined from the temperature rise. In this way very accurate plots of heats of adsorption against uptake can be obtained directly for both weakly and strongly bound sorbates.

Thermal desorption is less accurate but more widely available than microcalorimetry. It is of particular use for strongly chemisorbed species. The adsorbate is adsorbed onto the sorbent and allowed to reach equilibrium. Physisorbed species are then removed by gentle heating in a flow of an inert gas or under vacuum, before the sample is heated (usually by a programmed linear temperature ramp). From thermal desorption experiments of this kind (TPD) it is possible to determine the total chemisorbed amount. In addition, under conditions where the rate of evolution of the adsorbate is limited only by desorption from the adsorption site (and not diffusion along the pores) the activation energy of desorption energy can be determined by modelling desorption kinetics at different linear heating rates using the basic equation:

$$\frac{d\theta}{dt} = -\nu\theta^n \exp\left(-\frac{E_d}{RT}\right)$$

where θ is the fractional coverage of the surface, n is the order of the desorption reaction and v the frequency factor of desorption. General equations to estimate the heat of adsorption from thermal desorption data obtained at different heating rates have been derived.[62]

The chemisorption of basic molecules at acidic sites in microporous solids, for example, is also readily studied by thermal desorption to give details of acid site strength and type (Section 8.4.2.1). Not all chemisorption can be studied straightforwardly by TPD, however. For molecules such as alkenes, even moderate temperatures rapidly cause reaction of the adsorbed species, so that the experiment monitors reaction rather than desorption. The method becomes temperature-programmed reaction and requires chemical analysis of the products.

7.2.2.1 *Variation of the Heat of Adsorption with Uptake*

The variation of the heat of adsorption with uptake can take one of several forms depending on the relative strengths of contributing interactions. In general, the adsorbate-adsorbent interaction decreases with uptake, as the stronger adsorption sites are filled first whereas the adsorbate-adsorbate interactions increase with coverage, as more adsorbates are present. Figure 7.4 illustrates this schematically. For example, for adsorption on cationic zeolites, high heats at low coverage are associated with uptake at the most energetically favourable cation sites, with a gradual reduction in heat of adsorption at higher coverage as less favourable sites are filled. The adsorption of carbon dioxide on cationic zeolites (Figure 7.5) is of this type.[63] For adsorption of molecules on uncharged frameworks, however, the heat of adsorption can increase with loading, as the dispersive sorbate-sorbent interaction experienced by the first molecules is augmented by additional sorbate-sorbate interactions. Behaviour intermediate between these types is also observed. Maurin *et al.*[63] describe such a variation for the adsorption of carbon dioxide on different zeolitic faujasites. For Na-X, with a high concentration of cations, the heats decrease with loading, whereas for dealuminated Y, without cations present, the heats increase with loading. This is modelled well by simulation, by first deriving

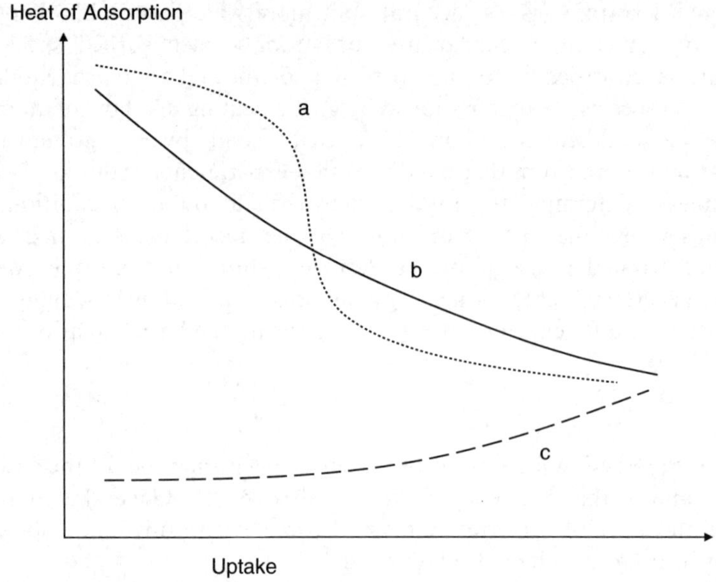

Figure 7.4 Schematic variation of heats of adsorption with uptake for the cases (a) with a fixed number of sites with high heats of adsorption, (b) with a heterogeneous distribution of adsorption sites with varying energy and (c) where there is weak adsorbate–adsorbent interaction and adsorbate–adsorbate interactions become increasingly important as loading increases.

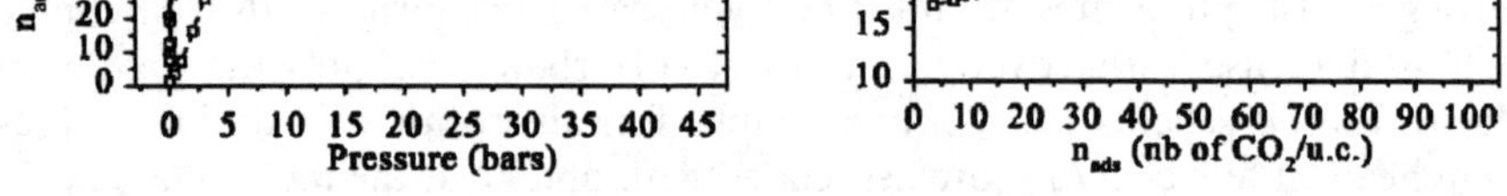

Figure 7.5 Adsorption isotherms (left) and enthalpies of adsorption (right) as a function of coverage for CO_2 on zeolites Na-X, Na-Y and dealuminated Y (experimental ○; and simulated □). [Reproduced from reference 63 with permission. Copyright 2005 American Chemical Society.]

interatomic potentials for framework-CO_2 and cation-CO_2 interactions by fitting the results of *ab initio* simulations, and then using Grand Canonical Monte Carlo methods based on the molecular mechanics approach using these interatomic potentials (Figure 7.5).

The adsorption of amines at Brønsted acid sites (Section 8.4.2.1) gives a constant heat of adsorption until all the stoichiometric amine-proton interactions have been achieved, followed by a rapid drop in the adsorption energy, and so is a good example of chemisorption.

7.2.3 Molecular Motion of Adsorbates

Whereas adsorption isotherms and microcalorimetry give information on the thermodynamics of adsorption (including the entropy of adsorption, which is related to the motion of adsorbed species), other techniques are required to measure molecular motion within the pores directly. This motion can take the form of either re-orientation (discussed below) or diffusional transport through the pores, which is addressed in Section 7.4.

Molecules adsorbed within micropores have motional modes and frequencies of re-orientation that depend on the geometric constraints of the pores and the strength of the adsorption interaction. Deuterium (^{2}H) wideline spectroscopy is one of the most powerful experimental methods to study this – quasi-elastic neutron scattering is another. Molecular dynamics simulations are also of great value, as described in Section 4.5.3, although the timescales involved only extend to nanoseconds, rather than the timescale of microseconds or greater that is accessible to NMR measurements.

7.2.3.1 NMR Studies of Motion

Motion of adsorbed molecules in static solids has the result of narrowing the resonances of all the NMR-active nuclei they contain. ^{13}C and ^{1}H MAS NMR resonances in solids, for example, are further narrowed by molecular motion. Deuterium NMR of static samples has particular advantages for studying molecular re-orientation[64] (rather than translational motion, which is better studied by pulsed field gradient NMR) and has been exploited to understand adsorption in zeolitic solids.[65,66] The ^{2}H nucleus is quadrupolar ($I = 1$) with quadrupolar coupling constants in the range of 140–220 kHz for most deuterated organic compounds. The lineshapes of its wideline powder patterns (collected without sample spinning) are readily observed using the quadrupole echo pulse sequence, and appear as so-called Pake doublets (Figure 7.6). Re-orientation of the C–D bonds due to molecular motion results in narrowing the powder pattern and changes in the lineshape. The lineshapes for different mechanisms of re-orientation depend on their geometry (molecular rotations, angular flips, *etc.*) and on their frequency, when this is in the 10^4–10^7 s^{-1} range. Slower motion than this is static on the NMR timescale whereas motion faster than this range is said to be at the fast limit of motion. Examples of ^{2}H lineshapes for deuterons undergoing different kinds of motion at the fast limit, for example those in methyl groups or aromatic rings, are given in Figure 7.6. An interesting improvement of the method is the so-called slow spinning MAS NMR technique, in which the spinning is not fast enough to remove the quadrupolar coupling, so that spectrum becomes a large number of spinning sidebands, the envelope of which has the same shape as the static lineshape. This has two advantages: not only is the signal intensity increased (being

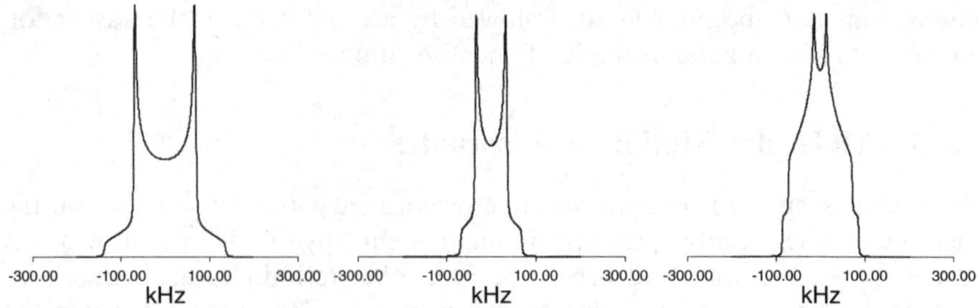

Figure 7.6 Simulated ^2H NMR spectral lineshapes at the fast limit of motion for C_6D_6 benzene (centre) undergoing motion around the molecule's C_6 axis and (right) for H_3C-C_6D_4-CH_3 *p*-xylene undergoing π flips around the molecule's *para* axis. These are compared with (left) the spectrum of static ^2H.

concentrated into the sidebands) but it is also possible to resolve signals from ^2H nuclei with different shifts. A similar effect is achieved by use of the quadrupolar Carr–Meiboom–Purcell–Gill pulse sequence.[67]

The static (or slow MAS) ^2H spectra can be simulated assuming modes and frequencies for independent mechanisms of re-orientation, and simulations proceed iteratively to match the spectra. Several computer codes are available to perform the simulation,[68] while some groups calculate spectra directly from analytical functions. ^2H wideline NMR studies have been applied to many adsorbate-microporous solid systems, including both physisorbed and chemisorbed species. The lineshape-matching process can sometimes be ambiguous, so that additional constraints on the possible mechanisms of motion, such as those provided by molecular dynamics or (time-averaged) by crystallography, are very helpful.

Combined ^2H NMR and lineshape analysis have unambiguously revealed the temperature-dependent motion of physisorbed species in a range of zeolite sorbate systems. For the adsorption of benzene in Na-X and Na-Y, for example, the preferred adsorption site is shown to be the sodium ion in the cationic forms, at which position the benzene is held at low temperatures, executing rotation around its six-fold axis.[69] Benzene is much more mobile in the pure silica faujasite, however, where it exhibits fully isotropic motion even at the lowest temperatures studied (155 K).[70] The motion of deuterated aromatics in the ZSM-5 structure, of relevance in understanding the zeolite's shape selectivity in adsorption and catalysis involving aromatic molecules, has also been studied by ^2H NMR.[71,72] *Para*-xylene in Na-ZSM-5, for example, is found at low loading to be strongly bound at low temperatures (the aromatic ring does not re-orient), whereas by 150 °C most of the molecules execute rapid re-orientation but are not translationally mobile on the NMR timescale. In H-ZSM-5 at 150 °C, however, *para*-xylene is much more mobile, and is consequently thought to move through the pore structure on the NMR timescale.

In a recent example in our own laboratory that demonstrates the effect of small structural changes in pore structure on the constrained motion of adsorbates, we studied the motion of d_6-benzene in the two aluminium

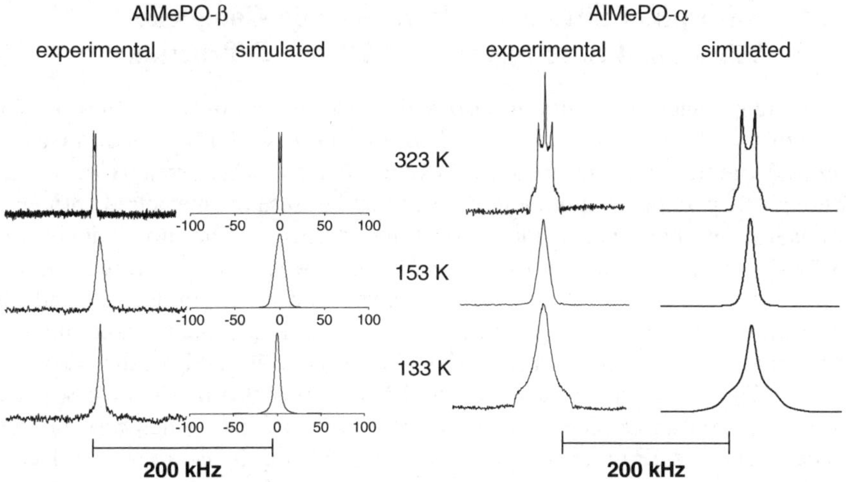

Figure 7.7 Variable temperature wideline ^{2}H NMR of d_6-benzene in AlMePO-β (left) and AlMePO-α (right). Making use of Molecular Dynamics simulations described in Section 4.5.3 (Figure 4.6), these spectra can be simulated by assuming isotropic motion of C_6D_6 in AlMePO-β (left) and π/3 flips around a C_6 axis in the more constrained channels of AlMePO-α (right). In each case the spectra are finally matched by varying the frequency of the re-orientations.[73] [Reproduced from reference 73 with permission. Copyright 2005 American Chemical Society.]

methylphosphonates, AlMePO-α and -β.[73] These polymorphs have closely similar structures, with the α-form possessing channels with a more acutely triangular cross section. This difference is sufficient that while benzene can tumble freely within the channels of β, giving a narrow lineshape, it is strongly constrained within the channels of α, giving rise to broader lineshapes (Figure 7.7). MD simulations support this interpretation and suggest the mechanism of re-orientation in α is mainly through 120° (2π/3) jumps, where the plane of the ring remains approximately parallel to the channel axis (Section 4.5.3). Using this model, the lineshapes can be modelled satisfactorily.

Chemisorbed species show much less re-orientational motion, due to their stronger bonding at the adsorption site.

7.2.3.2 Quasi-elastic Neutron Scattering

Quasi-elastic neutron scattering is another method that can give information on both the translational and also the rotational motion of adsorbates. Energy transfers of ±2 meV give rise to broadening of the spectra of scattered neutrons if the molecules diffuse over a timescale of 10^{-8} to 10^{-12} seconds, and this broadening can be analysed in terms of the molecular motion. Results for self-diffusivities obtained by this technique, for example, give good agreement with the methods of PFG NMR and Molecular Dynamics (Section 7.4.2).[74]

7.2.4 Adsorption Sites and Adsorbate-solid Complexes: Vibrational Spectroscopy, NMR and Diffraction

Spectroscopic measurements of adsorbed molecules can only be interpreted in terms of a molecular picture of the adsorption process. This is in contrast with thermodynamic quantities measured experimentally, which require no model, although they do give supporting information because the measured enthalpies of adsorption give direct evidence on bond strengths. The most widely used spectroscopic technique for the study of chemisorption is infrared spectroscopy, which probes the interaction between adsorbed molecules and the adsorption site by measuring the change in absorption frequency and intensity of absorbances from bonds within the adsorbed species and on the adsorbing surface. This is described in more detail below and also in Chapter 8. Other vibrational spectroscopies, such as Raman spectroscopy (which has different selection rules to IR) and inelastic neutron scattering (which has no selection rules), have also been used to give complementary information, but their use is much more limited. The entire internal surface of microporous solids is in principle accessible to adsorbed molecules, so that any bulk technique is applicable to the study of adsorption complexes. As a result, many other spectroscopies have been applied, including NMR, ESR and UV-visible spectroscopy (changes in electronic energy levels are a sensitive guide to symmetry and strength of interaction). Taken together with diffraction studies, a full picture of the adsorption complex can be established.

7.2.4.1 Infrared Studies

Infrared absorption occurs for vibrations that result in a change in dipole moment: for adsorbed molecules this may refer not only to those vibrations that are already active in the gas phase but also to those that become active upon adsorption, as is the case with non-polar diatomics such as hydrogen, oxygen and nitrogen. The rotational fine structure is of course lost upon adsorption, leaving characteristic vibrational bands. The intensity and frequency of the vibrations are modified by the adsorption, and the magnitude of these changes is very sensitive to the strength and type of the interactions. An important example is for the adsorption of pyridine on acid sites, where the pyridinium ion formed by proton transfer absorbs at clearly different frequencies from pyridine molecules adsorbed at Lewis acid sites (Section 8.4.2.2). The vibrational resonances of the adsorbent can also be strongly affected. This is most important for hydroxyl stretching frequencies for solids containing Brønsted acid sites.

In practice, the complete vibrational spectrum of an adsorbed molecule cannot be measured, because many of the vibrations occurring below around $1500\,cm^{-1}$ are obscured by the absorption spectrum of the host solid. In addition, it is often difficult to quantify the concentration of adsorbed species because the extinction coefficients are not well known. Nevertheless, infrared spectroscopy remains an extremely important tool, since it permits simultaneous study of both the adsorbate and the adsorption site (for example the protons at the Brønsted acid site) on a data collection timescale of seconds, and microporous solids are commonly transparent (above 1500 cm^{-1}) for IR. Spectra may be measured in transmission in thin self-supporting wafers, using Fourier Transform double beam spectrometers and sensitive detectors,[75] or, where scattering or low transmittance is a problem, in diffuse reflectance mode.

7.2.4.2 NMR Methods

Multinuclear solid state chemical shift NMR of ^{13}C, ^{15}N, ^{1}H and ^{129}Xe nuclei has been widely used to investigate the environments experienced by adsorbates, and NMR of protons and metal cations (such as aluminium) in framework and extra-framework positions reveals changes in environment of these sites in the solid upon adsorption of molecules. The specific application of NMR to the study of structure in adsorption is outlined below, whereas applications in diffusion are described in Section 7.4. The adsorption process can be followed either by observing changes at the adsorption sites or within the adsorbates. NMR is inherently a less-sensitive technique than infrared spectroscopy, particularly in the study of dilute spins such as ^{13}C, and data collection times on adsorbed hydrocarbons can reach hours.

For chemical shift studies the geometry of the experimental setup for magic angle spinning typically requires the controlled loading of a sample in a small glass ampoule, subsequently sealed and placed inside an NMR rotor. Examples of the effects on molecules that have been observed by NMR upon adsorption include the change of chemical shift of adsorbed xenon as a function of its environment and the adsorption of basic molecular probes on Brønsted and Lewis acid sites. The adsorbing solid itself can also undergo structural change upon adsorption, for example the environments of protons and metal cations are affected and changes in crystallographic symmetry frequently occur, particularly in aluminophosphate or pure silica frameworks. These are shown in the NMR by the changes in the multiplicity and chemical shifts of resonances in the ^{1}H, ^{27}Al, ^{29}Si and ^{31}P spectra. Recent developments include double resonance experiments (CPMAS, TEDOR and REDOR – see Section 3.4.1.1) in which the locations of ordered hydrocarbon within microporous solids can be determined precisely: the results can be complemented by single crystal X-ray diffraction studies, when single crystals are available; examples are given in Section 7.3.1. NMR is also able to detect the formation of reaction products within the solid which may be difficult to detect by other methods, and there has been much interest in measuring catalysis *in situ*, discussed further in Section 8.5.

7.2.4.3 Other Spectroscopies

X-ray absorption spectroscopy gives similar information to that obtained by NMR (local environment, redox state) for a very wide range of nuclei. For example, XANES and EXAFS can be used to monitor the change in coordination state of, for example, transition metal cations as molecules are adsorbed and become bound to them. Transition metal cations are usually not suitable for study by NMR, either due to their poor NMR characteristics, which is the case for titanium, or the presence of unpaired electrons, which result in rapid relaxation of the NMR signal and broaden and shift the resonances. Discussion of X-ray absorption studies of catalytic importance involving titanium and cobalt are given in detail in Chapter 9. X-ray edges of light elements such as carbon, nitrogen or oxygen that are usually present in adsorbates occur at very low ('soft') X-ray energies that are readily absorbed and so can only be measured in vacuum and from adsorbates at the external surface.

7.2.4.4 Diffraction Methods

Diffraction enables the time and space averaged location of molecules within the pores to be determined in those cases where the adsorbates are ordered over a long range at similar sites in each unit cell. Single crystal X-ray diffraction has the advantages of providing more accurate structural data than X-ray powder diffraction, but the combination of low scattering of typical adsorbates (hydrocarbons and small gases) and the difficulties of loading single crystals with known concentrations of adsorbates has limited the use of this method. Successful examples of its use include studies of aromatic hydrocarbons adsorbed within silicalite crystals.[76] Powder X-ray diffraction studies have also been possible for favourable cases. Examples include the location of xenon[77] and methylchloride[78] in zeolite Rho from laboratory X-ray data, through to the use of synchrotron X-rays to locate hydrocarbons in ZSM-5[79] and even the location of molecules of hydrogen in a microporous inorganic-organic framework.[80]

The high sensitivity of neutron diffraction to light elements that make up typical adsorbates (such as hydrocarbons, water, methanol, amines *etc.*) have made neutron diffraction particularly suitable for locating adsorbed molecules. The relative insensitivity of neutron diffraction means that single crystal neutron diffraction experiments are very rare, particularly for synthetic microporous solids, so that powder diffraction is the usual approach. In a typical neutron powder diffraction experiment, a known concentration of the adsorbate is loaded on the microporous solid in a vanadium can (vanadium is essentially transparent to neutrons). The diffraction experiment is typically performed at low temperatures, to 'freeze out' motion of the adsorbate. The location of carbon monoxide adsorbed on cobalt-exchanged zeolite A, in which the molecule adsorbs end-on, is one example.[81] Neither the adsorbate nor the solid should contain protons (^{1}H nuclei scatter neutrons incoherently and lead to very high background signals) so this requires the replacement of all protons in the solid by deuterium and the use of deuterated adsorbates. The locations of benzene in zeolite Na-Y,[82] pyridine in zeolite L[83] (Figure 7.8) and of D_2O in the aluminophosphate SAPO-34,[84] for example, have been determined in this way. Such studies permit a structural picture of adsorption to be obtained and provide invaluable experimental confirmation of computationally simulated minimum energy configurations.

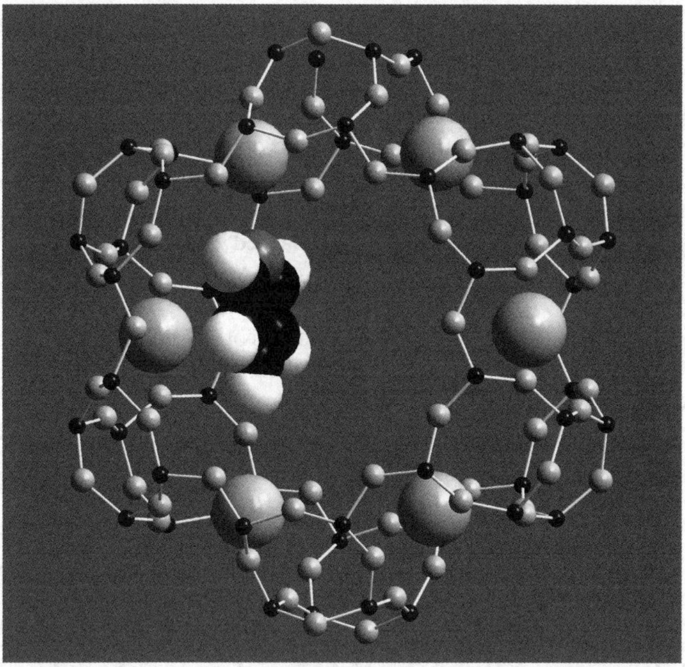

Figure 7.8 The minimum energy configuration of C_5D_5N pyridine coordinated to potassium cations in the channels of zeolite K-L, as determined from neutron powder diffraction data by Rietveld refinement.

7.2.5 Computer Simulation of Adsorption: General Lessons

The methods of computer simulation of adsorption (and diffusion) in microporous solids were described in Chapter 4: a summary is given in Table 4.1. These techniques are now sufficiently well developed for physisorption that thermodynamic properties can be predicted routinely for relatively simple adsorbates, once the structural details of the host are known. Molecular mechanics using standard forcefields are very successful for zeolitic systems, which take into account dispersive interactions satisfactorily, but it is also possible to use higher level calculations.

Progress has also been made in understanding the adsorption behaviour for hydrocarbons of some of the microporous solids with new framework chemistry and known structure, such as aluminophosphates and hybrid solids. For the latter category the important step is to parametrise interactions between both inorganic and organic groups in the framework and the adsorbates. First results suggest that, for MOFs related to the zinc terephthalate MOF-5, adsorption sites close to the inorganic clusters are the preferred ones for hydrogen and hydrocarbons. Challenges remain in understanding adsorption on hybrid solids, however. One of the main ones is to model the minimum energy structure of those materials which have very flexible frameworks such as structures related to the iron fumarate

MIL-88.[85] Neither molecular mechanical methods nor quantum mechanical approaches have so far been very successful. In the case of DFT methods, the main difficulty arises from a poor description of dispersive interactions.

Limitations in our understanding of physisorption also arise in interpreting adsorption of complex molecules and in predicting the behaviour of mixtures. In the first case, the difficulty lies in finding, even by Monte Carlo methods, reasonable molecular ensembles. Recent advances in the use of configurational bias methods of GCMC have, however, enabled the successful simulation of, for example, linear and branched hexanes in ZSM-5. These simulations suggest that whereas at low pressures branched alkanes are retained preferentially (because they retain more entropy upon adsorption due to their easier rotation), at high pressures the linear hexane is adsorbed preferentially. Such calculations shed light on the 'compensation effect' and 'inverse shape selectivity' observed in hydrocarbon transformations (Chapter 8).

The study of chemisorption requires the calculation of both dispersive and bonding interactions, because the interaction of the overall adsorbate with the framework must be considered as well as specific bond formation. For chemisorption, the electronic structure is important, so that quantum mechanical modelling is required.

7.3 Adsorption Sites and Interactions with Adsorbates

The adsorption of molecules by microporous solids is discussed here within six categories, according to the nature of the adsorption sites, and without considering adsorption by metal clusters introduced within the pores. Tables 7.3–7.5 give enthalpies of adsorption for examples within these categories, which vary from the interaction of neutral frameworks with small molecules to strong adsorption on acid or basic sites.

7.3.1 Neutral Frameworks

A number of molecular sieves can be prepared as defect-free, neutral frameworks. Traditional, inorganic examples include pure silicas and aluminophosphates; most of the organic-inorganic MOFs also fall into this category. In these cases the interactions of adsorbed molecules are dominated by dispersive interactions of the adsorbed molecules with the atoms of the framework and so can be modelled by interatomic potential models that have been parametrised according to experimental data. Because these inorganic systems are so well defined and well characterised, they make excellent model systems for calculations. From an

Table 7.3 Heats of adsorption of small non-polar molecules on silicalite.[5]

Framework	*Heats of adsorption of adsorbates ($kJ\,mol^{-1}$)*							
	CO	N_2	O_2	Ar	CH_4	CO_2	C_2H_6	SF_6
Silicalite	17	17.5	17.5	17.5	21	27	31	34

Table 7.4 Initial heats of adsorption of small molecules on cationic forms of ZSM-5 and X ($kJ\ mol^{-1}$).

	Silicalite MFI	*ZSM-5 (Si/Al =40)*[97]				*Zeolite X*[61]							
		H	Li	Na	K	SiO_2 FAU	Li	Na	K	Mg	Ca	Sr	Ba
CO	17	27	36	33	28								
N_2	16	18	27	25	23	15	23.6	19.2	14.1	30.1	26.5	26.0	20.8
O_2	16.3		17		17								
Ar	15.8			18			13	12	11	17	21.5	20.5	17

Table 7.5 Heats of adsorption of hydrocarbon on H-zeolites (kJ mol^{-1}).[89,105]

	C_3H_8	n-C_4H_{10}	i-C_4H_{10}	n-C_5H_{12}	i-C_5H_{12}	n-C_6H_{14}
Silicalite						70
H-ZSM-5 (35)	46	58	52	70	64	82
H-Mordenite (10)	41	50	52	59	61	69
H-Y (2.7)	31	39	40	46	46	53

applications point of view, neutral frameworks are of interest in shape selective adsorption and separation of hydrocarbons, in membranes, and in the adsorption and release of incompletely combusted hydrocarbons in auto-exhausts (see Section 7.5.2). Most microporous metal organic framework solids are neutral, but they show distinctive adsorption behaviour, particularly in terms of their flexibility, and this is discussed separately in Section 7.3.6.

Representative examples of heats of adsorption on the neutral inorganic framework of silicalite at zero loading are given in Table 7.3. These give information on the initial sorbent-sorbate interaction. For small non-polar gas molecules, the physisorption heats vary from 17 kJ mol^{-1} (CO, N_2, O_2, Ar) to 34 kJ mol^{-1} (SF_6).[5] Interactions of neutral frameworks with polar molecules such as water and ammonia with silicalite depend very strongly on whether appreciable defect hydroxyl content is present as a result of the sample preparation. For defect-free silicalite, heats of adsorption of ammonia may be as low as 12 kJ mol^{-1}, which is lower than the heat of liquefaction of ammonia, whereas other silicalite samples that are shown by IR studies to have appreciable hydroxyl content that can interact by H-bonding have much higher initial heats (60–80 kJ mol^{-1}).[86]

Interaction of hydrocarbons with neutral frameworks is of industrial as well as academic interest, particularly in separation and adsorption. For a linear alkane such as *n*-hexane, the energy of adsorption is higher in pure silica zeotypes that have higher framework densities (and smaller pores) due to closer framework-molecule contacts and stronger dispersive interactions. Figure 7.9 illustrates this for the heats of adsorption of *n*-hexane in pure silica polymorphs with one-dimensional channels of different pore sizes.[87] The adsorption of hydrogen on $AlPO_4$s of different pore size also show this strong inverse dependence on pore size, decreasing from 9 kJ mol^{-1} in a small-pore material to 4 kJ mol^{-1} for VPI-5.[88] Molar heats of adsorption of straight chain alkanes increase linearly with carbon number, the gradient of the increase depending inversely on the pore size (in the range 9–15 kJ mol^{-1} per additional -CH_2- group). Very similar trends are seen for interactions of non-polar molecules with neutral aluminophosphates[89] and indeed for isostructural SiO_2 and $AlPO_4$ adsorbents (such as $AlPO_4$-5 and SSZ-24) the curves of adsorption enthalpy versus uptake are superimposable.[90] For larger hydrocarbon molecules, steric factors become important. For example, for the adsorption of C6 alkanes on silicalite the adsorption enthalpy decreases as the degree of branching increases, due to steric effects.[91] These packing effects can also give rise to steps in the adsorption isotherms and lower uptakes for the bulkier molecules.

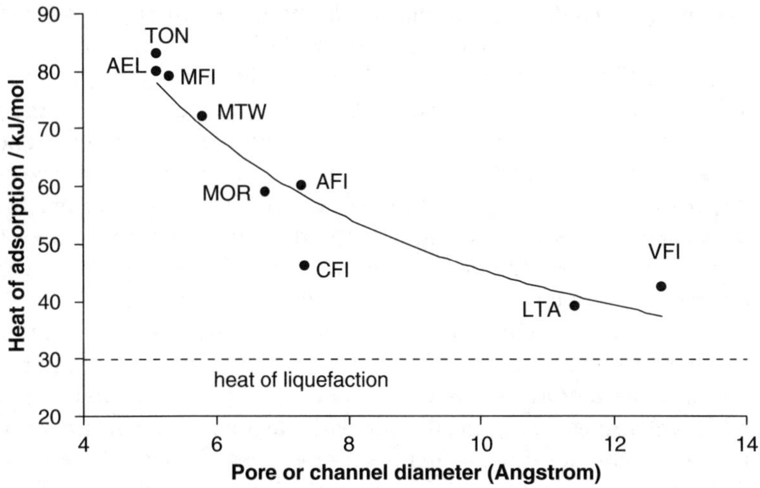

Figure 7.9 Experimentally determined heats of adsorption of n-hexane plotted against cage size of channel diameter for a series of pure silica zeolite polymorphs and aluminophosphates.

Structural determination of adsorbed molecules in neutral frameworks is also of interest. The location of molecules of *para*-xylene in silicalite (MFI) has been particularly well studied, because of the relevance to separation processes used to obtain the valuable *para*-isomer from mixtures of xylenes from the refinery. Two approaches stand out: the single crystal diffraction studies of van Koningsveld[76,92] and the MAS NMR studies of Fyfe.[93,94] The silicalite/p-xylene complex exhibits two slightly different orthorhombic symmetries, depending on the loading. At <4 molecules per unit cell, 'low loading,' the symmetry is Pnma (with 12 independent T-sites) whereas with loadings of from 6–8 molecules per unit cell, 'high loading,' it transforms to $P2_12_12_1$ (with 24 T-sites).[93] Using single crystal diffraction, van Koningsveld solved the structure of the high-loaded sample, where molecules are located both in the straight channels and in the intersections of straight and zig-zag channels (Figure 7.10). Fyfe used this data to help in the development of an NMR methodology to solve the structures of pure silica zeolite-adsorbate complexes. (Only pure silica zeolites exhibit sufficient resolution between different T-site signals.)

The NMR method makes use of measurements of the dependence of ^1H to ^{29}Si cross polarisation efficiency as a function of CP contact time to determine the distances between different silicon atoms in the ordered complex to H atoms in *para*-xylene molecules in the pores. The experiment has to be performed where the molecules show little translational motion, and under conditions where any molecular re-orientation is well understood (from ^2H wideline NMR, for example – see Section 7.2.3). To give additional information, the experiments were performed separately on selectively deuterated d_4- and d_6-*p*-xylene molecules. Plots of the ^{29}Si signal intensity from different crystallographic T-sites against the contact time show an increase to a maximum value, the steepness of which depends on the proximity of the ^{29}Si nuclei to protons of the adsorbed *p*-xylene (the more efficiently polarised ^{29}Si are closer to more ^{1}H nuclei). The CP experiment is based on the through-space heteronuclear ^{1}H-^{29}Si interaction that shows inverse third power dependence on the internuclear distance, so that distances can be determined accurately. The ^{29}Si resonances can be assigned to sites in the host structure of the framework by INADEQUATE ^{29}Si-^{29}Si experiments[95] (the structure and therefore the ^{29}Si shifts of particular sites vary strongly upon adsorption). Furthermore, the re-orientational motion of the protons is known (-CH_3 groups rotate at all temperatures, the aromatic ring is either stationary or undergoes 180° flips in these experiments). Therefore the location of the methyl and ring protons can be established, essentially by modelling (*via* 'triangulation') using appropriate software. In a *tour de force* of solid state NMR,[94] Fyfe *et al.* show that, if CP MAS NMR intensities of ^{29}Si in silicalite with adsorbed *p*-xylene are measured as a function of contact time, at temperatures where the adsorbate molecules show slow or no translational motion, then the location of the adsorbates can be determined with precision. After confirming the feasibility of the method against van Koningsveld's data for the high-loaded structure, the method was used to determine the low energy location of *p*-xylene at low loading. The INADEQUATE ^{29}Si-^{29}Si NMR spectrum of the low-loaded sample is given in

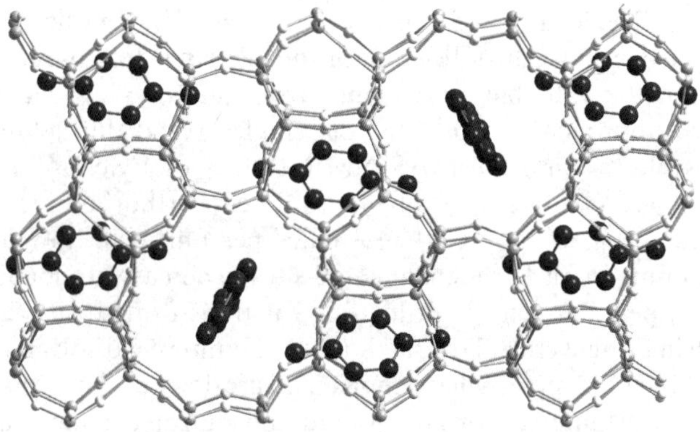

Figure 7.10 The location of *para*-xylene in silicalite/ZSM-5 in samples loaded with 4 molecules per unit cell (low loading) and 8 molecules per unit cell (high loading).[76,92–94] The adsorption isotherm of *p*-xylene on MFI shows a step at 4 molecules per unit cell, as the structure changes space group from Pnma (low loading) to $P2_12_12_1$ (high loading). In the high-loaded sample, the structure of which was solved by single crystal diffraction, the xylene molecules occupy sites in both straight and sinusoidal channels (top). In the low-loaded sample, the structure of which was determined first by NMR methods, the ^{29}Si signals had first to be assigned by INADEQUATE ^{29}Si–^{29}Si 2D NMR (middle) before ^{1}H–^{29}Si variable contact time CP MAS NMR could be used to triangulate the position of the xylene molecules: these occupy sites along the straight channels (below). Only C atoms of *p*-xylene are shown. [Reproduced from reference 95 with permission. Copyright 1991 American Chemical Society.]

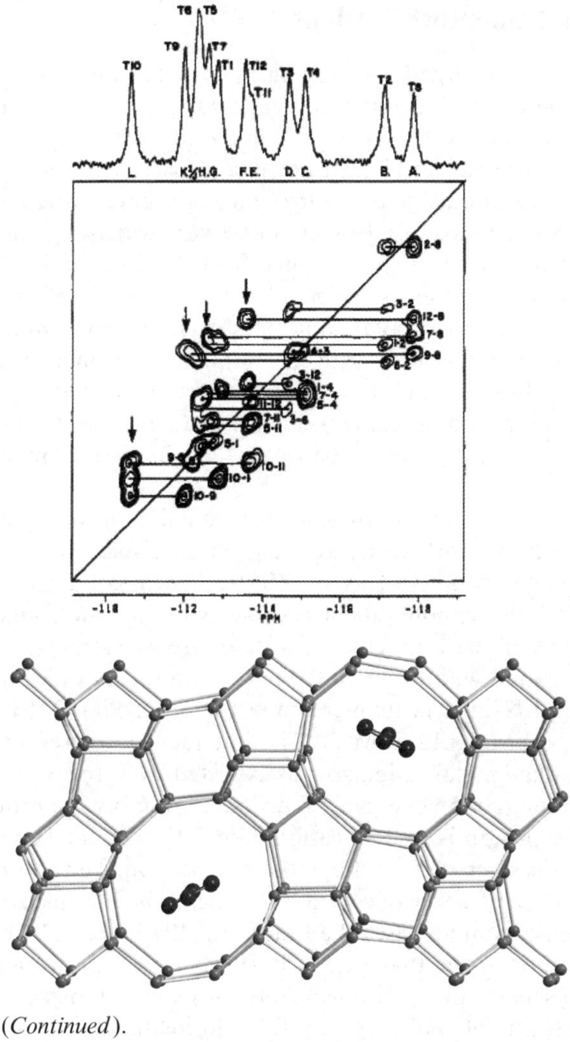

Figure 7.10 (*Continued*).

Figure 7.10, which enables ^{29}Si peak assignment – vital to the subsequent analysis of CP MAS NMR data. These studies show that the location of the first *p*-xylene molecules to adsorb at low loading is in the centre of the straight channels: subsequent single crystal XRD has confirmed this. A more detailed exposition of the method, and a description of its application to locate *para*-dichlorobenzene in silicalite, has also been given.[96]

The determination of the position of adsorbed molecules by XRD and/or NMR is only feasible at low temperatures (room temperature and below), where the effects of translational motion are negligible, and limited to pure silica zeolites, with T-sites resolved in the ^{29}Si MAS NMR, but it does establish a deeper understanding of the energetics of hydrocarbon adsorption as a function of structure and loading.

7.3.2 Extra-framework Cations

Upon dehydration, charge-balancing cations in zeolites lose their water of hydration and adopt well-defined positions where they coordinate to framework oxygens. In most cases, they are unable to satisfy their optimum coordination, and so present unshielded positive charge to molecules entering the pores. The electrostatic fields that surround such 'bare' cations can be of the order of $10\,V\,nm^{-1}$. Heats of adsorption of a very wide range of molecules on such sites have been measured and range from $15–30\,kJ\,mol^{-1}$ for argon and simple homonuclear diatomics, where charge-induced dipole interactions are present, to $90\,kJ\,mol^{-1}$ for polar molecules such as water where charge-permanent dipole interactions dominate. The total interaction includes terms from dispersive interactions, from interactions between cations and induced dipoles and between cations and permanent dipoles and permanent quadrupoles, and similar electrostatic interactions between the charged framework and the adsorbates.

Instructive examples of the dependence of initial heats of adsorption on cation type and framework structure are given in Table 7.4.

The alkali metal cation forms of ZSM-5 are a good starting point for discussion.[97] All the cations are accessible and the low framework charge results in them being well dispersed. Furthermore, the strengths of interaction are readily compared with those on the pure silica form of the structure, silicalite. For CO and N_2, the heats increase in the order SiO_2-MFI < H-ZSM-5 < K-ZSM-5 < Na-ZSM-5 < Li-ZSM-5. The interaction therefore depends inversely on the alkali metal cation size, as expected from the inverse dependence of electrostatic interaction energy with distance. Notably the proton form does not act as if the proton is a small cation: interaction of the H-form of zeolites with adsorbates is discussed further in the next section. The adsorption of O_2 is affected little by the presence of cations, because of the low quadrupole moment of the molecule, and in all cases the heat is *ca.* $17\,kJ\,mol^{-1}$. This difference of heat of adsorption from that of N_2 is the basis of the room temperature pressure swing separation of nitrogen from air over cation-exchanged zeolites. Over cationic forms of zeolite X, careful calorimetry has also shown similar dependence on cation size of the adsorption enthalpies of, for example, N_2 and Ar (Table 7.4).[61] Among the alkali metal forms, N_2 binding enthalpy increases in the sequence KX < NaX < LiX; for alkali earth metal forms, BaX < SrX < CaX < MgX. The trends are similar for argon, except that there is a maximum in adsorption enthalpy for the Ca-X. These results are in line with trends in the hardness of the cations. In addition to changing the enthalpies of adsorption, the exchange of larger cations for smaller cations can also change the available space and the effective window size, as described previously for K-A (3A) compared to Na-A (4A), and exchanging (fewer) divalent cations for monovalent cations will result in greater sorption capacity and can result in larger effective pore size (as seen for Ca-A (5A)).

For well-characterised systems such as zeolite X, there are also cases of mixed cation materials where the degree of ion exchange is found to affect sorption capacities because the cation siting changes. Replacing sodium cations by lithium cations to give Na,Li-X has little effect on the adsorption capacity or N_2/O_2 selectivity until a critical level of exchange (which depends on the framework composition) is exceeded, even though lithium cations are expected to interact more strongly with the adsorbed nitrogen.[98] Above this critical level of Li exchange both the adsorption capacity and N_2/O_2 selectivity rapidly increase. This occurs because in competition for cation sites between the first lithium ions to be introduced and the original sodium cations, the accessible cation sites of type SIII (in the supercages, and therefore accessible to adsorbed molecules) are all filled by sodium ions. Only above 75% exchange (for zeolite X, Si/Al=1.2) are there too few sodium cations to fill these sites, so that some must be occupied by lithium cations, which then strongly influence the adsorption properties. Many such tunes can be played in tailoring the adsorption properties of zeolite molecular sieves.

The interaction of extra-framework cations with polar molecules is strong. Interactions with halocarbons (industrially relevant because zeolites are used to dry and remove water and acid degradation products of chlorofluorocarbons used as refrigerants) have been shown by Mellot *et al.*[99] to include terms from van der Waals interactions ($ZeO^- \cdot\cdot$ Cl–C), from electrostatic interactions ($M^+ \cdot\cdot$ Cl–C) and H-bonding ($ZeO^- \cdot\cdot$ H–C). The presence of chlorine in the hydrocarbons strongly enhances van der Waals and electrostatic interactions due to the greater polarisability and partial charge. Measured zero loading adsorption enthalpies of 53 kJ mol^{-1} on chloroform ($CHCl_3$) are therefore much higher than for methane (15.2 kJ mol^{-1}). Mellot *et al.* show that on going from siliceous zeolite Y *via* Na–Y (Si/Al = 3) to NaX (Si/Al = 1.2) increases the heat of interaction from 40 *via* 50 to 80 kJ mol^{-1}. In zeolite Na–Y the additional heat results from increased interaction of the chloroform with the charged framework, since few sodium cations are present in SIII supercage sites. The additional enthalpy increase in zeolite Na–X results because for this composition there are many more sodium cations and some of them occupy accessible SIII sites: these effects are well simulated computationally. Finally, the initial heat of water adsorption on zeolites such as CaNa–A, for example, is very high, *ca.* 90 kJ mol^{-1},[100] and explains their action as powerful drying agents.

A variety of methods have also been used to relate the observed bond strengths of cation–adsorbate interactions to the locations of cations within the pore space, the structure of the adsorption complex and its relation to the geometry of the pores. IR, NMR and diffraction are the most important experimental methods. Computational approaches have also been successful, although these have the added difficulty of modelling the location of the framework charge and cation locations.

Mid-IR spectroscopy (400–4000 cm^{-1}) has been shown by the group of Zecchina to be very informative when studying adsorption at different adsorption sites in zeolites.[101] The spectroscopy is performed at low temperatures using small diatomics such as CO, N_2 and H_2 as probe molecules that have suitable IR characteristics and that can gain access to the entire pore space. As well as being unreactive, these probe molecules have IR characteristics that enable their absorption spectrum when adsorbed to yield interpretable changes in vibration frequencies and intensities. CO is a particularly useful probe. The C=O stretch (2143 cm^{-1} in the free state) is readily observed, and important effects are seen in the vibrational spectra of CO adsorbed onto alkali metal cation forms of zeolites. CO usually interacts with cations at the carbon atom. The ν(CO) frequency depends on the electric field strength sensed at the charge centre, which for cations of a constant charge increases as the cation size decreases. The observed shifts of the carbonyl stretching frequency to higher wavenumbers are consistent with shifts expected from electrostatic fields in the region of 2–6 V nm^{-1}. These shifts result from the sum of a positive contribution from the cation and negative contributions from the framework oxygens and polarisation of the framework.

The high electrostatic fields are able to induce dipoles in homonuclear diatomics, such as hydrogen and nitrogen, and render the stretching modes infrared active. Upon adsorption of the N_2 molecule on alkali metal zeolites, for example, the N≡N stretch becomes IR active.[75] The vibration frequency of nitrogen is always observed to increase upon adsorption, which strongly suggests an end-on interaction of the nitrogen molecule with the alkali metal extra-framework cations. A comparison of the spectroscopy of the interaction of N_2 with cations of different sizes shows that the shift in the stretching frequency varies with the inverse of the square of the distance between the cation and the N_2 molecule (as it does for CO) so that there is a linear relationship between ν(N≡N) and ν(CO) and $1/(R_M + R_{ads})^2$ (Figure 7.11). This indicates the interaction is predominantly electrostatic.

Xenon-129 NMR has also been used as a suitable probe to investigate the electrostatic field that adsorbed xenon adatoms experience within zeolitic pores.[102] The chemical shift of adsorbed ^{129}Xe is a sensitive function of the electrostatic field that it experiences, as well as depending on collisions between

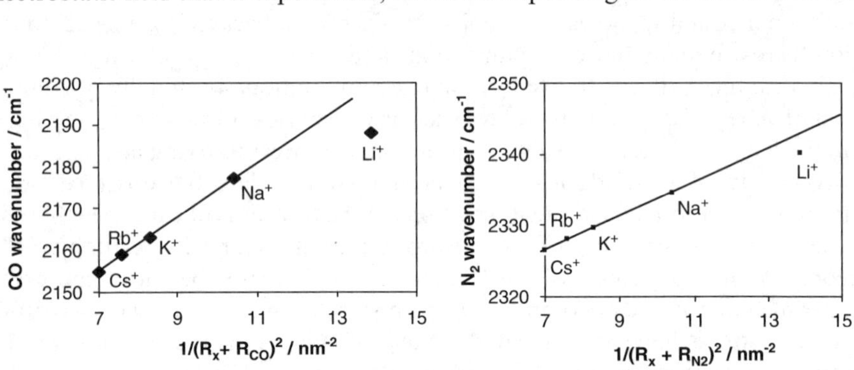

Figure 7.11 The wavenumbers of (left) the C=O stretch of carbon monoxide and (right) the N≡N stretch of dinitrogen adsorbed on a series of cation-exchanged zeolites, plotted against the inverse square of the cation··· adsorbate distance. A linear relationship suggests that electrostatic effects are dominant in determining the frequency shift. [Reproduced from reference 75 with permission. Copyright 1995 American Chemical Society.]

the Xe atoms with the framework and other Xe atoms. ^{129}Xe NMR is therefore very sensitive to both chemical composition and pore size. If the chemical shift δ is measured as the concentration of adsorbed xenon is varied, then the value of δ extrapolated to zero coverage is found to vary from 110 ppm for the medium-pore ZSM-5 to 58 ppm for zeolite Y, for example. In addition, if the xenon atoms adsorb in two different channel systems, between which the transport is slow, then two distinct signals are observed: the results of Ripmeester of xenon on mordenite at 240 K give two signals attributed to xenon in the main channels and the side pockets of this zeolite.[103]

The adsorption of basic molecules such as pyridine gives rise to characteristic shifts in the resonances that are readily interpreted in terms of specific coordination of the molecules to the cations through the nitrogen lone pairs, *via* a Lewis acid–Lewis base interaction. Indeed, pyridine acts as such a strong base that all acid centres (Lewis and Brønsted) interact stoichiometrically with the pyridine molecules. A strong base such as pyridine therefore distinguishes poorly between Lewis acid sites of different strength.

Diffraction studies have been used in a small number of cases to determine the minimum energy positions of adsorbed molecules within cationic zeolites. These difficult and relatively expensive experiments have been used to establish computational simulation as a reliable method to predict such interactions without recourse to experimentation. Crystallographic X-ray studies of the adsorption of small molecules such as CO and NO on cationic forms of zeolite A[81] are of particular relevance to study of cation–molecule interactions, as are neutron powder diffraction studies of the location of pyridine, coordinatively bound to potassium ions in zeolite K-L,[83] and of benzene, bound to sodium ions in zeolite Na-X.[82]

7.3.2.1 Transition Metal Cations

Transition metal cations can be introduced as charge-balancing cations within zeolites and aluminophosphates by ion exchange (Section 6.3.1) or by decomposition of transition-metal-containing complexes acting as structure-directing agents (Section 6.2.2). In addition to electrostatic interactions, transition metal cations often form chemical bonds with adsorbed molecules through d–σ or d–π interactions, acting as ligands. Such behaviour can be the first step in catalytic behaviour. As a result, a good deal of calorimetry and IR spectroscopy has been performed to investigate the nature of transition metal cation–molecule interactions.

Copper(I) cations are known to interact strongly with two nitric oxide molecules, and copper-containing zeolites are good catalysts for the reaction of removal of NO from auto-exhaust streams (Section 9.3.2). Interactions of probe molecules with Cu^+-ZSM-5 are therefore of interest.[101] Adsorption of CO as a probe molecule on Cu^+ results in ν(C=O) of 2157 cm^{-1}, lower than that expected on the basis of its cationic radius according to the relation established for adsorption on alkali metal cation exchanged zeolites.[104] The shift in the stretching frequency is the net result of σ donation through the weakly antibonding 5σ molecular orbital (which raises the ν(CO) frequency) and back donation to the antibonding 2π MO (which lowers ν(CO)). For N_2 adsorbed on Cu^+-ZSM-5, ν(N≡N) is *less* than that of free N_2, which can also be rationalised in terms of chemisorption. The orbitals involved in this case are the $3\sigma_g$ MO of N_2, which acts as an electron donor, and the empty $1\pi_g$, which is the electron acceptor. Both σ–donor and π–acceptor interactions weaken the N≡N bond, leading to lower ν(N≡N) wavenumbers. Notably, low temperature IR of NO adsorbed on Cu^+-ZSM-5 indicates that mono- and dinitrosyl complexes are formed that convert at room temperature into $Cu^{2+}NO_2^-$ species in a reaction that could have relevance for the activity of copper zeolites in NO decomposition. Many other metal complexes of CO and NO have been prepared on transition metal zeolites.

7.3.3 Structural Hydroxyls

Zeolitic solids can be prepared with very high concentrations of protons attached to the framework. These are readily observed by infrared and NMR spectroscopy (Chapter 3). The bridging hydroxyl sites (Si–OH–Al) in aluminosilicate catalysts are of fundamental importance in acid-catalysed reactions. They are much more strongly acidic than the terminal SiOH hydroxyls that are also present at defect sites at the external surfaces or within crystals that have lost framework cations (for example by dealumination). Structural hydroxyl groups in aluminophosphates also arise, due to substitutions of divalent cations for aluminium, silicon for phosphorus, or in aluminosilicate islands within the framework. The interaction *via* hydrogen bonding of the acid site with any adsorbed molecule capable of acting, even weakly, as a base, perturbs the O–H bonding, in a way that infrared spectroscopy readily recognises. The interaction of molecules at such Brønsted acid sites is the first step in acid-catalysed hydrocarbon transformations and is therefore discussed in detail in the next chapter.

At this point, however, it is worth considering the enthalpies observed for adsorption on zeolites containing structural protons. For simple diatomics such as N_2 and O_2, the presence of hydroxyl groups makes little difference compared to that in neutral frameworks. The strength of interaction of CO with H-forms is similar to that with K^+-forms, and CO adsorption is found to be a sensitive IR probe of microporous acids. Interaction of hydroxyl groups with strong bases leads to their protonation, whereas the adsorption of water or methanol on Brønsted acid sites at a loading of less than 1 molecule per site is at the limit between H-bonding and protonation. Adding further molecules results in protonation of the cluster (see Sections 8.4.2.2 and 8.4.2.3).

Interaction of normal and iso-alkanes on protonated zeolites H-ZSM-5, H-mordenite and H-Y have been studied in detail by Eder *et al.* by a combination of calorimetry and IR spectroscopy[105] (Table 7.5).

At low coverages the hydrocarbons are adsorbed *via* H-bonds at the Brønsted acid sites. This is shown by loss of intensity of the sharp bridging hydroxyl resonance and the appearance of a broader band at lower wavenumber. That chemisorption is involved is also indicated by a sharp drop in the heat of adsorption above loadings of 1 molecule per strong acid site (or 2 for *n*-alkanes on H-ZSM-5). The H-bonding involves induction of polarity in the alkane. The heats of adsorption of the *n*-alkanes, where there are no steric restrictions due to their shape, are higher in the order H-Y < H-MOR < H-ZSM-5 (opposite to what would have been expected on the basis of the aluminium and proton concentrations in these solids, which increase in the reverse order). This arises because the most important interactions are the short-range dispersion forces, which are higher for the medium-pore ZSM-5 than for the large-pore zeolites. In addition, the increase in adsorption enthalpy along the homologous *n*-alkane series is linear, and is steeper for H-ZSM-5 than for H-MOR or H-Y for the same reason. In fact, the increase of the adsorption enthalpy due to the presence of protons (compared to the pure silica forms, for example) is relatively small and constant. Comparing H-ZSM-5 to silicalite, for example, shows that the heat of adsorption on the aluminium-containing form is greater by only around 10 kJ mol^{-1}, regardless of the carbon number of the adsorbed *n*-alkane.

Similar trends of increasing heat of adsorption with C-number are seen for the iso-alkanes. For H-ZSM-5 their adsorption is less favoured than for the *n*-alkanes whereas they are more favoured in the large-pore solids. This results from steric hindrance in the ZSM-5. As a more marked consequence of the steric features, the packing of the iso-alkanes is much less efficient than of the *n*-alkanes in H-ZSM-5, so that the uptake of iso-butane is only around one half of that of *n*-butane. This ratio is much closer to 1 in large-pore mordenite (0.85) and H-Y (1). Still larger steric effects were noticed for the adsorption of bulky alkylbenzenes on H-ZSM-5, where initial heats of adsorption of isopropyl- and *n*-butyl-benzene are much lower (50 and 10 kJ mol^{-1}, respectively) than those observed for ethyl- and *n*-propylbenzene (*ca.* 80 kJ mol^{-1}).[106]

7.3.4 Framework Lewis Acid Sites

In some microporous solids, framework cations in tetrahedral sites are able to expand their coordination by chemisorbing small molecules while remaining within the framework, so that removal of the adsorbed species permits their return to tetrahedral coordination. The most important example of this is the titanium in titanosilicates such as TS-1. The titanium cations in the framework of ETS-10 are only able to adopt octahedral coordination, and are unable to show this kind of behaviour.

In the case of titanosilicate analogues to zeolites, the titanium has been unequivocally shown to adopt tetrahedral coordination upon dehydration of the as-synthesised material.[107–109] Upon adsorption of aqueous solutions of hydrogen peroxide, which is important in selective oxidations, UV-visible and EXAFS spectroscopies show that the titanium centre exhibits complex behaviour.[110,111] In the presence of excess water, the titanium loses its tetrahedral coordination and there is a modification of the first and second coordination shells of the titanium. The UV-vis spectrum of the yellow complex is characterised by a Ligand-to-Metal Charge Transfer band at 26 000 cm^{-1} characteristic of a side-on Ti-peroxo species with bidentate (η^2) geometry. Ti-peroxo species with bidentate (η^2) geometry are usually inactive in selective oxidation, however. As water is removed the colour is lost, but can be regenerated upon rehydration. It is thought that the species that remains upon dehydration is monodentate (η^1) end-on hydroperoxide (Scheme 7.1), which is thought to be the active species in oxidation. Notably the same species (end-on hydroperoxide) is generated upon dosing anhydrous H_2O_2 onto TS-1, *via* decomposition of $KH_2PO_4.H_2O_2$, and can be hydrated to the side-on peroxo complex (Scheme 7.1).[110,111] X-ray absorption spectroscopy at the Ti K-edge reinforces these conclusions: the pre-edge peak, high in the case of tetrahedral titanium, is reduced on addition of aqueous hydrogen peroxide as the side-on species is coordinated, but slightly increases as water is lost and the end-on peroxide species is formed. Since bidentate peroxo species are not active in oxidation reactions, the equilibrium between these two configurations has direct implications for the activity of TS-1 in a range of selective oxidation reactions (Chapter 9).

Tetrahedral silicate frameworks also appear to be able to include small amounts of tin, which is then able to change coordination from tetrahedral to octahedral. Like titanium, variable coordination tin imparts activity in Lewis-acid-catalysed reactions (Section 9.2.2).

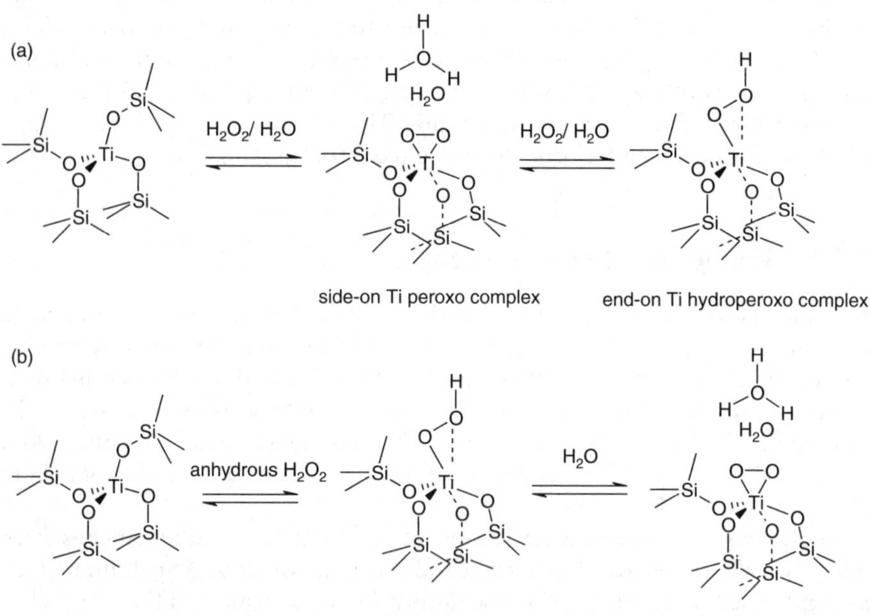

Scheme 7.1 Mechanisms for generation of peroxo and hydroperoxy species on titanium, postulated on the basis of X-ray absorption and UV-visible spectroscopy.[110,111]

Similar variable coordination behaviour is observed for metal cations in framework positions in aluminophosphates. Reversible hydration can result in five- or six-fold coordinated aluminium and marked changes in the crystallographic symmetry. The ability of cobalt in the framework of aluminophosphates to change oxidation state and coordination is thought to be important in the selective oxidation behaviour of these solids (Section 9.2.3). Related studies of the adsorption of the Lewis base acetonitrile onto a cobalt-exchanged aluminophosphate demonstrated that framework cobalt(II) can increase its coordination to six with two nitrogen atoms of adsorbed acetonitrile molecules.[112] These effects are also observed for the framework cations in many metal-organic framework compounds, such as the copper trimesate HKUST-1, the chromium trimesate MIL-100 and the nickel N,N′-piperazinebismethylenephosphonate described in Chapter 2, in which the transition metal cations can adopt different coordinations as bound water molecules are lost.

7.3.5 Basic Sites

There has also been interest in the potential of porous solids as shape selective basic catalysts (Section 9.3.5). The most easily prepared base forms of zeolites are the alkali metal cationic forms, which are easily handled but weakly basic. They possess framework oxygen sites that are electron donating and possess the tendency to attract protons. Stronger basic sites can be introduced by impregnation of alkali metal salts followed by their subsequent thermal decomposition to give intra-zeolitic metals or metal oxides. Additional methods include the treatment of zeolites with ammonia at high temperature to give imides and nitrides. For mesoporous solids, by comparison, it is possible to covalently attach organic bases to the framework *via* the surface hydroxyls at low temperatures.

The adsorption of molecular probes, followed by infrared or NMR spectroscopy and thermal desorption studies, is the most commonly adopted way to study basic sites. Carbon dioxide is frequently used in infrared studies, particularly of cationic zeolites with added alkali metals.[113] Chloroform is also suitable, since the interaction with the chlorine atom and subsequently the C–Cl stretching frequency is a measure of the basic strength. NMR studies of basic zeolites have concentrated on the use of ^{13}C containing probes such as methyl iodide and chloroform.[114] Addition of methyl iodide results in its decomposition and the formation of methoxy groups at framework oxygens.[115] The shift of the methyl carbon is expected to be related to the basicity of the framework, that is the tendency of the framework oxygens to donate electrons – ^{13}C MAS NMR of methoxy groups prepared in this way shows a clear distinction between basic zeolites, such as Cs-X, and acidic zeolites, such as H-ZSM-5.

For cationic forms of zeolites, the observed trends in (mild) basicity are clear. Zeolites with higher aluminium contents in the framework are more basic than high silica zeolites. Similarly, the basicity of zeolites exchanged with cations increases in the order $Na < K < Rb < Cs$.

7.3.6 Metal–Organic Frameworks

Very many microporous metal organic frameworks have been prepared, some of which are described in Section 2.8. The most striking feature of this class of material is the very-large-pore volumes exhibited by examples such as MOF-5, MOF-177, IRMOFs and the MIL-n M^{3+} carboxylates, where M = Al, Fe, Cr, *etc*. Furthermore, recent PFG NMR experiments show that the diffusivity of hydrocarbons in the archetypal MOF-5 is much higher than in zeolites such as Na-X and only slightly lower than in the neat liquids at the same temperature,[116] which is promising for possible uses in large-scale adsorption and separation processes. The most distinctive adsorptive behaviour exhibited by some MOFs, by comparison with inorganic microporous solids, derives from their properties of solid state reactivity and structural flexibility.[117]

Some frameworks are able to take up molecules reversibly, as a consequence of bond-breaking and making events. This can occur *via* amorphous, intermediate phases which recrystallise upon readsorption, such as that observed for zinc benzenedicarboxylates[118] or through crystal–crystal transformations, as is observed in the uptake of benzene into the $Cu_2(pzdc)_2(bpy)$ framework (pzdc = pyrazine-2,3-dicarboxylate, bpy = 4,4′-bipyridine).[119] A structure taking up molecules *via* bond-breaking can in some cases take up molecules of dimensions larger than would be expected on the basis of the window size measured from the starting crystal structure.[120] In other structures the pore and window sizes can be increased greatly in the presence of adsorbing molecules because the conformation of the ligands change markedly in response to sorbate uptake. In the iron(III) fumarate MIL-88, for example, the cell volume is almost doubled between the anhydrous state and the form that has opened fully to accept adsorbate molecules.[85]

Adsorption isotherms of gases such as N_2 and CO_2 can show stepped uptake at very high pressures. This 'gating' behaviour results from pore opening as a structural response to the increased chemical potential of the adsorbate phase, rather than the effect of strong sorbate–sorbate interactions that is responsible for the Type V isotherm of the IUPAC adsorption isotherm types. For MIL-53(Cr) at 304 K, for example, a low pressure adsorption step thought to be due to uptake of CO_2 at the Cr–OH sites in the pores and accompanied by distortion of the lattice is followed by a step at higher pressure (*ca*. 7 bar) when the resulting Cr–OH · · OCO · · HO–Cr interaction is disrupted and the 'trellis-like' lattice opens up fully (Figure 7.12).[121]

Finally, strong irreversible hysteresis has been observed in H_2 adsorption/desorption measurements on $Ni_2(4,4'\text{-bipy})(NO_3)_4$ by the groups of Thomas and Rosseinsky,[122] even to very low desorption pressures. This results from 'kinetic trapping' of H_2 molecules which can then only be released by raising the temperature to 110 K and above.

In summary, the adsorption behaviour of MOFs is of interest because of their large pore volumes, and the novel properties of their organic-lined pores. Furthermore, the reactivity and flexibility of their structures gives rise to reversible uptake of H-bonding molecules with recrystallisation, 'breathing'

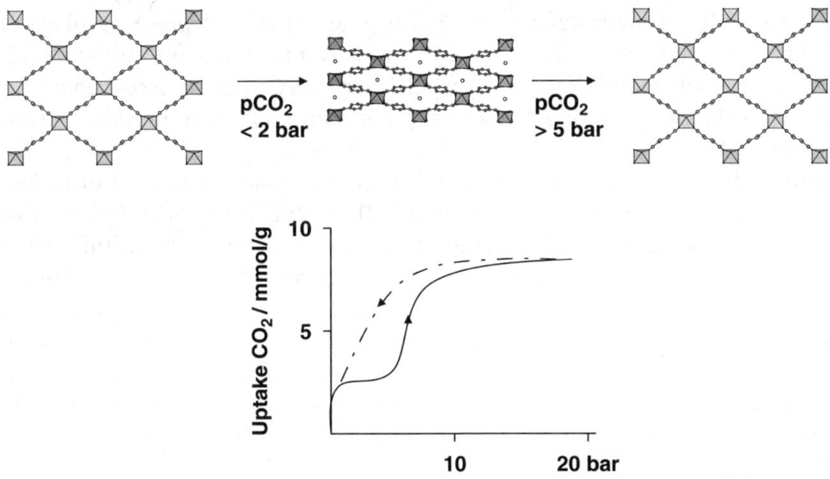

Figure 7.12 The chromium terephthalate MIL-53 ($CrOH(O_2CC_6H_4CO_2)$) is a very flexible structure. When heated to remove all adsorbed molecules the structure is very open (above left). Carbon dioxide adsorption at 304 K proceeds stepwise (below). At pressures below *ca.* 2 bar, the structure adapts by contraction of the channels to coordinate CO_2 molecules chemisorbed at the structural hydroxide group, and this structure remains stable until pressures higher than 2 bar, when it opens to take large volumes of CO_2 into the channels.

of the framework as a response to adsorbate uptake and 'gating' and 'kinetic trapping'. Potential advantages in applications in gas storage are discussed further in Chapter 10.

7.4 Diffusion

7.4.1 Introduction: Self-diffusion and Transport Diffusion

Diffusion of adsorbate molecules throughout the pore space of microporous solids is an essential step in many applications of microporous solids and determines their utility and selectivity in applications. Whereas the thermodynamics of the adsorption determines the equilibrium situation, the kinetics of an adsorptive or catalytic process is controlled by the diffusion rates. This is exemplified in their use in shape-selective catalysis, where molecules must reach and leave active sites distributed through the crystallites and therefore products that diffuse faster will be enriched in the molecular mix leaving the solid.

Diffusion in microporous solids occurs by activated jumps along the pore channels or across cages within the structure. The rate of diffusion over 'short' distances, of the order of the unit cell repeat, is determined by the frequency of re-orientation of molecules into configurations that permit motion, the strength of interaction with the framework, the distance between adsorption sites and the presence of other molecules in the pores. Structural defects, including stacking faults, growth defects and grain boundaries, the presence of species blocking the channels (such as coke in catalytic processes) and ultimately the external crystallite surfaces can affect diffusion rates on the macroscopic length scale. Effects on both length scales (nano- and macro-) influence overall diffusion rates.

Before discussing the results of measurements or simulations of diffusion, I give below a brief summary of the terminology and types of diffusion. Comprehensive treatments of the quantitative aspects of diffusion within porous solids are available elsewhere.[123] In general terms, there are two kinds of diffusion regimes: equilibrium and non-equilibrium. In the first regime, uptake achieves thermodynamic equilibrium, with the chemical potential of the sorbed species being equal to the chemical potential of the adsorbing gas. For non-equilibrium diffusion, there is net sorbate transport from regions of higher chemical potential to regions of lower chemical potential.

Starting from Fick's first law,

$$J = -D(q)\frac{\partial q}{\partial z}$$

where J is the sorbate flux, $\partial q/\partial z$ the concentration gradient and $D(q)$ the diffusion coefficient at a given uptake level, q, in $m^2 s^{-1}$. For the equilibrium regime at equilibrium, we consider the diffusion of marked, or 'tracer', molecules designated with an asterisk, so that

$$J^* = -\mathscr{D}(q)\frac{\partial q^*}{\partial z}$$

where $\mathscr{D}$ is then the self-diffusivity.

For non-equilibrium diffusion, the flux can be expressed in terms of a chemical potential gradient

$$J = Lq\frac{d\mu}{dz}$$

where Lq is the usual form of the constant of proportionality and $d\mu/dz$ is the chemical potential gradient. From this, inclusion of the expression for chemical potential of an adsorbate and comparison with Fick's law gives the transport diffusivity D,

$$D = LRT\frac{d\ln p}{d\ln q}$$

where p is the partial pressure of the adsorbate.

The transport diffusivity and self-diffusivity are related by the Darken relation,

$$D = \mathscr{D}(q)\frac{d\ln p}{d\ln q}$$

Using this relation, it is possible to compare self-diffusion coefficients and transport diffusivities measured or calculated under equilibrium or non-equilibrium conditions. The relationship of partial pressure and uptake is obtained from adsorption isotherms.

7.4.2 Experimental Methods and Simulations

Techniques for measuring diffusion are usually divided into two categories: those that measure microscopic diffusion, on the shorter, molecule-related length scale, and those that measure macroscopic effects of diffusion, such as pressure changes, mass uptake rates or chromatographic effects. These two types of measurements can give different results, with macroscopic diffusion coefficients typically being one or two orders of magnitude lower than those predicted on the basis of microscopic investigations, and the apparent discrepancy has given rise to extensive discussion.

Methods to measure microscopic diffusion aim to determine the time dependence of the root mean square displacement of molecules of an adsorbate, typically under equilibrium conditions. Most such measurements therefore measure 'self-diffusion', with measurements typically on the scale of nanometres. Typical methods of this kind include Pulsed Field Gradient (PFG) NMR[124] and Quasi Elastic Neutron Scattering (QENS),[74] both methods operating under conditions of macroscopic equilibrium. Pulsed field gradient NMR studies enable the self-diffusion rates of adsorbed molecules to be measured by following the position of proton-containing molecules within a magnetic field gradient. The experimental setup is more complicated than for chemical shift NMR studies, since a probe with a strong magnetic field gradient is required. In addition, diffusion constants taken from the analysis of molecular dynamics simulations also treat diffusion on this length scale. There is a high degree of agreement between methods that determine diffusion coefficients on this scale.

Most applications, however, rely on diffusion over length scales $>1\ \mu m$. Although PFG NMR measurements can also be made on this scale (where they give much smaller diffusion constants than those made at shorter root mean square displacements), more commonly used, 'macroscopic' methods include zero length chromatography (ZLC),[125] gravimetric measurement of uptake rates, frequency response (FR) studies[126] and interference microscopy and FTIR microscopy using single crystals.

In the frequency response technique the volume is periodically increased and decreased in square waves, and the response of the pressure measured and analysed in terms of Fick's laws under equilibrium conditions.[126] It has the twin advantages of being applicable over a wide frequency range and being able to distinguish between independent kinetic processes. As a result, it appears to show close agreement with microscopic processes, because it can resolve the effects on different length scales.

Interference optical microscopy is a recent and highly attractive experimental method[127,128] to study diffusion processes on an intermediate length scale of microns as a single crystal is exposed to molecules in the gas phase

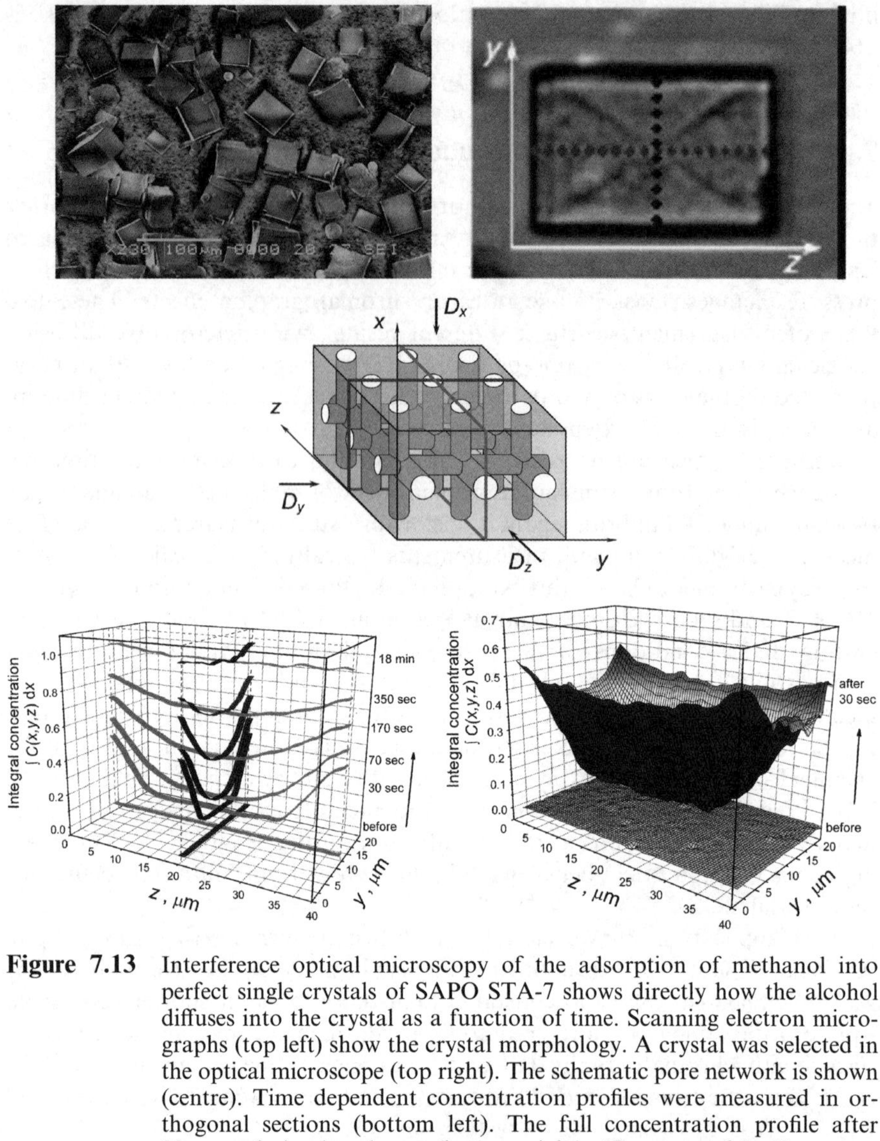

Figure 7.13 Interference optical microscopy of the adsorption of methanol into perfect single crystals of SAPO STA-7 shows directly how the alcohol diffuses into the crystal as a function of time. Scanning electron micrographs (top left) show the crystal morphology. A crystal was selected in the optical microscope (top right). The schematic pore network is shown (centre). Time dependent concentration profiles were measured in orthogonal sections (bottom left). The full concentration profile after 30 seconds is also shown (bottom right). (Courtesy of P. Kortunov, D. Tsoulakis, J. Karger, Leipzig). STA-7 has three-dimensional pore connectivity, with different pores along different directions, so that analysis reveals the dependence of diffusion rate on pore size.

(Figure 7.13).[128,129] It is conceptually simple, because it uses the phase difference that results from the presence of adsorbed molecules in the pores to quantify their concentration, and so directly measures the distribution of adsorbates within the crystal. Under the correct conditions, the method can be applied to determine non-equilibrium diffusion coefficients to be evaluated at different points in each crystal and as a function of different concentrations. The spatial resolution is good ($0.5 \times 0.5\,\mu m^2$) and it can be used to measure diffusion under non-equilibrium conditions, for example to follow the ingress of adsorbate into a crystal from the vapour phase as a function of time. It is therefore possible to obtain transport diffusivities as a function of loading at non-equilibrium conditions. A number of examples of the use of Interference Microscopy have been completed: these include the measurement of the diffusion coefficient of methanol at room temperature on Na,Ca-A[127] and SAPO STA-7.[130] For Na,Ca-A, the measured diffusivity by this method (*ca.* $10^{-13}\,m^2\,s^{-1}$) is similar to that through the larger 8MR pores in STA-7, and twice as large as that through the smaller 8MR pores along *x* and *y* directions in STA-7. Infrared microscopy can also be used to follow the ingress of molecules with a characteristic absorbing band, but the available resolution is much less using that technique ($20 \times 20\,\mu m^2$).

The evidence from these methods suggests that lower values of diffusion constants are obtained from methods which are sensitive to factors important over longer distances, and so indicate that defects can have a controlling effect in diffusional rates.[131] This is shown clearly in interference optical microscopy studies of methanol adsorbing into crystals of CrAPO-5, where the hexagonal prismatic crystals are found to possess large sectors that are inaccessible to adsorbate molecules.[128] The existence of barriers to molecular adsorbates at the external surfaces of zeolites has also frequently been reported[132] and can be visualised directly by comparative interference microscopy on as-prepared and also surface-etched ZSM-5 crystals.[131]

7.4.3 Examples

There are many studies of molecular diffusivities in microporous solids, particularly on silicalite and zeolites X and Y. Molecular shape is, as expected, found to critically affect the diffusivities of alkanes. For example, among hexanes diffusing through silicalite, diffusivities vary over three orders of magnitude (10^{-12}–$10^{-15}\,m^2\,s^{-1}$) going from linear to double-branched molecules (Figure 7.14).[133] Furthermore, aromatic molecules are found to diffuse much faster through silicalite than cycloalkanes of similar molecular mass.[134]

Within one-dimensional channels, where adsorbate molecules are too large to move past one another, single file diffusion occurs.[135] In single file diffusion, motion occurs *via* coordinated motion of assemblies of molecules along the channels. Where more than one set of channels is present and they are interconnected, molecular 'traffic control' takes place. In the silicalite structure at temperatures below *ca.* 450 K, for example, diffusion of *para*-xylene is thought, on the basis of Molecular Dynamics and Frequency Response studies, to occur much more readily along the straight channels than along the sinusoidal channels. Remarkably, diffusion of *para*-xylene is much faster than that of the smaller benzene,[136] which is attributed to the reluctance of benzene to lose its rotational entropy (in the form of free rotation at the intersections) during restricted diffusion along the connecting channels. In addition,

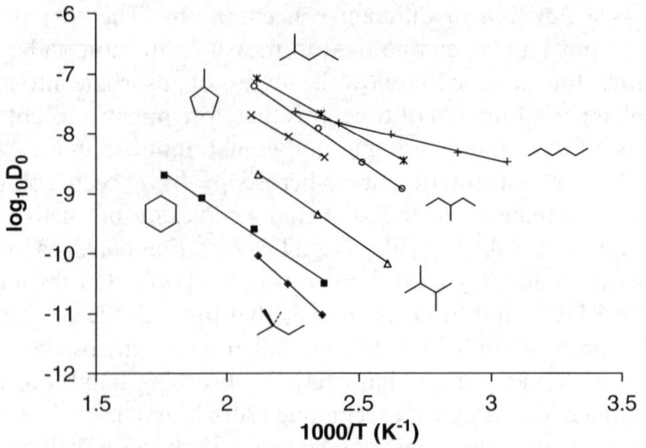

Figure 7.14 Arrhenius plots of temperature-dependent diffusivity coefficients of C6 alkanes within silicalite indicate that diffusion is increasingly hindered in the medium-pore channels as the degree of branching increases. Cyclohexane also diffuses very slowly. [Reproduced from reference 133 with permission. Copyright 1995 American Chemical Society.]

cycloalkanes can only diffuse through the straight channels, because their access to the sinusoidal channels is severely restricted due to repulsion from the channel walls. Molecular diffusivity effects are also important in the conversion of ethyl- and propylaromatics to toluene and xylene in zeolites that contain both medium- and large-pore channels. Whereas the larger (less valuable) alkylbenzenes reside preferentially in the large pores, where they react to give a mixture of products, the more valuable xylene and toluene products diffuse faster through the medium pores. In zeolites with high aluminium contents, and therefore high concentrations of charge-balancing cations, molecular diffusion is controlled by jumps between strong adsorption sites on the cations, as shown by combined MD and diffusion studies of benzene on zeolite X. Structural intergrowths and surface barriers also have an important effect on diffusion rates in crystals.[131]

7.5 Applications of Adsorption

The range of structure types and the tunable chemistry of zeolites makes them ideal for applications in adsorption, purification and separation.[98] The preparation of pure gases or the removal of impurities from gas mixtures is a very common requirement in metallurgical, energy-related and medical technologies and in scientific research. Two different process types can be identified: purifications, in which low levels of unwanted gases must be excluded, and separations, where much larger gas volumes must be adsorbed and removed. These have very different engineering requirements. In the first, two fixed beds can remain in place for an extended period of time, the first adsorbing impurities while the second, arranged in parallel, is degassed by heating in a flow of a

Table 7.6 Applications of zeolites in adsorption and separation.

Separation process	*Description of process*
Drying of solvents, hydrocarbon vapour streams and refrigerants	Na-A is typically used for drying, because of its high capacity for water adsorption and narrow pore openings.
Removal of other impurities (H_2O, HCl)	There is a need for specially designed adsorbents for halocarbon refrigerant fluids, to avoid decomposition of the zeolite.
Drying of syngas ($CO + H_2$) produced by reforming natural gas	Once CO_2 is largely removed by adsorption in basic solution, residual CO_2 and H_2O are removed by molecular sieves to <1 ppm
Hydrogen purification	CO_2 and N_2 removed non-cryogenically over molecular sieves
Air separation	H_2O, CO_2 and N_2 removed from air by non-cryogenic pressure swing adsorption over LiX zeolite, leaving $> 95\%$ O_2
Hydrocarbon trapping during engine cold starting	High silica zeolites adsorb unburnt hydrocarbons and desorb them as the engine and the catalytic converter warm up
Separation of normal from branched alkanes	Shape selective Ca-A zeolite (5A) adsorbs linear but not branched hydrocarbons
Separation of para-xylene from other xylenes and ethylbenzene	Shape selective MFI type zeolites (silicalite) adsorb *para*- but not *ortho*- or *meta*-xylenes. FAU type zeolites are also effective for this separation under simulated moving bed conditions

carrier gas, and these are then switched. In the second process, a continual process of adsorption and desorption from large adsorbent beds is performed by strong variation in pressure (Pressure Swing Adsorption, or PSA). In this way the least strongly adsorbed component of the gas mixture can be strongly enhanced. Adsorptions can also be performed cryogenically, with or without an adsorbent, but improved performance is offset against increased cost. Typical examples of applications of zeolite molecular sieves in adsorption and separation are given in Table 7.6.

7.5.1 Drying and Impurity Removal

The strong adsorption of water on extra-framework cations makes zeolites ideal drying agents. In particular, zeolite A has high cation content, large pore volume and small pores that prevent access to molecules larger than linear alkanes. Zeolite A, either as Na- or Ca- forms (4A or 5A) is therefore widely used in drying organic solvents and gas streams.

Zeolites are also used in the non-cryogenic purification of hydrogen by the adsorption of water, CO_2, N_2 and hydrocarbons and in the removal of residual CO_2 and H_2O from syngas mixtures prepared by reforming natural gas according to the reaction $CH_4 + H_2O \rightarrow CO + 3H_2$. These processes are important because hydrogen and syngas are widely used in petrochemical and chemical processes and for the synthesis of bulk chemicals such as methanol, alcohols and hydrocarbons *via* the Fischer–Tropsch process. Carbon dioxide removal by zeolites is also used in the purification of natural gas streams that contain small amounts of carbon dioxide as a natural impurity. (Higher contents are currently removed using liquid amine based absorption/regeneration technology.)

7.5.2 Air Separation

The separation of nitrogen and oxygen in air is performed both cryogenically and also by pressure swing adsorption at ambient temperatures, using pressures between zero and a few bar. Where pressures below 1 bar are used to accelerate desorption, the process is known as vacuum swing adsorption (VSA). Most separation of air to nitrogen and oxygen is performed cryogenically without adsorbents, accounting for around 80% of the total. Indeed, for very high purity and high volumes, this is the preferred technology. Nevertheless, pressure swing adsorption (PSA) can be run without recourse to cryogenic technology and so for applications that do not require high throughput or very high purity, adsorptive processes possess real advantages.[98]

To obtain nitrogen from air, adsorptive processes make use of microporous carbon as the molecular sieve, whereas for technologies giving oxygen, zeolites are used. In particular, zeolite LiX (Si/Al = 1) is the commercial adsorbent of choice. In both cases the adsorbent takes up the gases other than the one being produced much more strongly. The reason for the difference in adsorbent chosen for the desired gas lies in the different molecular mechanisms of adsorption. Properties of the gases are given in Table 7.7.

Microporous carbons act as true molecular sieves. The pore size of these is sufficiently small to adsorb the smaller oxygen and carbon dioxide molecules in preference to nitrogen, leaving nitrogen enriched (90–95%) in the product gases. When cationic zeolites are used, the nitrogen is adsorbed more strongly because the 'N_2 quadrupole–Li^+' and the 'induced N_2 dipole–Li^+' interactions are stronger than the corresponding interactions for O_2. As a result, the nitrogen is held more strongly and the product is enriched in oxygen to 95%, with much of the remainder being argon. PVA and VSA (Vacuum Swing Adsorption) methods account for around 5% of the world demand for oxygen.

The technology associated with pressure swing adsorption is well developed. Air at a pressure of several bar is brought into contact with large beds of LiX

Table 7.7 Physical properties of gases in air (2CLJQ potential model parameters).[36,137]

	Lennard-Jones σ/Å	*L-J energy/ kJmol^{-1}*	*Interatomic distance/Å*	*Quadrupole/ DÅ*	*Polarisability/ 10^{-25} cm^3*
N_2	3.321	0.291	1.0464	1.4397	17.4
O_2	3.106	0.359	0.9699	0.8081	15.8

adsorbent, preferentially adsorbing carbon dioxide and nitrogen and enriching the free gas in oxygen. The oxygen-rich gas is then contacted with fresh adsorbent, pressurised and the process repeated. The zeolite bed is isolated after each adsorption and the pressure reduced to 1 bar or below to allow the adsorbed gases to desorb and to regenerate the zeolite for the next cycle. No heating or cooling is required.

7.5.3 Hydrocarbon Adsorption and Separation

Zeolites are widely used to adsorb hydrocarbons, including unwanted volatile organic compounds (VOCs) such as solvent vapour and unburnt fuel in automobile engines. They are also used in refineries and petrochemical plants to separate and purify products on a large scale. The zeolites are chosen on the basis of cost, capacity, resistance to process conditions and, particularly in petrochemical applications, for their ability to act as true molecular sieves.

7.5.3.1 Adsorption

A recent example where hydrocarbon adsorption technology is becoming important is in auto-exhausts for gasoline engines. When the engine has been running and has reached its working temperature, any unburnt hydrocarbons are passed, together with the other pollutants, carbon monoxide and nitrogen oxides, over the auto-exhaust catalyst. This typically comprises platinum and rhodium supported on an alumina-coated ceramic monolith, although some cation-exchanged zeolites are also used. In the so-called three-way catalyst, nitric oxide acts as an oxidant for carbon monoxide and unburnt hydrocarbons so that the final exhaust contains only carbon dioxide, nitrogen and water. Under the conditions of a cold start, however, before the engine has warmed up, the unburnt hydrocarbons (and other pollutants) pass over a catalyst below its working temperature and conversion of pollutants will not occur. To prevent unwanted exhaust during this cold start regime, the nitrogen oxides, carbon monoxide and unburnt hydrocarbons must be trapped upstream of the catalyst and only released once it has achieved a temperature at which it is active. For the hydrocarbons, high silica zeolites such as Beta, US-Y or ZSM-5 are suitable adsorbents, because they are stable at the conditions of elevated temperatures, acidity and high water contents that prevail in the working engine. The performance of the novel high silica SSZ-33 in this application has also been reported.[138] The hydrocarbons are adsorbed at low temperatures and released as the temperature increases. Different materials, including metal oxides, are responsible for low temperature trapping of the other pollutant components.

7.5.3.2 Separation

Zeolites also have a key role as hydrocarbon adsorbents in the context of petrochemical production, in the separation of specific compounds or classes of compounds from complex mixtures that derive from the refining of crude oil fractions. Zeolites are able to separate components that have very similar boiling points that cannot be separated by fractional distillation, but whose different molecular dimensions enable them to be separated by molecular sieving. Two important examples of this kind are the separation of normal and iso-paraffins and the resolution of different isomers of xylene.

Small-pore zeolites such as zeolite A (usually as Ca-A, or 5A) are highly effective in adsorbing normal alkanes while not permitting branched alkanes access to the internal pore space. The reason for this is readily understood when comparing the molecular dimensions with the zeolite pore size. Removal of *n*-alkanes from gasoline fraction hydrocarbons increases the octane number for fuel applications. (This is higher for branched, cyclic and aromatic hydrocarbons.) In the MOLEX process, for example, linear paraffins are removed from a liquid hydrocarbon mixture by a zeolite adsorbent and desorbed using a liquid desorbent using UOP's SORBEX technology, which simulates a moving bed of adsorbent with a counter current flow of liquid feed (www.uop.com).

The separation of linear and branched alkanes is also of importance in the process known as dewaxing, in which the removal of normal alkanes makes the product hydrocarbon less viscous and reduces the so-called pour point temperature. Such processes can be combined with catalytic isomerisations to optimise the value of oil fractions (Chapter 8). Linear paraffins are also separated using a zeolite-based process from kerosene fractions to give reactants for the synthesis of linear alkylbenzene sulfonate anionic surfactants, which are both cost effective and biodegradable.

Medium-pore zeolites, such as those with the MFI structure, are particularly important in the separation of mixtures of monoaromatic hydrocarbons. *Para*-xylene is of great importance as a precursor to terephthalic acid and thereby to terephthalate polymers. However, ethylbenzene and the three isomers of xylene (*ortho*-, *meta*- and *para*-) cannot be separated on the basis of their boiling points. The slightly smaller kinetic diameter of *para*-xylene and its higher resultant diffusion rate in silicalite result in it being taken up with great specificity from such a mixture, and permits 99.9% *para*-xylene to be obtained. The other isomers that are left can subsequently be re-isomerised to give more *para*-xylene, or separated further (*meta*-xylene is the precursor to isophthalic acid, another important monomer). In fact, in addition to MFI type zeolites, cation-exchanged forms of synthetic faujasites are also used[139] in the separation of *para*-xylene from mixtures of xylenes using a modification of the SORBEX simulated moving bed technology referred to earlier. Examples of the industrial processes are those of UOP (the PAREX process[140]) and IFP (the ELUXYL process.[141])

7.5.4 True Molecular Sieving for Small Molecules

Whereas the separation of nitrogen and oxygen from air over cationic zeolites results from preferential adsorption, rather than molecular sieving, the well-defined pore structures of microporous framework solids do permit the separation of small gas molecules with similar size in the 3–4 Å region. Diffusivity of propane in the novel zeolite ITQ-12, for example, is shown to be around 10 000 times faster than that of propene at 30 °C.[142] Indeed, many small pore silicas show potential for the pressure swing separation of these two gases, currently run using energy intensive cryogenic distillation.

Some frameworks undergo framework distortions, for example upon cation exchange and migration and/or dehydration, which can be made use of to tune the pore size and modify the molecular sieving properties. For example, Kuznicki *et al.* demonstrate that by tailoring the degree of dehydration of the strontium-exchanged form of the titanosilicate ETS-4 (up to *ca.* 350 °C, above which the solid begins to lose crystallinity) the size of the 8MR window can be closely controlled over a range of 3.9–4.4 Å, so that effective separations of methane from ethane, nitrogen from methane or oxygen from nitrogen can be achieved. This fine-tuning has been called the 'molecular gate' effect.[143]

7.6 Summary

The ability of microporous solids to act as high-capacity molecular sieves has long been exploited in a wide range of applications in adsorption and separation. The electrostatic interactions of the traditional cationic forms of aluminosilicates are well suited for the uptake of polar molecules (such as H_2O) and are also able to separate oxygen from air. The development of microporous solids with varied chemistry has enabled adsorption and diffusion properties to be finely tuned for particular technologies. Pure silica zeolite polymorphs such as silicalite have particular importance, because they enable separation on the basis of a different range of polarity and on molecular size: the absence of aluminium in the framework also prevents the presence of unwanted acidity, so adsorbed hydrocarbons do not undergo any catalytic transformation.

Adsorption remains an important and fast-moving technology, as shown by new applications such as the storage and release of hydrocarbons during cold engine startup in auto-exhaust catalysis. Furthermore, there are potential future applications in hydrogen storage, where the huge adsorption capacities of the newly discovered MOF materials could be of great use, and in the storage and controlled release of small molecules of physiological importance. For larger molecules, such as complex drug molecules or proteins, mesoporous solids extend the working range in uptake, storage and release. Current developments in these areas are discussed in the final chapter.

Most of the above uses rely on physisorption rather than strong chemical bond formation. Chemisorption, however, is of particular importance in heterogeneous catalysis, where it is a necessary precursor to reaction. Chemisorption sites include Brønsted and Lewis acid sites as well as coordinatively unsaturated transition metal cations, either within the framework or as charge-balancing cations outside the frameworks. Their catalytic activity is discussed further in the following Chapters 8 and 9.

References

1. C. Baerlocher, W. M. Meier and D. H. Olson, *'Atlas of Zeolite Framework Types'*, Elsevier, Amsterdam, 2001.
2. J. J. Pluth and J. V. Smith, *J. Am. Chem. Soc.*, 1983, **105**, 1192.
3. J. J. Pluth and J. V. Smith, *J. Am. Chem. Soc.*, 1980, **102**, 4704.
4. A. Corma, F. Rey, J. Rius, M. J. Sabater and S. Valencia, *Nature*, 2004, **431**, 287.
5. J. A. Dunne, R. Mariwals, M. Rao, S. Sircar, R. J. Gorte and A. L. Myers, *Langmuir*, 1996, **12**, 5888.
6. M. E. Davis, C. Montes, P. E. Hathaway, J. P. Arhancet, D. L. Hasha and J. M. Garces, *J. Am. Chem. Soc.*, 1989, **111**, 3919.
7. M. A. Camblor, A. Corma and S. Valencia, *Chem. Commun.*, 1996, 2365.
8. A. Corma, M. Diaz-Cabanas, J. Martinez-Triguero, F. Rey and J. Rius, *Nature*, 2002, **418**, 514.
9. T. Blasco, A. Corma, M. J. Diaz-Cabanas, F. Rey, J. Rius, G. Sastre and J. A. Vidal-Moya, *J. Am. Chem. Soc.*, 2004, **126**, 13414.
10. M. Yoshikawa, P. Wagner, M. Lovallo, K. Tsuji, T. Takewaki, C. Y. Chen, L. W. Beck, C. Jones, M. Tsapatsis, S. I. Zones and M. E. Davis, *J. Phys. Chem. B*, 1998, **102**, 7139.
11. J. L. Paillaud, B. Harbuzaru, J. Patarin and N. Bats, *Science*, 2004, **304**, 990.
12. A. Corma, M. J. Diaz-Cabanas, J. L. Jorda, C. Martinez and M. Moliner, *Nature*, 2006, **443**, 842.
13. S. T. Wilson and E. M. Flanigen, *ACS Symp. Ser.*, 1989, **398**, 329.
14. P. A. Wright, M. J. Maple, A. M. Z. Slawin, V. Patinec, R. A. Aitken, S. Welsh and P. A. Cox, *J. Chem. Soc. Dalton Trans.*, 2000, **8**, 1243.
15. X. D. Zou, T. Conradsson, M. Klingstedt, M. S. Dadachov and M. O'Keeffe, *Nature*, 2005, **437**, 716.
16. S. M. Kuznicki, *US Patent*, 1990, **4**, 938–989.
17. P. Brandao, A. Philippou, A. Valente, J. Rocha and M. Anderson, *Phys. Chem. Chem. Phys.*, 2001, **3**, 1773.
18. N. Guillou, Q. Gao, P. M. Forster, J.-S. Chang, M. Nogues, S.-E. Park, G. Férey and A. K. Cheetham, *Angew. Chem. Int. Ed.*, 2001, **40**, 2831.
19. K. Maeda, Y. Kiyozumi and F. Mizukami, *J. Phys. Chem. B*, 1997, **101**, 4402.
20. S. R. Miller, P. A. Wright, C. Serre, T. Loiseau, J. Marrot and G. Férey, *Chem. Commun.*, 2005, 3850.
21. H. Li, M. Eddaoudi, M. O'Keeffe and O. M. Yaghi, *Nature*, 1999, **402**, 276.
22. H. K. Chae, D. Y. Siberio-Perez, J. Kim, Y. Go, M. Eddaoudi, A. J. Matzger, M. O'Keeffe and O. M. Yaghi, *Nature*, 2004, **427**, 523.
23. A. C. Sudik, A. P. Côté, A. G. Wong-Foy, M. O'Keeffe and O. M. Yaghi, *Angew. Chem. Int. Ed.*, 2006, **45**, 2528.
24. C. Serre, F. Millange, C. Thouvenot, M. Nogues, G. Marsolier, D. Louer and G. Férey, *J. Am. Chem. Soc.*, 2002, **124**, 13519.

25. K. Barthelet, J. Marrot, G. Férey and D. Riou, *Chem. Commun.*, 2004, 520.
26. G. Férey, C. Serre, C. Mellot-Draznieks, F. Millange, S. Surblé, J. Dutour and I. Margiolaki, *Angew. Chem. Int. Ed.*, 2004, **43**, 6296.
27. G. Férey, C. Mellot-Draznieks, C. Serre and F. Millange, *Acc. Chem. Res.*, 2005, **38**, 217.
28. Q. M. Wang, D. M. Shen, M. Bulow, M. L. Lau, S. G. Deng, F. R. Fitch, N. O. Lemcoff and J. Semanscin, *Micropor. Mesopor. Mater.*, 2002, **55**, 217.
29. D. N. Dybtsev, H. Chun and K. Kim, *Angew. Chem. Int. Ed.*, 2004, **43**, 5033.
30. K. S. Park, Z. Ni, A. P. Côté, J. Y. Choi, R. D. Huang, F. J. Uribe-Romo, H. K. Chae, M. O'Keeffe and O. M. Yaghi, *Proc. Natl. Acad. Sci. USA*, 2006, **103**, 10186.
31. J. A. Groves, S. R. Miller, S. J. Warrender, C. Mellot-Draznieks, P. Lightfoot and P. A. Wright, *Chem. Commun.*, 2006, 3305.
32. M. J. Kim and R. Ryoo, *Chem. Mater.*, 1999, **11**, 487.
33. J. S. Beck, J. C. Vartuli, W. J. Roth, M. E. Leonowicz, C. T. Kresge, K. D. Schmitt, C. T. W. Chu, D. H. Olson, E. W. Sheppard, S. B. McCullen, J. B. Higgins and J. L. Schlenker, *J. Am. Chem. Soc.*, 1992, **114**, 10834.
34. D. W. Breck, '*Zeolite Molecular Sieves*', Wiley, 1974.
35. J. Stoll, J. Vrabec and H. Hasse, *J. Chem. Phys.*, 2003, **119**, 11396.
36. J. Stoll, J. Vrabec and H. Hasse, *AIChE J.*, 2003, 49, 2187; J. Vrabec, J. Stoll and H. Hasse, *J. Phys. Chem. B*, 2001, 105, 12126.
37. M. Cook and W. C. Conner, in '*Proc. 12th Int. Zeol. Conf.*', Ed. M. M. J. Treacy, B.K. Marcus, M. E. Bisher, J. B. Higgins, Mat. Res. Soc., Warrendale, Pennsylvania, 1999, 409.
38. Y. Traa and J. Weitkamp, in *'Handbook of Porous Solids'*, Ed. F. Schuth, K. S. W. Sing and J. Weitkamp, Wiley-VCH, New York, 2002, 1015.
39. K. G. Strohmaier and D. E. W. Vaughan, *J. Am. Chem. Soc.*, 2003, **125**, 16035.
40. S. Brunauer, L. S. Deming, W. E. Deming and E. Teller, *J. Am. Chem. Soc.*, 1940, **62**, 1723.
41. S. J. Gregg and K. S. W. Sing, '*Adsorption, Surface Area and Porosity*', Academic Press, 1982.
42. S. Brunauer, P. H. Emmett and E. Teller, *J. Am. Chem. Soc.*, 1938, **60**, 309.
43. M. M. Dubinin and E. D. Zaverina, *Zh. Fiz. Khimii*, 1949, **23**, 1129.
44. M. M. Dubinin, *Carbon*, 1989, **27**, 457.
45. G. Horvath and K. Kawazoe, *J. Chem. Eng. Jpn.*, 1983, **16**, 470.
46. A. Saito and H. C. Foley, *Micropor. Mater.*, 1995, **3**, 531.
47. N. Dufau, N. Floquet, J.-P. Coulomb, P. L. Llewellyn and J. Rouquerol, *Stud. Surf. Sci. Catal.*, 2001, **135**, 2824.
48. L. J. Song, Z. L. Sun, H. Y. Ban, M. Dai and L. V. C. Rees, *Phys. Chem. Chem. Phys.*, 2004, **6**, 4722.
49. Y. Lee, T. Vogt, J. A. Hriljac, J. B. Parise and G. Artioli, *J. Am. Chem. Soc.*, 2002, **124**, 5466.

50. Y. Lee, T. Vogt, J. A. Hriljac, J. B. Parise, J. C. Hanson and S. J. Kim, *Nature*, 2002, **420**, 485.
51. J. M. Thomas and W. J. Thomas, *'Principles and Practice of Heterogeneous Catalysis'*, VCH, 1996.
52. B. C. Lippens, B. G. Linsen and J. H. de Boer, *J. Catal.*, 1965, **4**, 319.
53. G. D. Halsey, *J. Chem. Phys.*, 1948, **16**, 931.
54. E. P. Barrett, L. G. Joyner and P. H. Halenda, *J. Am. Chem. Soc.*, 1951, **73**, 373.
55. A. H. Janssen, A. J. Koster and K. P. de Jong, *Angew. Chem. Int. Ed.*, 2001, **40**, 1102.
56. A. H. Janssen, A. J. Koster and K. P. de Jong, *J. Phys. Chem. B*, 2002, **106**, 11905.
57. C. M. Yang, B. Zibrowius, W. Schmidt and F. Schuth, *Chem. Mater.*, 2004, **16**, 2918.
58. J. Fan, C. Z. Yu, L. M. Wang, B. Tu, D. Y. Zhao, Y. Sakamoto and O. Terasaki, *J. Am. Chem. Soc.*, 2001, **123**, 12113.
59. O. Terasaki, Z. Liu, T. Ohsuna, H. J. Shin and R. Ryoo, *Microsc. Microanal.*, 2002, **8**, 35.
60. P. I. Ravikovitch and A. V. Neimark, *Langmuir*, 2002, **18**, 1550.
61. P. L. Llewellyn and G. Maurin, *Compt. Rend. Chim.*, 2005, **8**, 283.
62. J. P. Joly and A. Perrard, *Langmuir*, 2001, **17**, 1538.
63. G. Maurin, P. L. Llewellyn and R. G. Bell, *J. Phys. Chem. B*, 2005, **109**, 16084.
64. M. J. Duer, *'Solid State NMR Spectroscopy: principles and applications'*, Oxford University Press, New York, 2002.
65. R. R. Eckman and A. J. Vega, *J. Phys. Chem.*, 1986, **90**, 4679.
66. D. F. Shantz and R. F. Lobo, *Topics In Catalysis*, 1999, **9**, 1.
67. F. H. Larsen, H. J. Jakobsen, P. D. Ellis and N. C. Nielsen, *Chem. Phys. Lett.*, 1998, **292**, 467.
68. M. S. Greenfield, A. D. Ronemus, R. L. Vold, R. R. Vold, P. D. Ellis and T. E. Raidy, *J. Magn. Reson.*, 1987, **72**, 89.
69. S. M. Auerbach, L. M. Bull, N. J. Henson, H. I. Metiu and A. K. Cheetham, *J. Phys. Chem.*, 1996, **100**, 5923.
70. L. M. Bull, N. J. Henson, A. K. Cheetham, J. M. Newsam and S. J. Heyes, *J. Phys. Chem.*, 1993, **97**, 11776.
71. I. Kustanovich, D. Fraenkel, Z. Luz, S. Vega and H. Zimmermann, *J. Phys. Chem.*, 1988, **92**, 4134.
72. I. Kustanovich, H. M. Vieth, Z. Luz and S. Vega, *J. Phys. Chem.*, 1989, **93**, 7427.
73. J. Gonzalez, R. N. Devi, P. A. Wright, D. P. Tunstall and P. A. Cox, *J. Phys. Chem. B*, 2005, **109**, 21700.
74. H. Jobic, *Curr. Opin. Solid State Mater. Sci.*, 2002, **6**, 415.
75. F. Geobaldo, C. Lamberti, G. Ricchiardi, S. Bordiga, A. Zecchina, G. T. Palomino and C. O. Arean, *J. Phys. Chem.*, 1995, **99**, 11167.
76. H. van Koningsveld, *J. Mol. Catal. A-Chem.*, 1998, **134**, 89.

77. I. Gameson, P. A. Wright, T. Rayment and J. M. Thomas, *Chem. Phys. Lett.*, 1986, **123**, 145.
78. I. Gameson, T. Rayment, J. M. Thomas and P. A. Wright, *J. Phys. Chem.*, 1988, **92**, 988.
79. J. B. Parise, J. A. Hriljac, D. E. Cox, D. R. Corbin and V. Ramamurthy, *J. Chem. Soc., Chem. Commun.*, 1993, 226.
80. Y. Kubota, M. Takata, R. Matsuda, R. Kitaura, S. Kitagawa, K. Kato, M. Sakata and T. C. Kobayashi, *Angew. Chem. Int. Ed.*, 2005, **44**, 920.
81. J. M. Adams and D. A. Haselden, *J. Solid State Chem.*, 1984, **55**, 209.
82. A. N. Fitch, H. Jobic and A. Renouprez, *J. Phys. Chem.*, 1986, **90**, 1311.
83. P. A. Wright, J. M. Thomas, A. K. Cheetham and A. K. Nowak, *Nature*, 1985, **318**, 611.
84. L. Smith, A. K. Cheetham, R. E. Morris, L. Marchese, J. M. Thomas, P. A. Wright and J. Chen, *Science*, 1996, **271**, 799.
85. C. Mellot-Draznieks, C. Serre, S. Surblé, N. Audebrand and G. Férey, *J. Am. Chem. Soc.* , 2005, **127**, 16273.
86. V. Bolis, C. Busco, S. Bordiga, P. Ugliengo, C. Lamberti and A. Zecchina, *Appl. Surf. Sci.*, 2002, **196**, 56.
87. E. N. Gribov, G. Sastre and A. Corma, *J. Phys. Chem. B*, 2005, **109**, 23794.
88. S. H. Jhung, H. K. Kim, J. W. Yoon and J. S. Chang, *J. Phys. Chem. B*, 2006, **110**, 9371.
89. J. Janchen, H. Stach, L. Uytterhoeven and W. J. Mortier, *J. Phys. Chem.*, 1996, **100**, 12489.
90. F. Eder and J. A. Lercher, *J. Phys. Chem.*, 1996, **100**, 16460.
91. C. L. Cavalcante and D. M. Ruthven, *Ind. Eng. Chem. Res.*, 1995, **34**, 177.
92. H. Van Koningsveld, F. Tuinstra, H. van Bekkum and J. C. Jansen, *Acta Crystallogr.*, 1989, **B45**, 423.
93. C. A. Fyfe, H. Strobl, G. T. Kokotailo, G. J. Kennedy and G. E. Barlow, *J. Am. Chem. Soc.*, 1988, **110**, 3373.
94. C. A. Fyfe, A. C. Diaz, H. Grondey, A. R. Lewis and H. Forster, *J. Am. Chem. Soc.*, 2005, **127**, 7543.
95. C. A. Fyfe, Y. Feng, H. Grondey, G. T. Kokotailo and H. Gies, *Chem. Rev.*, 1991, **91**, 1525.
96. C. A. Fyfe and D. H. Brouwer, *J. Am. Chem. Soc.*, 2006, **128**, 11860.
97. S. Savitz, A. L. Myers and R. J. Gorte, *Micropor. Mesopor. Mater.*, 2000, **37**, 33.
98. T. R. Gaffney, *Curr. Opin. Solid State Mater. Sci.*, 1996, **1**, 69.
99. C. F. Mellot, A. K. Cheetham, S. Harms, S. Savitz, R. J. Gorte and A. L. Myers, *J. Am. Chem. Soc.*, 1998, **120**, 5788.
100. J. Janchen, D. Ackermann, H. Stach and W. Brosicke, *Solar Energy*, 2004, **76**, 339.
101. A. Zecchina and C. O. Arean, *Chem. Soc. Rev.*, 1996, **25**, 187.
102. J. Fraissard and T. Ito, *Zeolites*, 1988, **8**, 350.
103. J. A. Ripmeester, *J. Am. Chem. Soc.*, 1982, **104**, 289.

104. G. Spoto, A. Zecchina, S. Bordiga, G. Ricchiardi, G. Martra, G. Leofanti and G. Petrini, *Appl. Catal.*, 1994, **38**, 150.
105. F. Eder, M. Stockenhuber and J. A. Lercher, *J. Phys. Chem. B*, 1997, **101**, 5414.
106. R. Schumacher and H. G. Karge, *J. Phys. Chem. B*, 1999, **103**, 1477.
107. D. Gleeson, G. Sankar, C. R. A. Catlow, J. M. Thomas, G. Spano, S. Bordiga, A. Zecchina and C. Lamberti, *Phys. Chem. Chem. Phys.*, 2000, **2**, 4812.
108. T. Blasco, M. A. Camblor, A. Corma and J. Perez-Pariente, *J. Am. Chem. Soc.*, 1993, **115**, 11806.
109. T. Blasco, M. A. Camblor, A. Corma, P. Esteve, J. M. Guil, A. Martinez, J. A. Perdigon-Melon and S. Valencia, *J Phys Chem B*, 1998, **102**, 75.
110. F. Bonino, A. Damin, G. Ricchiardi, M. Ricci, G. Spano, R. D'Aloisio, A. Zecchina, C. Lamberti, C. Prestipino and S. Bordiga, *J. Phys. Chem. B*, 2004, **108**, 3573.
111. C. Prestipino, F. Bonino, S. Usseglio, A. Damin, A. Tasso, M. G. Clerici, S. Bordiga, F. D'Acapito, A. Zecchina and C. Lamberti, *Chem. Phys. Chem.*, 2004, **5**, 1799.
112. P. A. Barrett, G. Sankar, R. H. Jones, C. R. A. Catlow and J. M. Thomas, *J. Phys. Chem. B*, 1997, **101**, 9555.
113. F. Yagi, H. Tsuji and H. Hattori, *Micropor. Mater.*, 1997, **9**, 237.
114. D. K. Murray, J. W. Chang and J. F. Haw, *J. Am. Chem. Soc.*, 1993, **115**, 4732.
115. D. K. Murray, T. Howard, P. W. Goguen, T. R. Krawietz and J. F. Haw, *J. Am. Chem. Soc.*, 1994, **116**, 6354.
116. F. Stallmach, S. Groger, V. Kunzel, J. Karger, O. M. Yaghi, M. Hesse and U. Muller, *Angew. Chem. Int. Ed.*, 2006, **45**, 2123.
117. A. J. Fletcher, K. M. Thomas and M. J. Rosseinsky, *J. Solid State Chem.*, 2005, **178**, 2491.
118. M. Edgar, R. Mitchell, A. M. Z. Slawin, P. Lightfoot and P. A. Wright, *Chem-Eur J.*, 2001, **7**, 5168.
119. R. Matsuda, R. Kitaura, S. Kitagawa, Y. Kubota, T. C. Kobayashi, S. Horike and M. Takata, *J. Am. Chem. Soc.*, 2004, **126**, 14063.
120. A. J. Fletcher, E. J. Cussen, T. J. Prior, M. J. Rosseinsky, C. J. Kepert and K. M. Thomas, *J. Am. Chem. Soc.*, 2001, **123**, 10001.
121. S. Bourrelly, P. L. Llewellyn, C. Serre, F. Millange, T. Loiseau and G. Férey, *J. Am. Chem. Soc.*, 2005, **127**, 13519.
122. X. B. Zhao, B. Xiao, A. J. Fletcher, K. M. Thomas, D. Bradshaw and M. J. Rosseinsky, *Science*, 2004, **306**, 1012.
123. J. Karger and D. M. Ruthven, '*Diffusion in Zeolites and Other Microporous Materials*', Wiley and Sons, 1992.
124. J. Karger and S. Vasenkov, *Micropor. Mesopor. Mater.*, 2005, **85**, 195.
125. S. Brandani, M. A. Jama and D. M. Ruthven, *Chem. Eng. Sci.*, 2000, **55**, 1205.
126. L. J. Song and L. V. C. Rees, *Micropor. Mesopor. Mater.*, 2000, **35–6**, 301.

127. U. Schemmert, J. Karger and J. Weitkamp, *Micropor. Mesopor. Mater.*, 1999, **32**, 101.
128. E. Lehmann, C. Chmelik, H. Scheidt, S. Vasenkov, B. Staudte, J. Karger, F. Kremer, G. Zadrozna and J. Kornatowski, *J. Am. Chem. Soc.*, 2002, **124**, 8690.
129. O. Geier, S. Vasenkov, E. Lehmann, J. Karger, U. Schemmert, R. A. Rakoczy and J. Weitkamp, *J. Phys. Chem. B*, 2001, **105**, 10217.
130. P. Kortunov, P. A. Wright, J. Karger and M. Castro, *in preparation.*
131. C. Chmelik, P. Kortunov, S. Vasenkov and J. Karger, *Adsorption*, 2005, **11**, 455.
132. J. Wloch, *Micropor. Mesopor. Mater.*, 2003, **62**, 81.
133. C. L. Cavalcante and D. M. Ruthven, *Ind. Eng. Chem. Res.*, 1995, **34**, 185.
134. J. Karger and D. M. Ruthven, in *'Handbook of Porous Solids'*, Ed. F. Schuth, K. S. W. Sing and J. Weitkamp, Wiley-VCH, New York, 2002, 2089.
135. V. Kukla, J. Kornatowski, D. Demuth, I. Gimus, H. Pfeifer, L. V. C. Rees, S. Schunk, K. K. Unger and J. Karger, *Science*, 1996, **272**, 702.
136. L. J. Song, Z. L. Sun and L. V. C. Rees, *Micropor. Mesopor. Mater.*, 2002, **55**, 31.
137. W. Jia and S. Murad, *J. Chem. Phys.*, 2004, **120**, 4877.
138. S. P. Elangovan, M. Ogura, M. E. Davis and T. Okubo, *J. Phys. Chem. B*, 2004, **108**, 13059.
139. V. Cottier, J.-P. Bellat, M.-H. Simonot-Grange and A. Méthivier, *J. Phys. Chem. B*, 1997, **101**, 4798.
140. D. B Broughton, R. W. Neuzil, J. M. Pharis and C. S. Brearley, *Chem. Eng. Prog.*, 1970, **66**, 70.
141. D. Pavone and G. Hotier, *Oil Gas Sci. and Technol.*, 2000, **55**, 437.
142. P. A. Barrett, T. Boix, M. Puche, D. H. Olson, E. Jordan, H. Koller, M. A. Camblor, *Chem. Commun.*, 2003, 2114; D. H. Olson, X. B. Yang and M. A. Camblor, *J. Phys. Chem. B*, 2004, **108**, 11044.
143. S. M. Kuznicki, V. A. Bell, S. Nair, H. W. Hillhouse, R. M. Jacubinas, C. M. Braunbarth, B. H. Toby and M. Tsapatsis, *Nature*, 2001, **413**, 652.

CHAPTER 8

Microporous Solid Acid Catalysts and their Applications

8.1 Introduction

Solid acid catalysts play a crucial role in the petrochemical industry, where they have largely replaced liquid acids in hydrocarbon transformations. They are also finding increasing use in the production of commodity organic chemicals and the syntheses of fine chemicals. The acid forms of aluminosilicate zeolites are pre-eminent among them, because their unique combination of thermal stability, pore geometry and acid site strength enables good control of product selectivity in complex reactions.

As well as possessing a similar fundamental chemistry to acids in solution, including the classes of reaction that they catalyse, solid acids possess special features.[1] In particular they are convenient to handle and they may be used at high temperatures. Their acid strength is controlled to a high degree by the arrangement and chemical composition of the atoms in the local environment of their acid sites, which may be Brønsted or Lewis in type (as described in the previous chapter). The site geometry is dictated in zeolites and related microporous solids by their long-range crystal architecture, which underlines why these solids are so attractive for study – their catalytic activity is controlled directly by their structure, which is readily measured by diffraction and spectroscopy and modelled computationally.

The concentration, types and strength of accessible acid sites have been studied by experimental and computational methods. Spectroscopic and adsorption measurements using probe molecules have been used to establish a structural and thermodynamic basis for acid site behaviour while catalytic test reactions give information on the effect of pore geometry and acid site strength on activity and selectivity in reactions of direct relevance to applications. The controlling role of structural chemistry in the acid behaviour of microporous solids has been established from these data, while a theoretical basis is being established by quantum mechanical modelling.

Although the acid site strength of microporous solids is important in their use as catalysts, it is not their most important attribute – other solids have stronger acidity and for certain reactions are preferred for that reason. For microporous solids, it is the product selectivity that they impart, particularly in complex hydrocarbon transformations, that makes them so valuable. This shape selectivity, as it is generally termed, is a consequence of the geometry of the channels and cages from which the active sites are accessed. In a manner analogous to that of enzymes, the combination of active site, local environment and the available pathways to and from the particles' external surface controls the product selectivities, often to product distributions well away from gas phase thermodynamic equilibrium.

Solid acids catalyse a wide range of conversions, varying from reactions of molecules containing heteroatoms such as oxygen or nitrogen, which generally occur under mild conditions, through reactions involving alkenes and aromatics, to the catalytic cracking of saturated hydrocarbons at elevated temperatures.[2] In this chapter the catalytic behaviour of microporous solid acids is discussed according to reactant type. The reason why microporous solids, particularly zeolites, have been so widely studied is that several of these catalytic reactions have been scaled up into important industrial applications in petroleum refining and petrochemical conversions.[3] Details of some of the more important processes are given in this chapter. Zeolitic solid acids also show potential utility in a range of fine chemicals syntheses, particularly for the synthesis of flavours and fragrances, and in the preparation of precursors in pharmaceuticals and liquid crystal chemicals. The application of porous solid acids in industrial catalysis that started with their use in fluidised catalytic cracking back in the 1950s is therefore highly successful, with a high degree of interplay between industry and academia. The level of understanding of the structural control of acid catalysis in these materials rivals any structure–property relationship in materials chemistry. Moreover, the story is not finished, and the prospects for development of new materials and reactions for a changing (if relatively mature) market remain encouraging.

8.2 The Chemistry of Acid Catalysis

Solid acids accelerate reactions *via* the same general mechanism as acids in solution.[4] For Brønsted acid catalysis, the key step is protonation, which lowers the activation energy barrier to further reaction. Ease of protonation decreases in the order: oxygen- and nitrogen-containing compounds > alkenes and aromatics > alkanes, as the molecules' basicity decreases. As a consequence, reactions involving oxygen-containing compounds (such as esterification and etherification) or alkenes (such as isomerisation, oligomerisation and alkylation of aromatics) proceed at lower temperatures than catalytic conversions of alkanes. Lewis acids, which accept electron density, reduce activation energies by polarising molecules to make them more reactive, and often catalyse the same types of reaction as Brønsted acids. In general, Brønsted acid catalysis over microporous materials is the better understood.

In Brønsted acid catalysis of reactions involving heterocompounds, the important first step is protonation of the heterogroup, making the molecule more reactive. Protonated carbonyl groups are more susceptible to nucleophilic attack, for example, whereas other positively charged species can act as electrophilic reagents. Many excellent general texts describe acid catalysis of this sort.[4] For zeolitic solid acids, examples of catalysed reactions of this type include isomerisations such as the Beckmann rearrangement and a range of nucleophilic substitution and addition reactions, including etherification and dehydration of alcohols, reactions of ammonia with alcohols and addition and elimination reactions on carbonyl-containing compounds (see later). A scheme for Brønsted acid-catalysed nucleophilic addition is shown (Scheme 8.1) in which protonation of the carbonyl oxygen activates the carbonyl group to nucleophilic substitution. Lewis acid catalysts also activate the carbonyl group, by interacting with the carbonyl oxygen and enhancing the polarity of the bond.

Much of the mechanistic understanding of acid catalysis of hydrocarbon reactions stems from work on superacids.[4] Superacids are stronger proton donors than concentrated sulfuric acid and usually consist of the acid form of a highly stable anion. They are sufficiently strong proton donors to protonate very weak bases, including alkenes, in solution. The resulting carbenium ions have lifetimes that are sufficiently long for their chemistry to be studied. The reactivities of carbenium ions are therefore well understood. Selected reactions of carbenium ions are given in Table 8.1 and include rearrangement, scission, hydride abstraction, oligomerisation and alkylation. The product distributions of reactions proceeding through carbenium ion pathways are strongly influenced by the relative stabilities of the different cations, which increase in the order primary < secondary < tertiary. The energy difference between each of these is *ca.* 60–70 kJ mol^{-1}.[4] The generation of products directly from carbenium ions results from the return of the proton to the acid site and desorption of the resulting olefin or by hydride abstraction to give alkanes. The elementary steps of Table 8.1 (or closely related ones) are thought to be responsible for hydrocarbon reactions such as isomerisation, alkylation and cracking.

Microporous solids are not superacidic. As a result, the only unambiguous observations of carbenium ions on zeolites are of strongly stabilised aromatic and cyclic species observed by solid state NMR, such as cyclopentenyl, indanyl and pentamethylbenzenium ions observed during *in situ* studies of the reaction of methanol (see Sections 8.5.1 and 8.6.1). The nature of the intermediates

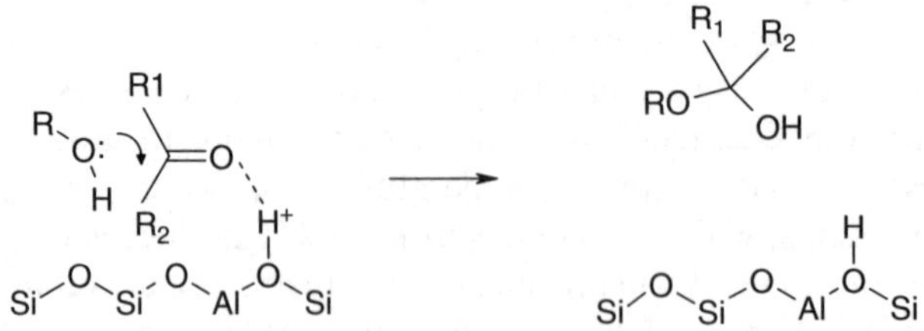

Scheme 8.1 Acid-catalysed nucleophilic substitution.

Table 8.1 Selected reactions of carbenium ions relevant to acid catalysis over solid acids.

Reaction	*Associated carbenium ion reaction*	*Hydrocarbon transformation*
1,2-Alkyl shift		Branching isomerisation
Aromatic alkyl shift	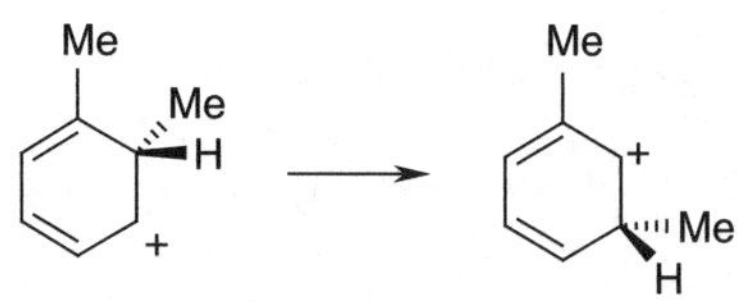	Xylene isomerisation
Hydride abstraction		Hydrogen transfer
β-scission		Cracking
Addition to aromatics	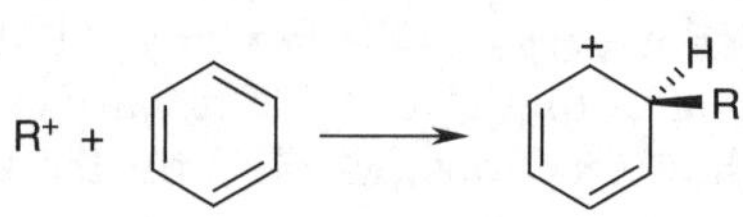	Alkylation of aromatics
Addition to alkenes	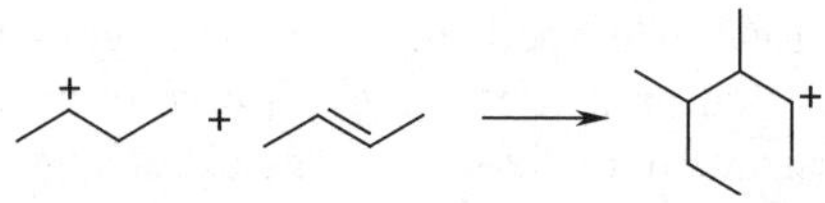	Oligomerisation
Addition to alkanes	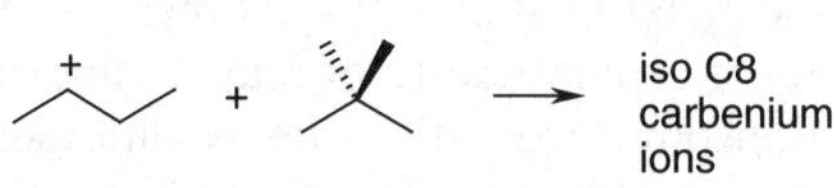	Alkylation of alkanes

formed by interaction with protons at the surfaces of solid acids in most hydrocarbon reactions therefore remains open to debate. However, quantum mechanical calculations suggest that the most likely reaction mechanisms pass through transition states similar to carbenium ions, which are formed by protonation of alkenes and stabilised by interaction with framework oxygen atoms (as discussed further in Section 8.5). Carbenium ion chemistry is therefore thought to be relevant over microporous solid acids.

Scheme 8.2 Protonated cyclopropane pathway proposed for carbenium ion branching isomerisation.

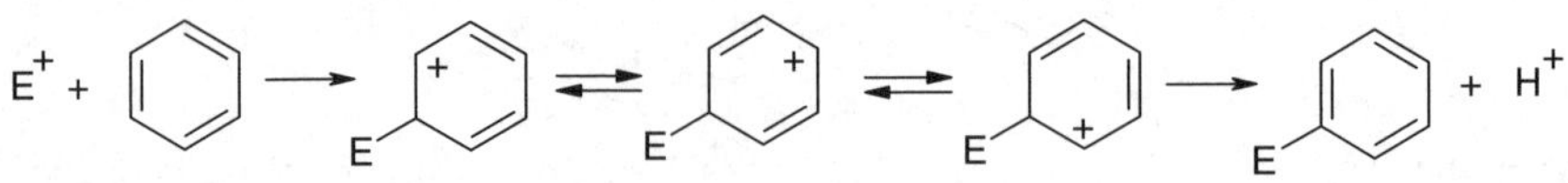

$E^+ = R^+, RCO^+$, etc

Scheme 8.3 Electrophilic aromatic substitution.

Carbenium ions can rearrange by hydride shifts (in which the charge moves along the framework), alkyl shifts (in which the degree of branching remains the same but its position changes) and branching isomerisation (where the degree of branching increases – sometimes described as skeletal isomerisation). Whereas hydride shifts are facile, branching isomerisation is thought to proceed *via* a protonated cyclopropane ring (Scheme 8.2) and requires a higher activation energy. For cases such as the branching isomerisation of an *n*-butyl cation to the iso-butyl cation, the mechanism passes through a primary carbenium ion and so requires a more strongly acidic catalyst than skeletal isomerisation of *n*-pentyl cations.

Carbenium ions can also undergo β-scission to yield alkenes and another, smaller, carbenium ion. Routes that give stable tertiary carbenium ions in the products are favoured. These reactions, which form the key step in cracking reactions, result in an increase in entropy and are favoured at high temperatures. The reverse reaction, the reaction of carbenium ion and alkene, is therefore favoured at lower temperature. Such oligomerisation reactions rapidly propagate at low temperatures to give polymeric species. Carbenium ions are also able to abstract hydride ions from other hydrocarbons, such as alkanes, to generate other carbenium ions in a chain process. Finally, they can alkylate aromatics, or, with more difficulty, alkanes. Alkylation and other zeolite-catalysed reactions of aromatics are of major importance.

Electrophilic substitution onto aromatic rings is catalysed by strong acids, which may donate protons directly or generate positively charged electrophilic species by protonation, hydride abstraction or cleavage of polar groups. Examples include carbenium ions, as mentioned above, and also protonated alcohols and acylium ions. Substituted aromatics are observed to undergo acid-catalysed alkyl shift isomerisation, *via* protonated intermediates. These intermediates can also react with other substituted aromatics to give transalkylation or addition products. A general mechanism is depicted below for electrophilic substitution (Scheme 8.3). The rate of reaction depends strongly on the presence and position of substituents on the ring, although in zeolitic catalysts the product distribution is also determined by shape selective effects.

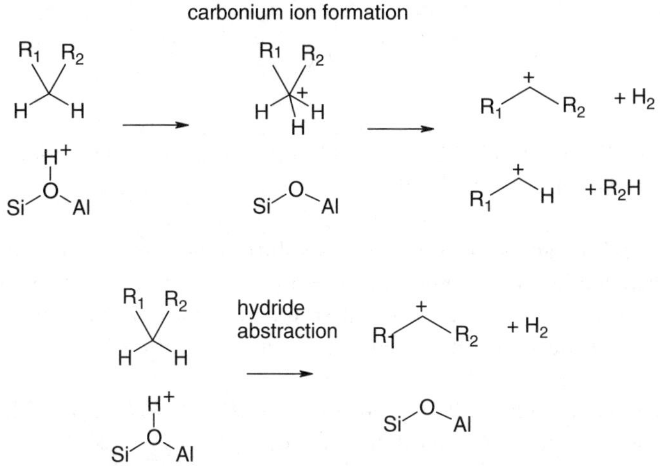

Scheme 8.4 Postulated routes to alkane activation over solid acids.

The mechanism of alkane activation on zeolites remains open to discussion. The initial step is thought to be either (1) protonation to yield a highly unstable and short-lived carbocation-like transition state containing 5-coordinate carbon (a carbonium ion, analogous to those inferred to be generated by interaction with superacids[5]) followed by loss of hydrogen or an alkane to give a carbenium-ion-like transition state, or (2) the direct generation of a carbenium-ion-like transition state by hydride abstraction (Scheme 8.4 and Section 8.7.2.5). Either way, the subsequent reaction is through reactions typical of carbenium ions.

8.3 General Features of Solid Acids

8.3.1 Acid Site Type, Concentration and Strength

Many features of acid sites in solids closely parallel those in solution. Materials that are Brønsted acids (proton donors) and Lewis acids (electron acceptors) are known. Typically, Brønsted acid sites in mixed metal oxides occur where the oxygen atoms of a hydroxyl attached to a metal atom of one kind are coordinating (acting as a Lewis base) to atoms of a different kind, resulting in the proton becoming more acidic (Scheme 8.5). Lewis acid sites are usually associated with incompletely coordinated cations that can accept electrons. The criteria for solid acidity are widely met by mixed metal oxides and oxo-salts, the varied chemistry of which gives rise to solids with acidities up to that of concentrated sulfuric acid and, in some cases, beyond.

In general terms, Brønsted acidity is defined in terms of the equilibrium between an acid, HA and a base, B:

$$HA + B \Leftrightarrow HB^{+} + A^{-}$$

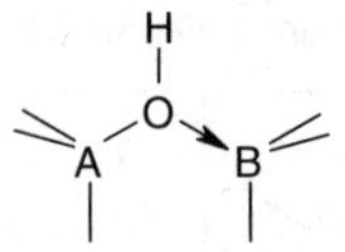

Scheme 8.5 Enhancement of acidity of AOH group by coordination of oxygen atom to adjacent metal cation B.

The strength of acids in non-aqueous solution can be determined by investigating their tendency to donate protons to bases of known pK_a values which have different colours in the protonated and unprotonated forms. UV-visible spectroscopy of the solution is then able to give the ratio of these forms of the indicator molecules.

For the indicator bases the equilibrium $B + H^+ \Leftrightarrow BH^+$ is established. The Hammett acidity value, H_o,[6] is then defined:

$$H_o = pK_{BH} - \log_{10}\frac{[BH^+]}{[B]} \quad \text{where} \quad K_{BH} = \frac{[H^+][B]}{[BH^+]}$$

pK_{BH} is increasingly negative for weaker bases: p-nitroaniline, $pK_{BH} = 1.0$; p-nitrotoluene, $pK_{BH} = -1.4$; 2,4-dinitrotoluene, $pK_{BH} = -13.8$. Adding an indicator to an acid, the ratio $[BH^+]/[B]$ will approach unity if the Hammett acidity H_o is equal to pK_{BH} (from the above equation). H_o is then approximately given by the pK_{BH} of the weakest base that the acid is able to protonate. For strong acids of different acid strength the indicator will be in large part protonated if $pK_{BH} < H_o$, so that a weaker base (with lower pK_{BH}) will be required to distinguish between them. By definition, superacids have an acidity stronger than concentrated sulfuric acid, and on the Hammett scale, $H_o(H_2SO_4) = -12$.

The use of Hammett indicators and the H_o value is therefore a reasonable approach to extend concepts of pH and pK_a to strong acids in non-aqueous solutions. However, application to microporous solid acids is problematical. On a practical level, the measurement of the intensity of UV-visible absorption on solids is difficult, although this can be circumvented by using NMR rather than UV-visible spectroscopy to measure the degree of protonation (Section 8.4.2.3). More seriously, the approach requires that the indicators have unhindered access to the protons, a situation that is not the case for large base molecules in any but the largest-pore solids. In addition, the energetics of formation of the protonated form, and therefore the position of equilibrium, is likely to depend to a large extent on interactions with the framework atoms in the vicinity of the acid site. Furthermore, the acid strength relevant to catalysis on microporous solids is not usually that in solution, but rather that in gas/solid reactions. Consequently, alternative techniques have been developed that are more applicable to solids and that enable their acidities to be compared with other solids' acidity and with those of mineral acids with known H_o values.

Nevertheless, the concept of Hammett acidity can be of some use. Bases with the values of -11.4 adsorbed on the large-pore zeolite H-Y, for example, are

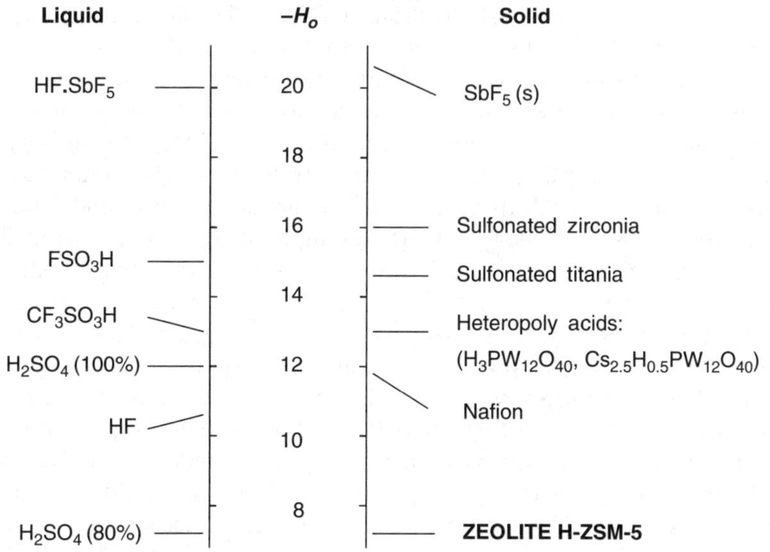

Figure 8.1 Comparison of the acidity of selected liquid and solid acids on the Hammett acidity scale. $H_o = -12$ for a superacid. (See text for an explanation of H_o.) [Reproduced from reference 1 with permission. Copyright 2002 American Chemical Society.]

not protonated, so the zeolite is not superacidic. In addition, some idea of the relative strengths of different types of solid acid can be obtained from a comparison of estimated H_o values (Figure 8.1).

The most obvious differences between solid and liquid acids are in their physical properties. Solids can be heated, which enhances the rate of proton transfer reactions which are slow at room temperature, can be used in solid-liquid and solid-gas reactions and can readily be separated from reactants and products. One of their limitations, however, is that the catalyst can become covered in strongly adsorbed by-product, or at high temperatures by carbonaceous residue, 'coke', resulting in deactivation. In this case, the utility of the catalyst may ultimately be determined by how readily it can be regenerated.

8.3.2 Microporous Solids as Acid Catalysts

The acid forms of aluminosilicate zeolites have found wider use as acid catalysts than any other materials.[7,8] Their outstanding utility derives from their relatively high acid strength, their high hydrothermal stability, their ability to impart shape selectivity to product distributions and the reproducibility with which they can be synthesised and modified. Each of these advantages stems directly from their crystalline structure. The two basic types of acid site types in microporous solids are Brønsted, which are protons located at bridging sites (Si–O–Al in zeolites, M–O–P in aluminophosphates) and Lewis, usually incompletely coordinated metal cations (especially aluminium in zeolites) in framework or extra-framework positions. Of these, the chemical structure and behaviour of the Brønsted sites is much better understood.

As described previously, zeolites are usually converted to the protonic form by calcining ammonium-exchanged or, in the case of templated varieties, the as-prepared materials. However, not all zeolites can be prepared as the acid form in this way, since zeolites with very high framework aluminium contents become unstable to hydrolysis once ammonia is removed and lose their crystallinity. Zeolites A, X and L, for example, cannot be prepared in the protonic form. Zeolites with moderate or high Si/Al framework ratios can usually be prepared as acids, and forms of H-Y that are widely used in catalysis can be prepared by processes involving the modification of framework composition (Chapter 6). This process can result in migration of some framework aluminium into extra-framework sites, and increase of the Si/Al ratio of the framework without loss of crystallinity and the generation of secondary mesoporosity that improves molecular transport to and from active sites in the microporous regions (Chapter 6). These ultrastabilised solids possess both Lewis and Brønsted acid sites and are the catalysts of choice where reaction and regeneration conditions are severe.

Catalytic properties of the acid forms of most zeolites have been reported, either in the open literature or in patents. In practice, though, only a handful are used in large-scale industrial applications of acid catalysis, including zeolites Y, mordenite, ferrierite, Rho and the high silica materials ZSM-5 and Beta. Many of the more recently discovered high silica solids require relatively expensive organic templates for their synthesis, and have made a more modest impact so far because of their relatively high cost. The zeolite MCM-22, which is prepared using a simple cyclic amine, hexamethyleneimine, is one of the more promising of the new materials for this reason. Among the germanium-containing silicates, many of which have novel and catalytically attractive structures, there is sufficient stability to perform test reactions, but ultimate applicability will require similar frameworks to be made in aluminosilicate form.[9]

Analogues of aluminosilicate zeolites, in which the place of aluminium is taken by other trivalent cations such as B^{3+}, Ga^{3+} or Fe^{3+}, are readily prepared. These give solids with closely similar structure but with weaker acid sites that are also more susceptible to removal of the heteroatom from the framework, for example in the presence of water or steam. Such weakly acidic solids can be selective catalysts for molecular rearrangements that give unwanted by-products in the presence of stronger acid catalysts. The Beckmann rearrangement is such a reaction (see Section 8.6.1).

Few of the other families of microporous solids can compete with zeolites in acid catalysis. The aluminophosphates display a great variety of novel structure types, and Brønsted acidity can be introduced through substitution during synthesis of framework aluminium by divalent metal cations or of phosphorus by silicon. Subsequent removal of the organic template gives the acid form. Great interest was aroused by the discovery of the extra-large-pore VPI-5, the 18-ring channels of which are larger than those of any zeolite. The attractive possibility of performing industrially significant acid-catalysed transformations of hydrocarbons with large kinetic diameters in VPI-5 is out of reach, however, because VPI-5, like many of the larger-pore aluminophosphates, lacks hydrothermal stability. Nevertheless, some of the aluminophosphate-based solids do show useful acidic catalytic performance. Metal and silicon-substituted forms of the small-pore $AlPO_4$-34 and $AlPO_4$-18, the medium-pore $AlPO_4$-11 and $AlPO_4$-31 and the large-pore $AlPO_4$-5 and $AlPO_4$-36 are the most promising. In particular, SAPO-34 has been found to be an active catalyst in the methanol-to-olefins reaction, suitable preparations giving high selectivities to ethene and propene, and SAPO-11 and SAPO-31 have been shown to be attractive for shape selective reactions such as catalytic isodewaxing, in which straight chain alkanes are selectively isomerised to monobranched rather than multibranched isomers, giving hydrocarbon mixtures with more convenient viscosities.[10]

Among the other families of microporous solids, the titanosilicate ETS-10 shows poor structural stability in the acid form, and is more active as a basic catalyst.[11] Most of the open framework metal phosphates other than aluminium phosphates lose crystallinity upon template removal (some nickel phosphates excepted). The solid acidity among MOF-type microporous solids remains largely unexplored, but their inherent instability to high-temperature regeneration in air means that they will not be important for high-temperature hydrocarbon transformations.

8.3.3 The Competitors: Other Classes of Solid Acids

In selecting a solid acid to catalyse any given reaction, zeolites are not the only candidates. Many solids fulfil the general requirement for solid acidity and some possess favourable attributes of enhanced acid strength and resistance to deactivation. Some of the more important varieties are discussed below.

Some ion-exchange resins can be prepared in the protonic form. Organic resins include 'Nafion®' in which strongly acidic perfluorinated sulfonate groups are attached to a cross-linked polymer which can take up molecules from solution which can gain access to the acid centres. The acid sites are strong, but the use of these materials is limited to reactions in solution and to relatively low temperatures. The polymer backbone has low thermal stability, so that regeneration of blocked catalyst is not possible. Despite these limitations, ion-exchange resins are widely used in catalysis,[12] particularly for reactions such as hydration and etherification that proceed rapidly at low temperatures.

Many mineral acid catalysts that are active in homogeneous catalysis can be 'heterogenised' by supporting them on inorganic oxides.[1] Strongly acidic heterogeneous catalysts are prepared by supporting Brønsted acids such as trifluorosulfonic (triflic) acid, sulfuric acid and phosphoric acid, and Lewis acids such as BF_3 and SbF_5, on high-surface-area oxides such as silica, alumina and zirconia. In particular, supported phosphoric acid on silica is widely used for the hydration of ethylene and the production of cumene (isopropylbenzene) by alkylation, and $BF_3/\gamma\text{-}Al_2O_3$ and H_2SO_4/ZrO_2 possess acidic sites that enable them to perform reactions that other solids are not acidic enough to catalyse. Supported acids are difficult to characterise, however, and are highly dependent on the methods and materials used in their preparation. Furthermore, although an improvement on homogeneous systems, their preparation requires the handling of corrosive acids. Nevertheless, they currently offer the best alternative to zeolites for reactions that require very strong and even superacidity, or where deactivation due to coking within the confined zeolite pores is a severe problem, such as the alkylation of alkanes.

Solids that are mixtures of two or more metal oxides, such as $SiO_2\text{-}Al_2O_3$ and $Al_2O_3\text{-}P_2O_5$, typically possess acid sites stronger than those found in the pure oxides. Amorphous silica-aluminas were among the first solid acids employed in the petrochemicals industry and are still used in hydrocarbon cracking. They possess both Brønsted and Lewis acid sites, but they are not strongly acidic, and so have been largely superseded by zeolites in reactions requiring high acid strength and thermal stability or shape selectivity. Sol-gel routes in the presence of micelle-forming surfactants can prepare well-ordered mesoporous silicas and silica-aluminas (the so-called MCM-41 type materials). These have revived interest in this field, because solids with regular pores with dimensions of 2–10 nm, and with surface areas in excess of $1000\,m^2\,g^{-1}$, can be prepared. Although these do not show the high thermal stability and acid strength required for cracking catalysts, they do permit acid-catalysed transformations of molecules larger than 8 Å in dimension that take place under milder conditions.[7]

Heteropolyoxometallate compounds are a class of metal salts that may be prepared in the acid form and used as homogeneous and heterogeneous acid catalysts. The anions of these salts are formed by the condensation of at least two different kinds of oxo-anions. Heteropolyanions with the Keggin structure consist of a central PO_4 tetrahedron surrounded by twelve WO_6 octahedra ($PW_{12}O_{40}{}^{3-}$). Their negative charge is balanced by cations, including hydrated protons. A very wide range of elements can be substituted for the tetrahedral, octahedral and charge-balancing cations, and the compounds can be strongly and even (in the case of $(Cs,H)_3PW_{12}O_{40}$),[1] superacidic. As solids, the packing of the heteropolyanions can lead to microporosity and enhanced adsorption of small molecules. Other hydrated heteropolyoxometallate compounds can absorb polar molecules and become 'pseudo-liquid' acid catalysts. More generally, they can be supported on inorganic solids to give high-surface-area acid catalysts – suitable, for example, for the industrial hydration of ethylene to ethanol, where the combination of high acidities and pore structure of zeolites results in unwanted by-products, such as oligomeric alkenes.

8.4 Measurement of Acid Site Concentration and Strength in Microporous Solids

As discussed earlier, establishing acid site strengths and concentrations *via* spectroscopic measurements through the use of organic bases of different strength (Hammett Indicators) is not strictly applicable to microporous solids, although the Hammett scale is a convenient comparative index. As a result, a number of alternative approaches have been taken to determine the concentration, type and strength of acid sites within these solids. Most of these methods give relative scales of acidity, but it is broadly possible to 'calibrate' these acidities to those observed for other acids. The methods fall into three broad categories: direct measurements, measurement of interaction with 'unreactive' probe molecules and the use of catalytic test reactions.

8.4.1 Direct Observation of Brønsted Acid Sites

The acid sites of porous solids are distributed evenly over their internal surfaces. As a consequence, they are amenable to study by bulk structural methods. Brønsted acid sites may be observed by proton NMR, vibrational spectroscopy and, in favourable cases, by neutron diffraction. Furthermore, Al-H distances can be determined by NMR spectroscopy[13] and, as described in Chapter 3, the local environment of aluminium ions associated with Brønsted acid sites can be measured by aluminium EXAFS.[14,15] The structural picture for the dehydrated acid sites on H-ZSM-5 is given in Figure 8.2.

^1H MAS NMR spectra of zeolitic solid acids are able to distinguish between terminal silanol protons, protons of hydroxyl groups attached to extra-framework aluminium and Brønsted acid protons.[16] The dipolar proton couplings are strong, so the resonances are quite broad, and in addition the chemical shift range of Brønsted protons is narrow (3.8–5.2 ppm), so that it is difficult to distinguish between different types of bridging hydroxyls. Measurements of chemical shift are therefore not very useful in the estimation of acid site strength. Semi-quantitative estimates of the concentrations of acid sites are available from 'spin counting' experiments, however.

Infrared spectroscopy of the O–H stretching vibration is more sensitive to differences between hydroxyl group environments, but uncertainty in the extinction coefficient makes quantification difficult. Also, there is not a strong

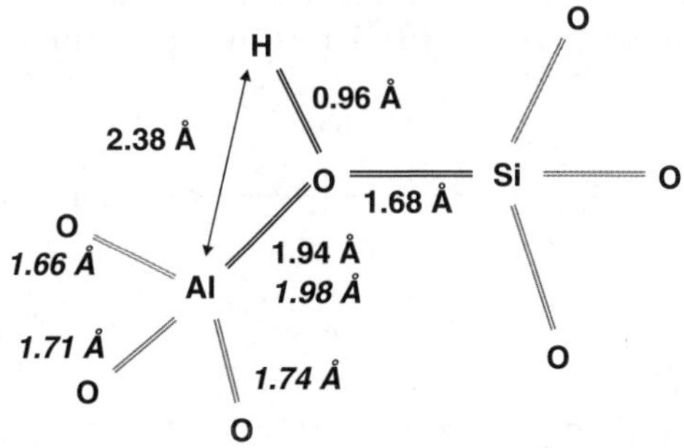

Figure 8.2 Geometry of the Brønsted acid sites of ZSM-5, as determined by NMR (NMR-derived dimensions in normal text) and EXAFS (italics) spectroscopies.[13–15]

correlation between the site acidity and the frequency, which can be strongly influenced by location, where interaction with the framework (*via* H-bonding, for example) reduces the frequency without indicating a high acid strength. IR is therefore best used in conjunction with basic probe molecules in studies of acidity.

Neutron diffraction has provided important quantitative information on the location and abundance of protons in crystallographically simple solids such as H-Y[17] and H-SAPO-34,[18] both of which have only four crystallographically distinct possible locations for bridging hydroxyls. The Al-OH-Si grouping is determined to be planar. However, the structural picture given crystallographically is an average, does not distinguish between tetrahedral cations, is not generally applicable to more complex or disordered materials and does not give a direct measure of the acid site strength.

8.4.2 Interaction of Brønsted Acid Sites with Probe Molecules

In order to study the interactions of molecules at these sites without the complication of subsequent reaction, much use has been made of probe molecules with different basicities, and therefore with different strengths of interaction at acid sites. The interactions have been monitored by many different techniques, including microcalorimetry and TPD, particularly of amines, and infrared spectroscopy and solid state NMR of the adsorbate-sorbate complex. This work has primarily aimed to establish a measure of acid strength that can be linked to catalytic activity in hydrocarbon transformations.

8.4.2.1 Microcalorimetry and TPD

There is growing evidence that there are not straightforward links between the local proton affinity of a bridged Si–O–Al bond and catalytic reaction rates. This can be explained by reference to the thermodynamic considerations of Farneth and Gorte[19] who present the thermochemical cycle shown in Scheme 8.6.

This describes the protonation of an adsorbed molecule B and the attendant formation of an ion pair at the zeolite's internal surface (**1**). The protonation is known from IR studies to occur with bases such as ammonia or pyridine. The

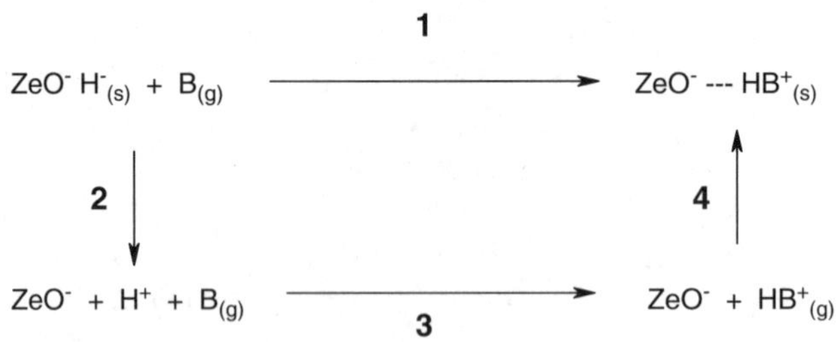

Scheme 8.6 Thermochemical cycle associated with the protonation of a base at an acidic site.[19]

process (**1**) can be thought to be made up of three steps: (**2**) removal of the proton from the zeolite (the reverse of the proton affinity of the zeolite acid site), (**3**) the gas phase proton affinity of the base (for which values are available in tabulated form) and (**4**) the enthalpy of interaction of the resultant protonated base with the negatively charged framework. The clear implication is that the measured heat of adsorption (**1**) will depend on the ion pair interaction with the zeolite framework as well as on the acid site strength. In other words, the environment of the acid site and the molecular structure of the base will determine the interaction energy. The measured order of adsorption strengths among a selection of zeolites can in some cases depend on the probe base molecule that is used. In fact, microcalorimetric studies suggest that there is little variation in heats of adsorption as a base (such as ammonia or a pyridine) is adsorbed on all acid sites in the proton forms of ZSM-5, ZSM-12, Y and mordenite.[20,21] This is inconsistent with large variations in Brønsted acid site strength between crystallographically distinct acid sites in these zeolites. Typical values for heats of adsorption of ammonia and pyridine on Brønsted acid sites are 145 and 200 kJ mol^{-1}, respectively.

With these provisos, microcalorimetry has clarified important details of acid sites. It is a highly accurate way to determine the acid site density, as the heat of adsorption drops off sharply once all accessible sites have been occupied (Figure 8.3). Furthermore, linear plots of the heats of adsorption against the gas phase proton affinities of series of chemically similar amines suggest that the interaction with ion pairs of similar bases (such as small amines, for example) are closely similar for the same zeolite. Estimation of the ion pair interaction energies can be used to give values for the proton affinity of a zeolite framework.

Thermal desorption studies of adsorbed molecules can be used to investigate acid site strength. In a typical experiment, the activated acid form is contacted with an amine (ammonia, pyridine, *etc.*) and then purged with an inert gas to remove weakly adsorbed species. Subsequent heating at a constant rate results in the evolution of the amine giving a peak with a maximum at a temperature of 300–600 °C. The desorbed amount is then a direct measure of the amount of acid sites, whereas the temperature at which the desorption rate is a maximum is a function of the acid site strength. If the effects of diffusion and transport can be neglected, the activation energy and, indirectly, the heat of desorption can be estimated as described in Section 7.2.2. More than one desorption peak is often observed, indicating the presence of different sets of acid sites with different strengths. The type of acid sites involved can be determined by combining TPD with infrared spectroscopy of the adsorbed phase (when pyridine is the base that is used as a probe – see below).

8.4.2.2 Infrared Spectroscopy

Whereas calorimetry measures a heat of adsorption, infrared spectroscopy is able to follow changes in bonding within probe molecules of different basicities adsorbed at Brønsted acid sites. The work of Zecchina's group in Turin, in

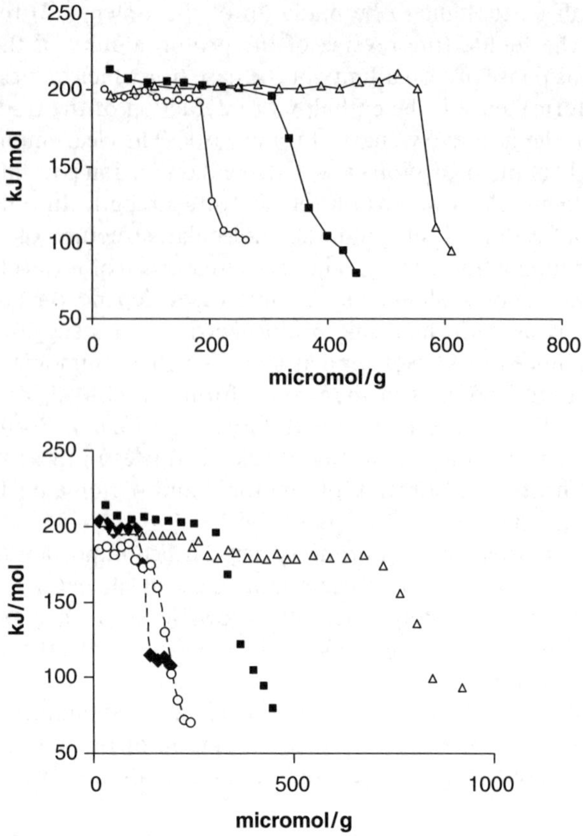

Figure 8.3 (Above) Heats of adsorption by calorimetry of pyridine on acid forms of ZSM-5 with different acid site contents, as measured by TPD/TGA measurements with isopropylamine, plotted as a function of loading (○, 180 μmol g^{-1}; ■, 360 μmol g^{-1}; Δ, 600 μmol g^{-1}). (Below) Heats of adsorption of pyridine on acid forms of different zeolites with different acid site concentrations, measured as a function of loading (◆ ZSM-12, 120 μmol g^{-1}; ○ Y, 160 μmol g^{-1}; ■ ZSM-5, 360 mol g^{-1}; Δ Mordenite, 800 μmol g^{-1}). [Data from references 20 and 21.]

particular, beautifully illustrates the utility and versatility of this approach.[22–24] The types of probe molecules used to study Brønsted acid sites vary from simple homonuclear diatomics, such as dinitrogen, which are bound very weakly, through diatomics such as carbon monoxide and slightly more complex molecules such as methanol and water, to more strongly basic molecules such as ammonia and amines. These have gas phase proton affinities that range from weak (495 kJ mol^{-1} for N_2) to very strong (928 kJ mol^{-1} for pyridine). The strength of interaction varies from weak hydrogen bonding (inducing dipoles in simple homonuclear molecules such as H_2 or N_2 at low temperature), through weak bases that form hydrogen-bonded adducts (H_2O, MeOH, CH_3CN) to complete proton donation to basic nitrogen atoms (protonating ammonia and pyridine, for example).

The interaction between Brønsted acid site and basic molecule can be followed by IR spectroscopic measurements on both the adsorbate and the Brønsted hydroxyl group, typically working at adsorbate:acid site ratios of one or less. In the absence of adsorbates, bridging hydroxyl acid sites give resonances at *ca.* 3250–3660, 990–1055 and 325–420 cm^{-1}, corresponding to the stretching mode and in- and out-of-plane bending vibrations, respectively (Figure 8.4).[23] In some cases different proton sites can be resolved, particularly if they have very different sites within the framework. The bending vibrations

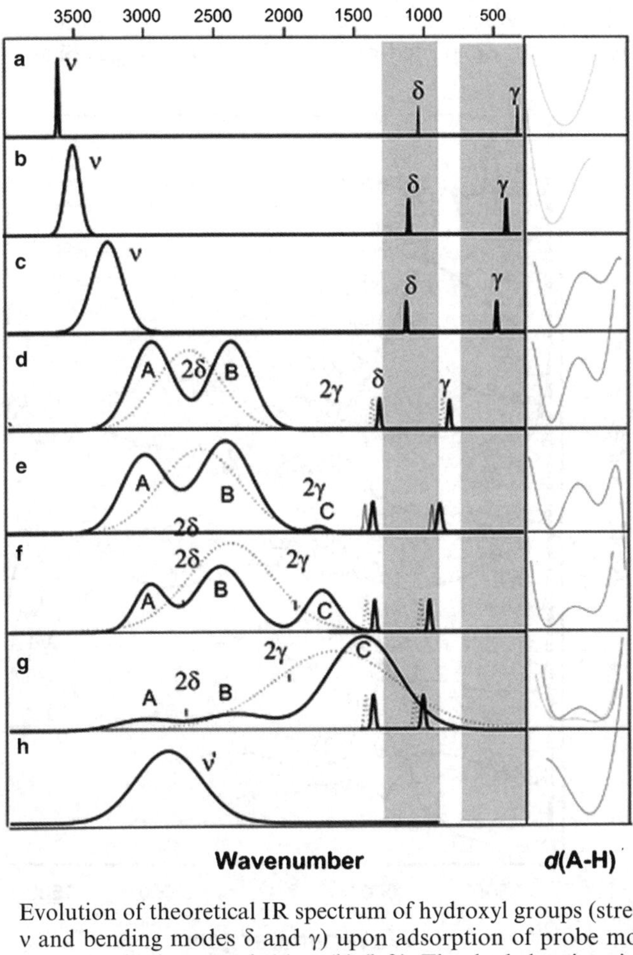

Figure 8.4 Evolution of theoretical IR spectrum of hydroxyl groups (stretching mode ν and bending modes δ and γ) upon adsorption of probe molecules with increasing basic strength (a) to (h) (left). The shaded regions indicate those parts of the spectrum dominated by framework absorption bands. The energy-distance plot of the proton between the acidic oxygen atom (A) and the adsorbed base is also shown (right). [Figure reproduced from reference 23 with permission. Copyright 1997 American Chemical Society.]

occur in the same regions of the spectrum as IR-active framework vibrations and are obscured. Interaction of molecules with the Brønsted site by hydrogen bonding or *via* protonation affects these absorptions. For molecules that are weakly adsorbed, such as H_2 or N_2, the H-bonding interaction is manifested as a small shift in the OH stretching frequency (ν(OH)) to lower frequency (lower wavenumber) while the bending vibrations, which are expected to shift to higher wavenumbers, remain obscured. In some cases the hydroxyl frequencies remain unaffected, indicating that the adsorbed molecules cannot gain access to the protons. This is the case for protons located within the sodalite cage of zeolite Y, for example. Adsorption also affects the vibrational spectra of the adsorbates, which are without rotational fine structure (Figure 8.5).[23] H-bonding interaction is sufficient to induce a dipole in adsorbed homonuclear diatomic molecules, which then become IR-active with a vibrational frequency that is higher than that expected of the molecule in the gas phase.

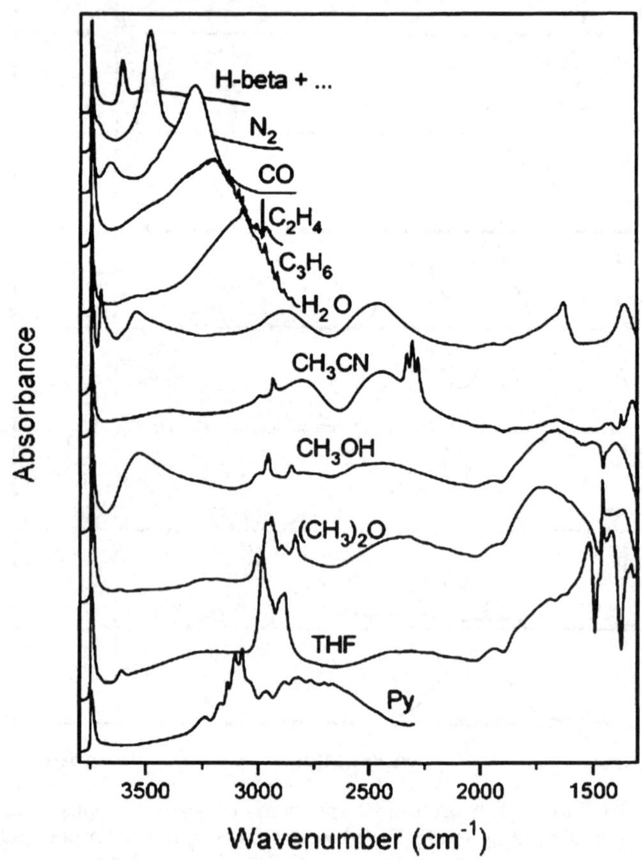

Figure 8.5 Experimental spectra showing the effect of adsorbing bases of increasing strength (downwards) on the acid hydroxyl groups of zeolite Beta for comparison with Figure 4.4. [Figure reproduced from reference 23 with permission. Copyright 1997 American Chemical Society.]

Adsorption of slightly more strongly basic probe molecules, such as CO, which forms a 1:1 adduct with the Brønsted proton with an interaction energy of around 30 kJ mol^{-1}, results in a negative shift in the ν(OH) stretching frequency and a positive shift in the ν(CO) mode. The negative shift of ν(OH) is around ten times the magnitude of the positive shift of ν(CO). Indeed, this makes IR study of the adsorption of CO a sensitive means of measuring acid strength, since the effect of the acid proton on the adsorbate is measurable over the complete range from slightly acidic terminal silanols (SiOH) to superacidic sites. The shift (to higher frequency) of the ν(CO) resonance is therefore a guide to the site's acidity. For example, upon adsorption of CO on H-ZSM-5, a strong acid, the Brønsted OH frequency changes from 3610 cm^{-1} (free) to 3275 cm^{-1} (Δν: − 345 cm^{-1}) and the ν(CO) frequency shifts from 2138 cm^{-1} to 2171 cm^{-1} (Δν: + 33 cm^{-1}). These shifts are among the largest measured for a zeolite, indicating that H-ZSM-5 is one of the strongest zeolitic acids. As well as shifts in the ν(OH) and ν(CO) frequencies, the bending vibrations also shift. This results in the bending resonances moving out from the envelope of framework vibrations and becoming more easily resolved.

The adsorption of alkanes on Brønsted sites also gives 1:1 hydrogen-bonded species, with the interaction being between the protonic charge and the induced dipole on the alkane. The adsorption enthalpy of hydrocarbons increase monotonically with size, as the dispersion forces involved increase. The adsorption of alkenes is stronger, as some electron density is transferred from the double bond. At all but the lowest temperatures reaction will then occur for all alkenes except ethene.

As progressively more basic molecules are adsorbed, the shift in the ν(OH) frequency increases. As well as the fundamental modes of the OH vibrations, there are also overtones (second harmonics). In the absence of adsorbates, or where the adsorption strength is weak, these are of very low intensity and are unimportant. As the interaction between the Brønsted site and the basic probe molecules increases, however, the energy of the overtones of the bending vibrations (which is increasing) approaches that of the fundamental mode of the stretching frequency (which is decreasing) leading to the phenomenon of Fermi resonance.[25] This causes the strong increase in the intensity of the resultant combination ν mode (with up to three components), which changes the appearance of the spectrum for molecules such as water, methanol and acetonitrile. In most zeolite examples 1:1 adsorption of molecules such as those onto the acid site gives strongly hydrogen-bonded adducts, with characteristic spectra.[26] By comparison, addition of these weak bases at low coverage to the acid forms of heteropolymetallates and Nafion gives spectral features consistent with a situation where the proton transfer is near to occurring, indicating that heteropolymetallates and Nafion are stronger acids than zeolites.[24] As more molecules of adsorbates are added, however, protonated complexes of several hydrogen-bonded molecules are thought to form as the ion pair is further stabilised.[24–27] This appears now to be consistent with the data from quantum mechanical modelling and diffraction on water on H-SAPO-34 discussed in Chapter 4.

The adsorption of strong bases such as ammonia or pyridine on Brønsted sites results in proton donation. The protonated bases remain within the solid, themselves showing hydrogen bonding to the framework. The infrared spectrum then resembles that of the ammonium (or pyridinium) cation. The pyridinium ion has a characteristic infrared spectrum with absorbances at *ca.* 1540 and 1490 cm^{-1}. Strong bases of this kind have very high enthalpies of interaction, and titrate any acid sites that are present. As such, they are not sensitive probes of acid site strength.

8.4.2.2.1 The Modified Bellamy–Hallam–Williams Plot for Solid Acids. For basic probe molecules that show relatively weak interactions, the shifts of ν(OH) at the Brønsted sites are a guide to the acid strength, but the magnitude of the proton–base interaction, which determines the shift, is also affected by the repulsive negative charge on the framework, which differs for zeolites with different Si/Al ratios. In a refinement of the method, the frequency shift ν(OH) of the terminal silanols for the zeolite is also measured, and acts as a kind of internal standard. For several acidic zeolites it has been shown that the shifts in the stretching frequencies of Brønsted and silanol vibrations that are observed as probe molecules of different basicity are linearly correlated,[23] and the slope of this line (Figure 8.6[23]) can be used to establish a relative spectroscopic scale of Brønsted acidity for protonic zeolites. This is similar to

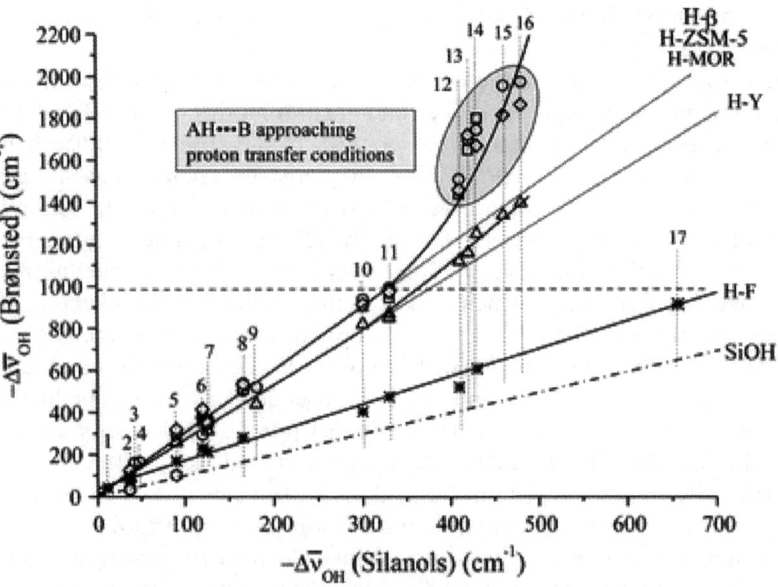

Figure 8.6 Modified Bellamy–Hallam–Williams plot of shift in OH frequency of acidic hydroxyl groups of different zeolites against the shift in frequency of framework silanol hydroxyls as probe molecules of different basic strengths (represented by numbers 1–17) are adsorbed. See reference for details. [Figure reproduced from reference 23 with permission. Copyright 1997 American Chemical Society.]

the Bellamy–Hallam–Williams[28] approach to evaluate the relative acidity of different molecules in solution. For the zeolites examined, the following order has been found: H-Y < H-mordenite, H-Beta, H-ZSM-5.

Related infrared spectroscopic studies of weak bases such as methanol and water have also been performed on the silicoaluminophosphate SAPO-34 (CHA structure type) and its zeolitic analogue SSZ-13. SAPO-34 is known to possess acidity that is among the highest found in aluminophosphate-based solids. Nevertheless, SAPO-34 is found to possess weaker acidity than the isostructural zeolite.[29]

8.4.2.2.2 Extra-framework Aluminium as an Important Example of Lewis Acid Sites. One of the most important types of extra-framework cation found in zeolites, particularly in catalytic applications, is aluminium that has left lattice positions as a result of hydrolysis (for example during steaming, or ultrastabilisation). The Lewis acidity of extra-framework aluminium sites is an important source of acidity in reactions such as catalytic cracking, where it is thought to have an effect in enhancing the strength of neighbouring Brønsted acid sites or to have an important role in hydrogen abstraction in subsequent catalytic steps. Adsorbed basic probe molecules such as ammonia or pyridine interact with extra-framework aluminium *via* coordination of the nitrogen lone pair. Coordinated pyridine gives characteristic IR absorptions at *ca.* 1490 and 1450 cm^{-1}, compared to absorptions at 1540 and 1490 cm^{-1} for the pyridinium ion. For zeolites of medium-pore size or larger, IR spectroscopy of pyridine adsorption is a powerful approach to measuring the relative abundance of Brønsted and Lewis acid sites. Furthermore, the strength of the interactions of bases such as pyridine with the different acid sites can be measured by combining infrared spectroscopy with thermal desorption. An example is shown in Figure 8.7, in which the types and acid site strengths in ZSM-5 are compared with those in the magnesium aluminophosphates MgAPO-36 and DAF-1.[30] Acid site strengths are in the order MgAPO DAF-1 < MgAPO-36 < H-ZSM-5.

8.4.2.3 NMR

Solid state NMR strongly supports conclusions on acid site-molecule interactions made on the basis of IR. The addition of pyridine, for example, can be studied by ^{1}H, ^{13}C or ^{15}N (for isotopically enriched pyridine). H-ZSM-5 and H-Y, for example, treated with C_5D_5N, give ^{1}H shifts at δ 15.5–19.5 ppm, characteristic of protonated pyridinium ions ($C_5D_5NH^+$). There is associated reduction in the signals from bridging hydroxyl protons, including those originally in smaller cages in H-Y, which indicates their mobility. ^{15}N MAS NMR of pyridine adsorbed on protonic forms of zeolites at low loading gives shifts characteristic of pyridinium ions. Ammonia has been used as a basic probe molecule for Brønsted sites, ^{1}H MAS NMR giving chemical shifts characteristic of ammonium ions (δ 6.8–7.4) upon adsorption.[16] The basic molecule trimethylphosphine has also been used to characterise acid sites by

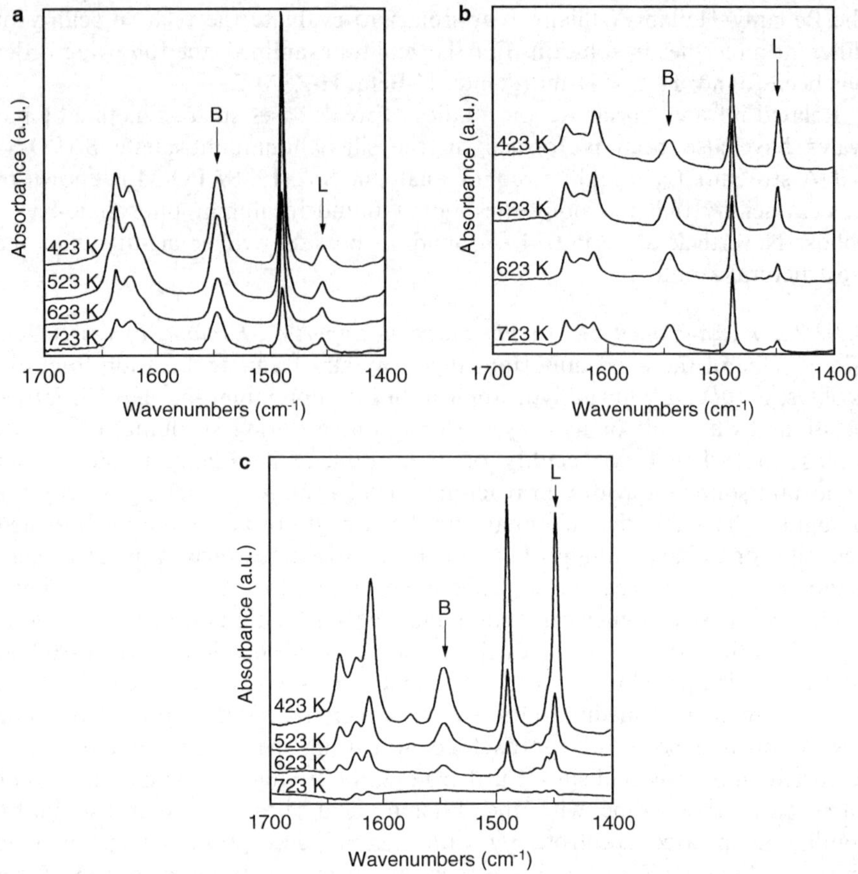

Figure 8.7 Temperature programmed desorption/IR of pyridine adsorbed on (above left) H-ZSM-5, (above right) calcined MgAPO-36 and (below) calcined MgAPO DAF-1. All absorbances are from chemisorbed pyridine: resolved bands from pyridine adsorbed on Brønsted sites (B) and Lewis sites (L) are annotated.

NMR spectroscopy due to the favourable NMR properties of the ^{31}P nucleus (spin 1/2, 100% abundance, high sensitivity and well-resolved spectra) and the fact that the phosphorus is directly involved in bonding at the acid site.[31] The ^{31}P chemical shift of TMP at Brønsted sites (−3 ppm) is easily told apart from that at Lewis sites (−32 to −58 ppm). Chemically, though, the molecule is less readily handled than amines, and prone to oxidation.

The ability of NMR to determine whether protonation of adsorbed basic probe molecules has occurred can also be made use of for Hammett indicators that have access to acid sites within the pores. This is useful because colour changes on solid samples are not readily converted into relative proportions of protonated and unprotonated forms. The NMR approach is more reliable in these cases: p-Nitroaniline-$^{15}NH_2$, for example (pK_{BH} 0.99) is observed by ^{15}N MAS NMR to be protonated, whereas p-nitrotoluene ($pK_{BH}=-11.4$) is

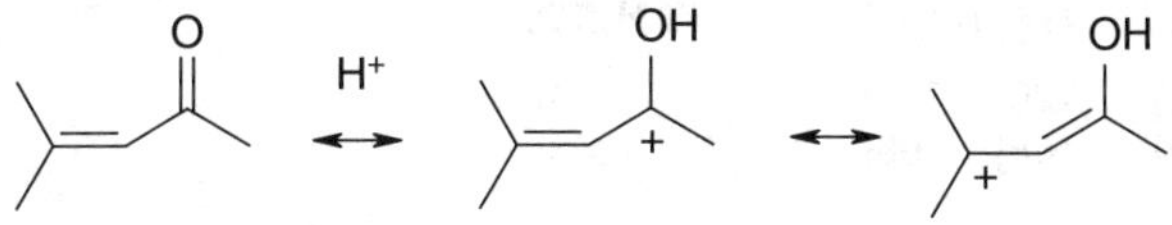

Scheme 8.7 Isomerism of protonated mesityl oxide.

observed to remain unprotonated.[32] Studies of this kind confirm that H-Y is not superacidic. The variation of chemical shift of nuclei in adsorbed species can also be used to give a scale of acidity, once the theoretical basis for the variation is established. Examples are the ^{13}C resonance of C2-^{13}C-labelled acetone adsorbed on bridging hydroxyls, which is shifted from 10–20 ppm from free acetone, and the ^{13}C shift of mesityl oxide (formed by dimerisation of acetone on the zeolite), which is shifted by up to 35 ppm from the free molecule, due to the formation of resonance-stabilised protonated forms, where the position of equilibrium depends on the acid strength (Scheme 8.7).[32]

The acid strength of solids relative to sulfuric acid solutions, for example, can be estimated by comparing the observed values of the chemical shifts on the solids with chemical shift values of the same molecules in acid solutions of different strengths. Using this approach, high silica zeolites such as ZSM-5 have acidities similar to that of 80% sulfuric acid, and are definitely not superacidic.[32]

The measurement of ^{13}C MAS NMR spectra on microporous solid acids which have reactive molecules adsorbed on them is discussed in detail in later sections, where the importance of *in situ* NMR in determining reaction mechanisms of acid catalysis is discussed.[33] At this point it is sufficient to consider the NMR evidence on which species are stable within the pores. One of the first points to make is that adsorption of alkenes such as 2-^{13}C-propene on a protonic zeolite gives a ^{13}C signal that can be assigned to framework-bound alkoxy-species (with characteristic shift, 150 ppm downfield of free propene, and spinning sideband pattern) but not free carbenium ions ($C_3H_7^+$), which are known from studies in superacids to have a chemical shift of 320 ppm. Some carbenium ions have been observed in zeolites, however, as a result of reactions. These include resonance stabilised ring and aromatic species such as indanyl and benzenium species (Figure 8.8). Theoretical calculations on these and other species indicate that zeolites are able to protonate molecules with proton affinities greater than *ca.* 875 kJ mol^{-1}, but that simple carbenium ions are not stable. Combining and relating information available from probe molecules, Haw came up with a scheme (Figure 8.8) to compare acid strengths on a range of scales, including the Hammett H_o scale.[32] This illustrates that zeolites are not superacidic and that is unlikely that free carbenium ions could be intermediates in reactions.

8.4.3 Acid Catalysis: Activity Measurement by Catalytic Test Reaction

Whereas spectroscopic measurements give indirect evidence for the acid site strength and their likely performance in acid-catalysed reactions, measurements

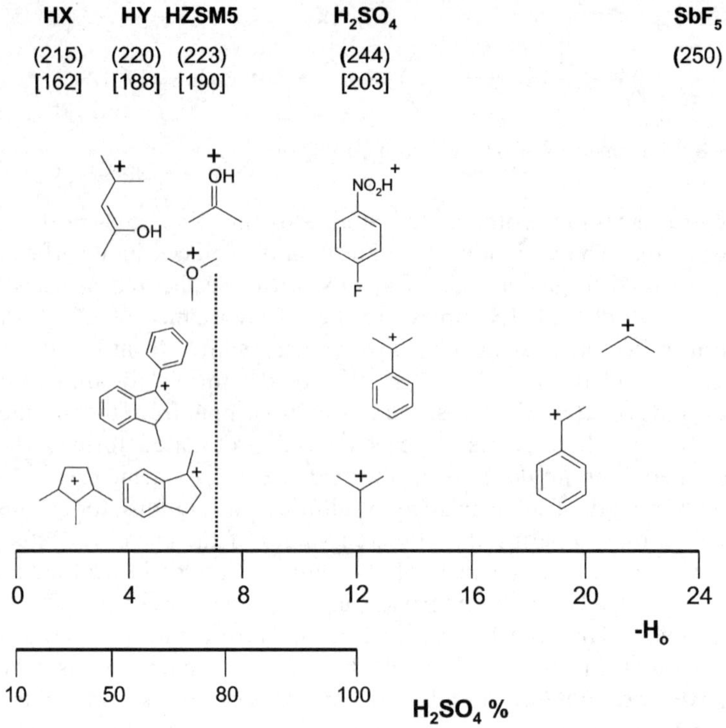

Figure 8.8 Comparison of acid strength of different zeolites with those of sulfuric acid and the Hammett indicator scale (see text), on the basis of the observation of (i) different protonated molecules at the acid sites and (ii) the ^{13}C NMR chemical shifts of mesityl oxide (·) and acetone [·]. [After the Scheme of Haw, reference 32 with permission. Copyright 1996 American Chemical Society.]

of catalytic activity in standardised hydrocarbon conversion reactions give direct information. A wide variety of test reactions have been devised, to be able to compare the performance of different microporous solid acids. Whereas some of these have been developed specifically to give information on the relevant pore geometry (and are described in Section 8.7), others measure the activity more directly. Examples of this kind are hexane and isobutane cracking.

Hexane cracking is a typical test reaction used by Mobil in characterising acid catalysts. The Alpha value, often quoted in patents, is the ratio of the catalytic cracking activity of the zeolite related to that of a highly active silica-alumina cracking catalyst. Details of the Alpha test are given in US Patent No. 3,354,078. Isobutane cracking was suggested by McVicker *et al.*[34] as a probe of acid strength that is applicable to both medium- and large-pore zeolites, because isobutane is a relatively small molecule. The relative cracking rates on different solids, calculated at low conversions and normalised to acid site density, can readily be compared, for example on Arrhenius plots (ln Rate *vs.*

1/T). Furthermore, all the products of cracking can readily be identified by gas chromatography, and the ratio of alkanes to alkenes is a sensitive index of the acid site strength. This is thought to be because carbenium-like transition states formed on weaker sites decompose and desorb as olefins whereas on stronger acid sites they are more likely to exist for longer and undergo bimolecular hydride shift reactions with molecules in the pores, leading to the generation of alkanes.

Care should be taken in interpreting the activation energies of hydrocarbon cracking obtained from Arrhenius plots, which can include effects of diffusion, adsorption and secondary reaction. Values taken at low conversion, avoiding the effects of secondary reaction, are in principle the easiest to interpret. Measurements of this kind for different hydrocarbons show that although the rates at elevated temperatures are similar, different Arrhenius energies are obtained for different homologous alkanes, which is at odds with the expected similarity in the mechanism. In fact, the observed Arrhenius energy also includes a component from adsorption, because for larger alkane molecules, more are adsorbed at lower temperature so that the rate of product formation is enhanced (since more molecules are present at the active sites).

$$E_{act}(\text{observed}) = E_{act}\ (\text{intrinsic}) + \Delta H_{ads}$$

The apparent activation energy is therefore lower than the real value, and the measured activation energies and pre-exponential factors for alkanes of different mass tend to compensate to give similar rates (the 'Compensation Effect').[35] For *n*-hexane cracking over a series of high silica zeolites working at the same conversion, and showing similar product distributions, Babitz *et al.*,[36] for example, showed that the differences in apparent activation energies (and rates) could be attributed simply to different heats of *n*-hexane adsorption onto the different solids and not to differences of acid site strength. Higher heats of adsorption are observed on the medium-pore ZSM-5 than zeolite Y, for example, and the greater resultant concentration of reactants in the pores leads to higher reaction rates.

8.5 Mechanisms of Acid Catalysed Reactions from *in situ* NMR and Quantum Mechanical Calculations

The product distributions of acid-catalysed reactions over acidic zeolites have long been interpreted in terms of the reactions of short-lived carbenium ion intermediates in line with observed reactions in superacid solutions. Information from NMR studies and theoretical calculations has, since the early 1990s, indicated that a different interpretation is required. Alkoxy species bound to the framework are the observed intermediates in many of these reactions, rather than carbenium ions, and carbenium-ion-like species, strongly stabilised by interaction with the framework, are postulated high-energy transition states. In addition, the observation of a reactive hydrocarbon pool is gaining acceptance as an important part of the mechanism in reactions such as the conversion of methanol. *In situ* MAS NMR studies, more than any other experimental technique, are changing our view of acid catalysed reactions, and quantum mechanical molecular modelling is now sufficiently well developed that the possible mechanisms can be assessed theoretically.

In situ NMR[37,38] has been performed in three different ways: using sealed samples and *via* 'pulse-quench' and continuous flow (CF) techniques. To shorten data collection times, ^{13}C-labelled reactants are used. The sealed ampoule technique is experimentally the simplest. Typically, a sample is loaded inside a quartz tube, which is the correct size to fit snugly inside an MAS NMR rotor, activated and loaded with the required amount of reactant before the glass tube is sealed off (keeping the solid and adsorbate/reactant cool while sealing). The crucial requirement is that the glass ampoule remains symmetrical when sealed and enables the ampoule + rotor assembly to be spun at tens of kHz. For *in situ* studies the sample can be heated at different temperatures and the heated sample, including reactants and products, analysed by NMR. A modification of this method has been developed by Munson *et al.*, in which a sample is pre-treated and a reactant adsorbed before being loaded within a rotor (and heated as desired).[39,40] This removes the difficulties associated with sealing the glass and gets around the associated problem of spinning a rotor plus ampoule assembly.

In general, the 'sealed sample' approach constrains the experiment to following catalysis run in batch mode with products remaining in contact with the catalyst over the entire reaction period. Heterogeneous catalysis, however, is much more likely to be run in flow mode, in which the reactant stream is passed over a catalyst bed and the products are removed after a single pass. As a result, there is a strong incentive to make *in situ* MAS NMR measurements on catalysts operating in continuous flow mode. The group of Haw has developed a pulse-quench method by which a catalyst operating under flow conditions in a microreactor is rapidly quenched and analysed at ambient or sub-ambient temperatures by MAS NMR.[41,42] This set-up enables the products to be monitored at the point where the sample is taken, so that the appearance of products in the microreactor stream can be related to the observation of species, possibly reaction intermediates, adsorbed on the catalyst.

A different approach has been taken by Hunger,[38,43] who has developed a technique to perform MAS NMR *in situ* during reaction in continuous flow mode. This requires an ingenious apparatus design, which enables the introduction of reactants into a heated rotor spinning at the magic angle in an NMR magnet. In addition, modifications enable sampling of the product stream by gas chromatography. In this way the species on the catalyst under continuous flow can be measured. Using the same apparatus, possible variations on the experiment include switching the isotopic composition of reactants and step-wise removal of the reactant flow. These enable the response of the system to be followed under transient conditions, and so give information on the reactivity of adsorbed species, so that spectator species can be distinguished from reaction intermediates, which is important in determining the reaction mechanism.

The power of these *in situ* methods to determine the mechanism of acid-catalysed reactions is best illustrated with examples of each of the approaches. One of the first reactions to be studied by heating a sealed sample was the reaction of prop-2(^{13}C)-ene on the acid form of zeolite Y.[33] This showed important differences from the reaction of propene over super-acids. No carbenium ions or scrambling of ^{13}C were observed over the zeolite and the reactive intermediates appeared to be alkoxy species bound to the framework. Another important result of heating sealed ampoules was obtained by Anderson and Klinowski on the methanol to gasoline (MTG) reaction over H-ZSM-5.[44,45] The reaction of methanol over microporous solids is discussed in detail in later sections, where the details of the mechanism are elaborated. Using a combination of 1D and 2D ^{13}C MAS NMR methods,[46] Anderson and Klinowski were able to compare the products of reaction measured by NMR analysis of adsorbed molecules inside the pores with those present in the gas phase outside the zeolite, and observed important differences. Among the aromatic products in the gas phase, for example, were *m*- and *p*-xylene, 1,2,4-trimethylbenzene and toluene but no tetramethylbenzenes, whereas *o*- and *p*-xylene and 1,2,4,5-tetramethylbenzene were the most abundant species inside the pores. Among the trimethylbenzenes neither 1,2,3- nor 1,3,5-trimethylbenzene (kinetic diameters, σ, 6.4 and 6.7 Å, respectively) were found in the gas phase although they were seen in the adsorbed phase, whereas 1,2,4-trimethylbenzene (σ 6.1 Å) is seen in both. These clear differences between molecular distributions in adsorbed and gas phases represents a direct observation of shape selectivity, illustrating the controlling role of diffusivity of product hydrocarbons out of the 10MR channels of the zeolite.

The study of the reaction of methyl halides over cationic zeolites, performed using the set-up of Haw, is another example of an *in situ* NMR study of conversions taking place in batch mode.[47] Methyl halides decompose over cationic zeolites to give ethylene and other hydrocarbons, with the reaction being faster over more basic (nucleophilic) zeolites. NMR indicates the reaction proceeds *via* reactive methoxy species bound to the framework (observed with a characteristic signal, δ 58 ppm).

The application of *in situ* NMR to catalysts working under flow conditions is a powerful approach that is of direct relevance to acid catalysis. The *in situ* study of methanol conversion, and the role of the reactive hydrocarbon pool, is described later. The methylation of aniline on H-Y using ^{13}C-labelled methanol as a reactant and with a reactant ratio of MeOH:aniline of 4 : 1 has been studied by Hunger using this method.[48] At low temperatures (below 200 °C) methanol reacts to give DME and methoxy groups on the framework, whereas above this temperature aniline is alkylated, giving a mixture of protonated *N*-methylanilinium (N-MA), *N*,*N*-dimethylanilinium (NN-DMA) and *N*,*N*,*N*-trimethylanilinium ions (NNN-TMA) (see Scheme 8.8), as well as toluidine species alkylated on the aromatic ring. The relative amounts of the methylated species vary with the temperature of the reaction. At 200 °C the NNN-TMA species is the most abundant species, whereas at higher temperatures the others become favoured. In order to determine the reactivity of the NNN-TMA, and determine whether it is in dynamic equilibrium with the other products, a

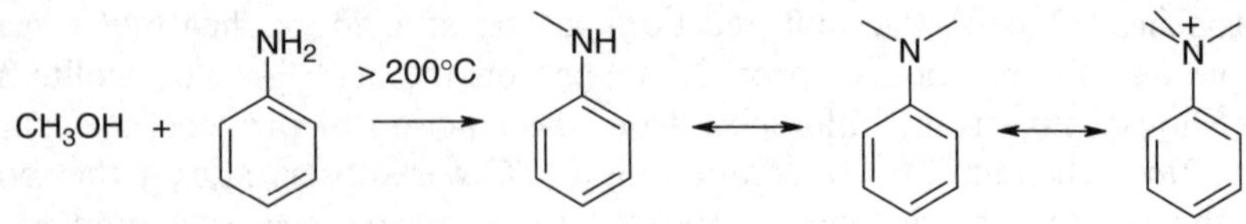

Scheme 8.8 Methylation of aniline over a solid acid.

stopped flow experiment was performed, in which the reactant flow was stopped and replaced with an inert gas flow as the sample is cooled to room temperature, leaving only NNN-TMA ions in the zeolite. Subsequent heating above 200 °C (in a sealed system) resulted in conversion of the NNN-TMA to the other species, showing the NNN-TMA is indeed reactive and part of the reaction mechanism. Analysis of the transient effects of sharp changes in reactant composition by *in situ* NMR in this way is likely to be very powerful in understanding reaction mechanisms over microporous solid acids.

Theoretical calculations of the stability of intermediates of acid-catalysed reactions (see Chapter 4) have proceeded in parallel with experimental observations, and provide important additional insight. Nicholas and Haw, for example, performed DFT calculations on the carbenium ions they had identified by *in situ* NMR and compared their energies with those of the parent alkenes.[49] In this way the proton affinities (PAs) were calculated. Comparison of calculated and measured PAs of different carbenium ions gives an estimate of the required PA for the formation of stable carbenium ions (875 kJ mol^{-1}). Taking calculations a step further, quantum mechanical calculations have also been performed for the possible transition states for a range of acid-catalysed conversions of hydrocarbons over zeolites. Kazansky has considered double-bond isomerisation in alkenes, skeletal isomerisation of alkenes, olefin cracking and oligomerisation and alkane cracking.[50] In agreement with NMR, the calculations indicate that free simple carbenium ions are not stable intermediates and that surface alkoxides are found to be the stable species. Carbenium-ion-like species are in fact the transition states, so that mechanisms based on carbenium ions still appear to be relevant. For example, skeletal isomerisation is still thought to go through a protonated cyclopropane ring, and protolytic cracking of alkanes is thought to go *via* a carbonium-ion-like transition state stabilised by interaction with the zeolite surface. Kazansky estimates the activation energy for protolytic cracking of butane to be 240 kJ mol^{-1}. This is higher than that observed (170 kJ mol^{-1}), the difference being ascribed to the energy of adsorption of butane in the pores.

8.6 Trends in Performance of Microporous Solid Acids

An enormous volume of research on the behaviour of microporous solid acids, including spectroscopic studies of probe molecules and the measurement of their activities in catalytic reactions, has established the following features of the variation of Brønsted acid strength:

(1) Acid site strengths of microporous solids vary, depending on the framework composition, but are not superacidic.
(2) The strongest acid sites are present in crystalline aluminosilicates: acid sites of other metallosilicates (B, Ga, Fe), substituted aluminophosphates or amorphous silica-aluminas are weaker.
(3) Within a single zeolitic structure type that can be prepared with a wide range of Si/Al ratios, acid site strength varies with chemical composition below Si/Al of *ca.* 6–10.
(4) Studies of high silica zeolites indicate that the activity for acid-catalysed reaction depends linearly on the framework aluminium concentration.
(5) Acid site strength shows some variation between high-silica solids with different structures.

8.6.1 The Role of Chemical Composition

The acid strengths of some microporous solids were compared with those of liquid acids and other classes of solid acid in Figure 8.1. The high-silica zeolites H-ZSM-5 and mordenite are among the most acidic of microporous solids, but do not achieve the high acid strengths of solids such as sulfated zirconia or heteropolyacids.

The chemical composition of the immediate environment of the bridging hydroxyl plays an important role in determining the acid strength. In silicates, Al–O–Si linkages can give rise to strong acids, whereas isomorphous replacement of Al by B, Ga or Fe in the same structure gives much weaker sites. The reason for the weak acidity of borosilicates derives from the fact that in the protonated form boron adopts trigonal coordination with oxygen, as shown by ^{11}B NMR (Chapter 4). In this geometry it has little effect on removing electron density from the silanol oxygen, and therefore the proton remains weakly acidic.

Chemical composition affects acid site strength in another way. Acid site strength within a given medium or low-silica zeolite structure, such as zeolite Y, increases with Si/Al ratio to a maximum value, where it remains at a constant level. This is attributed to an increase in overall electronegativity of the framework surrounding the active site as silicon replaces aluminium in framework sites around the bridging hydroxyl. Specifically, decreasing the aluminium content reduces the number of aluminium atoms in the second-nearest-neighbour shell of the aluminium associated with the bridging hydroxyl. The increased local electronegativity imparts a greater ionic character to the hydroxyl group and therefore a stronger Brønsted acidity. The Sanderson electronegativity, σ, takes account of these compositional changes and indices of acidity (shifts in IR frequencies upon adsorption of weak bases, rates of catalytic reactions) correlate well with σ.[51] Above a certain value of the framework Si/Al ratio (*ca.* 10 for zeolite Y) the acid strength changes little. At these compositions almost all the aluminium atoms associated with bridging hydroxyls possess no aluminium atoms in their second-next-nearest-neighbour tetrahedral cation environment.[1] Aluminium atoms further away in the structure therefore appear to have little effect, so that in the protonic forms of zeolites with Si/Al ratios such that the aluminium atoms are well spaced in the framework the catalytic activity is observed to vary linearly with the aluminium concentration. For example, Haag and co-workers showed that the rate of hydrocarbon cracking over H-ZSM-5 is directly proportional to the concentration of aluminium in the framework.[52]

Among the other microporous solid acids, SAPOs and MAPOs tend to possess acidities that are weaker than those found for aluminosilicates. Not all substituted aluminophosphates are weak acids, however. Magnesium-substituted $AlPO_4$-36, for example, gives a strong performance in acid catalysed alkane cracking.[30,53,54]

8.6.2 The Role of Local Framework Structure

Crystalline silicates possess stronger acidities than amorphous silica-aluminas of similar chemical compositions. This arises from the geometric constraints placed on the Si–O–Al bond angle in structures ordered on the long range. Quantum mechanical studies of acid sites using carefully chosen framework 'fragments' indicate that the deprotonation of SiOAl bridges requires less energy when the angle is larger and consequently protons on bridges with larger crystallographic angles are more acidic. For amorphous silica-aluminas the framework can relax to give bond angles with less associated strain. For crystalline zeolites some of the bridging bond angles are constrained to be higher than these lowest energy configurations, so they have higher associated acidities. The more acidic zeolites are found to have a wider range of TÔT angles (ZSM-5, 137–177°; mordenite, 143–180°) than less acidic zeolites (zeolite Y, 138–147°) in general agreement with theoretical predictions.[1,55] Care should be taken not to oversimplify correlation of rates of reaction with structure in terms of acid site strength, however, because it is likely that they will depend on the local environment of the pores around the active site rather than the isolated properties of the sites alone. Notably, the calorimetry of the group of Gorte suggests that all Brønsted sites in ZSM-5 interact with amines with very similar heats of adsorption.[56] Furthermore, the presence of extra-framework aluminium can strongly affect the rate in some reactions, so that only very carefully characterised samples can be directly compared.

8.6.3 The Role of Lewis Acidity

Most working zeolite catalysts are likely to have both Brønsted and Lewis acidity, the latter resulting from aluminium, either as extra-framework cations or incompletely coordinated within a dehydroxylated framework. Strong deviation from spherical symmetry can result in quadrupolar aluminium being NMR-invisible,[57] except at the highest magnetic field,[58] but Lewis acid sites are readily observed by IR spectroscopy of adsorbed probe molecules such as ammonia and pyridine. Steamed zeolite Y that contains extra-framework aluminium has been shown to be highly active in cracking reactions. Several explanations for this effect have been proposed and are discussed in Section 8.7.2.5.

In addition, rare earth cations such as La^{3+} are often included in zeolite cracking catalysts to enhance reaction rates, and a variety of metal-exchanged zeolites are active for acid-catalysed reactions of carbonyl compounds, among others. In one detailed study, however, Karge *et al.*[59] showed that La^{3+} zeolites without hydroxyl groups are not active for ethylbenzene disproportionation, suggesting that, in this case at least, the La^{3+} does not function as an active site. In all such metal-containing compounds, one role of the cation will certainly be to increase adsorption during reaction, due to increased interactions with the reactants, and this will accelerate the reaction rate.

8.6.4 The Role of Pore Structure: The Origin of Shape Selectivity

Most acid sites of microporous solids are located within channels and cavities that are in the range 4 to 10 Å. Some acid sites do exist at the external surfaces but the fraction of these is negligible unless the catalyst particles are very small. The unique pore structures of these solids therefore strongly influence their catalytic behaviour. Most importantly for their use, they introduce a selectivity into product distributions that depends on the shapes of reactant and product molecules and of transition states in bimolecular reactions (Figure 8.9).[60] In addition, the degree of coke formation during high-temperature reactions of hydrocarbon molecules can be influenced by the space available. Finally, the confinement effects of pores – high electrostatic fields and reactants brought into close contact – have a strong influence on reaction rates of bimolecular reactions, including intermolecular hydride transfer and alkylation.

8.6.4.1 Shape Selectivity

The most straightforward cause of shape selectivity is the discrimination between molecules on the basis of their diffusion rates through the channels or cage windows. Microporous solids act as true molecular sieves, because the well-defined pores are able to select molecules on the basis of differences in dimensions of 0.1 Å or less. Examples of strong molecular sieving effects include the selection of normal alkanes over branched ones by small-pore solids and the selection of *para*-substituted over *ortho*- and *meta*-substituted aromatics over medium-pore zeolites. This type of selectivity according to molecular diffusion rate may act on both reactant and product molecules. The much faster dehydration of *n*-butanol compared to isobutanol over Ca-A demonstrated by Frilette and Weisz is the classic example of *reactant diffusion*

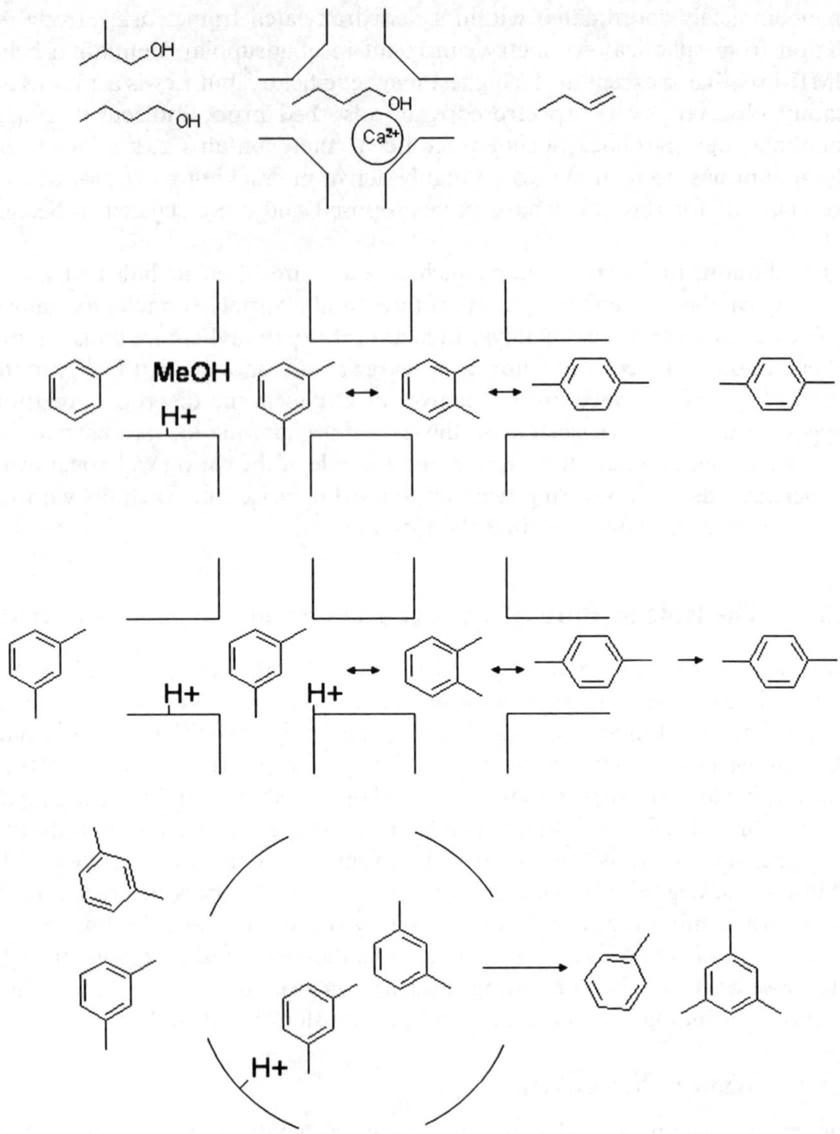

Figure 8.9 Shape selectivity in zeolite pores. From top to bottom: reactant selectivity in alkanol dehydration; diffusion selectivity in alkylation of aromatics; diffusion selectivity in xylene isomerisation; transition state selectivity in xylene isomerisation.

shape selectivity.[61] (The source of the acidity is the protons resulting from the polarisation of water by the calcium ions ($Ca^{2+} + H_2O \rightarrow Ca^{2+}OH^- + H^+$.)) Similarly, normal alkanes are cracked preferentially over more slowly diffusing multibranched alkanes over small- and medium-pore zeolites. This is utilised in refining processes.

Similarly, *product diffusion selectivity* may act on the products of a reaction that are formed within the catalyst, so that the fastest diffusing products are able to leave the catalyst preferentially whereas the bulkier remain behind, where they can react further. The direct observation of the difference in relative abundance of aromatic molecules in adsorbed and gas phases during the MTG process over H-ZSM-5 by MAS NMR (Section 8.5) is a good example of this. The effect of product diffusion control on product selectivity has been well demonstrated by the use of *in situ* ^{13}C NMR under reaction conditions,[44,45] where the adsorbed hydrocarbon mix of the methanol-to-gasoline reaction over H-ZSM-5 is demonstrated to have a distribution much closer to thermodynamic equilibrium than the product, due to strongly different diffusion rates out of the pores. Where equilibria exist within the pores, the effect of diffusion selectivity is to enrich the product mixture in the faster-diffusing product to levels that can be far above equilibrium concentrations. A well-studied example of this is the alkylation of aromatic molecules within zeolites, when the product stream is enriched to well above equilibrium levels in the faster diffusing *para*-substituted isomer, which is the most valuable one.

Transition state selectivity is another way in which the pore structure may change the product distribution from that expected upon reaction over an acid site of similar strength located on an open surface. The transition state in a reaction is the activated state through which reaction proceeds. This will adopt a favourable configuration with respect to the atoms of the microporous solid. When there are constraints of space, it may be that the transition state cannot adopt a favourable configuration or is too large to be accommodated within the pores. In such a case the reaction pathway through that transition will be retarded or blocked.

The isomerisation of xylenes within H-ZSM-5 provides the textbook example of transition state shape selectivity (Table 8.2). Xylenes isomerise on zeolite H-Y to an equilibrium distribution of *o*-, *m*- and *p*-xylenes within the large cavities. Under the same conditions they also undergo disproportionation to give toluene and a mixture of trimethylbenzenes. This (unwanted) side reaction proceeds *via* a bimolecular reaction. Over the medium-pore H-ZSM-5, however, only xylenes are observed, enriched in the *para*-isomer by product shape selectivity. Little or no disproportionation is observed. This is a result of transition state selectivity, since there is no space within the ZSM-5 structure for the bimolecular transition state of the disproportionation reaction.[62]

Table 8.2 Products of *m*-xylene isomerisation and disproportionation reactions.[1]

Zeolite	*Pore system window; connectivity*	*p-/o- selectivity*	*Isomerisation/ disproportionation*
ZSM-5	10 MR; 3-D	2.9	33
ZSM-48	10 MR; 1-D	2.4	13
ZSM-12	12 MR; 1-D	1.0	3
H-Y	12 MR; 3-D	1.0	1.5

A similar effect is observed in the coking behaviour of zeolites. For hydrocarbon reactions within zeolites, oligomerisation and side reactions invariably occur, resulting in carbon-rich oligomers being synthesised in the pores. Ultimately, this leads to blocked pores and consequent deactivation. Regeneration of the catalytic activity is only possible by burning out the deposits in air at high temperatures. This occurs extremely rapidly in cracking reactions over zeolite Y, where the pores become blocked after a few seconds, and is also observed over large-pore zeolites in a range of other reactions. Coking is markedly slower over medium-pore zeolites however: H-ferrierite is active over long periods in butene isomerisation, for example, whereas zeolite Y rapidly deactivates and similar effects are seen in the methanol-to-gasoline reaction. This arises because it is difficult for extended aromatic sheets to grow within the available space in the medium pores.

Finally, an interesting effect is observed in the product distribution of alkanes prepared by the catalytic hydrocracking over zeolites of high-molecular-weight straight-chain hydrocarbons to smaller alkanes. The more valuable products of these reactions are branched alkanes, because they possess higher octane numbers than normal alkanes. Experimentally, branched alkanes are found to be favoured over the faster-diffusing linear alkanes, an effect that was named '*inverse shape selectivity*'. Furthermore, the observed ratio of branched/normal cracked products is found to vary with the effective pore size of the zeolitic catalyst used, reaching a maximum where the effective pore size is around 7 Å. The explanation for this effect remains a matter of debate, but it appears that thermodynamic as well as kinetic factors determine the product distribution.[63–66]

8.6.5 Catalytic Test Reactions: Information on Pore Geometry

Shape selectivity therefore plays an important role in determining product distributions, and this feature is exploited in a range of industrial processes to give high yields of desired compounds. Conversely, standardised reactions, through the molecular makeup of their products, can be used to give information on the pore geometry of solids.[67] The criteria of a suitable test reaction are that it can be operated readily with easy product analysis and that it should not be strongly dependent on particle size or time-on-stream, so that the results of different laboratories can be compared directly. Furthermore, it should give molecular indices (ratios of products or relative rates of reactant conversion) that vary steadily with pore size over reasonable ranges. Such test reactions can be used to establish details of the origin of shape selectivity in microporous solids of known structure (for example in comparison with molecular modelling of reaction rates) or used to estimate the pore geometry of new materials with unknown structure.

The initial reaction used by Mobil (the results of which are quoted in most of their patents of new materials) is the competitive cracking of an equimolar mixture of *n*-hexane and 3-methylpentane under standard conditions.[68] In the absence of spatial restrictions the rate of cracking of branched isomers is faster than for linear species, but as the pore size is reduced the reverse is true. The Mobil researchers therefore defined a constraint index (CI) that depends on the rates of cracking of these isomers as an empirical indicator of pore size.

$$\text{Constraint Index (CI)} = \frac{\log(\mathit{fraction}\ n - C6\ \mathit{remaining})}{\log(\mathit{fraction}\ 3 - Me - C5\ \mathit{remaining})} \approx \frac{\log k_{n-C6}}{\log k_{3-Me-C5}}$$

The faster cracking of the branched isomer where no space constraints exist is attributed to its passing through a secondary carbenium ion-type transition state, whereas the linear alkane must crack *via* a higher-energy primary carbenium ion. Medium-pore zeolites give CIs in the range 4–10 (*e.g.* ZSM-5 gives CI values of 7–8) whereas large-pore zeolites give values of 2 or less (*e.g.* zeolite Y gives a CI of around 0.3). Detailed modelling studies suggest that within medium-pore zeolites the cracking of the branched isomer is inhibited by reactant transport shape selectivity and by the need for a bulkier transition state. There are limitations to the CI as an index of the pore size, however.[69] The reaction over large-pore solids is strongly affected by coking, so that the cracking rates change strongly with time-on-stream, and there is little differentiation in the large- or extra-large-pore ranges. Furthermore, the value tends to average the effects of pore size and cage diameters in zeolites with channels and cages.

The rapid coking during cracking reactions, particularly over large-pore zeolites that do not restrict intra-pore coke formation, remains a drawback for their use as test reactions. One route around this is to load the catalysts with precious metals, and perform bifunctional hydrocracking or hydroisomerisation over them, because the combination of dehydrogenation/hydrogenation and high hydrogen pressures permit reactions at lower temperature and with slower deactivation. Bifunctional catalysis is an important reaction in hydrocarbon refining and has been used with considerable success in test reactions. The usual mechanism is dehydrogenation of the alkane followed by formation of a carbenium-ion-like transition state, skeletal isomerisation and/or β-scission, return of a proton to the catalyst, alkene desorption and hydrogenation to alkane product. High hydrogen pressures ensure that low concentrations of alkene are present so that the likelihood of oligomerisation of alkenes is reduced. The product diffusion is strongly affected by reactant and product transport rates, together with shape selective effects on the transition states that can form.

The hydroisomerisation of *n*-decane is one such test reaction where a modified constraint index CI* is defined as the ratio of the formation of 2-methylnonane to that of 5-methylnonane.[70] CI* is found to increase over the range 1–10 as the pore size decreases from large to medium pore (for Y, CI* is *ca.* 1, whereas for ZSM-5, CI* is *ca.* 10). However, like CI, it gives relatively little differentiation between large-pore zeolites, and explanation for the origin of the CI* values remains ambiguous.

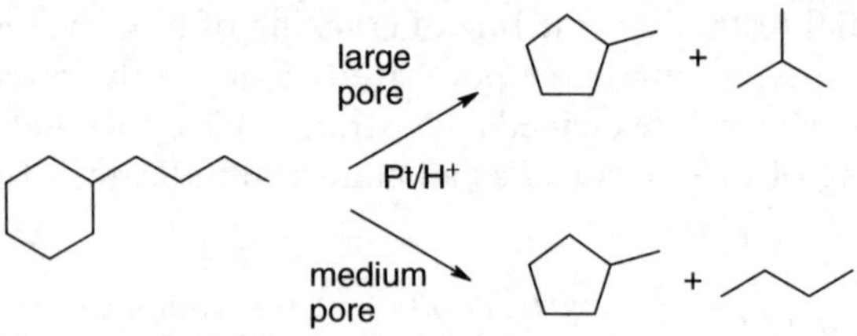

Scheme 8.9 Hydrocracking of butylcyclohexane: shape selectivity in bifunctional zeolites.

Hydrocracking of bulky cyclic alkanes is found to give a sensitive index in the large-pore region.[71] For example, the hydrocracking of butylcyclohexane in the absence of spatial constraints proceeds almost entirely to methylcyclopentane and isobutane, whereas shape selectivity that occurs as pore size decreases forces the reaction to go *via* alternative processes, resulting in the formation of *n*-butane (Scheme 8.9).[67]

The Spaciousness Index (SI) is defined as the ratio of iso- to *n*-butane in the product,

$$\text{SI (hydrocracking of butylcyclohexane)} = \frac{iso - C4}{n - C4}$$

This index is found to be very sensitive to changes in pore size. For example, for zeolites Y and Beta, SI is around 20, but for ZSM-5 it is around 2.

Combinations of these test reactions, and others like them, can therefore be used to indicate the likely pore geometry of a new microporous solid acid and suggest its probable performance in reactions where shape selectivity plays a controlling role.

McVicker *et al.* have recently suggested the acid-catalysed ring contraction of methylcyclohexane as a test reaction able to give information on acid site density and strength and also on pore size, *via* the product distribution.[72] The authors show that the reaction is sensitive to acid site density, acid site strength and pore size, and that the effects of the three variables are separable. The reaction is run under hydrogen and in the presence of a hydrogenation/dehydrogenation catalyst to prevent rapid deactivation. Methylcyclohexane reacts to give a mixture of isomers of alkylcyclopentanes: ethylcyclopentane plus three dimethylcyclopentane (DMCP) isomers (Scheme 8.10).

One of the advantages of the reaction is that it gives a manageable number of reaction products and the reaction pathways are understood. Reaction to ethylcyclopentane is *via* a low energy tertiary carbenium ion transition state whereas generation of the doubly branched dimethylcyclopentanes must take place *via* a higher energy protonated cyclopropane ring-containing intermediate. As a result, high ethylcyclopentane selectivities (70%) at low conversion are given by weak acid sites, whereas high dimethylcyclopentane contents (50–70%) are favoured by stronger acids. In addition, the *trans*-1,2-DMCP is bulkier than the *trans*-1,3-DMCP, so that product ratios of *trans*-1,2-DMCP/*trans*-1,3-DMCP are less than the equilibrium value at reaction conditions (1.9)

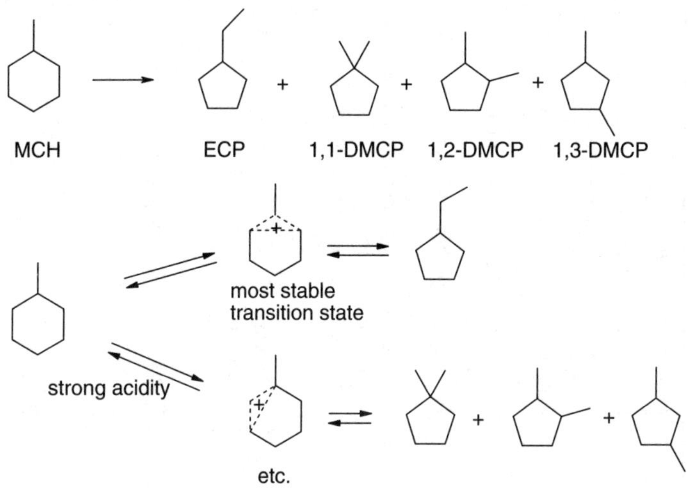

Scheme 8.10 Acid catalysed ring contraction of methylcyclohexane.

over medium-pore zeolites. Over ZSM-5, for example, this ratio is 0.02, whereas for larger-pore zeolites such as Y the value, 2.1, is close to the equilibrium value, indicating no product diffusion constraints.

In fact, the acid-catalysed ring contraction of alkylcyclohexanes to alkylcyclopentanes is potentially important as a reaction, because cyclopentanes are more readily ring-opened than alkylcyclohexanes over precious metal hydrogenolysis catalysts to give linear hydrocarbons. Such linear alkanes are suitable for diesel fuel.[73] Addition of an acidic function therefore enables both cyclopentanes and cyclohexanes to be transformed at a similar rate in this way.

8.7 Reactions over Microporous Solid Acids

Acid catalysts are used on a massive scale in oil-refining processes to give improved fuels and feedstock hydrocarbons for the solvent, polymer, pharmaceutical, additives and detergents industries.[3] Key reactions include isomerisations, alkylations, catalytic reforming of alkanes, cracking and those involving oxygenated hydrocarbons (Table 8.3). In many cases bifunctional catalysts, involving both acidic and metal catalytic species, are involved. There is also introduction of solid acid catalysis into highly selective fine chemical syntheses. Tanabe and Holderich[2] describe over 90 acid-catalysed processes that are either exploited commercially or likely to be commercialised. Zeolites have proved particularly important, and are used in around 60% of these processes. Where they are not applicable, due to limitations of pore size or acid site strength, other solids, such as ion-exchange resins, supported acids or heteropolyacids are likely to replace homogeneous catalysts.

Table 8.3 Industrial processes involving zeolitic solid acids.[2,3]

Industrial process	*Catalyst*	*Conditions*	*Product importance*
Synthesis of methylamines from methanol and ammonia	H-Rho, H-Mordenite	350–500 °C 15–30 bar	Important intermediates for solvents, insecticides, herbicides, pharmaceuticals and detergents
Alkylation of mono-aromatics with ethene, propene, *etc.*	H-ZSM-5	300–400 °C 20 bar	Ethylbenzene, ethyltoluene precursors for styrenes used for polymers, isopropyl benzene precursor for phenols
Isomerisation of xylenes	H-ZSM-5	300–400 °C 20 bar	*Para*-xylene precursor for terephthalic acid, *etc.*
Skeletal isomerisation of *n*-butenes, *n*-pentenes	H-Ferrierite	400–500 °C *ca.* 1 bar	Isobutenes, isopentenes for synthesis of octane enhancers
Reforming gasoline to increase alkane branching	Pt, Re on Al_2O_3, zeolites	450–550 °C 15–70 bar	Improved octane number of fuels
Hydrocarbon cracking	Y-zeolites ZSM-5 and Beta zeolite additives SiO_2-Al_2O_3	400–500 °C *ca.* 1 bar	More valuable lower molecular weight fractions
Hydrocracking of heavy fractions of petroleum	Pt/H-zeolites	270–450 °C 80–200 bar H_2	Useful fractions from heavy residue
CYCLAR process Dehydro cyclodimerisation of propane and butane to aromatics	Ga/H-ZSM-5 (see Section 9.3.1)	> 500 °C, 1–10 bar	Valuable aromatics
Beckmann rearrangement of cyclohexanoneimine to ε-caprolactam	Silicalite	vapour phase	Precursor for nylon-6

8.7.1 Acid Catalysis with Reactants that Contain Heteroatoms

Reactions involving molecules that contain oxygen and nitrogen and are relatively easy to protonate can be catalysed at low temperatures and by acid catalysts of moderate strength. The etherification of isobutene and isopentene with methanol to methyl tertiary butyl ether (MTBE) and tertiary amyl methylether (TAME), for example, is readily carried out over the acid forms of ion exchange resins at *ca.* 100 °C with extended lifetimes. These ethers have useful octane blending characteristics for gasoline, and have been much in demand as replacements for lead compounds, although they have more recently been discovered to have environmental drawbacks and are themselves being phased out in favour of other additives. The hydration of ethylene is more demanding, requiring stronger acidity and elevated pressure, and it is likely that heteropolyacids, either supported or in their own phase, will be able to replace supported phosphoric acid as the catalyst of choice. Zeolites are not favoured for this reaction, however, because alkenes tend to oligomerise within their pores. For many reactions involving heteroatom-containing reactants, however, zeolites or zeotypes have been found to be suitable catalysts, and in these cases the choice of catalyst has been determined by issues of required acid strength, shape selectivity and, for large reactants, accessibility of active sites and ease of product desorption.

8.7.1.1 Hydration

For the reasons outlined above, some typical acid-catalysed reactions, such as hydration and etherification, may be better performed over non-microporous acid catalysts, but microporous acids have found uses in this area. Asahi, for example, have established the zeolite-catalysed hydration of cyclohexene as a commercial process,[74] where in a two-phase reaction mixture (aqueous and non-aqueous layers) the H-ZSM-5 catalyst stays in the aqueous phase but adsorbs enough cyclohexene, because of its relative hydrophobicity, that the reaction proceeds in the zeolite pores. This has the advantage over the previously used cyclohexene/sulfuric acid system that the aqueous layer is not acidic and corrosive. Furthermore, the medium-pore structure impedes etherification to dicyclohexyl ether and the highly siliceous zeolite has long-term stability in boiling water.

8.7.1.2 Methanol-to-hydrocarbon Conversion

The conversion of methanol to hydrocarbons is the most studied reaction of oxygenates over microporous solids, for both commercial and academic reasons. Methanol can be generated from syngas over copper- and zinc-based catalysts using the ICI process, and syngas can be prepared from methane, which is a readily available resource. Under the correct economic conditions, methanol conversion reactions can provide an important route to higher hydrocarbons. The originally disclosed Mobil process[75] concerned the reaction of methanol over ZSM-5 to give a mixture of alkanes and aromatics similar to high-octane gasoline. This was operated in New Zealand for a number of years (the MTG process) but the high aromatics content of the product is no longer acceptable for gasoline. More recently, the conversion of methanol to light olefins over the small-pore aluminophosphate SAPO-34 (the UOP/Hydro MTO process[76]) has excited interest as a source of raw material for the production of polyolefins, and the first world-scale MTO commercial project is currently underway.

The methanol conversion reactions have been studied extensively. The hydrocarbon product distribution depends strongly on the zeotype catalyst used and on the contact time in the reactor. The first products are light olefins. For medium- or large-pore zeolites secondary reactions give rise to hydrocarbons with higher molecular weight as the contact time is increased, and deactivation due to the build up of larger hydrocarbon species in the zeolitic pores is also an important issue for large-pore solids. Most recently, the organic species formed within the pore space have been postulated to play a central role in the catalytic mechanism, which remains a matter of debate. In fact, the reaction occurs *via* five stages or regimes as shown in Scheme 8.11, most of which are well understood, and most discussion centres on the formation of the first C–C bonds.[77]

The first stage is the reaction of methanol over an acid site to give dimethylether and water. Modelling of this dehydration step suggests that it occurs *via* an S_N2 reaction between a protonated methanol molecule and a second, physisorbed methanol molecule, the transition state of which is stabilised by its interaction with the zeolite framework (Section 4.6.2). Experimental studies suggest that an alternative mechanism can operate at higher temperatures, by which methoxonium species are formed on the zeolite acid sites by water loss, and that these species can then react with physisorbed methanol. A second stage of the reaction is the so-called 'kinetic induction period' in which there are few hydrocarbon products in the product stream, and there is a build up of methoxy species on the catalyst. This precedes the third stage, which includes

	CH_3OH	⟶	$CH_3OCH_3 + H_2O$
Induction period	Methoxy species	⟶	Reactive Hydrocarbon Pool
C-C bond formation	Reactive Hydrocarbon Pool	⟶	C_2H_4, C_3H_6
	C_2H_4, C_3H_6	⟶	Oligomerisation, etc.
		⟶	Deactivation by pore filling

Scheme 8.11 Steps in the reaction of methanol over microporous solid acids.

the first C–C bond formation to give light olefins (ethene and propene). Subsequently, if pore shape and acidity permit, these light olefins can react further to give higher hydrocarbons in a fourth step. Finally, as hydrocarbons formed within the pores reduce access of molecules to and from the active sites, deactivation occurs. This is very rapid over SAPO-34 but much slower over ZSM-5, which deactivates slowly. This is attributed to the SAPO-34 cavities bound by 12 MRs allowing coke formation, whereas the narrower 10 MR channels of ZSM-5 inhibit formation of polyaromatic coke precursors.

A plot of the products of the reaction of methanol over H-ZSM-5 as a function of contact time is shown in Figure 8.10. Light olefins formed in the first step grow by oligomerisation or addition of methanol and dehydration: these larger species are then able to crack, skeletally isomerise and undergo cyclisations and intermolecular hydride transfers that give rise to alkanes and aromatics. The ZSM-5 structure contains sufficient space (for example in the channel intersections) for cyclisation and hydride transfers, and the 10 MR pores permit aromatics and branched alkanes to leave the pores. The final product distribution over different zeolites is therefore strongly controlled by the pore geometry. For larger-pore solids, such as Y or Beta, products of higher molecular weight are formed (briefly) but the pores are rapidly blocked and the catalyst deactivated. For the medium-pore ferrierite, where the one-dimensional channel structure does not have sufficient space for cyclisations or bimolecular hydride transfers, the products are dominated by linear butenes and pentenes, with few aromatics.[77] Finally, the small-pore SAPO-34 gives only

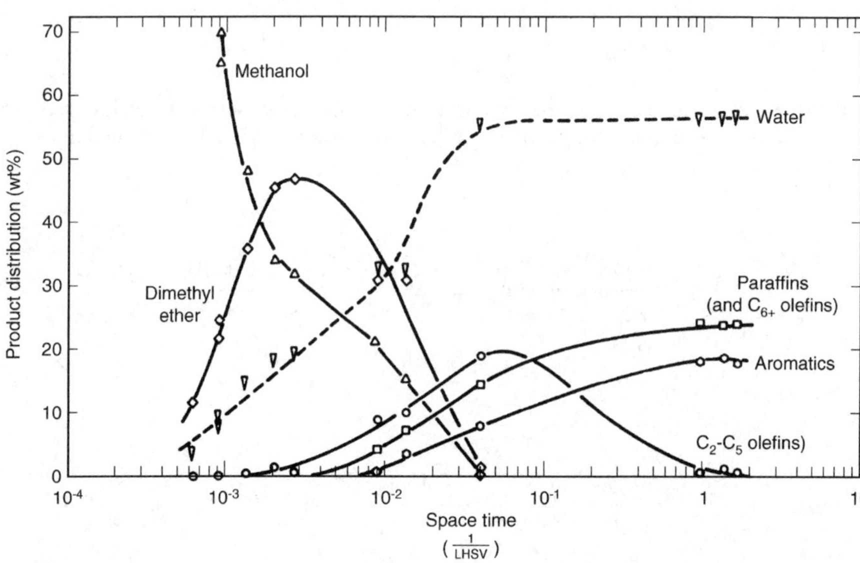

Figure 8.10 Variation of product distribution with contact time for the reaction of methanol to hydrocarbons over zeolite H-ZSM-5. [Figure reproduced from reference 75 with permission. Copyright 1977 Elsevier.]

ethene and propene as products, with traces of higher hydrocarbons. This arises in part because the structure consists of cages linked only by small-pore 8 MR windows, which only permit small linear alkenes to pass. The weaker acidity of the SAPO also plays a role, since it does not promote olefin oligomerisation.

The key mechanistic step in the reaction is the formation of the first C–C bond. Compelling evidence that this proceeds *via* organic species held within the pores has been obtained by the group of Haw through the use of *in situ* 'pulse quench' ^{13}C MAS NMR.[77] They have shown that the generation of light alkenes only occurs once a hydrocarbon pool of polymethylated benzenes or cyclic carbenium ions is established within the pores (Scheme 8.12).

Whereas cyclic carbenium ions (**a**, **b**) are stabilised by the more strongly acidic zeolites such as ZSM-5, only neutral cyclic species (such as **c**) are observed on the milder acid SAPO-34. Quantum mechanical modelling has been used to investigate the likely geometry of such species at the Brønsted acid sites and different reaction mechanisms have been proposed by which ethene and propene can be generated from these methylated aromatics within the pores. In one mechanism, the exocyclic methylation mechanism[77] (Scheme 8.13), small alkenes are removed from the aromatic rings, which remain the same size. In the 'paring' mechanism (as proposed for reactions over zeolite Beta[78]: Scheme 8.14) the aromatic C6 ring contracts, loses an external alkyl group as an alkene and

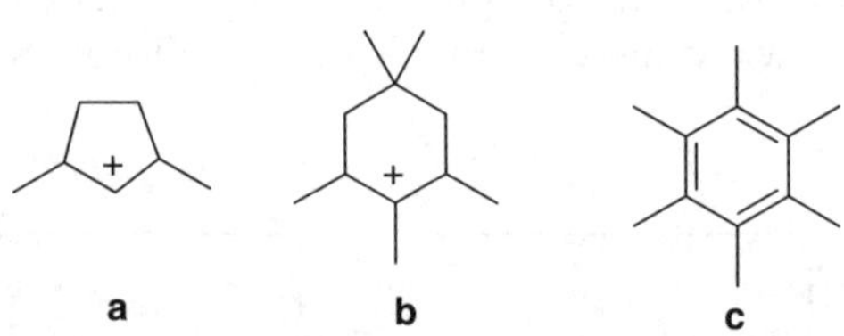

Scheme 8.12 Cyclic carbenium ions and neutral species observed on acidic zeolites and aluminophosphate solid acids during the reaction of methanol.

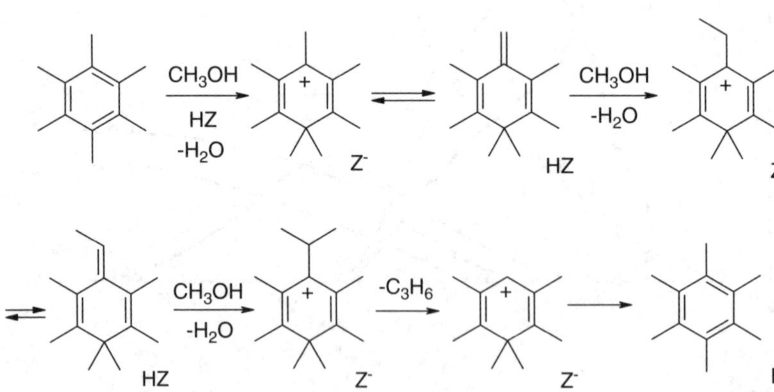

Scheme 8.13 The exocyclic methylation mechanism, by which substituents on the aromatic ring are formed and removed without the ring changing size.

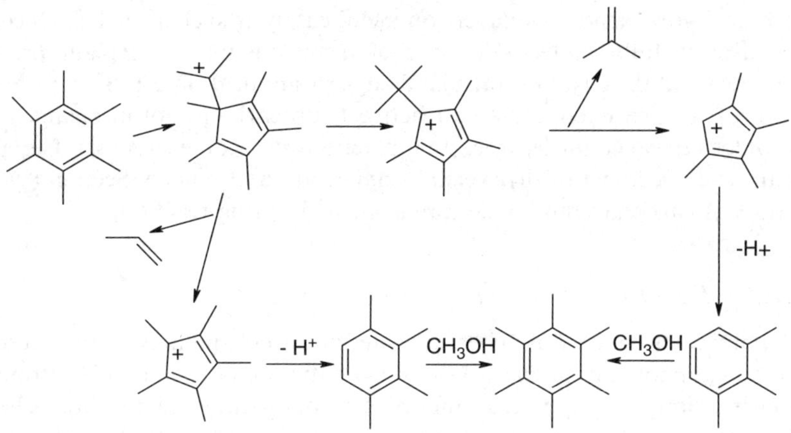

Scheme 8.14 The 'paring' mechanism, by which polymethylated aromatics give alkenes.

expands back to a C6 ring. The exact structural composition of the carbenium ions will depend on the space available in the pores – SAPO-34 has cages large enough to hold hexamethylbenzene, for example, and larger carbenium ions are formed in zeolite Beta than in ZSM-5.[79]

Whatever the exact details, the mechanism involves reaction of methanol with the methylated intermediates, resulting in loss of ethene and propene by stripping methyl and methylene groups off the rings. Isotope labelling evidence indicates that methyl groups from both the reactant methanol, and the hydrocarbon pool end up, scrambled, in the products. In addition, MAS NMR on the reaction performed under continuous flow conditions over ZSM-5, but with a sharp change in the isotopic composition of the methanol, has shown that around one half of the hydrocarbon pool is directly involved in generating products.[80] Haw describes the reaction as proceeding *via* 'supramolecular' inorganic-organic complexes, where the adsorbed organic species acts as a scaffold for the synthesis of light hydrocarbons, avoiding high-energy, direct mechanisms. Quantum mechanical calculations confirm that direct mechanisms have very high energies, whereas routes *via* polymethylated intermediates are energetically favourable.

The nature of the reactions in the induction period are of interest, since they must take place on exposure of the freshly prepared acid form of the zeotype catalyst to the reactants, and give rise to the active hydrocarbon pool. Although direct reactions have long been envisaged to be responsible for this first C–C bond forming event, Haw *et al.* consider that the observed features might also be explained by the presence of small amounts of impurities (such as ethanol) in the methanol feed.[77] In any case, the acid sites are responsible for the initial generation of the hydrocarbon pool.

The demonstration that a reactive hydrocarbon pool is of importance in the MTO and MTG reactions bears striking resemblance to the observed reactivity of reactive hydrocarbon overlayers on metal catalysts such as nickel. Once the mechanism is fully understood, it should be possible to explain product selectivities and deactivation rates and suggest an ideal pore topology for the reaction. It is also likely that the reactive hydrocarbon pool mechanism will have wider relevance for hydrocarbon reactions on zeolite catalysts. Certainly the catalytic selectivity of the skeletal isomerisation of butene (Section 8.7.2.1) increases as carbonaceous species are deposited during reaction.

8.7.1.3 Reaction of Alcohols with Ammonia

Methylamines are important chemical intermediates[3] and may be produced by alkylating ammonia using methanol over zeolitic catalysts containing Brønsted or Lewis acidity – a typical example of alcohol-ammonia reactions. When zeolites containing Brønsted acid sites are used as catalysts the ammonia is rapidly protonated and reacts with the methanol (or methoxy species) *via* a bimolecular dehydration step similar to that involved in etherification. The methylammonium ions that are formed are too strongly bound to be removed at the reaction temperatures, so that additional ammonia is required to scavenge the methyl groups or displace the methylamine from the zeolitic acid sites. Quantitatively, dimethylamine is in highest demand, with monomethylamine second, but trimethylamine is thermodynamically favoured. By running the reaction over forms of the small-pore zeolites Rho and ZK-5, DuPont have commercialised a process in which migration of the larger trimethylamine out of the pores is strongly hindered and dimethylamine yields are obtained that are well above those expected at thermodynamic equilibrium.[81]

In Section 8.5 the *in situ* ^{13}C MAS NMR of the methylation of aniline over H-Y was described. Methanol, DME and methoxy groups are all possible alkylating agents for the aniline, which gives *N*-methyl, *N,N*-dimethyl, *N,N,N*-trimethyl and C-methylated cations on the internal surface, the relative amounts of which depend on the temperature of the reaction. These then give products by deprotonation and desorption.

8.7.1.4 Rearrangements and Fine Chemicals Synthesis

Beckmann Rearrangement. The Beckmann rearrangement of cyclohexanone oxime to ε-caprolactam (Scheme 8.15) is an important step in the synthesis of ε-caprolactam, the monomer for nylon-6. Traditional reaction conditions involve the use of sulfuric acid as a catalyst for this conversion, and require

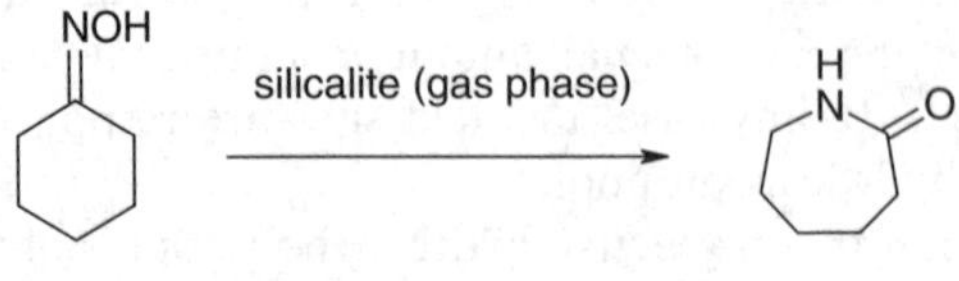

Scheme 8.15 Beckmann rearrangement of cyclohexanone oxime.

neutralisation of this acid once the reaction is complete, resulting in waste ammonium sulfate. Gas phase catalytic routes are therefore attractive, particularly where they can be developed in conjunction with improved routes to the oxime described in Chapter 9.

Following careful studies using a variety of zeolites, it was found that weakly acidic materials such as B-ZSM-5[82] and silicalite are found to be excellent catalysts for this reaction, giving selectivities in excess of 90% at high space velocities. Amorphous silica and silica-alumina display lower selectivity. The lifetime of the zeotype catalysts is increased by adding diluents such as water, toluene or acetonitrile to the feed, presumably preventing blocking of the acid sites, which are thought to be on the external surfaces of highly ordered crystallites. This reaction now forms part of a much-improved route to nylon-6, described in detail in Chapter 9.

8.7.1.4.1 Fine Chemicals Synthesis. Zeolitic solid acids have been shown to be promising catalysts for the synthesis of fine chemicals[83,84] that are of widespread use as, for example, flavours and fragrances. Some examples of transformations that have been performed with high selectivity are illustrated in Scheme 8.16, and include an acid-catalysed rearrangement of α-pinene oxide (**1**) to give campholenic aldehyde (**2**), used in sandalwood-like fragrances, and the acetalisation of phenylacetaldehyde (**3**) with glycerol to give acetals with a hyacinth fragrance (**4, 5**). Many acetals of this general kind find application as fragrances, and Climent *et al.* report the use of zeolitic solid acids for the synthesis of vanilla and blossom orange scented molecules.[85]

These reactions promise to replace current homogeneous routes that use strong mineral acids and metal salts with environmentally friendly routes without toxic waste. The chemical nature of the products (which may be large, polar molecules) can require that large-pore, hydrophobic solids be used as

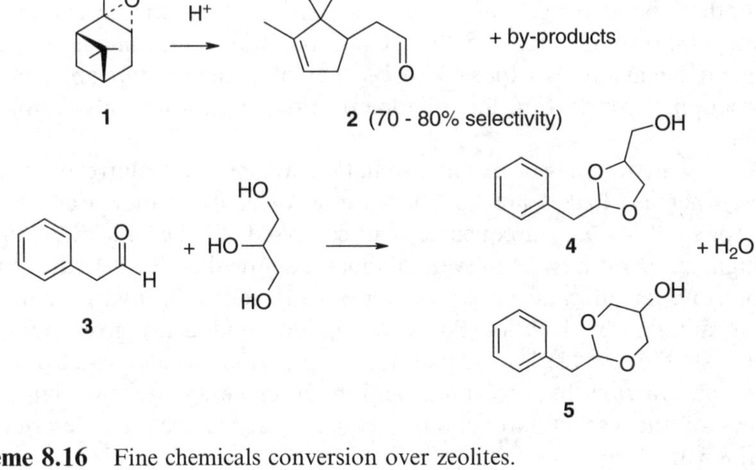

Scheme 8.16 Fine chemicals conversion over zeolites.

catalysts, to enable easy access to active sites and ready desorption from the pores.[86] Ultrastabilised Y is therefore of great utility in such reactions. For conversions involving molecules larger than *ca.* 8 Å, even-larger-pore solids are required, and it may be that mesoporous solids can play an important role here. The review of De Vos and Jacobs describes selected zeolite-catalysed reactions of interest in fine chemical conversions of this type.[87]

8.7.1.4.2 Dynamic Kinetic Resolution. Another typical acid-catalysed reaction is the racemisation of chiral alcohols, due to inversion at the chiral carbon. This can actually be made use of in the formation of enantiopure compounds, by dynamic kinetic resolution using an enzyme, such as a lipase, that catalyses enantioselective esterification in an organic medium. By coupling zeolite Beta-catalysed interconversion of benzylic alcohol enantiomers with enzyme-catalysed esterification of only one of the enantiomeric alcohols, almost complete conversion to enantiopure ester can be achieved.[88]

8.7.2 Hydrocarbon Conversions

The most important catalytic applications of zeolites are as solid acids in hydrocarbon transformations such as isomerisations, alkylations and cracking. Figure 8.11 gives a schematic view of the processes expected to be important in petroleum refining in 2010,[89] indicating those that employ a zeolitic acid catalyst: cracking, isomerisation, oligomerisation and bifunctional conversions including hydrocracking and reforming. As stated before, the associated reaction mechanisms are interpreted largely in terms of carbenium ion chemistry, *via* carbenium-like transition states. In fact, after adsorption of alkenes on Brønsted sites at low temperatures and upon increasing the temperature the main adsorbed species is the alkoxy species. The reactions are thought to go through transition states that can be interpreted in terms of partially charged carbenium ions interacting with both the bridging oxygen atom of the acid site and another, basic oxygen of the zeolite lattice. A schematic energy level diagram is shown in Figure 8.12.[90] Once the transition state is achieved, subsequent behaviour is expected to be that of a carbenium ion, including isomerisation, hydride transfer, oligomerisation, cyclisation, alkylation and scission.

Alkenes, and to a lesser extent aromatics, are highly reactive over acidic zeolites, whereas alkanes are not, because as weak bases they are not readily protonated. As a consequence, acid-catalysed alkane cracking requires very high temperatures. However, alkanes can readily be dehydrogenated over precious metal catalysts and alkenes can readily be hydrogenated, so that the admixture of Pt or Pd with zeolitic solid acids gives bifunctional catalysts that enable low temperature conversions of alkanes. Reforming of gasoline fraction hydrocarbons and hydrocracking are two important examples of the use of bifunctional zeolite catalysts that are described in Section 8.6.3.

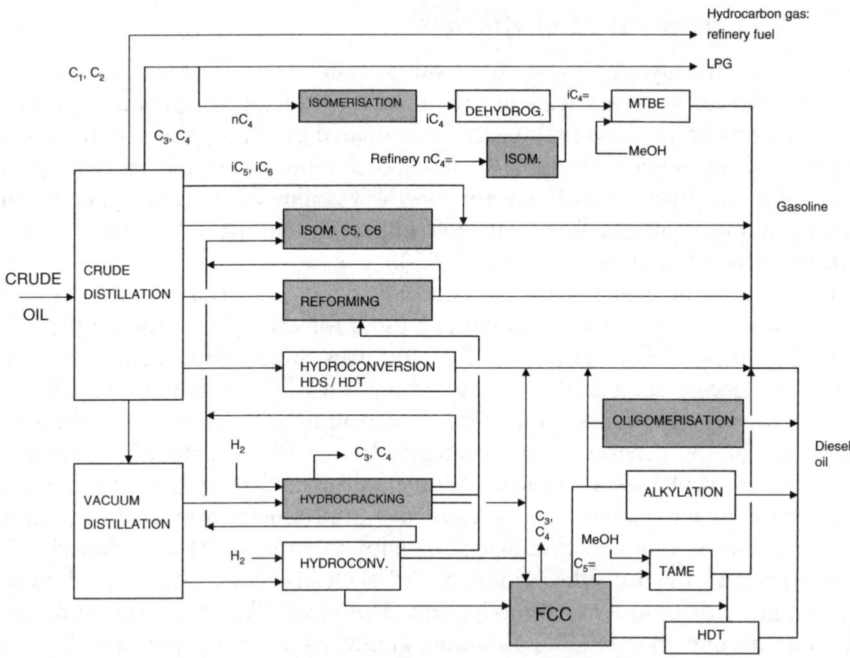

Figure 8.11 Refinery flow scheme showing the processes for treatment of the different fractions of crude oil. The main processes that use zeolite catalysts are indicated by shading. (HDS-hydrodesulfurisation, HDT-hydrotreating, FCC-fluidised catalytic cracking) [Adapted from reference 89 with permission. Copyright 1997 Wiley-VCH Verlag GmbH & Co. KGaA.]

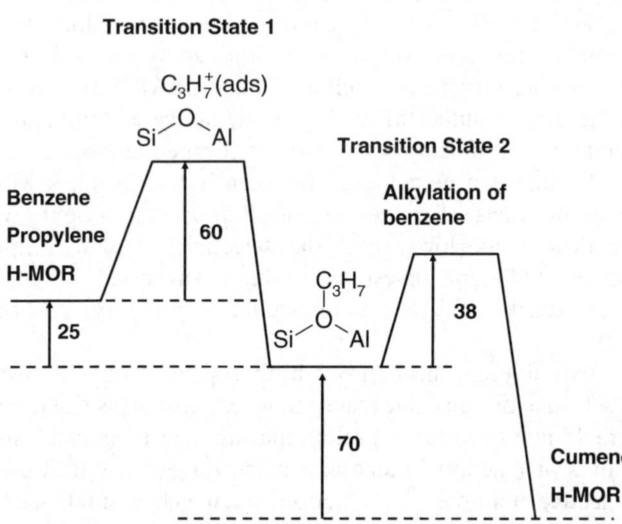

Figure 8.12 Energy level diagram for the alkylation of benzene with propylene, over the acid form of mordenite, as calculated by *ab initio* methods, and illustrating that the transition state is 'carbenium-ion-like'.

8.7.2.1 Isomerisation of Alkenes

Weak acids readily catalyse hydride shifts, resulting in double-bond isomerisation, whereas stronger acids are required for skeletal isomerisation because the reactions proceed *via* high-energy protonated cyclopropane ring transition states. Skeletal isomerisation of alkenes is one important step in increasing the branching of saturated hydrocarbons in the gasoline fraction and thereby the octane number of the fuel – this is important in the reforming reaction catalysed by bifunctional catalysts (8.6.3).

Direct skeletal isomerisation of normal alkenes is also important in the formation of isopentenes and isobutene used, for example, in the synthesis of TAME and MTBE ethers used as fuel additives. Because the isomerisation of *n*-pentene occurs *via* a secondary carbenium ion while formation of isobutene (2-methylpropene) occurs *via* a primary carbenium ion, stronger acid sites are required for the generation of isobutene. In the 1990s there was a surge in demand for MTBE as a gasoline additive, and therefore an increase in isobutene requirement. Isobutene is present in the products from cracking, along with *n*-butenes, but was insufficient to supply demand. Much research was therefore performed on the acid-catalysed skeletal isomerisation of *n*-butenes (but-1-ene and *cis*- and *trans*-but-2-ene) to isobutene. The ideal catalyst for this reaction is one that enables thermodynamic equilibrium between the four isomers to be set up at low temperatures (which favour isobutene) whilst not catalysing reactions (such as oligomerisation and cracking) that give rise to a wide range of products and low selectivity to C4 olefins.

Although H-ZSM-5 is an active catalyst for the reaction, it is very unselective, with its high acidity and spacious channel intersections favouring a high percentage of bimolecular interactions and, as a consequence, unwanted by-products (Figure 8.13).[91] The best catalysts for this reaction are found to be the acid forms of medium-pore zeolites and zeotypes with medium-pore one-dimensional pore structures, such as Theta-1, SAPO-31 and in particular ferrierite.[92] These have sufficient acidity to be active at temperatures as low as 300 °C, while not permitting secondary or competing reactions. Over these catalysts, equilibrium can be achieved between butene isomers with less than *ca.* 5% loss to products of higher or lower molecular weight. As a bonus, the catalysts deactivate slowly and the selectivity actually improves with time on stream. Different investigators have suggested both bimolecular (oligomerisation-cracking)[93] and unimolecular[94] pathways for the reaction (Scheme 8.17).

In fact, the two different mechanisms both appear to operate over ferrierite, with the less selective bimolecular reaction, which also gives rise to propene and pentene as cracking by-products, predominating over fresh catalysts, at higher partial pressures of reactant butene and when longer contact times are used. The more selective unimolecular reaction, by which a single carbenium ion rearranges *via* a protonated cyclopropane ring-containing C4 intermediate, is favoured at shorter contact times, lower butane concentrations and over catalysts that have been on stream longer. ^{13}C labelling experiments, in which

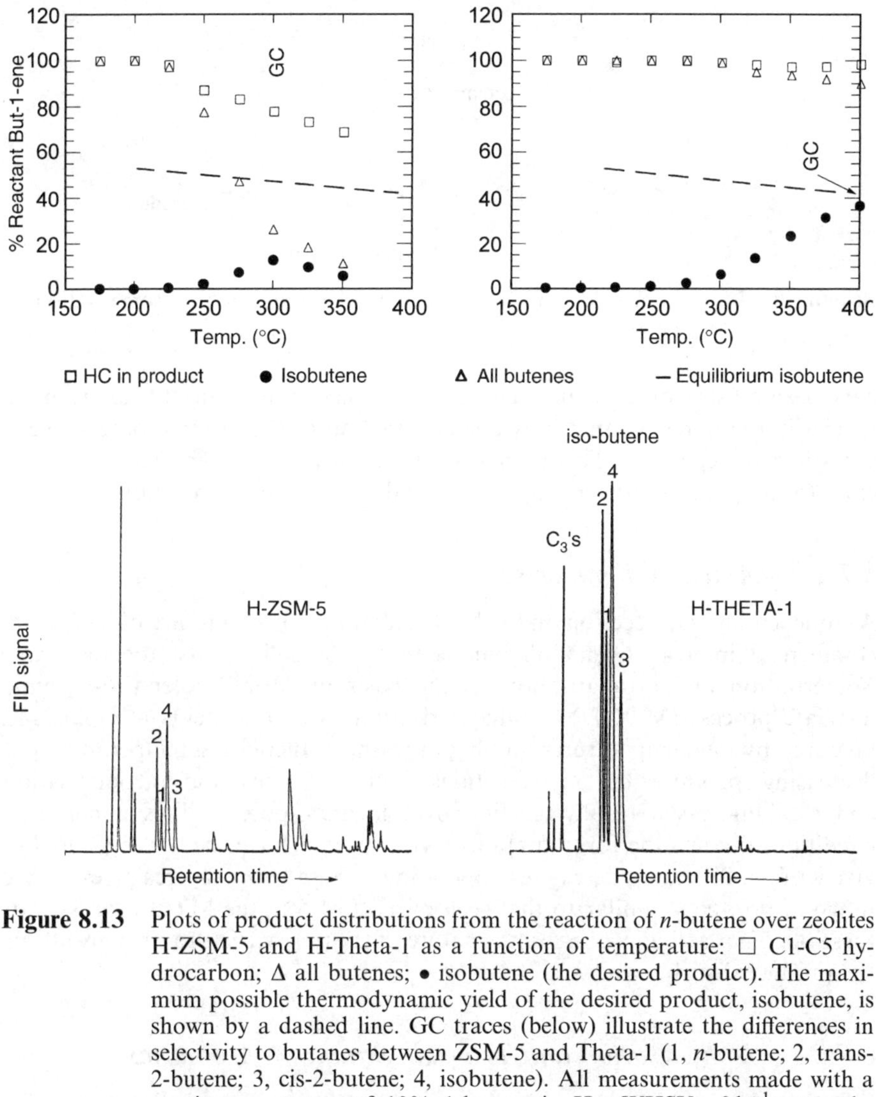

Figure 8.13 Plots of product distributions from the reaction of *n*-butene over zeolites H-ZSM-5 and H-Theta-1 as a function of temperature: □ C1-C5 hydrocarbon; Δ all butenes; • isobutene (the desired product). The maximum possible thermodynamic yield of the desired product, isobutene, is shown by a dashed line. GC traces (below) illustrate the differences in selectivity to butanes between ZSM-5 and Theta-1 (1, *n*-butene; 2, trans-2-butene; 3, cis-2-butene; 4, isobutene). All measurements made with a continuous stream of 10% 1-butene in He, WHSV = 3 h^{-1}, over the activated catalysts.

n-butene containing a single ^{13}C atom is used as reactant, support these models.[95] Under non-selective reaction conditions, product isobutene molecules have two, one or no ^{13}C atoms, indicating a dimerisation-cracking mechanism, whereas under selective conditions the isobutene product molecules still have one ^{13}C atom, indicating a unimolecular mechanism at an acid site. This also explains the observation that reaction selectivity increases on catalysts where the acid sites are further apart, and bimolecular reactions are less likely. The highest selectivity for this reaction is observed for catalysts that

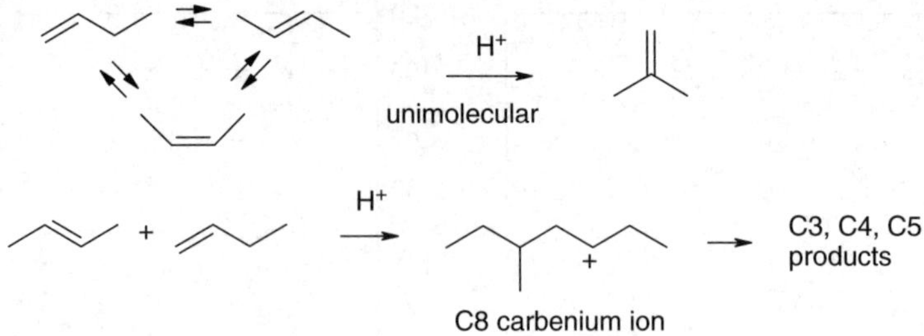

Scheme 8.17 Unimolecular and bimolecular mechanisms for the isomerisation of butenes.

have been on stream for some time and have considerable amounts of aromatic species in the pores, so that only a small fraction of the original pore space is accessible to reactants. This points to an important role for organic cations close to the pore mouths as important catalytic sites in the reaction.[96]

8.7.2.2 Alkene Oligomerisation

As touched on in the section on the MTO reaction, polyolefins are of increasing commercial interest. Olefin oligomerisation over solid acids, together with isomerisation and aromatisation, is the basis of Mobil's olefin-to-gasoline distillate process (MOGD).[97] Oligomerisation over solid acids is rapid, and favoured by low temperatures and high pressures. Increasing temperature and decreasing pressure favours reactions such as cyclisation, aromatisation and cracking. As a result, reaction over medium-pore zeolites under mild conditions favours the formation of polyolefins which can be hydrogenated to give jet fuel, for example, whereas operating at high temperatures gives a wide range of products, similar to that seen over ZSM-5 in the MTG process. For these oligomerisation processes, the nature of the pore structure strongly affects the product distribution.

8.7.2.3 Isomerisation and Transalkylation of Aromatics

Substituted aromatics are essential chemical feedstocks. Among the xylenes, for example, *p*-xylene is in great demand as a precursor to terephthalic acid, a polyester building block. The *para*-isomer is therefore more valuable than the *o*- and *m*-xylenes, so there is a powerful incentive for conversion of *o*- and *m*-xylene to *p*-xylene. Isomerisation over solid acids occurs readily as a result of alkyl shift reactions of the carbenium-ion-like transition state. The initial protonation occurs by interaction of the Brønsted acid site with the aromatic π system, by an electrophilic addition. Over non-microporous solid acids, at high conversion, xylenes are produced at their thermodynamically determined ratios, which favour the *meta* rather than the *ortho* or *para* isomers. In addition, unwanted transalkylation reactions occur, giving rise, for example, to toluene and trimethylbenzenes. Zeolite catalysts can be much more selective.

When zeolitic solid acids are used for the isomerisation of xylenes, the product distributions are found to depend strongly on the pore geometry. Large-pore zeolites give distributions similar to those over amorphous silica-aluminas, but the use of medium-pore zeolites gives *p*-xylene concentrations above equilibrium values and suppresses transalkylation reactions.[1,62] These advantages arise from shape selectivity due to the pore structure. Transalkylation proceeds *via* a bimolecular reaction that involves a bulky transition state and the space restrictions within medium-pore solids hinder its formation – an example of transition state selectivity. Furthermore, once a distribution of isomers is formed within the pores, the diffusion of the *para* isomer through the 10MRs is orders of magnitude higher than for the other isomers, and because diffusion is more rapid than isomerisation, its concentration in the product is favoured by product diffusion shape selectivity.

In a related process, the conversion of ethyl and propyl-substituted benzenes to more valuable xylenes, it is found that zeolites that contain interconnected 10- and 12MR channel systems, such as NU-87 and ITQ-23, show high selectivities to *p*-xylene. It is thought that while the larger aromatics enter and react within the large pores, the smaller xylenes can diffuse and leave the solid *via* the 10MR channels, leading to efficient and selective conversion.[98]

8.7.2.4 Alkylation of Aromatics

Solid acid catalysts have largely replaced the use of Friedel–Crafts reagents, such as $AlCl_3$, in the large scale alkylation of aromatics. The Friedel–Crafts route is stoichiometric rather than catalytic, produces much waste and involves the use of highly corrosive reagents. Furthermore, the concentrations of different compounds in the product are determined by equilibrium thermodynamics, and therefore less valuable isomers and by-products are also formed in the process. Zeolitic solid acids are not only true catalysts and easy to handle, but they also impart a high degree of selectivity to product distributions through transition state and product diffusion shape selectivity (Scheme 8.18).

Examples of commercial processes that use zeolitic catalysts for the alkylation of aromatics include the Mobil-Badger ethylbenzene process[3,99] and the

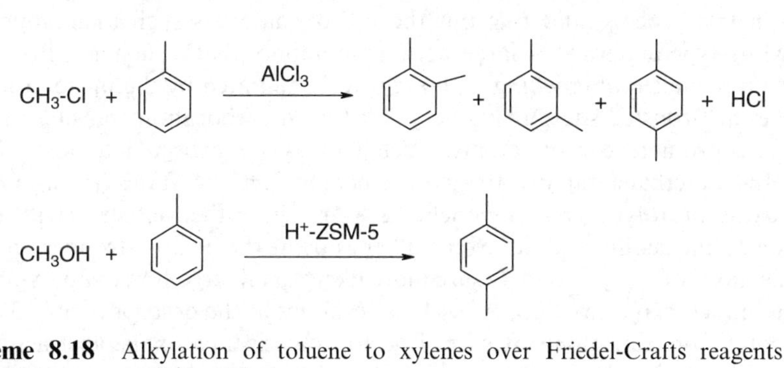

Scheme 8.18 Alkylation of toluene to xylenes over Friedel-Crafts reagents and zeolitic catalysts.

TAM process for toluene alkylation with methanol to give *p*-xylene.[100] In the ethylbenzene process, yields of 99% ethylbenzene (a precursor to styrene, the monomer of polystyrene) can be produced using ethene as the alkylating agent and H-ZSM-5 as the catalyst. A combination of high benzene-to-ethene ratios and shape selectivity strongly inhibits the formation of diethylbenzene by-product. In the TAM process, methanol is the alkylating agent and pore-size modification of ZSM-5 together with product diffusion selectivity ensures that up to 97% of the xylene product is the *para* isomer.

The selective alkylation of biphenyl and naphthalene is also possible, and is of commercial interest because 4,4′-diisopropylbiphenyl and 2,6-diisopropyl-naphthalene are used in the synthesis of speciality chemicals, for example for application as liquid crystal polymer materials for displays. For these larger molecules, large-pore zeolites such as mordenite give the best performance for isopropylation of naphthalene, introducing shape selectivity to linear (or more linear) dialkylated products over dialkylated products.[101]

8.7.2.5 *Alkane Transformation over Solid Acids: Cracking*

Catalytic cracking of high-molecular-weight hydrocarbons to more valuable compounds of lower molecular weight is a major application of solid acids. At present, more than 40% of oil conversion is performed using catalysts based on zeolites. The heavy hydrocarbon fraction left after the gas-oil fraction has been removed at low pressure includes alkanes, naphthenes (cyclic alkanes) and aromatics, the relative amounts of which depend on the geological origin of the oil. This fraction has to be cracked to more valuable fractions, such as the gasoline fraction, and to give alkenes and aromatics for use as petrochemical feedstocks. Cracking to low value dry gas (methane and ethane) or to carbon-aceous residues (coke) has to be avoided.

Advances in reactor design, such as the introduction of fluidised bed reactors in which a catalyst is on-stream for only a few seconds before it is stripped of hydrocarbons, removed, regenerated in air at 700 °C and recirculated, have long been of major importance in this reaction.[4,102] These chemical engineering requirements place strenuous requirements on the thermal stability of catalysts used in this process, and rule out the use of materials such as mesoporous MCM-41-type materials or large-pore aluminophosphates such as VPI-5.

The mechanism of cracking is thought to be initiated by the protonation of alkanes at Brønsted sites, giving very short-lived carbocation transition states with 5-coordinate carbon atoms. Such species are expected to lose H_2 or methane or ethane rapidly to give carbenium-ion-like states *via* a process known as protolytic cracking (Scheme 8.4). These then either (1) return a proton to the zeolite and desorb as alkenes or, if the acid is stronger and the alkene more strongly bound, (2) are more likely to undergo β-scission to give an olefin in a cracking reaction. As well as resulting in the desorption of alkenes, carbenium ion transition states can lead to isomerisation or the abstraction of hydride ions from other alkanes to give new alkane-carbenium ion transition state pairs. This bimolecular chain reaction generates cracked products and is favoured by strong acidity and high alkane concentrations in the pores.

Experimental evidence of cracking of *n*-butane over H-ZSM-5 at a range of conversions supports this model, so that the monomolecular reaction dominates at low conversions and the bimolecular mechanism becomes more important as conversion increases.[103] In particular, at low conversion the complementary products of breakdown of the carbonium ion intermediate (H_2 + butanes, methane + propylene and ethane + ethylene) are all 1:1. Additional models for the initial step include hydride abstraction over Lewis acids and protonation of trace levels of alkenes within the feed. In any case, the preferred reaction products are those of transformations *via* carbenium-ion-like transition states, rather than of protolytic cracking, which gives dry gas. Among the other hydrocarbon fractions, naphthenes undergo hydride abstraction leading to aromatics and faster rates of coke formation, whereas within the aromatic fraction, cracking of long alkyl side chains gives useful product.

Cracking catalysts include a combination of zeolite Y and smaller quantities of other zeolites. The large pores of zeolite Y permit access of molecules of free diameters up to *ca.* 8 Å to the acid sites. Commercially available Y is prepared by ion exchange of the as-prepared sodium form (Si/Al = 2.4–2.8) into which acidity is introduced by ion exchange with rare earth cations (such as La^{3+}) and/or by ammonium ions. Heating and steaming leaves acidic protons and also extra-framework aluminium cations, so that the catalysts contain Brønsted and Lewis acid sites. Catalytic activity increases as the Si/Al ratio is increased up to a value of 6–8 as the site strength increases and then decreases as the aluminium content is further decreased. Lowering acid site density further also results in increased production of dry gas, whereas the use of steamed zeolite Y increases the rate of hydride abstraction and improves selectivity to more valuable liquid products.

The catalytic effect of steaming Y to give ultrastable Y is very marked, and the activity for cracking can be increased by up to two orders of magnitude. This has been attributed variously to the effects of special sites, generated by interaction between bridging hydroxyl groups and extra-framework aluminium cations, the increased surface area generated by the introduction of secondary mesoporosity and by the effect of Lewis acidity in initiating the reaction by hydride transfer (thought to be least likely). According to Kung *et al.*,[104,105] the rate enhancement is best accounted for in terms of the different cracking mechanisms that operate. These include the monomolecular and bimolecular routes described above, plus an 'oligomerisation' route, in which adsorbed and protonated oligomeric species are important, rather like the reactive hydrocarbon pool in the methanol-to-hydrocarbon conversions. Whereas the monomolecular reaction, which dominates at low conversion, depends mainly on acid site strength, the other mechanisms are strongly dependent on the concentration of alkenes or oligomers within the pores. Since reactions of alkenes (and oligomers) are orders of magnitude faster than protolytic cracking, features of the catalyst that promote associated bimolecular reactions will have a major enhancing effect on the overall reaction rate. Additional surface area and so greater adsorption and more available diffusion pathways are produced by steaming and therefore enhance the rate.

As well as reducing molecular weight, the cracking process can be tuned to give different product distributions, according to demand. The addition of smaller amounts of other zeolites such as Beta or ZSM-5 results in an increase in branching and increased levels of useful isobutene and isopentenes while keeping a good gasoline selectivity.[106]

8.7.3 Bifunctional Catalysis

8.7.3.1 Alkane Isomerisation: Reforming

Although isomerisation of alkanes does occur over acid zeolites at high temperatures, the selectivity is low because of the competing cracking reactions: *n*-alkane isomerisation therefore requires an alternative approach. Given that alkenes are readily skeletally isomerised by acid catalysts and that *n*-alkanes are easily dehydrogenated (and alkenes hydrogenated) over precious metals such as platinum and palladium, bifunctional catalysts (solid acid/Pt or Pd) are ideal for alkane branching reactions important, for example, in gasoline reforming.

The aim of gasoline reforming is to increase the octane number of gasoline, important in its use as a fuel, without reducing its average molecular weight. This is achieved by increasing the degree of branching and its content of aromatics, although environmental demands are currently driving down the tolerance to aromatics in fuel. Reforming catalysts typically consist of platinum supported on an acidic oxide, such as alumina. Fine-tuning of the Pt/Al_2O_3 catalyst has involved alloying the metal to reduce hydrocracking and chlorinating the alumina to strengthen the acid strength. A very clear account of the process is given by Gates.[4]

Platinum supported on zeolites such as mordenite is also active[107] and the isomerisation of longer-chain alkanes is achieved with such catalysts, under mild conditions. The essential details of the process are illustrated in Scheme 8.19, with the metal function being to establish equilibrium between alkanes, hydrogen and a low level of olefins. The olefin content must be sufficient to allow isomerisation but sufficiently diluted by alkanes to prevent rapid oligomerisation and coking of the catalyst. The performance of the bifunctional catalyst is improved by the introduction of secondary mesoporosity in the mordenite catalyst, to enhance molecular transport between active site and gas stream, and by the generation of extra-framework aluminium sites that promote Lewis acid catalysed reactions.

8.7.3.2 Hydrocracking

Bifunctional zeolitic catalysts are also used for hydrocracking, in which the heaviest fraction of crude oil is cracked in the presence of hydrogen to give

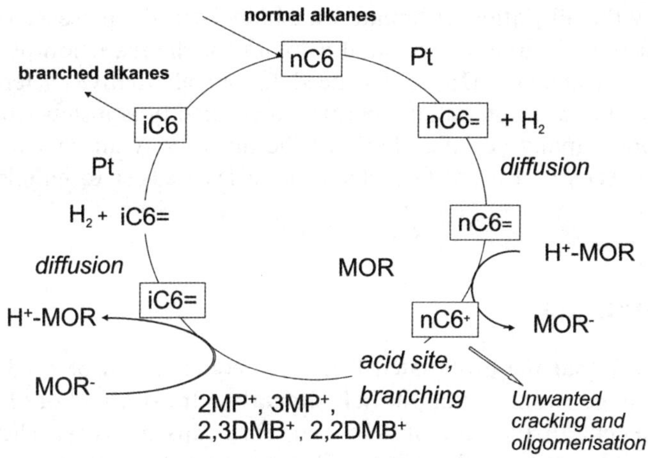

Scheme 8.19 Hydroisomerisation of *n*-hexane over bifunctional zeolite catalysts.[107]

more valuable products. In hydrocracking of the very heavy fraction of crude oil, acid catalysts loaded with precious metals are used in the presence of *ca.* 100 bar H_2 to give lighter products *via* cracking and hydrogenation, enabling maximum value to be extracted from each barrel of oil. By using bifunctional catalysts it is possible to crack large hydrocarbons at lower temperatures (350–450 °C) than those required for protolytic cracking, because the olefins generated in the mix are more reactive than alkanes. The hydrogen content of the heavy residue is lower than that of the desired product, so the addition of hydrogen is achieved by reactions such as the hydrogenation of aromatics, hydrodecyclisation, hydrodealkylation of aromatics and hydrocracking of alkanes. The acid sites catalyse skeletal isomerisation and cracking. Zeolite Y is in commercial use for hydrocracking, giving better activity, liquids yield and thermal stability than amorphous silica-aluminas. The secondary mesoporosity present in steamed, ultrastabilised Y is important to accelerate the diffusion of the large molecules present in the heavy fraction to gain access to the active sites.

8.7.4 Limitations of Microporous Solid Acid Catalysts

Although zeolitic solid acids have replaced supported mineral acids in many reactions, they cannot be used universally. For some reactions their acidity is too strong and their pore structure is readily blocked (etherification, alkene hydration); for others they are not strong enough acids. An example where zeolites have not been applied successfully is in the alkylation of alkanes with alkenes. Hydrocarbon fuels with high octane numbers are in constant demand for the majority of automobile engines, and branched alkanes of the gasoline fraction possess the required properties. Isoalkanes such as these are currently prepared by the alkylation of butane with butenes in the presence of HF and H_2SO_4, but there is an environmental demand for the reduction of the use of these acids, particularly HF, and a need for an alternative, heterogeneous, catalyst. Zeolites are not suitable because their acidity is insufficiently strong and the pores rapidly become blocked. The most likely alternatives are supported acids such as BF_3/Al_2O_3, which are at least easier to handle than the liquid acids.

8.8 Summary

The discovery that the protonated forms of zeolites could be used as active, stable and shape selective catalysts in hydrocarbon transformations has been of immense benefit to the refining and petrochemicals industry. The need for optimised microporous acid catalysts will continue as fuel specifications change and the requirements of the chemicals market shift. The likely growth in demand for synthetic fuels, including diesels, is one expected trend that could involve zeolite catalysts. Diverse feedstock chemicals and fine chemicals synthesis involving zeolite catalysts are also being developed.

As a result of this importance, great efforts have been made to understand the interplay between structure and chemistry to produce optimised acid catalysts for processes such as cracking, alkylation and isomerisation. It is now well established that zeolites are not superacidic, so that the apparent carbenium ion controlled conversions are thought to pass through carbenium-ion-like transition states stabilised by the zeolite framework. For methanol-to-hydrocarbon reactions, elegant *in situ* NMR has demonstrated that a reactive hydrocarbon pool that forms within the pores is observed to be responsible for the formation of the first C–C bonds, and it is likely that reactive hydrocarbon intermediates have a greater role in acid-catalysed reactions than previously spelt out.

The importance of the pore structure in exerting shape selective control over product distributions is well established. ZSM-5 type zeolites, for example, have a special role in selective alkylations and isomerisations of monoaromatics *via* diffusion and transition state selectivity. There is growing appreciation that subtle 'inverse shape selective' effects are also important, for example in hydroisomerisations, and that reactions and separations can be controlled by entropic as well as enthalpic effects. New structures could offer additional catalytic possibilities: the large-pore high silica material ITQ-21, which possesses a three-dimensional channel system with interconnected channels, could assist in the (hydro)cracking of larger oil molecules, and an aluminosilicate version of the 18 MR germanosilicate zeotype ITQ-33 would have distinct advantages for the cracking of large molecules – but the expense of the structure-directing agents required in their synthesis counts against them for bulk, lower value products such as fuel and petrochemicals. No crystalline solids with much larger pore systems have been prepared, and mesoporous solids, which do have the desired pore sizes, do not have the necessary strong acidity or hydrothermal stability for cracking applications.

As well as the crystal structure of the microporous catalysts, the secondary mesoporosity is also important, because molecular transport to and from the active sites is favoured in these materials. In steamed Y the mesoporosity and extra-framework aluminium results in a very active catalyst for cracking. Designed hierarchical structures, in which nanoparticles of zeolites are joined together to and connected by a secondary mesopore system for the same reason are discussed further in Chapter 10.

References

1. A. Corma, *Chem. Rev.*, 1995, **95**, 559.
2. K. Tanabe and W. F. Holderich, *Appl. Catal. A*, 1999, **181**, 399.
3. K. Weissermel and H.-J. Arpe, *'Industrial Organic Chemistry'*, VCH, 1993.
4. B. C. Gates, *'Catalytic Chemistry'*, John Wiley, 1991.
5. G. A. Olah, *Angew. Chem. Int. Ed.*, 1973, **12**, 173.
6. L. P. Hammett and A. J. Deyrup, *J. Am. Chem. Soc.*, 1932, **54**, 2721.
7. A. Corma, *Chem. Rev.*, 1997, **97**, 2373.
8. J. A. Rabo, *Cat. Rev.-Sci. Eng.*, 1981, **23**, 293.
9. A. Corma, M. J. Diaz-Cabanas, J. L. Jorda, C. Martinez and M. Moliner, *Nature*, 2006, **443**, 842.
10. S. J. Miller, *Micropor. Mater.*, 1994, **2**, 439.
11. A. Philippou and M. W. Anderson, *J. Catal.*, 2000, **189**, 395.
12. G. Gelbard, *Ind. Eng. Chem. Res.*, 2005, **44**, 8468.
13. D. Freude, J. Klinowski and H. Hamdan, *Chem. Phys. Lett.*, 1988, **149**, 355.
14. R. W. Joyner, A. D. Smith, M. Stockenhuber and M. W. E. van den Berg, *Phys. Chem. Chem. Phys.*, 2004, **6**, 5435.
15. J. A. van Bokhoven, A. M. J. van der Eerden and D. C. Koningsberger, *J. Am. Chem. Soc.*, 2003, **125**, 7435.
16. M. Hunger, *Solid State Nucl. Mag. Reson.*, 1996, **6**, 1.
17. M. Czjzek, H. Jobic, A. N. Fitch and T. Vogt, *J. Phys. Chem.*, 1992, **96**, 1535.
18. L. Smith, A. K. Cheetham, L. Marchese, J. M. Thomas, P. A. Wright, J. Chen and E. Gianotti, *Catal. Lett.*, 1996, **41**, 13.
19. W. E. Farneth and R. J. Gorte, *Chem. Rev.*, 1995, **95**, 615.
20. C. Lee, D. J. Parrillo, R. J. Gorte and W. E. Farneth, *J. Am. Chem. Soc.*, 1996, **118**, 3262.
21. D. J. Parrillo and R. J. Gorte, *J. Phys. Chem.*, 1993, **97**, 8786.
22. A. Zecchina and C. O. Arean, *Chem. Soc. Rev.*, 1996, **25**, 187.
23. C. Paze, S. Bordiga, C. Lamberti, M. Salvalaggio, A. Zecchina and G. Bellussi, *J. Phys. Chem. B*, 1997, **101**, 4740.

24. A. Zecchina, G. Spoto and S. Bordiga, *Phys. Chem. Chem. Phys.*, 2005, **7**, 1627.
25. G. Herzberg, '*Infrared and Raman Spectra of Polyatomic Molecules*', van Norstrand, New York, 1945.
26. A. Zecchina, S. Bordiga, G. Spoto, D. Scarano, G. Spano and F. Geobaldo, *J. Chem. Soc. Farad. Trans.*, 1996, **92**, 4863.
27. G. Mirth, J. A. Lercher, M. W. Anderson and J. Klinowski, *J. Chem. Soc. Farad. Trans.*, 1990, **86**, 3039.
28. L. J. Bellamy, H. E. Hallam and R. L. Williams, *Trans. Farad. Soc.*, 1958, **54**, 1120.
29. S. Bordiga, L. Regli, C. Lamberti, A. Zecchina, M. Jorgen and K. P. Lillerud, *J. Phys. Chem. B*, 2005, **109**, 7724.
30. P. A. Wright, C. Sayag, F. Rey, D. W. Lewis, J. D. Gale, S. Natarajan and J. M. Thomas, *J. Chem. Soc. Farad. Trans.*, 1995, **91**, 3537.
31. J. H. Lunsford, P. N. Tutunjian, P. J. Chu, E. B. Yeh and D. J. Zalewski, *J. Phys. Chem.*, 1989, **93**, 2590.
32. J. F. Haw, J. B. Nicholas, T. Xu, L. W. Beck and D. B. Ferguson, *Acc. Chem. Res.*, 1996, **29**, 259.
33. J. F. Haw, B. R. Richardson, I. S. Oshiro, N. D. Lazo and J. A. Speed, *J. Am. Chem. Soc.*, 1989, **111**, 2052.
34. G. B. McVicker, G. M. Kramer and J. J. Ziemiak, *J. Catal.*, 1983, **83**, 286.
35. C. E. Ramachandran, B. A. Williams, J. A. van Bokhoven and J. T. Miller, *J. Catal.*, 2005, **233**, 100.
36. S. M. Babitz, B. A. Williams, J. T. Miller, R. Q. Snurr, W. O. Haag and H. H. Kung, *Appl. Catal. A*, 1999, **179**, 71.
37. M. Hunger and J. Weitkamp, *Angew. Chem. Int. Ed.*, 2001, **40**, 2954.
38. M. Hunger, *Micropor. Mesopor. Mater.*, 2005, **82**, 241.
39. E. J. Munson, D. K. Murray and J. F. Haw, *J. Catal.*, 1993, **141**, 733.
40. T. Xu and J. F. Haw, *Top. Catal.*, 1997, **4**, 109.
41. P. W. Goguen, T. Xu, D. H. Barich, T. W. Skloss, W. G. Song, Z. K. Wang, J. B. Nicholas and J. F. Haw, *J. Am. Chem. Soc.*, 1998, **120**, 2650.
42. T. Xu, D. H. Barich, P. W. Goguen, W. G. Song, Z. K. Wang, J. B. Nicholas and J. F. Haw, *J. Am. Chem. Soc.*, 1998, **120**, 4025.
43. M. Hunger, M. Seiler and A. Buchholz, *Catal. Lett.*, 2001, **74**, 61.
44. M. W. Anderson and J. Klinowski, *Nature*, 1989, **339**, 200.
45. M. W. Anderson and J. Klinowski, *J. Am. Chem. Soc.*, 1990, **112**, 10.
46. M. W. Anderson and J. Klinowski, *Chem. Phys. Lett.*, 1990, **172**, 275.
47. D. K. Murray, J. W. Chang and J. F. Haw, *J. Am. Chem. Soc.*, 1993, **115**, 4732.
48. W. Wang, M. Seiler, I. I. Ivanova, U. Sternberg, J. Weitkamp and M. Hunger, *J. Am. Chem. Soc.*, 2002, **124**, 7548.
49. J. B. Nicholas and J. F. Haw, *J. Am. Chem. Soc.*, 1998, **120**, 11804.
50. V. B. Kazansky, *Catal. Today*, 1999, **51**, 419.
51. W. J. Mortier, *J. Catal.*, 1978, **55**, 138.

52. W. O. Haag, R. M. Lago and P. B. Weisz, *Nature*, 1984, **309**, 589.
53. S. T. Wilson and E. M. Flanigen, *ACS Symp. Ser.*, 1989, **398**, 329.
54. D. B. Akolekar, *J. Chem. Soc. Farad. Trans.*, 1994, **90**, 1041.
55. E. Kassab, K. Seiti and M. Allavena, *J. Phys. Chem.*, 1988, **92**, 6705.
56. D. J. Parrillo, C. Lee and R. J. Gorte, *Appl. Catal. A*, 1994, **110**, 67.
57. C. P. Grey and A. J. Vega, *J. Am. Chem. Soc.*, 1995, **117**, 8232.
58. C. A. Fyfe, J. L. Bretherton and L. Y. Lam, *J. Am. Chem. Soc.*, 2001, **123**, 5285.
59. H. G. Karge, S. Ernst, M. Weihe, U. Weiss and J. Weitkamp, *Stud. Surf. Sci. Catal.*, 1994, **84**, 1805.
60. S. M. Csicsery, *Zeolites*, 1984, **4**, 202.
61. P. B. Weisz, V. J. Frilette, R. W. Maatman and E. B. Mower, *J. Catal.*, 1962, **1**, 307.
62. J. A. Martens, J. Pérez-Pariente, E. Sastre, A. Corma and P. A. Jacobs, *Appl. Catal.*, 1988, **45**, 85.
63. J. F. M. Denayer, R. A. Ocakoglu, I. C. Arik, C. E. A. Kirschhock, J. A. Martens and G. V. Baron, *Angew. Chem. Int. Ed.*, 2005, **44**, 400.
64. J. F. M. Denayer, R. A. Ocakoglu, K. De Meyer and G. V. Baron, *Adsorption*, 2005, **11**, 49.
65. M. Schenk, S. Calero, T. L. M. Maesen, L. L. van Benthem, M. G. Verbeek and B. Smit, *Angew. Chem. Int. Ed.*, 2002, **41**, 2500.
66. T. L. M. Maesen, S. Calero, M. Schenk and B. Smit, *J. Catal.*, 2004, **221**, 241.
67. L. Forni, *Catal. Today*, 1998, **41**, 221.
68. V. J. Frilette, W. O. Haag and R. M. Lago, *J. Catal.*, 1981, **67**, 218.
69. S. I. Zones and T. V. Harris, *Micropor. Mesopor. Mater.*, 2000, **35–6**, 31.
70. J. A. Martens, M. Tielen, P. A. Jacobs and J. Weitkamp, *Zeolites*, 1984, **4**, 98.
71. J. Weitkamp, S. Ernst and R. Kumar, *Appl. Catal.*, 1986, **27**, 207.
72. G. B. McVicker, O. C. Feeley, J. J. Ziemiak, D. E. W. Vaughan, K. C. Strohmaier, W. R. Kliewer and D. P. Leta, *J. Phys. Chem.*, 2005, **109**, 2222.
73. G. B. McVicker, M. Daage, M. S. Touvelle, C. W. Hudson, D. P. Klein, W. C. Baird, B. R. Cook, J. G. Chen, S. Hantzer, D. E. W. Vaughan, E. S. Ellis and O. C. Feeley, *J. Catal.*, 2002, **210**, 137.
74. T. Okuhara, *Chem. Rev.*, 2002, **102**, 3642.
75. C. D. Chang and A. J. Silvestri, *J. Catal.*, 1977, **47**, 249.
76. J. Q. Chen, A. Bozzano, B. Glover, T. Fuglerud and S. Kvisle, *Catal. Today*, 2005, **106**, 103.
77. J. F. Haw, W. G. Song, D. M. Marcus and J. B. Nicholas, *Acc. Chem. Res.*, 2003, **36**, 317.
78. M. Bjorgen, U. Olsbye, D. Petersen and S. Kolboe, *J. Catal.*, 2004, **221**, 1.
79. W. G. Song, J. B. Nicholas, A. Sassi and J. F. Haw, *Catal. Lett.*, 2002, **81**, 49.

80. M. Seiler, W. Wang, A. Buchholz and M. Hunger, *Catal. Lett.*, 2003, **88**, 187.
81. D. R. Corbin, S. Schwarz and G. C. Sonnichsen, *Catal. Today*, 1997, **37**, 71.
82. G. P. Heitmann, G. Dahlhoff, J. P. M. Niederer and W. F. Holderich, *J. Catal.*, 2000, **194**, 122.
83. W. F. Holderich, J. Roseler, G. Heitmann and A. T. Liebens, *Catal. Today*, 1997, **37**, 353.
84. M. J. Climent, A. Corma, A. Velty and M. Susarte, *J. Catal.*, 2000, **196**, 345.
85. M. J. Climent, A. Corma and A. Velty, *Appl. Catal. A: Gen.*, 2004, **263**, 155.
86. A. Corma, *J. Catal.*, 2003, **216**, 298.
87. D. E. De Vos and P. A. Jacobs, *Micropor. Mesopor. Mater.*, 2005, **82**, 293.
88. S. Wuyts, K. De Temmerman, D. E. De Vos and P. A. Jacobs, *Chem.-Eur. J.*, 2004, **11**, 386.
89. G. Martino, P. Courty and C. Marcilly, in '*Handbook of Heterogeneous Catalysis*', Eds. G. Ertl, H. Knozinger, J. Weitkamp, VCH, 1997, p. 1801.
90. X. Rozanska and R. A. van Santen, in '*Computer Modelling of Microporous Materials*', Eds. C. R. A. Catlow, R. A. van Santen and B. Smit, 2004, Elsevier, Amsterdam, p. 165.
91. S. Natarajan, P. A. Wright and J. M. Thomas, *J. Chem. Soc. Chem. Commun.*, 1993, 1861.
92. H. H. Mooiweer, K. P. de Jong, B. Kraushaar-Czarnetzki, W. H. J. Stork and B. C. H. Krutzen, *Stud. Surf. Sci. Catal.*, 1994, **84**, 2327.
93. M. Guisnet, P. Andy, N. S. Gnep, E. Benazzi and C. Travers, *J. Catal.*, 1996, **158**, 551.
94. M. A. Asensi, A. Corma and A. Martinez, *J. Catal.*, 1996, **158**, 561.
95. P. Meriaudeau, R. Bacaud, L. N. Hung and A. T. Vu, *J. Mol. Cat. A.*, 1996, **110**, L177.
96. P. Andy, N. S. Gnep, M. Guisnet, E. Benazzi and C. Travers, *J. Catal.*, 1998, **173**, 322.
97. S. A. Tabak, F. J. Krambeck and W. E. Garwood, *AIChE J.*, 1986, **32**, 1526.
98. J. M. Serra, E. Guillon and A. Corma, *Stud. Surf. Sci. Catal.*, 2005, **158**, 1757.
99. P. J. Lewis and F. G. Dwyer, *Abstr. Pap. Am. Chem. Soc.*, 1977, **173**, 22.
100. W. W. Kaeding, C. Chu, L. B. Young, B. Weinstein and S. A. Butter, *J. Catal.*, 1981, **67**, 159.
101. J. Wang, J. N. Park, Y. K. Park, S. Y. Han, J. O. Baeg and C. W. Lee, *Catal. Today*, 2004, **97**, 283.
102. C. J. Plank, E. J. Rosinski and W. P. Hawthorne, *Ind. Eng. Chem. Prod. Res. Dev.*, 1964, **3**, 165.
103. H. Krannila, W. O. Haag and B. C. Gates, *J. Catal.*, 1992, **135**, 115.

104. B. A. Williams, S. M. Babitz, J. T. Miller, R. Q. Snurr and H. H. Kung, *Appl. Catal. Al.*, 1999, **177**, 161.
105. H. H. Kung, B. A. Williams, S. M. Babitz, J. T. Miller and R. Q. Snurr, *Catal. Today*, 1999, **52**, 91.
106. A. Corma, V. Gonzalez-Alfaro and A. V. Orchilles, *Appl. Catal. A: Gen.*, 1999, **187**, 245.
107. P. B. Koradia, J. R. Kiovsky and M. Y. Asim, *J. Catal.*, 1980, **66**, 290.

CHAPTER 9

Further Catalytic Applications of Microporous Solids

9.1 Introduction

The most important catalytic application of microporous solids is as solid acids in petrochemicals refining and in the synthesis of commodity and fine chemicals. In recent years, however, the potential for using their unique structural chemistry in other catalytic reactions has been investigated with some important successes. Most notably, the last twenty years has seen the development of microporous silicates bearing framework Lewis acid sites, such as titanium and tin, as important catalysts for a range of selective transformations of organic molecules.[1] Examples of Lewis-acid-catalysed reactions that have been performed with high selectivities, and that are an improvement on existing reactions, include selective oxidations (epoxidations, ammoximations, hydroxylation and alcohol oxidation over titanium-bearing zeolites, Baeyer–Villiger ketone oxidation over tin-bearing zeolite β) and the Meerwein–Ponndorf–Verley reduction of ketones (and the reverse Oppenauer oxidation of alcohols). Furthermore, aluminophosphates that contain transition metals such as manganese, iron and cobalt in framework positions have been shown to be active in the selective oxidation of alkanes using air as an oxidant *via* free radical mechanisms.

There has also been an enormous amount of work on the investigation of cation-exchanged zeolites as catalysts. Among these, zeolites containing the first row transition metal cations iron, cobalt and copper show real potential for use in environmental catalysis such as that required in automobile exhaust catalysis, particularly in the so-called lean burn deNO_x reactions (selective catalytic reduction). Other such zeolites that show specific activity for reactions of petrochemical and synthetic interest include those containing extra-framework gallium (dehydrocyclisation of small alkanes to give aromatics), iron and molybdenum-containing zeolites (oxidation of benzene, synthesis of aromatics from methane), palladium-bearing zeolites (active for the Heck reaction for C–C bond formation) and caesium-containing solids (as basic catalysts).

In all these solids, the catalytic activity derives from a combination of the high surface area and the unique environment of the species isolated within the pores. In principle, high surface area zeolitic solids can be used as highly thermally and hydrothermally stable supports for very many catalytic species, including metal and metal oxide nanoparticles, but the full range of such materials is outside the scope of this text. Special mention is made here of attempts to synthesise catalytic complexes within the pores of microporous solids *via* the so-called ship-in-a-bottle route. This has the attraction of preventing catalyst loss during catalytic processes, but is limited by the chemistry of the complexes introduced in this way and the space available for catalysis once the catalytic complex is in place. Recent approaches to the grafting of organometallic complexes onto mesoporous solids appear to show more general promise for catalyst preparation, their pore size enabling access to active sites of reactant molecules that are far larger than those that can enter crystalline microporous solids.

9.2 Framework Lewis Acids as Selective Oxidation Catalysts

9.2.1 Selective Oxidation over Titanosilicates

Crystalline titanium silicalite-1, with the same framework topology as ZSM-5 and known as TS-1, was patented in 1983 for application as a catalyst in the selective oxidation of organic substrates by hydrogen peroxide (H_2O_2).[2] Its performance improves upon that of amorphous TiO_2/SiO_2 solids, particularly in terms of selectivity and stability in the presence of H_2O. TS-1 has been found to be a versatile catalyst for hydrogen-peroxide-catalysed reactions such as alkene epoxidation, hydroxylation of aromatics, ammoximation and the oxidation of nitrogen- and sulfur-bearing organics.[3] Examples are shown in Scheme 9.1. In such reactions, the catalysed reaction competes with less selective radical reactions of H_2O_2 that has undergone homolytic fission. Two industrial applications, phenol hydroxylation and cyclohexanone ammoximation, have been demonstrated on a commercial scale, and there has been considerable progress towards an industrial process for the epoxidation of propylene. Academically, the main efforts have been towards the synthesis of new microporous titanium-containing silicates with pores larger than those of TS-1, establishing the mechanism of the activation of hydrogen peroxide and other peroxides, and investigating the utility of this new class of catalyst in a wide range of reactions.

9.2.1.1 Catalyst Preparation

TS-1 can be synthesised by hydrothermal methods using a variety of silica and titania sources, structure-directing agents and mineralisers. Alkali metal hydroxides cannot be used in the syntheses, however, because their presence results in the precipitation of separate titanate phases. It is only possible to

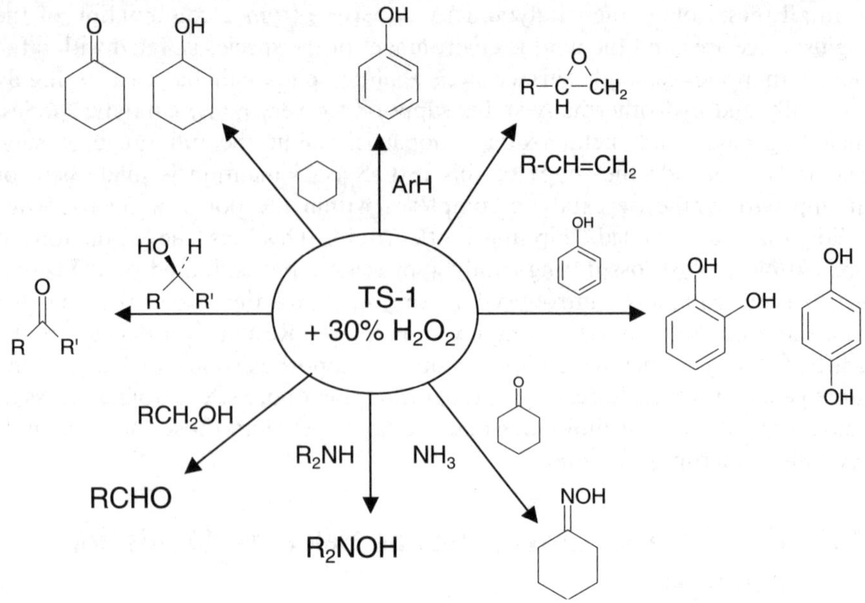

Scheme 9.1 Selective oxidation reactions catalysed by titanosilicalite-1 and aqueous hydrogen peroxide.

incorporate titanium in framework cation sites at loadings of up to 1–2 weight%, above which non-framework titanium species, including bulk TiO_2, result. It is important to avoid this, because unwanted side reactions occur over extra-framework titanium species. That the titanium is incorporated in framework positions is best confirmed by X-ray spectroscopy, because, in the calcined and dehydrated forms, such species adopt tetrahedral geometry, which gives characteristic XANES and EXAFS spectra (Section 3.4.2). The method of synthesis also determines whether any Brønsted acid sites are present in the final product. In most cases, acidity is to be minimised, to avoid side reactions (such as opening of epoxide rings, for example, in alkene epoxidation). Ideally, then, no aluminium should be present, and the density of silanol defects should be as low as possible.

TS-1 catalysts show high selectivity and activity for many conversions, but their medium-pore size precludes their use for conversions of large molecules. For example, cyclohexene is already too large to enter the pores and be epoxidised. For this reason, the incorporation of titanium into framework positions into larger-pore solids is an important research direction, with the same requirements of no aluminium, few framework defects and titanium present only in framework positions. The requirements for low acidity and only framework titanium favour the direct synthesis route using organic templates for such large-pore solids, and therefore high silica Beta, which has a three-dimensionally connected large-pore channel system facilitating high molecular diffusivities, is of particular interest. Different routes to Ti-Beta have been proposed and the group of Corma has successfully prepared a Ti-Beta that satisfies these requirements by direct synthesis in fluoride medium.[4] This is an important addition to the catalysts available for selective oxidations. Another approach has recently been suggested by Tatsumi, in which titanium species are introduced between the laminae of the zeolite precursor MCM-22(P) prior to a final condensation step[5] giving a large-pore solid with crystallinity[6] that is able to perform oxidation catalysis on larger molecules.

Many studies have aimed to prepare mesoporous solids with active titanium, either by direct synthesis or by post-synthetic modifications. Although some of the advantages of crystalline framework solids are lost, the larger pore size increases markedly the size limit of molecules that can be converted by this catalytic chemistry, and active and selective catalysts have been prepared.[7,8] In addition, such solids can be used with organic peroxides (such as *tert*-butylhydroperoxide) that do not produce water as an inhibiting by-product.

9.2.1.2 Mechanism of Hydrogen Peroxide Activation

The mechanism of the activation of H_2O_2 by TS-1 and related catalysts has been the subject of much research using spectroscopic and computational techniques. This has centred on the nature of the active site and its mode of reaction with H_2O_2, solvents and the organic substrates. Work to elucidate the structure of the active site has concentrated on the coordination chemistry of the titanium. X-ray and neutron diffraction studies, coupled with X-ray absorption, infrared and Raman spectroscopies, give evidence that most of the Ti(IV) in calcined TS-1, in the absence of any adsorbate molecules, is in tetrahedral coordination. Upon addition of one molecule of water, one of the Ti–OSi bonds is hydrolysed and the titanium adopts tetrahedral coordination as $Ti(OSi)_3OH$. Addition of a further water molecule gives rise to a pentacoordinated titanium.[9]

Similar behaviour is proposed for titanium upon the addition of aqueous hydrogen peroxide under catalytic conditions (Section 7.3.4). The framework titanium is thought by Prestipino *et al.*[10] to form complexes containing peroxide and hydroperoxidic species in equilibrium, depending on the concentration of water (Scheme 7.1). Corma *et al.* propose a similar arrangement, with hydrogen bonding between a hydroperoxide oxygen and the proton of the O-coordinated solvent (see Scheme 9.2).[11]

The hydroperoxide species is thought, on the basis of EXAFS spectroscopy and Quantum Mechanical modelling,[12,13] to coordinate end-on, through one

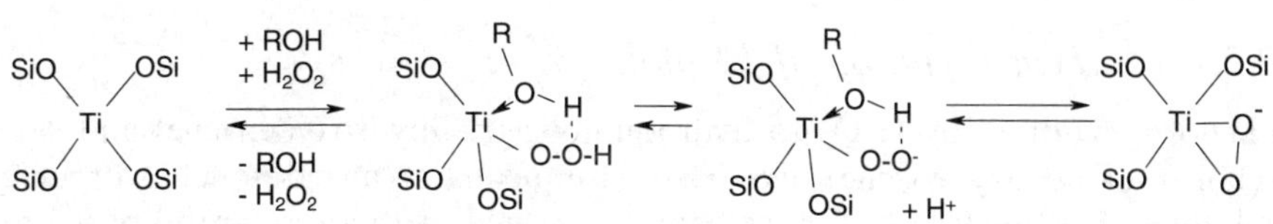

Scheme 9.2 An alternative scheme for the activation of H_2O_2 on framework titanium sites.[11]

oxygen, Ti-η_1-OOH. Side-on η_2 peroxide species are usually observed to be inactive in oxidation reactions, so that the complexed hydroperoxidic species are the active sites for further reaction. Olefinic groups of molecules small enough to gain access to the active sites of TS-1 undergo selective epoxidation by this catalyst, whereas bulkier molecules with C=C double bonds do not, indicating that the active sites are complexed species within the pores. Coordination of the peroxide to the titanium activates it by increasing its electrophilicity. The double bonds of alkenes are then able to abstract an oxygen atom. This leaves an oxygen atom at the titanium, which can be removed as water to regenerate the original titanium site.

The presence of excess water adsorbed within the pores of the catalyst acts to inhibit the catalytic reaction, presumably by favouring the inactive η_2 peroxide complex. Using H_2O_2, water is a by-product and cannot be avoided; in addition the catalysts typically use 30% H_2O_2/H_2O solutions. For this reason the best catalysts using this source of hydrogen peroxide are hydrophobic and contain few defects or aluminium atoms in the silicate framework. This has the added benefit of excluding acidic sites, which can cause unwanted side reactions such as ring opening of product epoxides. Also hydrocarbon reactants (such as olefins) are concentrated preferentially over water in the pores during reactions, favouring the active hydroperoxide complex.[10]

If catalysts with larger pores than those in TS-1 are used, then it is possible to use organic peroxides such as *tert*-butylhydroperoxide as the source of oxygen. This is expensive, but avoids the presence of water in the reaction and associated deactivation.

The use of crystalline microporous titanosilicates in selective oxidation reactions is highly attractive: there are potential routes to a range of useful oxygenated products, process conditions are mild, selectivities are high, and the main by-product of the reaction is water. The feasibility of commercialising such selective oxidations in the synthesis of bulk chemicals, however, is strongly influenced by the high cost of hydrogen peroxide, and requires the catalyst to be low in cost and highly selective in both the oxidation and in the use of hydrogen peroxide. Nevertheless, the environmentally friendly nature of the processes, particularly when compared with existing non-catalytic technologies, will become increasingly attractive. The main classes of selective oxidations that are catalysed by TS-1 are described below, including reference to three examples where processes have been or are being commercialised: hydroxylation of phenol, ammoximation of cyclohexanone and propylene epoxidation.[3]

9.2.1.3 Hydroxylation of Phenol

Phenol is oxidised by H_2O_2 to hydroquinone (*p*-dihydroxybenzene) and catechol (*o*-dihydroxybenzene) over TS-1 in aqueous or mixed aqueous/organic solutions (Scheme 9.1).[3] Both isomers are formed, with an *o*-/*p*-ratio of 0.5 to 1.3, depending on the conditions. No hydroxylation occurs at the *meta*-position (Scheme 9.3).

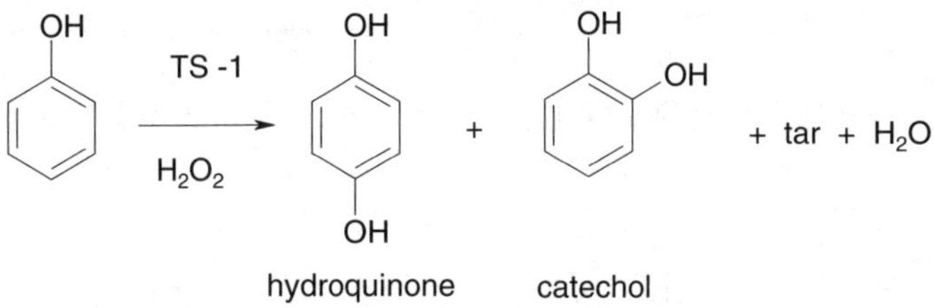

Scheme 9.3 TS-1 is a catalyst for the oxidation of phenol.

Both catechol and hydroquinone are widely used in photographic developers, and printing and reprographic applications, while hydroquinone is in demand as an intermediate in the synthesis of anti-oxidants, inhibitors, stabilisers, agrochemicals and dyes. The process has been operated commercially by EniChem since 1986, replacing the Brichima process which was based on radical hydroxylation. The titanosilicate-catalysed reaction has reduced the use of H_2O_2 and decreased the number of phenol separation and recycle steps required. The selectivity of the reaction to the desired products is improved by using TS-1 catalysts in which all of the titanium is in framework positions. The presence of extra-framework titanium species or bulk anatase promotes the decomposition of hydrogen peroxide and subsequent homolytic (radical) hydroxylation paths which are much less selective and increase the formation of tar as a by-product.

It should be noted that TS-1 is also active for the hydroxylation of benzene, where phenol can be prepared selectively at low conversions where consecutive reactions to the dihydroxybenzenes are minimised. Such a process could compare favourably with current multistep routes if acceptable selectivity can be achieved.[14]

9.2.1.4 *Ammoximation of Ketones*

The ammoximation of cyclohexanone to cyclohexanone oxime (Scheme 9.1) is an important industrial process, because the oxime is an intermediate for the formation of ε-caprolactam, the monomer for nylon-6, that is currently manufactured on a scale of 4 Mton *p.a.* TS-1 catalyses the ammoximation of ketones, including cyclohexanone, in liquid phase in the presence of ammonia and hydrogen peroxide. The reaction proceeds through hydroxylamine as an intermediate, which is generated at the active site *via* complexes of framework titanium with ammonia and hydroperoxide. The hydroxylamine then reacts with cyclohexanone molecules that are weakly adsorbed within the pores or can diffuse out into solution. The reaction is a general one, and can be applied to the ammoximation of a wide range of ketones.

An industrial plant using this reaction at 12 000 t/y has been operated by EniChem since 1994 as part of their new caprolactam process. The conversion and selectivity of the conversion of cyclohexanone exceeds 99% and the

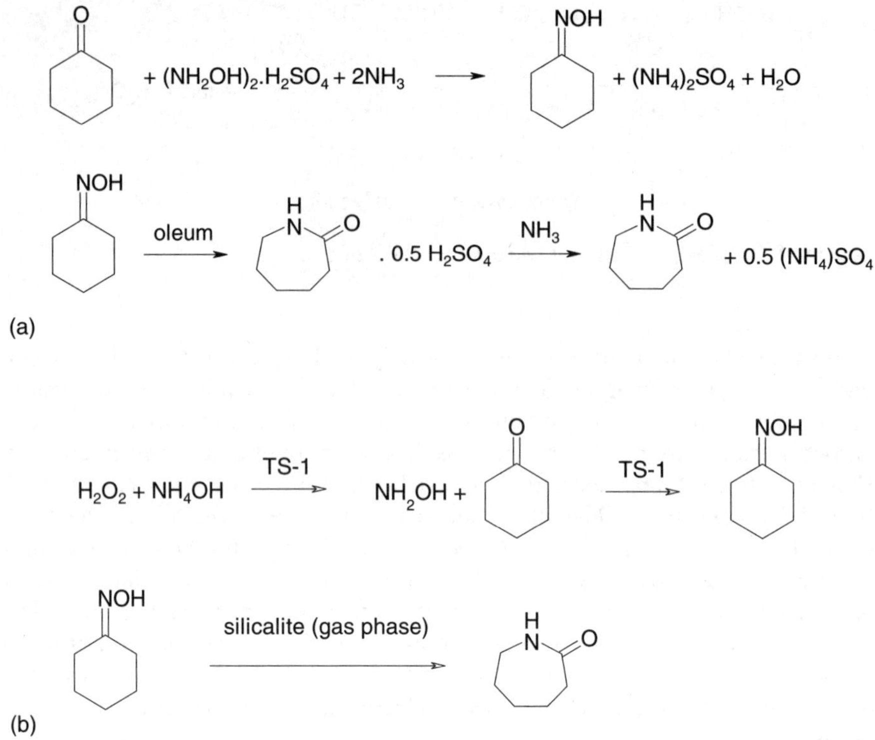

Scheme 9.4 **(a)** Conventional route to ε-caprolactam. **(b)** New EniChem/Sumitomo 2-stage synthesis of ε-caprolactam.

conversion of hydrogen peroxide is over 90%.[3] The process replaces a method in which hydroxylamine sulfate was first prepared and then used to prepare the oxime, giving stoichiometric quantities of unwanted ammonium sulfate as a by-product in the reaction. The new process therefore both reduces the capital cost of the project and is more environmentally benign. Taken together with the Beckmann rearrangement of cyclohexanone oxime to ε-caprolactam over weakly acidic MFI-type zeolites (described in the previous chapter), it is clear that zeolite catalysis offers an improved route to the nylon-6 precursor than the older technology, based on the Raschig process, which operated with corrosive reagents and generated very large volumes of unwanted ammonium sulfate by-products. It is therefore an excellent example of 'green' chemical technology (Schemes 9.4(a) and (b)). Further improvements may be expected (see Section 9.2.3).

9.2.1.5 Epoxidation of Alkenes

Epoxides are important feedstock chemicals. Currently the C_3-C_6 epoxides are prepared from the corresponding alkenes by the ARCO chlorohydrin process, which produces chlorinated organics and stoichiometric sodium chloride as

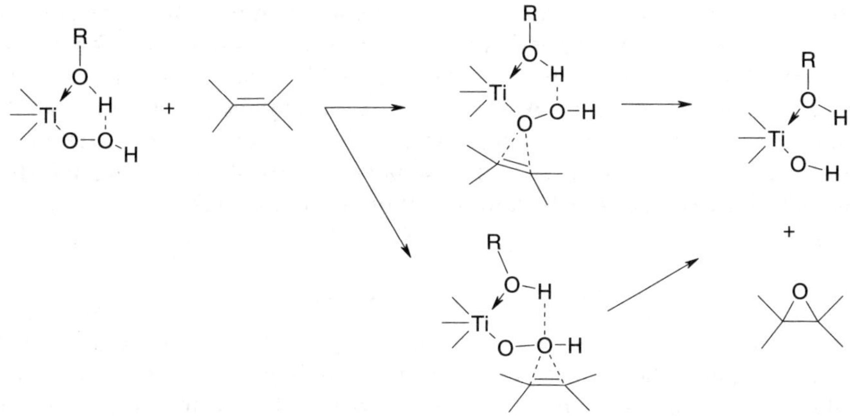

Scheme 9.5 Proposed mechanisms for the epoxidation of alkenes (adapted from reference 1).

unwanted by-products. TS-1 has been found to catalyse the epoxidation of alkenes using H_2O_2 with high selectivity to the desired epoxide and without toxic or corrosive by-products. The absence of products from radical epoxidations and the high stereospecificity of the conversion indicate the non-radical nature of the mechanism. The mechanism is best considered as an electrophilic oxidation in which activated oxygen is transferred to the alkene (Scheme 9.5).[11] The reaction rate is strongly affected by solvent, which is thought to be complexed at the active site during the catalytic reaction step. Methanol is the preferred solvent for reactions in TS-1. As described earlier, care must be taken to ensure that the titanosilicates do not possess appreciable acidity because the presence of Brønsted sites results in opening of the epoxide to give a diol.

EniChem have also designed a large-scale process for the epoxidation of propylene to propylene oxide over TS-1 in a slurry reactor that can be operated with high selectivity and for extended periods of time, with only water and minor amounts of oxidised organics such as propylene glycol and its monomethylethers as by-products. Ti-Beta is found to be less active than TS-1 for epoxidation of smaller alkenes, but for longer alkenes than *ca.* C_8 the enhanced molecular diffusivities possible through the large pores make it a better epoxidation catalyst.

9.2.1.6 *Alcohol Oxidation*

In the presence of TS-1 and dilute H_2O_2 under mild conditions, primary alcohols are oxidised to aldehydes, secondary alcohols to ketones and tertiary alcohols are converted to alkylperoxides. Where the alcohols possess a double bond, the double bond is epoxidised preferentially. The mechanism of alcohol oxidation requires activation of the C–H bond on the same carbon as the OH group. It is thought that alcohols of the type ROH coordinate to the titanium sites at the same time as the hydroperoxide group, as described for alcohol solvents during alkene epoxidation. For alcohol oxidation, a scheme is proposed by which concerted oxidation occurs *via* oxygen transfer to a similar cyclic intermediate, shown in Scheme 9.6. Other oxidations of heteroatom-containing molecules with H_2O_2 are also possible over titanosilicates, including the selective oxidations of N- and S-containing compounds.

9.2.1.7 *Direct Use of Molecular Oxygen*

The use of hydrogen peroxide as oxidant imparts a significant cost to these processes, particularly when the product is a bulk chemical of relatively low value. A further widening of the applicability of this and other selective oxidations using titanosilicates would be achieved if molecular oxygen could be used as the oxidant instead and the hydrogen peroxide generated *in situ*. Two possible ways to achieve this have been reported (Scheme 9.7). In one, an anthrahydroquinone is added as a co-catalyst to generate the hydrogen peroxide in a first step, and then reduced back to the starting form.

A second method involves adding a mixture of gaseous hydrogen and oxygen to a titanosilicate containing platinum or palladium. Hydrogen peroxide is produced *in situ* on the precious metal and is subsequently used at the titanium

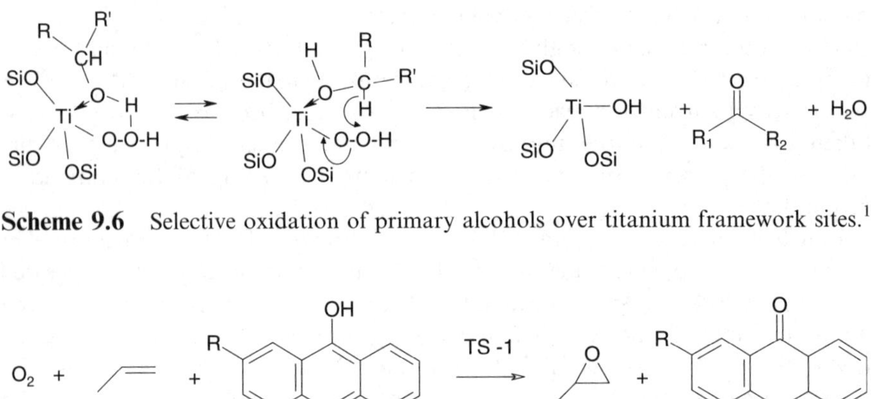

Scheme 9.6 Selective oxidation of primary alcohols over titanium framework sites.[1]

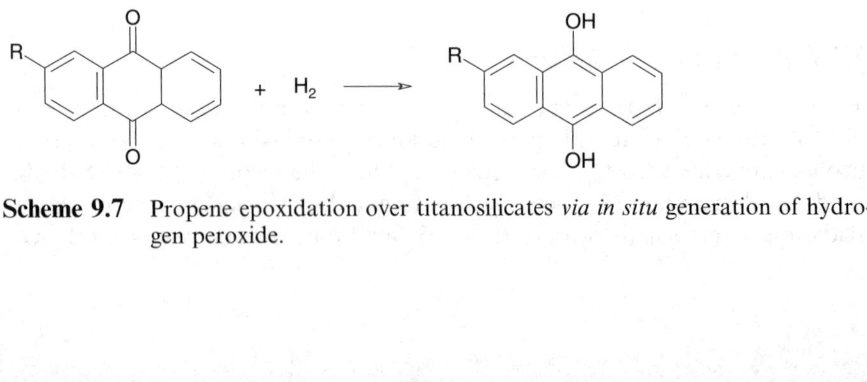

Scheme 9.7 Propene epoxidation over titanosilicates *via in situ* generation of hydrogen peroxide.

oxidation site. Although the conversions and selectivities observed in this route are much lower than those with direct addition of H_2O_2, the idea offers considerable promise.

9.2.2 Sn-Beta: a Versatile Catalyst for Baeyer–Villiger Oxidations and Meerwein–Ponndorf–Verley/Oppenauer Conversions

A further important development in the use of silica zeolite polymorphs containing Lewis acid sites in the framework in selective oxidation reactions was made by the group of Corma. They found that tin-containing zeolite Beta is a highly selective catalyst for Baeyer–Villiger oxidations, in which ketones are oxidised by hydrogen peroxide to give esters or lactones (Scheme 9.8).[15,16]

The Baeyer–Villiger reaction is an important conversion in organic chemistry and Sn-Beta provides an improved heterogeneous catalyst for it. Firstly, it permits the use of hydrogen peroxide rather than organic peroxides, which has benefits in terms of safety, by-products and cost. Although a range of other solids catalyse the reaction, including aluminium- and titanium-containing zeolites, they are less effective. Furthermore, when the reactant is an unsaturated ketone, the use of titanium zeolites favours epoxidation of the double bond rather than the Baeyer–Villiger oxidation. (Titanium-containing solids are excellent epoxidation catalysts, as described earlier.) To demonstrate the selectivity attainable by the Sn-Beta, Corma showed the following reaction results for the oxidation of dihydrocarvone with H_2O_2 (Scheme 9.9).[15]

The reaction mechanism over the tin Lewis acid sites is therefore quite different from that over titanium Lewis acid sites. On the basis of ^{119}Sn MAS NMR, the catalytic tin sites are thought to be within the framework. In the dehydrated solid the sites are tetrahedrally coordinated, and are able to bind with adsorbed molecules to become five- and six-fold coordinated. The Sn site activates the ketone rather than the hydrogen peroxide. The carbonyl oxygen from the ketone is thought to form a Lewis adduct with the framework tin, which is able to expand its coordination sphere as a so-called 'Criegee' adduct (Scheme 9.10).[15]

This coordination of the ketone at the Lewis acid site polarises the CO bond and activates the carbonyl carbon to nucleophilic attack by the hydrogen peroxide. By contrast, in the case of a titanium site, the Ti-OOH species is observed by UV-visible spectroscopy to be formed, and hydrogen peroxide is activated, and is then able to react selectively, for example to epoxidise C=C

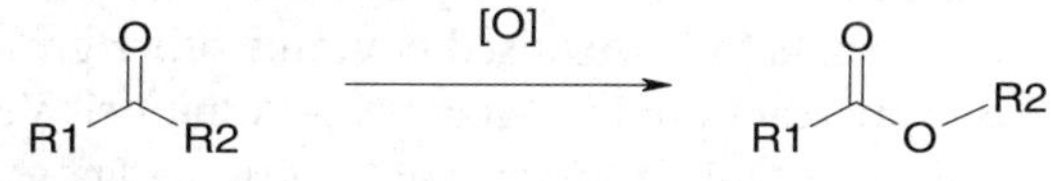

Scheme 9.8 Baeyer–Villiger oxidation of ketones to esters.

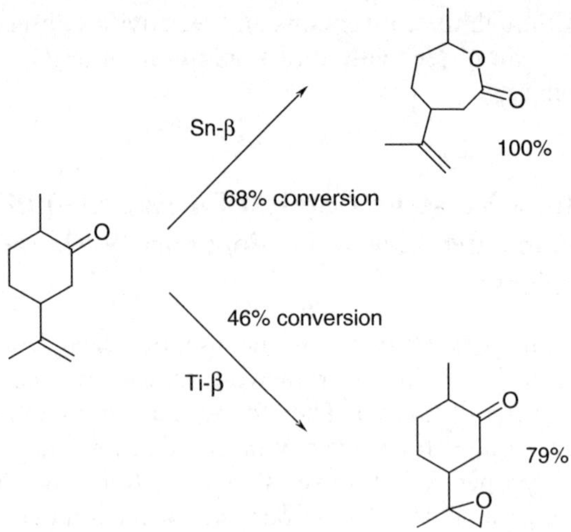

Scheme 9.9 The selective oxidation products of dihydrocarvone with H_2O_2 depend on the catalyst used: Ti-β catalyses epoxidation, whereas Sn-β catalyses reaction to the lactone.

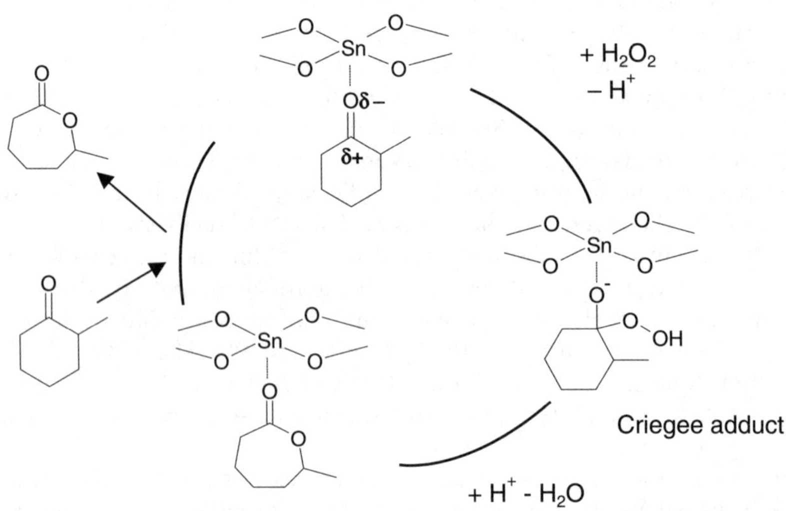

Scheme 9.10 Baeyer–Villiger oxidation over Sn-Beta. Modified from reference 15.

double bonds. Sn-Beta is therefore found to be a highly shape- and chemoselective catalyst for Baeyer–Villiger oxidation.

The Oppenauer oxidation of alcohols by ketones is a very selective oxidation reaction when the molecule to be oxidised contains other groups susceptible to oxidation. The opposite reaction, the Meerwein–Ponndorf–Verley reduction of ketones by alcohols is simply the reverse reaction. These conversions are catalysed by Lewis acids. These are typically metal *tert*-butoxides in solution,

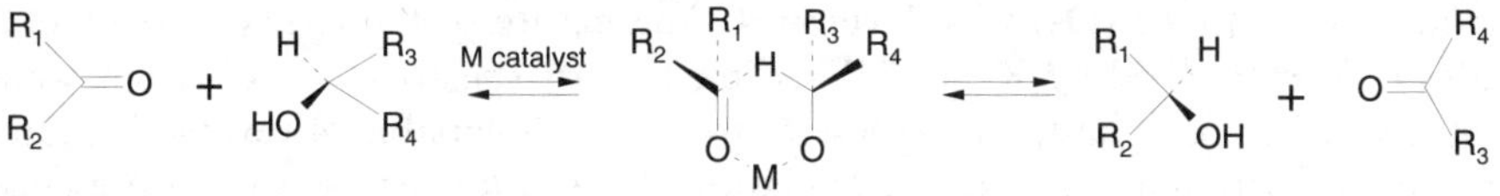

Scheme 9.11 Meerwein–Ponndorf–Verley and Oppenauer conversions over the Lewis acid site M.

but zeolites containing Lewis acid sites (examples being Sn-Beta and Ti-Beta) are also active for these conversions (Scheme 9.11).

9.2.3 Selective Oxidations over $AlPO_4$s Containing Redox-active Framework Cations

There is a strong incentive to oxyfunctionalise hydrocarbons because it opens the way to important chemical intermediates such as alcohols, ketones and carboxylic acids. Selective oxidations can be achieved using sacrificial oxidants such as hydrogen peroxide, but O_2 in air is the oxidant of choice for reasons of economy. Many current oxidations are operated industrially that involve complex multistep reactions, solvents and give unwanted by-products, so there is a clear need for clean and efficient catalytic alternatives.

For cyclohexane oxidation, for example, a reaction central to the generation of intermediates in routes to the synthesis of nylons-6 and -6.6 and urethanes (such as cyclohexanone and adipic acid), one industrial process uses molecular oxygen in the presence of cobalt or manganese salts as homogeneous catalysts in acetic acid solvent. The autoxidation of cyclohexane (and other alkanes) under these conditions proceeds *via* free radical chain reactions, where the first step is reaction of dioxygen to give the hydroperoxide. The metal cations do not activate the oxygen but rather coordinate the hydroperoxides and catalyse their decomposition, generating more radicals and accelerating the reaction.

It was therefore a logical step to measure the activity of microporous cobalt aluminophosphates as heterogeneous catalysts for these reactions, particularly because the cobalt was known to be able to exist as both Co(II) and Co(III) in tetrahedral framework sites. Initial attempts, for example by Kraushaar-Czarnetski *et al.*,[17] directly transferred the solids to existing homogeneous conditions, with acetic acid as a solvent. Although activity was observed, the cobalt was found to be leached from the framework by the acidic solvent, so that at least part of the reaction was homogeneously catalysed. Similar problems with metal leaching were also found for oxidations with the chromium-substituted $AlPO_4$-5 at around that time.

In 1999, however, the group of Thomas reported a series of selective oxidations of alkanes in which the redox metals were retained within the framework by working under solvent-free conditions and at sufficiently low conversions ($<10\%$) that only low concentrations of carboxylic acids were present.[18] Under these conditions, and by judicious choice of the framework type of the aluminophosphate and the substituted metal, a family of selective oxidations with air (O_2) was developed. The nature of the redox active metal sites was established by X-ray, UV-visible and IR spectroscopies.[19] They were established to be cobalt, manganese or iron cations capable of reversible redox behaviour between +2 and +3 oxidation states, with the associated presence or absence of protons on adjacent bridging M–O–P oxygen atoms as required for charge balance. Redox activity at the metal site was observed by XAS while IR revealed bridging hydroxyls were present. The reversibility of the redox process was demonstrated by alternate oxidations (in O_2) and reductions (in H_2). Furthermore, by careful examination of the effects of adding free radical initiators (which reduced induction times) and free radical scavengers (which inhibit the reaction) the reactions were shown to proceed *via* free radical mechanisms. By analogy with homogeneous oxidation reactions, these require a catalytic cycle between M(II) and M(III) oxidation states that involves the hydroperoxide as an intermediate. A selection of the selective oxidations using air reported for this class of solids[20] is shown in Scheme 9.12. The catalysts themselves are shown in Figure 9.1.

A reaction mechanism involving free radical steps and one-electron transfers on the redox active metals has been written explicitly by Moden *et al.*[21] for the oxidation of cyclohexane to cyclohexanol and cyclohexanone over substituted metalloaluminophosphates. The latter confirms that the generation of oxidation products bears strict relation to the number of framework cations able to cycle

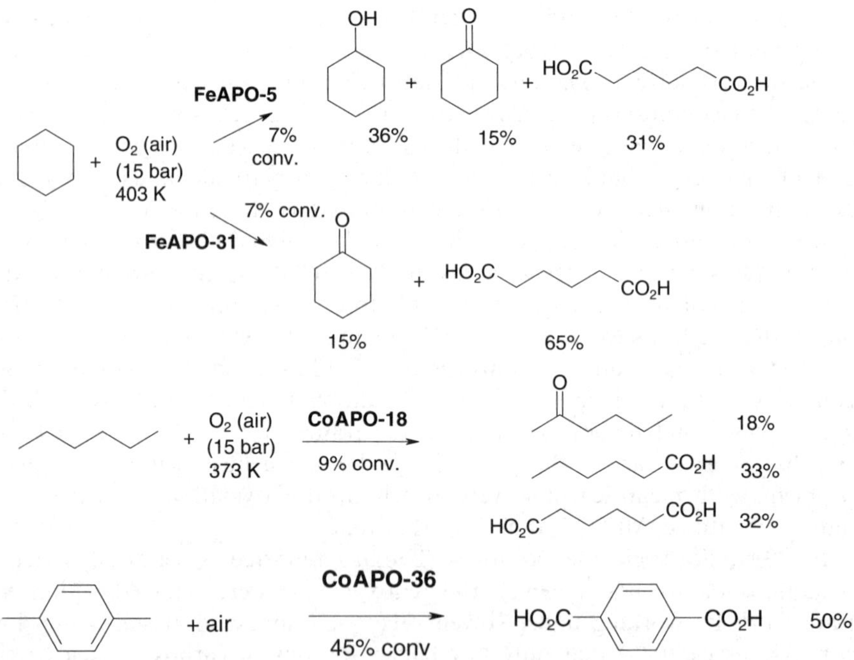

Scheme 9.12 Selective oxidations of hydrocarbons with molecular oxygen catalysed by transition metal-containing aluminophosphates.

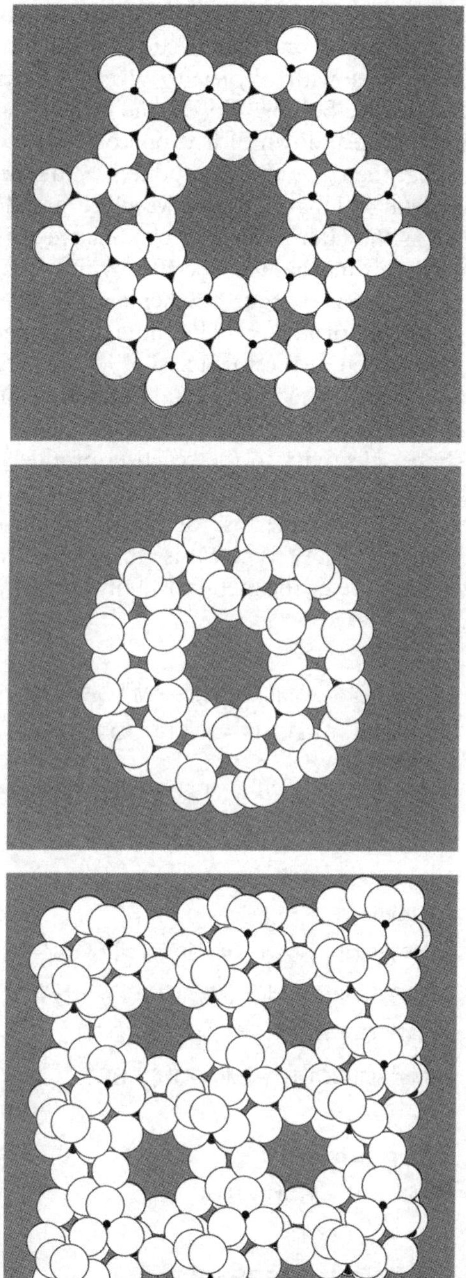

Figure 9.1 Framework structures of aluminophosphates-34, -18, -31, -5 and -36, with framework oxygen atoms shown with van der Waals radii. Once doped with cobalt, iron or manganese, these materials can act as shape selective oxidation catalysts, for example in the aerial oxidation of alkanes and aromatics.

reversibly between 2+ and 3+ states. The mechanism is based largely on that accepted for the homogeneously catalysed oxidation of alkanes with oxygen. Most aliphatic alkanes react with oxygen to give hydroperoxides, but these decompose only very slowly. Soluble Co(II) and Mn(III) salts accelerate the reaction by catalysing the breakdown of hydroperoxides *via* one-electron redox steps to give radicals that then give rise to products *via* the propagation steps of chain reactions. The metal cations shuttle between 2+ and 3+ oxidation states in the process (known as the Haber–Weiss cycle) Scheme 9.13.

Moden *et al.* re-write this mechanism specifically in terms of the framework metal sites and bridging hydroxyl groups (Scheme 9.14). The radicals take part in propagation steps to give products. (The radical intermediates are thought most likely to occur adsorbed at the surface sites of the catalysts.)

A potential advantage of using such metal cations in microporous solids, apart from the usual benefits of heterogeneous catalysts, is that the well-defined pores may impart shape selectivity to the reaction products by retaining bulk products, or inhibiting their formation by transition state selectivity. For cyclohexane oxidation, for example, use of the transition metal form of the large-pore $AlPO_4$-36 structure (ATS) gives rise to cyclohexanol and cyclohexanone as products, whereas use of the metal form of the medium-pore $AlPO_4$-31

$$M^{II} + ROOH \rightarrow M(HOOR) \rightarrow M^{III}(OH) + RO\cdot$$

$$M^{III} + ROOH \rightarrow M^{II} + ROO\cdot + H^{+}$$

Scheme 9.13 The Haber–Weiss cycle for catalysis of the reactions of peroxides.

Initiation

$$[Me^{2+}\text{-}OH] + ROOH \leftrightarrow [Me^{2+}\text{-}OH]\text{-}ROOH$$

$$[Me^{2+}\text{-}OH]\text{-}ROOH \rightarrow [Me^{3+}\text{-}O] + RO\cdot + H_2O$$

$$[Me^{3+}\text{-}O] + ROOH \leftrightarrow [Me^{3+}\text{-}O]\text{-}ROOH$$

$$[Me^{3+}\text{-}O]\text{-}ROOH \rightarrow [Me^{2+}\text{-}OH] + ROO\cdot$$

Propagation

$$RO\cdot + RH \rightarrow ROH + R\cdot$$

$$R + O_2 \rightarrow ROO\cdot$$

$$ROO + RH \rightarrow ROOH + R\cdot$$

Termination

$$ROO + ROO \rightarrow ROH + R{=}O + O_2$$

Scheme 9.14 The catalytic decomposition of peroxides over redox metal centres.[21]

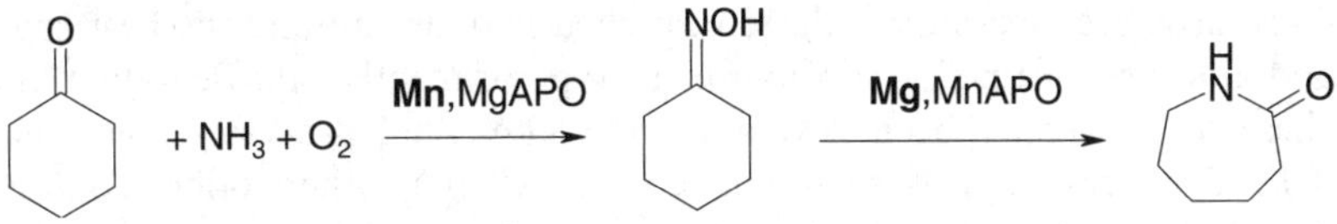

Scheme 9.15 Proposed one-step synthesis of ε-caprolactam over a bifunctional nanoporous solid. Modified from reference 23.

structure (ATO) gives a product distribution that is dominated by adipic acid.[22] This is thought to result because the narrower channels inhibit the release of cyclohexanol and cyclohexanone and the reaction proceeds further to the more mobile linear products, such as adipic acid. Selectivity is also observed in the aerial oxidation of linear alkanes.[18] If the reaction is performed over large-pore solids, *n*-alkanes are oxidised preferentially at carbon atoms at C2 and C3 positions in the chain, in accordance with the C–H bond strengths at these positions. If a small-pore structure such as CoAPO-18 is used, however, the product selectivity favours C1 oxyfunctionalised products. The synthesis of terminally oxidised alkanes would be of use for many applications, because linear terminal alcohols could be prepared from alkane feedstocks, rather than from α-olefins (*via* hydroformylation).

Further examples of selective oxidations using O_2 include the oxidation of *p*-xylene to terephthalic acid, Baeyer–Villiger oxidations of cyclic ketones to lactones using molecular oxygen and benzaldehyde as a sacrificial aldehyde[20] and catalytic epoxidation *via* a free radical route (rather than the electrophilic oxidation proposed for hydrogen peroxide mediated epoxidation over TS-1).[20]

A one-pot synthesis of ε-caprolactam over bifunctional MeAPOs has also been demonstrated in which a Mg,Mn(III)APO-5 is able to catalyse the synthesis of ε-caprolactam (Scheme 9.15) from cyclohexanone in the presence of air and ammonia alone. The transition metal is responsible for generation of hydroxylamine and the mild acidity associated with the metals in the aluminophosphate framework is sufficient to catalyse the Beckmann rearrangement.[23]

9.3 Catalysis over Extra-framework Metal Species

The combination of catalytically active metal clusters, metal cations or metal oxide species with microporous solids offers advantages of high dispersion and the possibility of introducing two catalytic functions in close proximity. Zeolites are generally favoured over other microporous solids because of their greater structural stability. Metal cations impart Lewis acidity, whereas metallic clusters, particularly of precious metals, are good catalysts for hydrogen transfer and hydrocarbon conversions: the example of platinum clusters in zeolite L acting in aromatisation is described below. Combining the function of metals with strong Brønsted acidity gives strong synergistic effects, as described in Chapter 8 for precious metal/zeolite bifunctional catalysts for hydrocracking and reforming. Very many combinations of metal cations or metal oxides with zeolites have been examined: those described here are among the best characterised, and are grouped according to the type of reaction they catalyse. Some of these have already been developed as commercial processes (such as the CYCLAR process that uses gallium zeolites) whereas others offer promise for future applications, particularly in environmental catalysis.

9.3.1 Transformation of Light Hydrocarbons into Aromatics

The best established reaction of this type is the conversion of light hydrocarbons such as propane and butane (for example as liquefied petroleum gas, LPG) to more valuable aromatics (benzene, toluene, xylenes) over gallium-bearing ZSM-5.[24] This reaction[25] has been developed jointly as the CYCLAR process by BP and UOP and is operated commercially. The catalyst preparation involves the impregnation of H-ZSM-5 with gallium salts followed by calcination of the mixture. The gallium added in this way exists as the oxide after calcination, probably outside the pore system, but under catalytic reaction conditions, which are strongly reducing, gallium migrates into the pores, probably as GaO^+ or Ga^+ (in a process reminiscent of reductive solid state ion exchange (Chapter 6)). Within the pores it can occupy extra-framework cation sites and can act as a Lewis acid. The gallium introduces dehydrogenation activity into the solid acid, probably through the activation of alkanes by hydride abstraction, and the pore size of ZSM-5 favours the generation of aromatic products. Gallosilicates prepared by direct crystallisation, in which the gallium begins in tetrahedral coordination, display similar catalytic behaviour, suggesting that the gallium leaves the framework positions under reaction conditions, and that this catalyst can best be considered as a bifunctional catalyst, similar to those described in Section 8.7.3.

In a similar process, platinum supported on the potassium-ion-exchanged zeolite L (Pt/K-L) has been found to be an excellent catalyst for the aromatisation of hexane and heptane. This is thought to be due to its ability to support small clusters of platinum metal that act as a 'molecular die'. The lack of acidity in this catalyst is important because it prevents side reactions, including cracking.[26–29]

The global abundance of methane in natural gas encourages the search for chemical routes for its conversion to more valuable higher hydrocarbons. Many attempts at selective oxidation to give higher hydrocarbons have been tried, but usually result in deep oxidation to carbon oxides and water. There is therefore much current investigation of the use of metal-loaded zeolites as dehydrogenation and dehydrocyclisation catalysts, and iron- and molybdenum-loaded catalysts have shown interesting results. Molybdenum supported on ZSM-5 (Mo/H-ZSM-5), for example, catalyses the formation of aromatics such as benzene from methane.[30] The active metal site of this catalyst is thought, largely on the basis of X-ray absorption spectroscopy, to be dispersed molybdenum carbide clusters that are generated *in situ*. These catalyse the dehydrogenation of methane to acetylene and the subsequent aromatisation of acetylene to aromatics.

9.3.2 Catalytic Removal of NO_x Species from Auto-exhaust and Power Plant Emissions

The formation of nitrogen oxides during the high-temperature combustion processes that take place in petrol and diesel engines and in stationary power generation plants and factories is a major potential pollution source, since NO_x species (NO, NO_2) are implicated in the production of photochemical smog and acid rain. The development of catalysts by which NO_x emissions can be strongly reduced has therefore been of great importance. In particular, the development of the three-way auto-exhaust catalyst (TWC) is a major achievement of modern chemistry.[31,32] In this process, the three pollutants from gasoline-fueled engines, unburnt hydrocarbons, carbon monoxide and nitrogen oxides (Figure 9.2),[33] are largely eliminated by passing the exhaust gases over catalysts that contain supported precious metals such as platinum, palladium and rhodium, together with a range of important co-catalysts. The effective action of the three-way catalyst is made possible by modern fuel injection systems that ensure that there is a stoichiometric mix of fuel and air, so that complete combustion would leave only CO_2 and H_2O in the exhaust stream. In practice, the high temperature equilibrium of N_2 and O_2 with NO and inefficiencies in the combustion result in the presence of the three pollutants in the exhaust stream. However, the reductants (CO and hydrocarbons) are present in the exact quantities needed to react with the NO_x species (and any unreacted oxygen). The TWC possesses all the necessary catalytic functions to ensure this occurs.

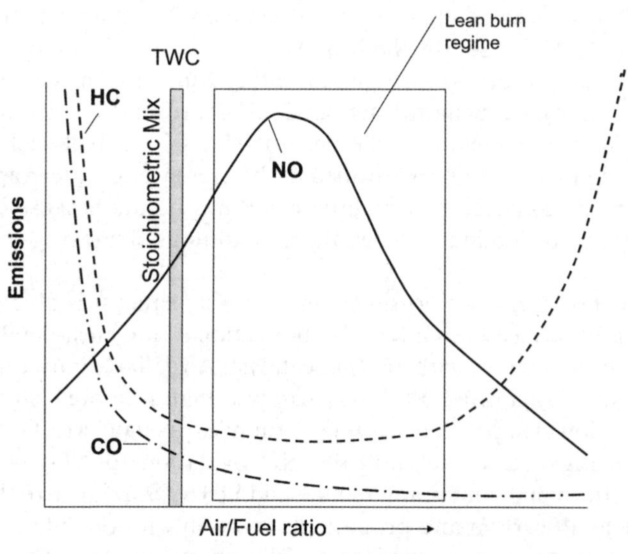

Figure 9.2 Schematic plot of the concentration of unburnt hydrocarbons, carbon monoxide and nitric oxide emissions in automobile engine exhaust as a function of the air : fuel ratio. The three-way catalyst (TWC) ensures removal of all these emissions only if the air : fuel ratio is stoichiometric.

$$CH_4 + xsO_2 + NO \xrightarrow{\text{metal-exchanged zeolite}} \tfrac{1}{2}N_2 + CO_2 + 2H_2O + xsO_2$$

Scheme 9.16 Selective catalytic reduction (SCR). (Idealised).

For reasons of fuel economy, however, there would be advantages of running petrol engines under so-called lean burn conditions, where there is an excess of oxygen over that needed for complete combustion. Under these conditions, the three-way catalyst is less effective in reducing NO_x emissions. In addition, diesel engines operate under lean burn conditions. In both cases, then, catalysts are required that enable the selective reduction of NO_x species by unburnt hydrocarbons and carbon monoxide in the presence of oxygen (Scheme 9.16). In other words, the catalysts must promote the reaction of the reducing species with NO_x species compared to their reaction with oxygen. The catalyst must be active at exhaust temperatures and in the presence of high partial pressures of H_2O. These are the requirements of a selective catalytic reduction (SCR) lean burn deNO_x catalyst.

The first major candidate for such a role was copper-exchanged ZSM-5 (Cu-ZSM-5).[34,35] Although it was subsequently found that this material suffered from poor hydrothermal stability and relatively low activity under working conditions, the discovery sparked much research into this reaction. It was determined, for example, that NO decomposition over Cu-ZSM-5 takes place over copper dimers that bridge two exchange sites, where the bridging oxygens on such dimers act as the active sites. The activity also appears to depend on the presence of Cu^+ species, so that the reaction involves cycling between Cu^+ and Cu^{2+}, where the Cu^+ adsorbs the NO. Many studies of the SCR of NO_x over zeolites containing transition metals (including Cu, Co, Mn, Fe and Ni) have been reported, which demonstrate gradual improvements in stability and activity.[32] Figure 9.3 illustrates the performance of a cobalt-exchanged high silica stilbite in this reaction, performed in the presence of water vapour.[36] The maximum in the conversion at high temperatures occurs because combustion reactions begin to dominate under these conditions, removing hydrocarbon reductant.

It may be, however, that no single phase zeolite catalyst is able to fulfil the role, and there has also been interest in bifunctional lean burn-deNO_x catalysts where the zeolite is mixed with another catalyst. A zeolitic catalyst may be used together with a supported Pt/Al_2O_3 catalyst that is more active for NO_x reduction at low temperatures. Other examples include Mn_2O_3/Ce-ZSM-5, where the manganese oxide catalyses NO oxidation to NO_2, which reacts rapidly over the metal-containing zeolite, and Pt/H-ZSM-5, where the acid sites activate the hydrocarbon and promote its reaction with the NO_x.

NO_x abatement is also required at stationary power sources. In these situations, ammonia is used as a reductant to selectively reduce the NO_x (and not the O_2) in the gas stream. As well as precious metal catalysts, zeolitic catalysts are commercially available for this reaction. Examples of successful

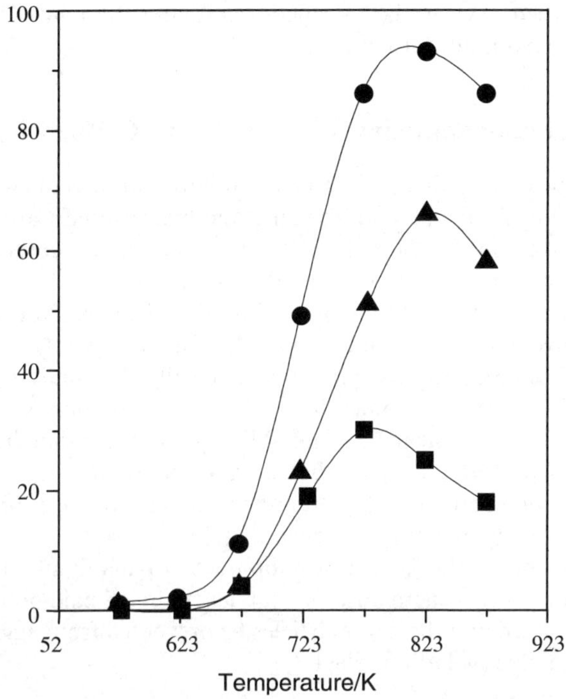

Figure 9.3 The efficiency of cobalt-exchanged high silica stilbite (Co-TNU-10) for the selective catalytic reduction of NO to N_2 with methane as a function of temperature at different inlet CH_4 levels of 2400 (■), 8000 (▲) and 16 000 (●) ppm. The reactions were run with a feed containing 1200 ppm NO, 2.6% O_2 and 10% H_2O at a GHSV of 14 000 h^{-1}. [Reproduced from reference 36 with permission. Copyright 2004 American Chemical Society.]

zeolite catalysts include Fe-Beta (active above 200 °C), Fe-ZSM-5 and mixtures of CeO_2 with H-ZSM-5.[37]

9.3.3 Selective Oxidation with N_2O over Fe-zeolites

As well as being active in $deNO_x$ catalysis, zeolites containing iron have been found to be good catalysts for the oxidation of benzene to phenol using N_2O. N_2O is available as a by-product from some industrial chemical processes and we have seen previously that a simple catalytic conversion of benzene to phenol would offer significant advantages over the current three-step cumene process. Fe-ZSM-5 synthesised with iron in framework positions has been found to be a promising catalyst. Upon activation by calcination, a fraction of the iron moves to extra-framework positions,[38] where it exists as isolated ions, small ferric complexes or highly dispersed oxide particles, as identified by a range of spectroscopies. These small complexes are thought to dissociate N_2O, giving a form of adsorbed oxygen, the so-called α-O, that is thought to be responsible for the selective oxidation of benzene.[39]

9.3.4 Palladium-containing Zeolites for C–C Bond Formation

Dispersed particles of palladium (and platinum) within zeolites are active in bifunctional catalysis, including reforming and hydroisomerisation and hydrocracking described in Chapter 8. Palladium-containing zeolites are also of interest as heterogeneous catalysts for the Heck reaction.[40,41] This is the Pd-catalysed arylation or vinylation of olefins of widespread use in C–C bond formation to give polyfunctional compounds of use as dyes, UV-absorbers and intermediates in fine chemicals syntheses. Typically the catalysis is performed homogeneously using organometallic palladium compounds, in which the palladium can cycle between Pd^0 and Pd^{II}, but zeolites that have been ion-exchanged with $[Pd(NH_3)_4]^{2+}$, or that contain nanoparticulate Pd^0, are also active. Dams *et al.* show that palladium-containing zeolites are heterogeneous Heck catalysts for the reaction of Scheme 9.17.

The requirement for the further development of palladium-containing zeolitic catalysts for more general use is that leaching of palladium in solution should be eliminated and the nanoparticles be prevented from agglomeration to catalytically inactive palladium black.

9.3.5 Base Catalysis

Whereas the appropriate forms of zeolites and related solids are widely used in acid-catalysed industrial processes, microporous solids are not currently of importance in commercial base-catalysed conversions. Instead, high-surface-area forms of alkali metal and alkaline earth metal oxides and hydroxides, often supported on alumina, fulfil the need for solid base catalysts. Nevertheless, interest remains in characterising basic sites in cationic zeolites and in developing routes to more strongly basic sites in microporous solids.[42] Routes to the latter include the introduction of metallic forms of alkali metals or nanoparticles of metal oxides and the partial replacement of amine groups at the sites of framework oxygen atoms. Porous solid bases have been shown to exhibit a varied catalytic chemistry, particularly for reactions such as dehydrogenations,

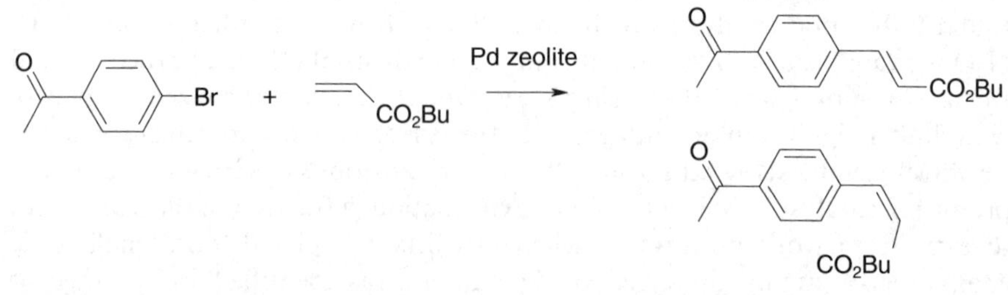

Scheme 9.17 The Heck reaction is catalysed over palladium zeolites.

double-bond isomerisations, C–C bond formation and methyl halide-to-hydrocarbon reactions. The last is probably responsible for the well-known laboratory precaution of not storing halogenated solvents over molecular sieve drying agents due to evolution of gases. Whereas most of these reactions remain predominantly of academic interest, the developing library of potential catalysts could develop into an important resource, particularly in fine chemicals synthesis.

A basic site can be described in terms of its ability to accept a proton (Brønsted base) or to donate electrons (Lewis base). In order for base-catalysed reactions to occur, the basicity must be sufficient to stabilise an anionic or polarised species with a significant negative charge that forms part of the catalytic cycle. For a typical base-catalysed C–C bond formation reaction (such as a Michael addition or the Knoevenagel condensation (Scheme 9.18)), the basic site stabilises the intermediate carbanion, which can then act as a nucleophile in the C–C bond formation. Surface sites on MgO are typical strong basic sites.

For an alkali–metal-exchanged zeolite the oxide ions on the framework bear a negative charge and are basic sites of moderate strength. The situation in such solids is complicated by the presence of alkali metal cations nearby, however, which act as Lewis acids (electron acceptors). A similar situation arises for the large-pore titanosilicate ETS-10, which also shows activity as a basic catalyst. In practice, therefore, 'basic' catalysis observed over such materials is the result of mixed acid/base behaviour as well as the enhanced molecular contacts promoted by adsorption within the pores. Care must also be taken when looking for basic catalysis to eliminate any strong acid sites (particularly Brønsted sites) that may be present, because acid catalysis is typically much faster than basic catalysis in these systems. This may be achieved by over-exchange with an excess of the alkali metal cations. In general the basicity of alkali metal cation-exchanged zeolites increases along the series Na < K < Rb < Cs, and high-aluminium-content zeolites are usually more strongly basic than high silica zeolites.

Typical base-catalysed reactions that occur over alkali–metal-exchanged zeolites include dehydrogenations, double bond isomerisations, side-chain alkylation of aromatics, conversion of methyl halides and a range of condensations. The reaction of alcohols over zeolites can be used to determine whether acid or base catalysis predominates. Whereas acid forms of zeolites catalyse dehydrations, leading to alkenes and the products of their subsequent reactions, basic sites catalyse dehydrogenations, leading to aldehydes and ketones.

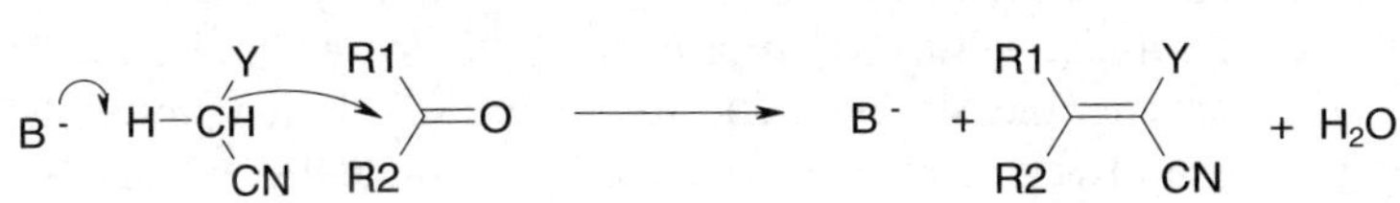

Scheme 9.18 The base-catalysed Knoevenagel is a useful test reaction for basicity. The electron withdrawing cyano group stabilises the intermediate carbanion to enable the reaction to proceed for bases of moderate strength.

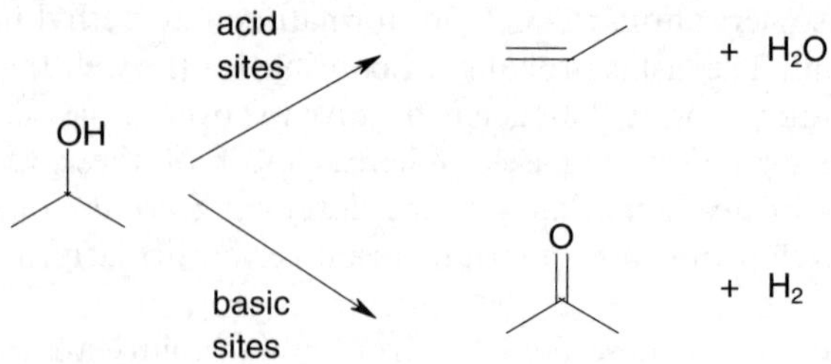

Scheme 9.19 The reaction of isopropanol is a diagnostic test for solid acid or basic sites.

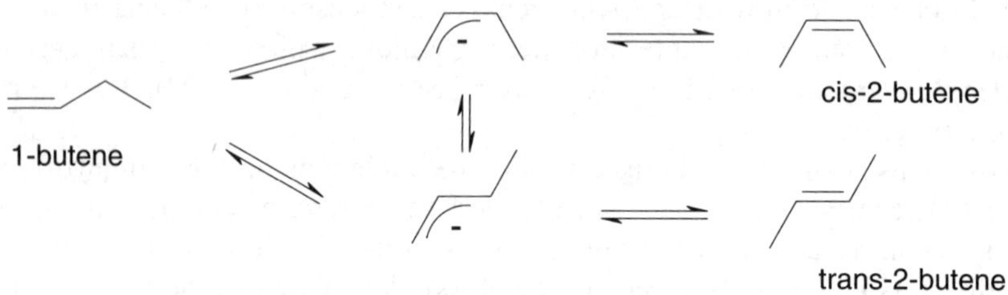

Scheme 9.20 Double-bond isomerisation on basic sites occurs *via* a carbanion intermediate.

The reaction proceeds by abstraction of a proton from the hydroxyl group, followed by formation of an alkoxy group on the alkali cation, hydride abstraction and hydrogen (H_2) liberation and subsequent formation of the aldehyde or ketone. The reaction of 2-propanol is commonly used as a test reaction in this regard: over acid catalysts propene is formed, over basic catalysts acetone is the product (Scheme 9.19).

Double bond isomerisation of alkenes is promoted over basic catalysts, where the carbon skeleton is retained, in contrast to the action of acid catalysts which are important catalysts for skeletal isomerisation. The difference lies in the mechanism, which for basic catalysts involves carbanions, rather than the carbenium ions that form in the presence of acids (Scheme 9.20).

Basic catalysts also show very different behaviour from acid catalysts for the alkylation of aromatics. Whereas acid catalysts promote alkylation of the aromatic ring, with high shape selectivity in the important case of ZSM-5 (Chapter 8), alkali metal zeolites catalyse side chain alkylation. In the case of the reaction of toluene with methanol over Cs-X, for example, the products include ethylbenzene and styrene. The side chain alkylation proceeds by the following base-catalysed steps, (i) formation of formaldehyde from methanol, (ii) activation of the toluene by polarisation of the methyl group (tending towards carbanion formation) and (iii) nucleophilic attack of the carbanion of toluene on the carboxyl group of formaldehyde. Side chain alkylation of aromatics is therefore a special case of aldol condensation. Reactions of this general sort, which result in C–C bond formation, are of considerable importance in the synthesis of fine chemicals (*e.g.* Michael addition, Knoevenagel condensation).

Alkali metal exchanged zeolites have also been shown to catalyse the breakdown of methyl halides, liberating hydrocarbons. *In situ* NMR studies by the group of Haw[43,44] suggest that the reaction proceeds by nucleophilic substitution by the framework oxygen atoms at the carbon atom of the halide. This gives methoxy species on the framework that subsequently combine. Basic framework oxygen anions and Lewis acidic cations appear to have complementary roles in the catalysis. Davis also reports the use of caesium oxide supported on a zeolite as a catalyst for the synthesis of 4-methylthiazole, an intermediate in the synthesis of thiabendazole, a fungicide.[45]

The basicity of alkali metal zeolites is relatively weak, so that several attempts have been made to prepare more strongly basic sites in zeolites. Direct inclusion of alkali metal clusters in the pores achieves this, but the resultant solids are highly reactive and impractical as useful catalysts. The partial nitridation of zeolites appears to be a more promising route. In this approach, high-temperature treatment with ammonia introduces NH_2 groups as defects in the structure (in place of defect OH groups) with retention of long-range order.[46] An alternative route to the preparation of ordered porous bases is the functionalisation of the internal surfaces of mesoporous silicas with organic amines or alkylammonium ions (see Chapter 6), where the associated hydroxyl ions can act as bases.[47] The ease with which large-pore solids can be prepared, and their versatility in accepting different organic bases offers much scope, particularly in the conversion of larger molecules.

9.4 Supported Metal Complexes in Porous Solids

9.4.1 Ship-in-a-bottle Type Catalysts in Zeolitic Solids

In Chapter 6 I described some of the synthetic approaches to the encapsulation of organic molecules and metal complexes within the pore space of microporous solids. The initial incentive for this was to prepare catalytic species held within inorganic particles that have the same activities in suspension as the complexes display homogeneously, but with the benefits of ease of separation and re-use that are typical of heterogeneous catalysis. This 'ship-in-a-bottle' catalytic concept as originally coined by Herron[48] has attracted a great deal of academic interest[49] and is reported in this section. The guiding concept of developing these catalysts has been to identify a catalytic species, typically a metal complex, and to assemble it within the large cages of zeolites such as X or Y, using either covalent or electrostatic bonding. It is then held in place even when in a solvent because it is larger than the windows connecting one cage to another and ultimately leading out of the crystallite. Other applications of this approach include in the preparation of photocatalysts or sensors.

Many of the first successful approaches concerned the preparation of metal complexes that resembled those observed at the active sites of metallo-enzymes, and particularly those enzymes involved in catalytic oxidation. The term 'zeozyme' was therefore introduced to cover this idea. Transition metal complexes of salen ligands in zeolites were the first 'ship-in-a-bottle' complexes prepared. These 'Schiff base' complexes are known as homogeneous catalysts for various reactions, depending on the complexed metal and the conditions. Zeolite-encapsulated Co(II) salen was first prepared by Herron[48] who demonstrated that the complex retained its ability to bind oxygen reversibly. Following the successful inclusion of transition metal salen complexes, the preparation of zeolites containing salen-based complexes known to be active for chiral epoxidation was a logical extension of the work. Schiff base complexes of Mn(III), for example, can be prepared using a chiral salen ligand synthesised by the condensation of a single enantiomer of a chiral primary diamine with a salicylaldehyde (Scheme 9.21).

In homogeneous reactions, complexes of this sort can achieve chiral alkene epoxidations because the freedom of approach of the pro-chiral alkene to the oxomanganese centre is controlled by the salicylidene groups. Particularly high enantioselectivities can be achieved in homogeneous epoxidations if the manganese complex of bis (2,4-di-*tert*-butylsalicylidene)-1,2-cyclohexanediamine (known as the Katsuki–Jacobsen catalyst) is used.* In 1997, the successful incorporation of a chiral salen ligand similar to the Katsuki–Jacobsen catalyst, but without the *tert*-butyl groups, was achieved by two groups, one using zeolite Y as the host,[50] the other zeolite EMT[51] (the hexagonal variant of Y, which contains larger supercages.) It was not possible to introduce the complex including *tert*-butyl groups for steric reasons. Both groups reported the catalysts to be active, but to show only moderate enantioselectivities in epoxidations (ee values of *ca.* 60% in the epoxidation of Z-methylstyrene, for example). The studies remain an important example of this approach, however.

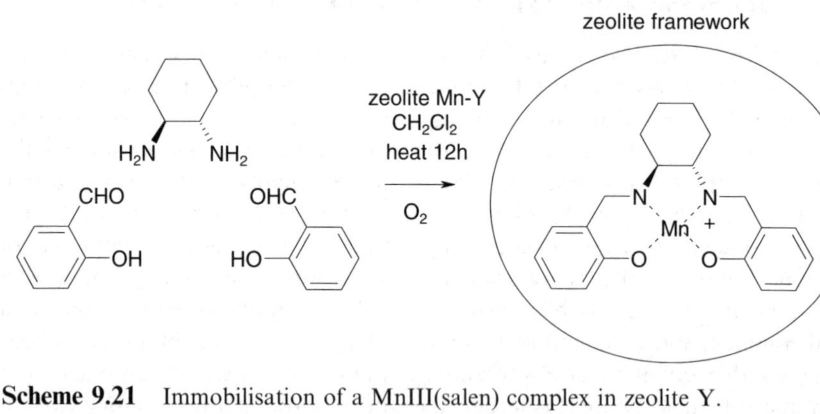

Scheme 9.21 Immobilisation of a MnIII(salen) complex in zeolite Y.

*Enantioselectivity of chiral reactions is described by an ee value, where ee = ([R] – [S])/([R] + [S]) × 100.

Zeolite-encapsulated metal salen complexes have also been used for reactions other than biomimetic oxidations: the Pd(II) salen complex, for example, has been used in the selective hydrogenation of carbonyl groups.[52]

Phthalocyanine complexes within zeolites have also been prepared by the 'ship-in-a-bottle method' (see Section 6.6), and have subsequently been investigated as selective oxidation catalysts, where their planar metal-N4 centres mimic the active sites of enzymes such as cytochrome P450, which is able to oxidize alkanes with molecular oxygen. Cobalt, iron and ruthenium phthalocyanines encapsulated within faujasitic zeolites are active for the oxidation of alkanes with oxygen sources such as iodosobenzene and hydroperoxides.[53] Following a similar route, Balkus prepared Ru(II)-perchloro- and perfluorophthalocyanines inside zeolite X and used these composites for the selective catalytic oxidation of alkanes (*tert*-butylhydroperoxide).[54] The introduction of fluorinated in place of non-fluorinated ligands increases the resistance of the complex to deactivation.

Manganese complexes are also well known as selective oxidation catalysts in metallo-enzymes, where the manganese ions may be present as mononuclear clusters or as dimers or complexes of higher nuclearity. There have therefore been intense efforts to synthesise inorganic analogues of these manganese centres and also to incorporate them into zeolites.[55] Among the most significant of these is that of the group of De Vos *et al.*, who incorporated triazacyclononane-based complexes of manganese in zeolite Y and used them in the selective epoxidation of alkenes.[56] Similar complexes are known to exhibit strong bleaching effects at low temperatures under homogeneous conditions using hydrogen peroxide as the oxidant. The trimethyltriazacyclononane ligand is more effective than triazacyclononane, giving rise to intra-zeolitic Mn^{III}-Mn^{IV} dimeric complexes that show selectivity for epoxidation over H_2O_2 decomposition. Catalytically active dimers can be formed within zeolites and have been characterised by ESR spectroscopy.[57]

There are also many examples of 'ship-in-a-bottle' catalysts that have no enzymatic analogues, including coordination complexes and organometallic complexes such as methyltrioxorhenium[58] and metal carbonyls. One noteworthy example where heterogeneous enantioselective catalysis has been achieved is the aziridination of styrene over Cu^{2+}-bis(oxazoline) complexes within the pores of zeolite Y[59] (Scheme 9.22). ESR indicates the Cu^{2+} is complexed in square planar geometry in the pores. Strong enhancement of the enantioselectivity of the reaction over the immobilised catalyst compared to that observed by homogeneous reaction is observed, and less-hindered bis(oxazolines) which gave little enantioselectivity homogeneously gave high ee values with the heterogeneous catalyst. This is attributed to a *containment* effect, where approach of the reactants to the active site in the zeolite is constrained. Such effects may find more general expression in other catalytic reactions.

The range of catalytic complexes that can be incorporated in zeolites is restricted by the limited space available to complex, substrate and reactant, however. As a result, mesoporous solids have increasingly been investigated as inorganic hosts, with covalent linkages rather than physical entrapment the

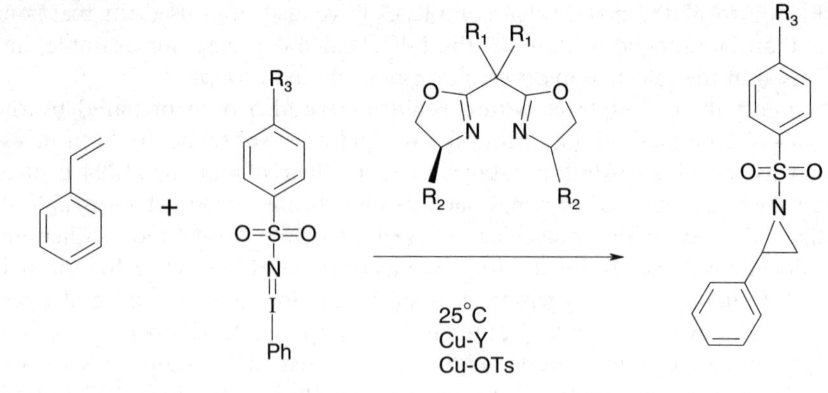

Scheme 9.22 Chiral aziridination of styrene over a chiral copper bis-oxazoline complex supported in zeolite Y (from reference 59).

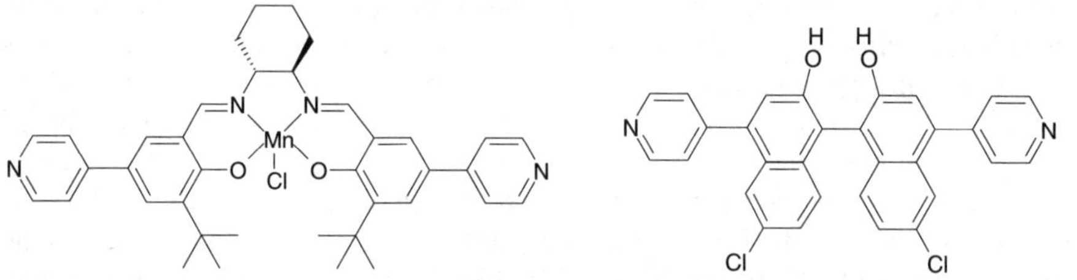

Scheme 9.23 Diamine ligands used as linkers in MOFs which also contain sites for catalysis: (left) ManganeseIII salen ligand, (right) dihydroxy-binaphthyl-bipyridine.

favoured mode of immobilisation. A full description is beyond the scope of this text, but some notable successes have been achieved. The regio- and enantio-selectivity of an amination reaction, for example, has been enhanced over that observed in solution or attached to an open silica surface by incorporating a ferrocene-based complex in MCM-41 of a suitable pore size.[60] This containment effect can be attributed to constraints on the approach of substrates to the active centre. Taguchi and Schuth describe other examples of catalysis over mesoporous solids in their review.[61]

Another approach to immobilising complexes into microporous solids is to prepare a metal organic framework using organic linkers that themselves include the catalytic sites. Typically, the ligand will have two 'primary' functional groups to bind to the metallocentric nodes, and will include a secondary functional group that is able to complex a metal site. This route is of particular interest to include chiral ligands that are known to act as enantioselective catalysts when operating homogeneously. Two examples of such ligands are shown in Scheme 9.23. The first is the Katsuki–Jacobsen salen ligand catalyst for enantioselective epoxidation,[62] the second a dihydroxy-binaphthyl-bipyridine that is able to coordinate titanium at the hydroxyl positions.[63]

The salen ligand can be introduced as a pillaring linker between porous layers of paddle wheel zinc dimers shown in Figure 2.32, linked within the layer by biphenyldicarboxylic acid linkers. The resulting microporous solid is found to retain the enantioselectivity of the ligand for chiral epoxidation, and is more stable, giving a higher total turnover number. The second ligand can be included into a cadmium chloride-based MOF, and the dihydroxy sites used to coordinate a titanium site that is active for the enantioselective conversion of aldehydes to secondary alcohols. In each case good enantioselectivities are achieved ($>80\%$).

9.5 Summary

Although the main applications of zeolitic solids in catalysis will continue to be as solid acids in the synthesis and transformations of petrochemicals and commodity chemicals they continue to be considered as catalysts and catalyst supports for a range of reactions of synthetic and industrial relevance. The most important of these are of titanium- and tin-containing solids in selective oxidations. Other well-studied reactions over zeolites include: light hydrocarbons-to-aromatics (Ga-zeolites); selective catalytic reduction of NO (transition metal exchanged zeolites); C–C bond formation (Pd zeolites); selective alkane oxyfunctionalisation with air (MAPOs, M=Mn, Fe, Co) and chiral catalysis over encapsulated chiral complexes.

Studies of the catalytic activity of MOFs are in their infancy with some encouraging results emerging in enantioselective catalysis. By contrast, mesoporous solids have already been studied extensively as catalytic supports, particularly of complexes too large to be encapsulated in zeolites. One of the most significant developments in this area is the observation that the constrained encapsulation of chiral catalysts in mesopores can raise the enantioselectivities of reactions well above those observed when the reaction is performed homogeneously.

References

1. A. Corma and H. Garcia, *Chem. Rev.*, 2002, **102**, 3837.
2. M. Taramasso, G. Perego and B. Notari, 1983, US Patent 4,410,501.
3. C. Perego, A. Carati, P. Ingallina, M. A. Mantegazza and G. Bellussi, *Appl. Catal. A*, 2001, **221**, 63.
4. T. Blasco, M. A. Camblor, A. Corma, P. Esteve, J. M. Guil, A. Martinez, J. A. Perdigon-Melon and S. Valencia, *J. Phys. Chem. B.*, 1998, **102**, 75.
5. W. B. Fan, P. Wu, S. Namba and T. Tatsumi, *Angew. Chem. Int. Ed.*, 2004, **43**, 236.
6. J. F. Ruan, P. Wu, B. Slater and O. Terasaki, *Angew. Chem. Int. Ed.*, 2005, **44**, 6719.

7. T. Maschmeyer, F. Rey, G. Sankar and J. M. Thomas, *Nature*, 1995, **378**, 159.
8. R. D. Oldroyd, J. M. Thomas, T. Maschmeyer, P. A. MacFaul, D. W. Snelgrove, K. U. Ingold and D. D. M. Wayner, *Angew. Chem. Int. Ed.*, 1996, **35**, 2787.
9. F. Bonino, A. Damin, G. Ricchiardi, M. Ricci, G. Spano, R. D'Aloisio, A. Zecchina, C. Lamberti, C. Prestipino and S. Bordiga, *J. Phys. Chem. B.*, 2004, **108**, 3573.
10. C. Prestipino, F. Bonino, S. Usseglio, A. Damin, A. Tasso, M. G. Clerici, S. Bordiga, F. D'Acapito, A. Zecchina and C. Lamberti, *Chem. Phys. Chem.*, 2004, **5**, 1799.
11. A. Corma, P. Esteve and A. Martinez, *J. Catal.*, 1996, **161**, 11.
12. G. Sankar, J. M. Thomas, C. R. A. Catlow, C. M. Barker, D. Gleeson and N. Kaltsoyannis, *J. Phys. Chem. B.*, 2001, **105**, 9028.
13. C. R. A. Catlow, S. A. French, A. A. Sokol and J. M. Thomas, *Phil. Trans. Roy. Soc. A*, 2005, **363**, 913.
14. L. Balducci, D. Bianchi, R. Bortolo, R. D'Aloisio, M. Ricci, R. Tassinari and R. Ungarelli, *Angew. Chem. Int. Ed.*, 2003, **42**, 4937.
15. A. Corma, L. T. Nemeth, M. Renz and S. Valencia, *Nature*, 2001, **412**, 423.
16. M. Renz, T. Blasco, A. Corma, V. Fornes, R. Jensen and L. Nemeth, *Chem.-Eur. J.*, 2002, **8**, 4708.
17. B. Kraushaar-Czarnetzki, W. G. M. Hoogervorst and W. H. J. Stork, *Stud. Surf. Sci. Catal.*, 1994, **84**, 1869.
18. J. M. Thomas, R. Raja, G. Sankar and R. G. Bell, *Nature*, 1999, **398**, 227.
19. J. M. Thomas, *Top. Catal.*, 2001, **15**, 85.
20. J. M. Thomas and R. Raja, *Chem. Commun.*, 2001, 675.
21. B. Moden, L. Oliviera, J. Dakka, J. G. Santiesteban and E. Iglesia, *J. Phys. Chem. B*, 2004, **108**, 5552.
22. M. Dugal, G. Sankar, R. Raja and J. M. Thomas, *Angew. Chem. Int. Ed.*, 2000, **39**, 2310.
23. J. M. Thomas and R. Raja, *Proc. Natl. Acad. Sci. USA*, 2005, **102**, 13732.
24. R. Fricke, H. Kosslick, G. Lischke and M. Richter, *Chem. Rev.*, 2000, **100**, 2303.
25. G. Giannetto, R. Monque and R. Galiasso, *Catal. Rev. Sci. Eng.*, 1994, **36**, 271.
26. S. J. Tauster and J. J. Steger, *J. Catal.*, 1990, **125**, 387.
27. R. J. Davis and E. G. Derouane, *Nature*, 1991, **349**, 313.
28. E. Mielczarski, S. B. Hong, R. J. Davis and M. E. Davis, *J. Catal.*, 1992, **134**, 359.
29. R. J. Davis, *Heter. Chem. Rev.*, 1994, **1**, 41.
30. W. P. Ding, S. Z. Li, G. D. Meitzner and E. Iglesia, *J. Phys. Chem. B.*, 2001, **105**, 506.
31. J. M. Thomas and W. J. Thomas, *'Principles and Practice of Heterogeneous Catalysis'*, VCH, 1996.
32. J. Kaspar, P. Fornasiero and N. Hickey, *Catal. Today*, 2003, **77**, 419.
33. M. Shelef, *Chem. Rev.*, 1995, **95**, 209.

34. M. Iwamoto, H. Furukawa, Y. Mine, F. Uemura, S. Mikuyira and S. Kagawa, *J. Chem. Soc. Chem. Commun.*, 1986, **16**, 1272.
35. S. Sato, Y. Yoshihiro, H. Yahiro, N. Mizuno and M. Iwamoto, *Appl. Catal.*, 1991, **70**, 1.
36. S. B. Hong, E. G. Lear, P. A. Wright, W. Z. Zhou, P. A. Cox, C. H. Shin, J. H. Park and I. S. Nam, *J. Amer. Chem. Soc.*, 2004, **126**, 5817.
37. K. Krishna, G. B. F. Seijger, C. M. van den Bleek and H. P. A. Calis, *Chem. Commun.*, 2002, 2030.
38. S. Bordiga, R. Buzzoni, F. Geobaldo, C. Lamberti, E. Giamello, A. Zecchina, G. Leofanti, G. Petrini, G. Tozzola and G. Vlaic, *J. Catal.*, 1996, **158**, 486.
39. V. I. Sobolev, A. S. Kharitonov, Y. A. Paukshtis and G. I. Panov, *J. Mol. Catal. A.*, 1993, **84**, 117.
40. M. Dams, D. E. De Vos, S. Celen and P. A. Jacobs, *Angew. Chem. Int. Ed.*, 2003, **42**, 3512.
41. M. Dams, L. Drijkoningen, B. Pauwels, G. Van Tendeloo, D. E. De Vos and P. A. Jacobs, *J. Catal.*, 2002, **209**, 225.
42. J. Weitkamp, M. Hunger and U. Rymsa, *Micropor. Mesopor. Mater.*, 2001, **48**, 255.
43. D. K. Murray, J. W. Chang and J. F. Haw, *J. Am. Chem. Soc.*, 1993, **115**, 4732.
44. D. K. Murray, T. Howard, P. W. Goguen, T. R. Krawietz and J. F. Haw, *J. Am. Chem. Soc.*, 1994, **116**, 6354.
45. M. E. Davis, *Micropor. Mesopor. Mater.*, 1998, **21**, 173.
46. M. Srasra, G. Poncelet, P. Grange and S. Delsarte, *Stud. Surf. Sci. Catal.*, 2005, **158**, 1811.
47. I. Rodriguez, S. Iborra, A. Corma, F. Rey and J. L. Jorda, *Chem. Commun.*, 1999, 593.
48. N. Herron, *Inorg. Chem.*, 1986, **25**, 4714.
49. A. Corma and H. Garcia, *Eur. J. Inorg. Chem.*, 2004, 1143.
50. M. J. Sabater, A. Corma, A. Domenech, V. Fornes and H. Garcia, *Chem. Commun.*, 1997, 1285.
51. S. B. Ogunwumi and T. Bein, *Chem. Commun.*, 1997, 901.
52. W. Kahlen, A. Janssen and W. F. Holderich, *Stud. Surf. Sci. Catal.*, 1997, **108**, 469.
53. N. Herron, *J. Coord. Chem.*, 1988, **19**, 25.
54. K. J. Balkus, A. K. Khanmamedova, K. M. Dixon and F. Bedioui, *Appl. Catal. A*, 1996, **143**, 159.
55. P. P. Knops-Gerrits, D. E. De Vos and P. A. Jacobs, *J. Mol. Catal. A*, 1997, **117**, 57.
56. D. E. De Vos, J. L. Meinershagen and T. Bein, *Angew. Chem. Int. Ed.*, 1996, **35**, 2211.
57. D. De Vos and T. Bein, *J. Am. Chem. Soc.*, 1997, **119**, 9460.
58. T. Bein, C. Huber, K. Moller, C. G. Wu and L. Q. Xu, *Chem. Mater.*, 1997, **9**, 2252.

59. S. Taylor, J. Gullick, P. McMorn, D. Bethell, P. C. B. Page, F. E. Hancock, F. King and G. J. Hutchings, *J. Chem. Soc., Perkin Trans. 2*, 2001, 1714.
60. B. F. G. Johnson, S. A. Raynor, D. S. Shephard, T. Mashmeyer, J. M. Thomas, G. Sankar, S. Bromley, R. Oldroyd, L. Gladden and M. D. Mantle, *Chem. Commun.*, 1999, 1167.
61. A. Taguchi and F. Schuth, *Micropor. Mesopor. Mater.*, 2005, **77**, 1.
62. S. H. Cho, B. Q. Ma, S. T. Nguyen, J. T. Hupp and T. E. Albrecht-Schmitt, *Chem. Commun.*, 2006, 2563.
63. C. D. Wu, A. Hu, L. Zhang and W. B. Lin, *J. Am. Chem. Soc.*, 2005, **127**, 8940.

CHAPTER 10

Advanced Applications and Current Trends

10.1 Introduction

Few classes of solids have found such important applications as zeolites. Development and improvement of their use in the established technologies of ion exchange, separation and catalysis continue in industrial companies around the world with an intensity that fluctuates according to the economic climate. Cleaner processes involving zeolites are progressively adopted as the balance between cost and environmental regulations shifts. At the same time there is intense investigation of potential applications of zeolites and their younger relatives, ordered mesoporous solids and hybrid organic-inorganic frameworks.

Among the furthest developed of these emerging applications is the preparation of thin films and membranes, which offer routes to surface modification for use in sensing and in new and more efficient separation technologies. In the field of electronics, mesoporous silica has recently been commercialised in semiconductor supports. The idea of using the pores of these materials as arrays of conducting nanowires (of metals or polymers) remains attractive. Possible applications in photonics may develop by making use of regular arrays of molecules aligned within the pores. Zeolitic solids are also finding a role as inert carriers in medical diagnosis and treatment. In the case of MOFs, the readily tailored and very high adsorption capacities, coupled in some cases with accessible low coordination metal sites or functional groups, suggest potential as adsorbents for gas storage and separation, in sensing applications or in catalysis.

In this chapter I will discuss how microporous framework solids are being developed for emerging applications and how the structures of ordered nanoporous solids are inspiring materials scientists in directions far away from the conventional uses of aluminosilicate zeolites.[1]

10.2 Fabricating Porous Solids for Developing Applications

10.2.1 Zeolite Membranes

Zeolite membranes have been actively studied for several years and the output of academic and patent publications has continued to grow steadily.[2,3] The goal of this work is to prepare high-quality ceramic membranes that may be applied in high-temperature industrial applications of separation and combined catalysis and separation. For all applications of this type, the two key aims are of high selectivity and high flux. In practice, this requires that the membrane should be as thin as possible while remaining defect-free. These strict requirements have posed difficult challenges to the manufacture of membranes that are only now being answered. Recently, though, there are indications that zeolite membranes could realise their commercial promise.

Zeolite membranes can be used in the separation of molecules that possess different diffusivities through the membrane (such as *para*- and *ortho*-xylene) or that are adsorbed with different strengths within the pores, and may either be selectively retained or selectively fill the pore spaces. The observed selectivities are found to vary considerably with partial pressures employed, however, decreasing sharply under conditions where adsorbed mixtures fill the pores to a high degree. Under these conditions, the transport of all molecules within the pores tends to be single file, and it is the relative concentrations of adsorbed molecules that are important in determining the exit fluxes, rather than the individual diffusivities. This is shown clearly by the selectivities observed for a zeolite membrane in contact with a mixture of methane and *n*-butane at room temperature (Figure 10.1).[2,4] The initial, transient, flux through the membrane is rich in methane, as expected from the higher diffusivity of the smaller molecule, but the equilibrium flux is rich in *n*-butane, as this is much more strongly adsorbed on the zeolite and essentially blocks the adsorption of methane. That adsorption plays an important role is underlined by running the separation at different temperatures: at a temperature of 150 °C the equilibrium flux is dominated by butane, whereas at much higher temperatures (above 350 °C) the flux is dominated by methane. This interplay between adsorption and diffusivity must be taken into account in any application of zeolitic membranes.

Cation-exchanged zeolites show a strong tendency to adsorb water compared to organic molecules, as described in Chapter 7. As a consequence, water can be preferentially removed from solvents. By the use of porous tubes coated in zeolite membranes, water can be removed from the solvent on one side of the membrane and evaporated from the other side. The Japanese Mitsui company, for example, have commercialised the first large-scale pervaporation plant that produces over $500\,l\,h^{-1}$ of solvents with less than 0.2 wt% water from simple alcohols containing 10 wt% water.[5] The plant makes use of over 100 individual sections of NaA zeolite membrane operating at 120 °C.

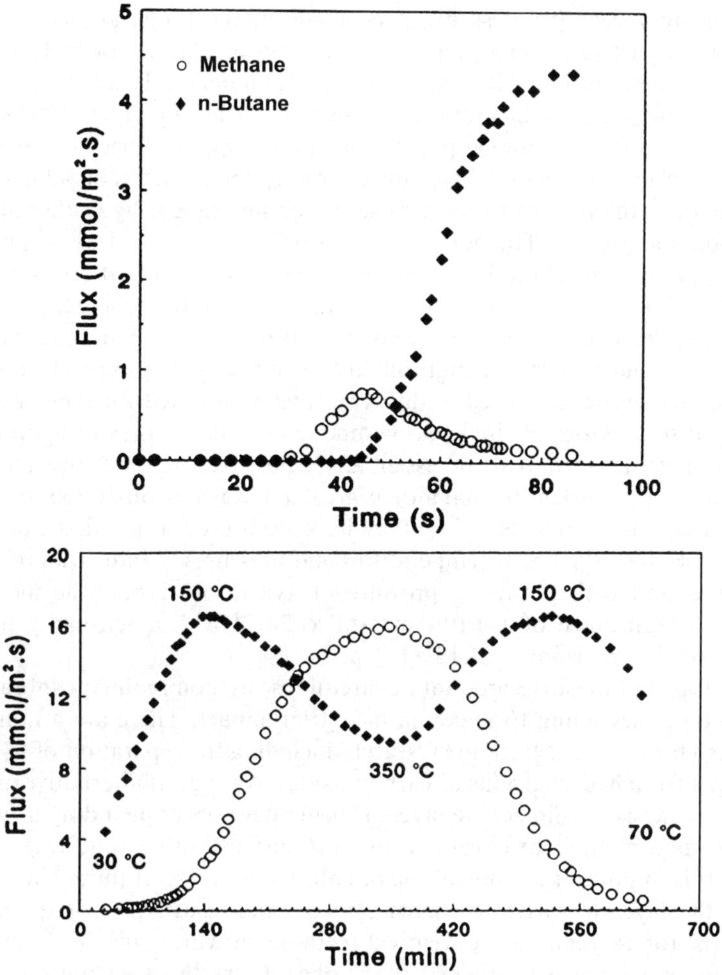

Figure 10.1 Measured flows through a silicalite membrane from a methane/*n*-butane 1 : 1 mixture. (Above) at 25 °C and (below) as the temperature is raised to 350 °C and then reduced to 70 °C. [Reproduced from reference 4 with permission. Copyright 1993, Elsevier.]

Pure silica microporous solids show no such strong affinity for polar molecules, and are of interest to separate hydrocarbon molecules of different shapes. In particular, the separation of xylene isomers is of great industrial significance, and since the relative diffusivities of the *para* isomer are known to be much higher than those of the *ortho* (and *meta*) isomers within silicalite, the preparation of silicalite membranes is an attractive target. If the aluminium content can be made vanishingly small, the membranes can be used for hydrocarbon separation at elevated temperatures without the effects of coke formation due to catalytic reaction.

Significant recent progress has been made in the preparation of silicalite membranes. One promising preparation method has been reported by Exxon Mobil in collaboration with the group of Hedlund at Luleå University in Sweden.[6] They prepare membranes consisting of silicalite layers 500 nm thick grown on an open and graded porous α-alumina support. Special care is taken to achieve a uniform zeolite layer by adsorbing pre-formed colloidal silicalite particles onto the porous support as seeds for subsequent hydrothermal crystallisation and growth. The porous support below the zeolite layer is protected against partial dissolution during the hydrothermal crystallisation step by pre-filling it with hydrocarbon wax. After crystallisation of the zeolite layer the wax is thermally removed to leave the support in its original, fully porous state. The method is found to prepare high quality membranes very reproducibly. The silicalite membranes prepared in this way have been tested for the permeation fluxes and selectivities of single gas components and mixtures of hydrocarbon isomers. They are found to be essentially defect-free and yet give molecular fluxes on to two orders of magnitude greater than previously reported, with comparable selectivities. Notably, they have been used in the first example of the separation of xylenes at temperatures and pressures of industrial relevance. At 400 °C and with a vapour pressure of xylenes of 1 bar, the membrane allowed a permeance of 1×10^{-7} mol $(m^2 sec Pa)^{-1}$ with a selectivity to *para*-xylene, the desired isomer, of 13 : 1.

It is clear that the large amount of careful research on zeolite membranes for separation is beginning to make an industrial impact. There are a number of attractive targets for membrane research, including the separation of hydrogen molecules from hydrocarbons or carbon oxides. A range of alternative methods for the synthesis of thin zeolite layers is being developed, including using laser ablation as a method to cover the support surface with zeolite fragments as seeds.[7] It is important to control the orientation and size of the zeolite crystals within the layer in order to increase fluxes and selectivities. The group of Tsapatsis, for example, has developed methods in which silicalite membranes can be prepared in which either (i) all the silicalite crystals are aligned with their *b* axes (and therefore all the straight channels) perpendicular to the surface[8] or (ii) the crystals are all aligned with their *a* axes (and therefore all the zig-zag channels) perpendicular to the surface.[9] This has been achieved by using a two-stage seed crystal deposition and subsequent crystallisation method, using two different structure-directing agents in the process (tetrapropylammonium and trimer TPA), one for each stage, and reversing the order in which the SDAs are added in the two steps – see Figure 10.2. This is an elegant example of how control of particle morphology can be used in a technological application.

It is therefore likely that coming years will see the application of zeolite membranes in the purification of solvents by per vaporisation, the ultrafiltration of macromolecules (such as dendrimer catalysts) from solutions and the separation of similar hydrocarbon isomers on the basis of molecular shape. In addition, the development of zeolite membranes for use in membrane reactors, in which a catalytic functionality is included into the zeolite, offers potential advantages. In principle, it should be possible to enhance yields in equilibrium

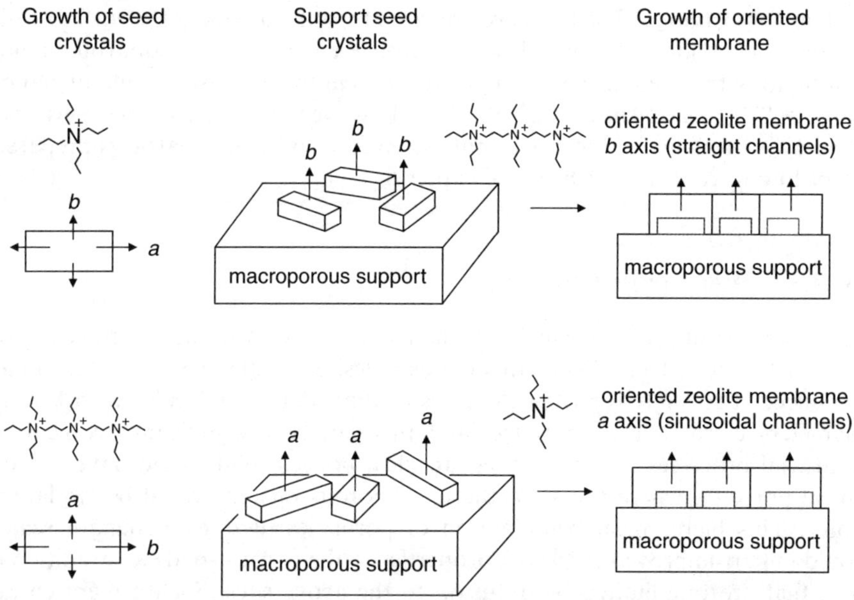

Figure 10.2 The orientation of crystal axes in a silicalite membrane can be controlled by a two-step process in which different structure-directing agents are used for (1) the growth of seed crystals, which lie down on the porous support, and (2) their subsequent growth to give an oriented membrane. (After Tsapatsis *et al.*, reference 9.)

limited reactors (by continuously removing product hydrogen, for example) or to isolate higher yields of intermediates in consecutive reactions. No industrial applications have so far been realised, however.

10.2.2 Thin films and Low Dielectric Constant Materials

Membranes are one class of thin films. If membranes provide a restrictive barrier to the free motion of molecules between phases, other thin film configurations possess a wider range of potential applications. One important new application of porous thin films is in the microelectronics industry, where the increase in the miniaturisation of circuits on microchips requires insulators with improved performance compared to the material that is currently used, dense silicon oxide. The electrical insulating properties of a material improve with decreased dielectric constant, k, so values lower than those of dense silicon oxide (k=3.9–4.2) are required. The lowest values of k are found in a vacuum, and gases have low k values, so porous silica with reasonable mechanical properties provides a sensible solution. Indeed, low k values ($k < 2.2$) have been demonstrated for porous materials, which can readily be prepared as thin films. One non-templated mesoporous silica, MesoELK (ELK=Extra Low k), has been developed by Air Products for applications in computing that require low k films, for example.[1] The films are prepared by spin coating of the surface with a silica precursor that hydrolyses to give a disordered, mesoporous solid. Continuous thin zeolite films prepared by spinning on suspensions of nanoparticles[10,11] (*ca.* 50 nm in size) also show very low dielectric constants (k=1.5),[12] providing a potential application for zeolite nanoparticles prepared according to routes described in Chapter 5.

10.2.3 Hierarchical Porosity

The pore size of zeolites, which are the most widely used microporous solids, extends to about 1 nm. While this imposes desired shape selectivity it also has the effect of strongly reducing the rates of diffusion of molecules with kinetic diameters close to the pore size, and this can be a significant problem in industrial processes. In these cases, an ideal porous solid would have hierarchical porosity, where access to the microporous regions would be *via* larger pores. This hierarchical combination of porosities of different length scales could impart improved transport properties to the porous solids for traditional uses that are rate-limited by diffusion to the active sites. Such a hierarchical distribution is observed in biology in the respiratory tract, where air enters the lungs *via* larger pipes and ends up at very fine branches, known as alveoli, where the exchange of oxygen and carbon dioxide takes place very rapidly.

In fact, the introduction of secondary mesoporosity in zeolites mordenite and Y during steaming (Section 6.2.3) enhances hydrocarbon diffusion and is an important feature of active catalysts for cracking and isomerisations, but the distribution of mesopores is not regular or hierarchical and rarely connected, so there can be micropore-mesopore-micropore sequences that inhibit rapid diffusion. There have therefore been many attempts to prepare hierarchically ordered porous solids. Most of these are based on the strategy of assembling and growing zeolite crystals (often nanocrystals) around a template that has dimensions on the meso scale (or larger) and which can be removed at a later stage to leave a secondary mesoporous (or macroporous) array of channels. To be of use, the connected assemblage of microporous solid must be self-supporting. This approach has been given the term 'exo-templating' by Schuth[13] – as opposed to the normal action of structure directing agents to template mesoporous solids 'from within', termed endo-templating. Many carbon-based species have been used as exo-templates, because they are stable at high temperatures and in chemically reactive solutions but can be removed by combustion, when inorganic species are stable. Suitable C-based exo-templates that have been used include carbon blacks, carbon nanocasts prepared from mesoporous silicas (Chapter 2), wood cells and carbon aerogels.[13] Schuth also describes an approach in which monoliths of silicalite are prepared *via* two stages of zeolite nanoparticle impregnation and growth into an organic aerogel made from resorcinol and formaldehyde in supercritical CO_2.[14] The organic aerogel is carbonised after the first stage of zeolitisation and removed by combustion after the second stage to leave microcrystalline silicalite with mesoporosity on the 3-4 nm scale as well as macroporosity. The monolithic silicalite displays both good mechanical strength and also enhanced activity in the Beckmann rearrangement of cyclohexanoneimine (Chapter 9) probably due to the greater accessible surface area of the meso-microporous composite *via* connected (but not ordered) mesopores.

Highly ordered and hierarchical microporous-macroporous solids have also been prepared. The method devised by the group of Stein[15] is a good example of exo-templating, and makes use of monodisperse colloidal polystyrene spheres 0.1-1 mm in diameter (Figure 10.3). These arrange into close-packed arrays upon settling (or under the action of a centrifuge) that stick together. The interstices of these arrays can be filled by zeolitic crystals which grow together to give a coherent microcrystalline and microporous framework. The polymer can then be removed to give a fully hierarchical macroporous/microporous composite, with interconnecting pore window diameters in the 10-100 nm range. It is also possible to transform macroporous skeletons of disordered silica to zeolites by steam-assisted conversion. Macroporous silica can also be prepared by a polymerisation process following phase separation.

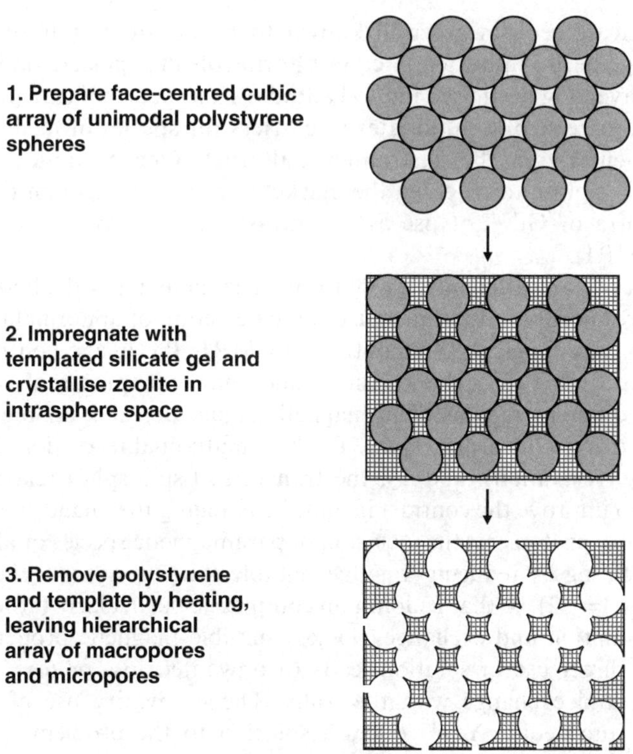

Figure 10.3 Exo-templating: the method of Stein[15] can be used to create hierarchical macroporous/microporous (or macroporous/mesoporous) silicas.

Resulting 'zeolitised' solid has appreciable microporosity and zeolitic microstructure, and shows enhanced cracking of large molecules.[16]

Another approach to introduce mesoporous channels to give better access of reactant molecules to the microporous regions is to assemble zeolite nanoparticles around micellar templates, in a modification of the standard route to mesoporous silicas. Reported examples include structures that possess walls made out of nanocrystals of zeolites such as Beta or ZSM-5. These composite solids possess enhanced hydrothermal stabilities and acidities compared to mesoporous solids with fully amorphous walls. The improved properties are attributed to the presence of the zeolite fragments, because zeolites are known to have higher acidity and hydrothermal stability than amorphous silica/aluminas.[17]

10.3 Using Ordered Porous Solids as Hosts in Functional Materials

10.3.1 Applications in Medicine: MRI Contrast Agents and Drug Delivery Agents

Aluminosilicate zeolites are well known to be non-toxic (although fibrous zeolites have been implicated in certain forms of lung cancer, *via* inhalation) and they have long been used in food additives and personal care products. As a result they are strong candidates as carriers for species of medical importance, particularly in the gastrointestinal tract. One example where their medical use is close to reaching the marketplace is as a medium for administering a form of Gd^{3+} of use as a contrast agent in Magnetic Resonance Imaging (MRI).

MRI is a powerful diagnostic tool for internal imaging of the body. Imaging of the gastrointestinal tract can assist in the detection of abnormalities, but the images typically display poor contrast. In MRI, the image contrast derives from distances between the densities and magnetic properties of protons present in different regions. The magnetic signal depends on the relaxation times T_1 and T_2 of the proton spins: T_1, the longitudinal relaxation time (due to 'spin-lattice' relaxation), and T_2, the transverse ('spin-spin') relaxation time. Strategies to improve the contrast include brightening the image by reducing T_1 times of water protons *via* the inclusion of paramagnetic species or alternatively image darkening by reducing T_2. One suitable contrast agent is the paramagnetic Gd^{3+} ($S=7/2$), so that gadolinium complexes, particularly those that show H_2O coordination and exchange, possess suitable magnetic properties. However, their direct use in solution leads to unwanted loss of toxic Gd^{3+} and interaction and exchange with the walls. The innovative use of Gd^{3+} ion-exchanged into zeolite Y[18,19] offers a solution to the problem – the zeolite matrix holds on strongly to the trivalent gadolinium cations (unless the pH drops below 2) while remaining structurally intact. The strength of the contrast enhancement depends on the rate at which water molecules from the solution outside the zeolite particle approach and leave the Gd^{3+} cations in the pores. This is determined by the cavity size (and water content) around the cations and by the window size. GdNaY exhibits favourable properties of ion exchange capacity, chemical stability and cavity and window size. A suspension of GdNaY, named 'Gadolite®' is in an advanced stage of testing as a positive contrast agent for diagnostic MRI. The material is found to have potential as a T_1 MRI contrast agent at low fields (< 60 MHz) and as a T_2 agent at higher fields.[20]

The potential of zeolites and porous solids as high capacity carriers for drug molecules has also been recognised. Whereas zeolites may have applications for the controlled release of small molecules of medicinal importance, such as nitric oxide,[21] mesoporous silica nanospheres that are sub-micron in particle size, with pore sizes of 2–10 nm, may be used for the targeted *in situ* delivery of larger drug molecules.[22,23]

10.3.2 Optical and Electronic Properties

As well as some of the previously described areas of emerging applications, there are several fields where microporous solids act as hosts for species that display remarkable physical properties, or combinations of properties. Some of the best studied of these have optical or electronic properties, including zeolite/laser dye composites,[24] photocatalysts (including organic and inorganic species),[25,26] luminescent lanthanide-containing microporous solids[27] and light-harvesting antenna assemblies.[28] As well as being of scientific interest, these investigations underline the use of crystalline microporous solids to organise functional species and to permit access of adsorbate molecules to functional centres. It is therefore likely that advanced material applications will evolve from an understanding of these physical phenomena, particularly in the fields of sensing and switching. In ongoing work on light-harvesting antenna materials,[28] systems based on zeolite L have been prepared and characterised in which light energy is absorbed by organic dye molecules organised in a constrained way within the one-dimensional channels and transferred to 'stop-cock' molecules at the channel mouths at the external crystal surface which can trap or inject excitation energy. Ultimately the aim is to couple these systems to external devices.

Research on photocatalysts encapsulated within zeolites is better developed.[25,26] The photocatalysts may be metal-organic complexes, organic cations or molecules or clusters of semiconductor particles such as TiO_2. In typical photocatalytic applications, illumination gives rise to excited or reactive species such as hydrogen peroxide or singlet oxygen that in the presence of other catalytic centres (framework titanium sites in the case of H_2O_2) react to give the desired products. Titania itself is an effective photocatalyst (for example for the decomposition of NO to N_2 and O_2) so that by isolating TiO_2 within the zeolite pores the cluster size and so the electronic band gap and optical absorption properties can be controlled.

10.3.3 Gas Storage and Polymerisation in Responsive Metal-organic Frameworks

In Chapter 7 the adsorption behaviour of some of the novel microporous metal-organic solids was discussed. The most remarkable differences from purely inorganic solids, such as zeolites, are the much higher pore volumes of solids such as MIL-100 and MOF-177 and the ability of some of these solids to interact *via* organic functional groups and/or to display structural response to adsorption. As a result, there is much current investigation of possible applications in gas storage and also of novel adsorption-based types of behaviour relevant to stimulus-driven sensing.

Hydrogen storage is of considerable interest as a pre-requisite technology for implementation of the 'Hydrogen Economy', which includes, for example, the use of hydrogen fuel cells for the generation of clean power. As a result, the hydrogen adsorption capacities of many of the new microporous metal-organic frameworks have been measured. It has been found that by working at low temperature (77 K) and high pressures (> 50 bar) the uptake of hydrogen gas can be as high as 7.5 wt% (gravimetric) and 34 g L^{-1} (volumetric) (measured on MIL-177).[29,30] The first of these values exceeds the gravimetric target set for 2010 for on-board automobile storage by the US Department of Energy (6 wt%) and the second approaches the corresponding volumetric target (45 g L^{-1}). However, the need for cryogenic storage and very high pressure suggests that an alternative application of hydrogen storage on these materials could be in large stationary tanks, if issues of long-term stability in the presence of impurities in the hydrogen can be addressed. In fact, for adsorption of hydrogen at room temperature, which is what is ultimately required, a stronger sorbate-H_2 interaction than available at the metal oxide clusters present in materials such as MOF-5 or -177 is required, and this area is the subject of intense current interest.

The storage of methane, another fuel gas, is more readily achieved at room temperature. Microporous carbons can take up 270 cm^3(stp) g^{-1} at 35 bar and 298 K, which meet the targets set by the US DoE for mobile storage. For comparison, IRMOF-6[29,31] (which is isoreticular with MOF-5, with an additional cyclobutane ring attached to the phenyl group) shows an uptake of 240 cm^3 g^{-1} under the same conditions. These values become competitive with (achieving *ca.* 70% of) the amounts stored in gas tanks at much higher pressures (200 bar). Other porous polymers made up of 2D-connected porous carboxylate layers linked by diamines show similarly high methane capacities.

Perhaps the most remarkable gas storage properties demonstrated for this type of system are in the uptake of acetylene by the coordination polymer $Cu_2(pzdc)_2(pyz)$,[32] where pzdc is pyrazine-2,3-dicarboxylate and pyz is pyrazine. The solid possesses one-dimensional channels of cross section 4×6 Å^2, into which uncoordinated oxygen atoms from carboxylate groups project. These oxygen atoms have basic character, and in the hydrated state coordinate 1 : 1 with adsorbed water molecules through H-bonding interactions. Comparing the adsorption properties for this polymer for carbon dioxide

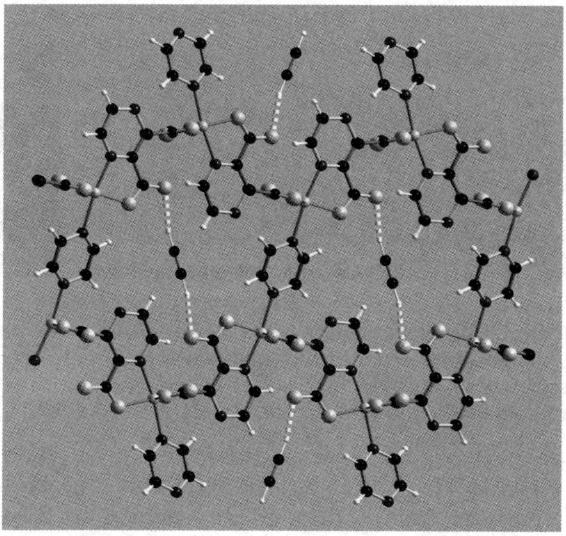

Figure 10.4 Acetylene storage at room temperature in the coordination polymer $Cu_2(pzdc)_2(pyz)$ is possible at a density of $0.43\,g\,cm^{-3}$, equivalent to a hypothetical state of acetylene at 41°MPa , 200 times the compression limit at room temperature. The crystal structure of the adsorbate-sorbent complex[32] shows the adsorption is strongly stabilised by hydrogen bonding (white dotted lines). (Carbon atoms black, nitrogen dark grey, oxygen light grey, hydrogen small white and octahedral copper bound to four oxygen atoms and two nitrogen atoms.)

and acetylene shows a remarkable difference: although the two gases have similar sizes, the acetylene is much more strongly sorbed (more than 25 times as much C_2H_2 as CO_2 is taken up at low pressures). This is due to the presence of H-bonding interactions between the H atoms of the C_2H_2 and the basic oxygen atoms (Figure 10.4). This H-bonding is demonstrated beautifully by electron density maps from the MEM-Rietveld X-ray diffraction approach (described briefly in Chapter 3). As a consequence of this tailor-made adsorption on a functional surface, acetylene can be stored in the pores at a density of $0.43\,g\,cm^{-3}$, equivalent to a hypothetical state of acetylene at 41 MPa at room temperature and 200 times the compression limit at room temperature, above which explosion occurs. Such polymers therefore offer opportunities for safe and efficient gas storage.

This ability of functional hybrid materials to show directional binding of arrays of guest molecules has also been exploited by Kitagawa to control polymerisation reactions within the pores.[33] If a more acidic acetylene such as the methyl-substituted acetylene methylpropiolate (MP) is adsorbed into the pores of $Cu_2(pzdc)_2pyz$ and allowed to stand, accelerated polymerisation to polyMP readily occurs (Figure 10.5). The basic oxygen atoms of the functional porous polymer deprotonate the acetylene and the resultant acetylide species initiate polymerisation to give *trans*-polyacetylene. This selectivity to the

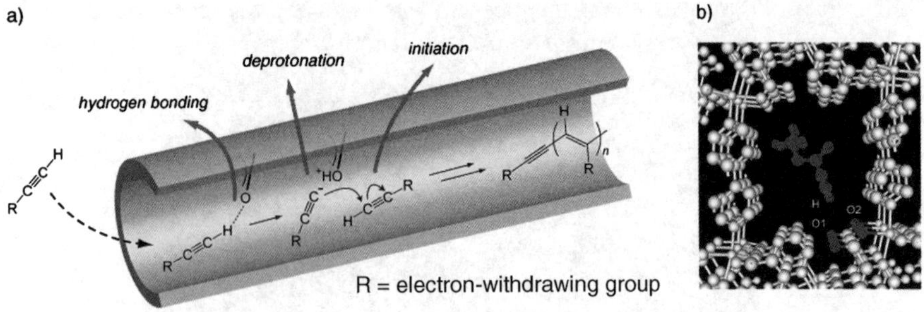

Figure 10.5 Controlled polymerisation of methylacetylenes in the pores of the coordination polymer $Cu_2(pzdc)_2(pyz)$ gives *trans*-polyacetylene due to the steric restrictions. [Reproduced from reference 33 with permission. Copyright 2006, Wiley-VCH Verlag GmbH & Co. KGaA.]

trans- rather than the *cis*-polymer is a natural consequence of the channel geometry. The polyacetylene product can be removed by solvent extraction. This is one of the first examples of the use of functionalised internal surfaces of porous coordination polymers for catalytic applications of a different type from those of zeolites. The same group have also demonstrated that styrene can be polymerised within a porous coordination polymer *via* a radical mechanism to give polystyrene with a narrower molecular weight distribution than that obtained from bulk polymerisation.[34]

The observation that some of the new MOF compounds, such as MIL-53 (Chapter 7) and other coordination polymers,[35] are highly flexible, and that for some adsorbates the porosity becomes accessible only at elevated adsorbate pressure (Figure 7.12), suggests that these materials could be used as sensors and actuators in response to increases in pressures. It may also be possible to use breathing solids in separations that are dominated by the effects of the fit of molecules within the flexible host.

10.4 Summary and Final Remarks

The earlier sections illustrate some of the emerging applications of microporous solids. Whether or not these achieve commercial success remains to be seen, but research in these areas continues to enhance understanding and engender excitement. Furthermore, research in microporous solids continues in parallel with that on mesoporous solids, which possess similar attributes of high surface area and well-defined pore sizes and which can easily be functionalised. These latter solids become the materials of choice when large adsorbate molecules become of interest. A thermally and hydrothermally stable, truly crystalline mesoporous solid remains elusive, but would serve to unify the fields. The area of porous coordination polymers, or MOFs, is at an early stage, and it will be fascinating to see how it grows, particularly if applications can be commercialised.

The overall aim of this text has been to introduce the researcher to the wide range of studies on these remarkable solids and wherever possible stimulate some of the next developments. I look forward to the results.

References

1. M. E. Davis, *Nature*, 2002, **417**, 813.
2. J. Caro, M. Noack, P. Kolsch and R. Schafer, *Micropor. Mesopor. Mater.*, 2000, **38**, 3.
3. J. Caro, M. Noack and P. Kolsch, *Adsorption*, 2005, **11**, 215.
4. E. R. Geus, H. van Bekkum, W. J. W. Bakker and J. A. Moulijn, *Micropor. Mater.*, 1993, **1**, 131.
5. Y. Morigami, M. Kondo, J. Abe, H. Kita and K. Okamoto, *Sep. Purif. Technol.*, 2001, **25**, 251.
6. J. Hedlund, J. Sterte, M. Anthonis, A. J. Bons, B. Carstensen, N. Corcoran, D. Cox, H. Deckman, W. De Gijnst, P. P. de Moor, F. Lai, J. McHenry, W. Mortier and J. Reinoso, *Micropor. Mesopor. Mater.*, 2002, **52**, 179.
7. T. Munoz and K. J. Balkus, *J. Am. Chem. Soc.*, 1999, **121**, 139.
8. Z. P. Lai, G. Bonilla, I. Diaz, J. G. Nery, K. Sujaoti, M. A. Amat, E. Kokkoli, O. Terasaki, R. W. Thompson, M. Tsapatsis and D. G. Vlachos, *Science*, 2003, **300**, 456.
9. J. Choi, S. Ghosh, Z. P. Lai and M. Tsapatsis, *Angew. Chem. Int. Ed.*, 2006, **45**, 1154.
10. L. Tosheva and V. P. Valtchev, *Chem. Mater.*, 2005, **17**, 2494.
11. M. Smaihi, E. Gavilan, J. O. Durand and V. P. Valtchev, *J. Mater. Chem.*, 2004, **14**, 1347.
12. Z. J. Li, C. M. Lew, S. A. Li, D. I. Medina and Y. S. Yan, *J. Phys. Chem. B.*, 2005, **109**, 8652.
13. F. Schuth, *Angew. Chem. Int. Ed.*, 2003, **42**, 3604.
14. W. C. Li, A. H. Lu, R. Palkovits, W. Schmidt, B. Spliethoff and F. Schuth, *J. Am. Chem. Soc.*, 2005, **127**, 12595.
15. B. T. Holland, L. Abrams and A. Stein, *J. Am. Chem. Soc.*, 1999, **121**, 4308.
16. Q. Lei, T. B. Zhao, F. Y. Li, L. L. Zhang and Y. Wang, *Chem. Commun.*, 2006, 1769.
17. Y. Liu and T. J. Pinnavaia, *J. Mater. Chem.*, 2002, **12**, 3179.
18. I. Bresinska and K. J. Balkus, *J. Phys. Chem.*, 1994, **98**, 12989.
19. S. W. Young, F. Qing, D. Rubin, K. J. Balkus, J. S. Engel, J. Lang, W. C. Dow, J. D. Mutch and R. A. Miller, *J. Mag. Res. Imaging*, 1995, **5**, 499.
20. E. Csajbok, I. Banyai, L. Vander Elst, R. N. Muller, W. Z. Zhou and J. A. Peters, *Chem.-Eur. J.*, 2005, **11**, 4799.
21. P. S. Wheatley, A. R. Butler, M. S. Crane, S. Fox, B. Xiao, A. G. Rossi, I. L. Megson and R. E. Morris, *J. Am. Chem. Soc.*, 2006, **128**, 502.
22. D. R. Radu, C. Y. Lai, K. Jeftinija, E. W. Rowe, S. Jeftinija and V. S. Y. Lin, *J. Am. Chem. Soc.*, 2004, **126**, 13216.

23. J. A. Gruenhagen, C. Y. Lai, D. R. Radu, V. S. Y. Lin and E. S. Yeung, *Appl. Spectrosc.*, 2005, **59**, 424.
24. O. Weiss, J. Loerke, U. Wustefeld, F. Marlow and F. Schuth, *J. Solid State Chem.*, 2002, **167**, 302.
25. A. Corma and H. Garcia, *Chem. Commun.*, 2004, 1443.
26. M. Anpo and J. M. Thomas, *Chem. Commun.*, 2006, 3273.
27. D. Ananias, A. Ferreira, J. Rocha, P. Ferreira, J. P. Rainho, C. Morais and L. D. Carlos, *J. Am. Chem. Soc.*, 2001, **123**, 5735.
28. G. Calzaferri, S. Huber, H. Maas and C. Minkowski, *Angew. Chem. Int. Ed.*, 2003, **42**, 3732.
29. A. J. Fletcher, K. M. Thomas and M. J. Rosseinsky, *J. Solid State Chem.*, 2005, **178**, 2491.
30. A. G. Wong-Foy, A. J. Matzger and O. M. Yaghi, *J. Am. Chem. Soc.*, 2006, **128**, 3494.
31. M. Eddaoudi, J. Kim, N. Rosi, D. Vodak, J. Wachter, M. O'Keefe and O. M. Yaghi, *Science*, 2002, **295**, 469.
32. R. Matsuda, R. Kitaura, S. Kitagawa, Y. Kubota, R. V. Belosludov, T. C. Kobayashi, H. Sakamoto, T. Chiba, M. Takata, Y. Kawazoe and Y. Mita, *Nature*, 2005, **436**, 238.
33. T. Uemura, R. Kitaura, Y. Ohta, M. Nagaoka and S. Kitagawa, *Angew. Chem. Int. Ed.*, 2006, **45**, 4112.
34. T. Uemura, K. Kitagawa, S. Horike, T. Kawamura, S. Kitagawa, M. Mizuno and K. Endo, *Chem. Commun.*, 2005, 968.
35. K. Uemura, R. Matsuda and S. Kitagawa, *J. Solid State Chem.*, 2005, **178**, 2420.

Subject Index